People, Plants,

People, Plants, and Genes

The Story of Crops and Humanity

Denis J. Murphy

OXFORD
UNIVERSITY PRESS

OXFORD
UNIVERSITY PRESS

Great Clarendon Street, Oxford OX2 6DP

Oxford University Press is a department of the University of Oxford.
It furthers the University's objective of excellence in research, scholarship,
and education by publishing worldwide in

Oxford New York

Auckland Cape Town Dar es Salaam Hong Kong Karachi
Kuala Lumpur Madrid Melbourne Mexico City Nairobi
New Delhi Shanghai Taipei Toronto

With offices in

Argentina Austria Brazil Chile Czech Republic France Greece
Guatemala Hungary Italy Japan Poland Portugal Singapore
South Korea Switzerland Thailand Turkey Ukraine Vietnam

Oxford is a registered trade mark of Oxford University Press
in the UK and in certain other countries

Published in the United States
by Oxford University Press Inc., New York

British Library Cataloguing in Publication Data

Data available

Library of Congress Cataloging in Publication Data

Data available

Typeset by Newgen Imaging Systems (P) Ltd., Chennai, India
Printed in Great Britain
on acid-free paper by
Antony Rowe, Chippenham

ISBN 978–0–19–920713–8   978–0–19–920714–5 (Pbk.)

10 9 8 7 6 5 4 3 2 1

This book is the story of the untold generations of agriculturalists who largely created the world as we know it—for both good and ill.
It is especially dedicated to the long-suffering people of Warka/Iraq, which was once one of the most important cradles of our civilization.
They surely deserve better.

Ad agricolis
Mundus noster fecistis
Dum aetas fugax

Contents

List of figures

List of tables

List of text boxes

Preface

This book has been a particularly challenging endeavour. My aim was to write a reasonably scholarly text that could also provide an accessible synthesis of up-to-date knowledge across some very diverse academic disciplines. It is aimed at a wide range of audiences, including anybody with an interest in how people and societies have evolved together with the crops upon which we now depend. While addressing a relatively broad spectrum of readers, it also seeks to deal with technical topics, from genetics to archaeology, in sufficient depth to satisfy most academic specialists. Such a balancing act is always difficult and there are inevitable simplifications and generalizations, especially when describing complex processes such as societal development or plant/human coevolution. In addressing other areas, such as molecular genetics or climatology, a scientific background would be an advantage for the reader but not absolutely essential to grasp the main points. As in the majority of academic discourse, some of the issues covered in the book are still vigorously disputed by experts. Examples include thorny topics such as human cognitive modernity and the impact of climatic change on societal development. In such cases, I have either remained neutral in the controversy or have explicitly agreed with a particular viewpoint, while drawing attention to the wider picture by citing alternative perspectives in the endnotes.

In order to meet the challenge of such wide-ranging and at times technical subject matter, the main text is supplemented by over 1200 detailed endnotes. These are linked in turn to a comprehensive bibliography of over 1460 citations, mostly from the peer-reviewed, primary literature. This should enable the interested reader to delve more deeply into the many complex and fascinating topics, many of them at the cutting edge of scientific discovery, that are perforce discussed more concisely in the main text. Wherever possible, I have provided web links to articles that are now available online. Many of the more enlightened scientific journals make their articles freely available on the Internet either immediately or within a year or so of initial publication. Such primary research articles are often surprisingly accessible to the interested layperson, and I recommend readers to consult at least a few examples. Secondary literature, for example scholarly reviews, government reports, conference papers, etc., is also often available on the Internet and can be a useful resource, especially for a more general reader or a technical specialist from a slightly different field. I have used relatively few 'tertiary' sources, such as popular magazines or newspapers, because while these tend to be more immediate in their content and often a 'good read', they are often less reliable, less accessible, and much more ephemeral in their Internet locations.

We often think about the history of humankind in terms of its 'progression' from a relatively simple and supposedly 'primitive' Palaeolithic past, to the sophisticated technological societies of today. It is normally assumed that one of the major defining features of this process was the 'invention' of agriculture a little over ten thousand years ago. One of my purposes here is to challenge this viewpoint and to present an alternative perspective based on a great deal of recent research, especially relating to human–plant interactions. Over the past decade or so, discoveries in fields as diverse as molecular genetics, palaeoanthropology, climatology, and archaeology, have immensely improved our understanding of human biological and societal development over the past two million years. Of course there are still many gaps in our knowledge of this complex process. Nevertheless, we are now beginning to appreciate more clearly how the course of human development has been modulated by a whole range of contingencies arising just as much (or sometimes more) from our biological and abiotic environments, as from internal societal factors.

The book is divided into four parts that cover the broad canvas of plant and human evolution, from 90 million years ago until the present day, and beyond into the medium-term future. In Part I, People and plants: two hundred millennia of coevolution, the three chapters are focussed mainly on the development of humankind from the emergence of *Homo sapiens* in Africa and its subsequent spread around the world. The interactions of early humans with the animals and plants upon which they depended were greatly affected by the hyper-variable climate of the Pleistocene Era. We will see that people in different regions interacted in many contrasting ways with plants and animals, and that in some cases these partnerships were as enduring and complex as agriculture has been. In a (very) few cases, human–plant partnerships became much more intimate, eventually favouring the evolution of different types of plant that were specifically adapted to growing in association with new forms of human management. These new management methods developed into what we now call agriculture and the new types of plant became our first crops. The first known case of plant domestication occurred about 12,000 years ago, at the village of Abu Hureyra in present day Syria. However, agriculture was neither inevitable nor necessarily enduring, and we will see how some societies either never adopted farming or later abandoned it in favour of more reliable and rewarding strategies of food acquisition.

In Part II, Crops and genetics: 90 million years of plant evolution, the focus switches to considering human–plant associations from the plant perspective. The four chapters in this section are probably the most technical in the book, dealing with plant genetics and its key role in enabling a few species to become domesticated into crops. Unlike humans, plant behaviour is solely determined by a combination of genetics and environment (i.e. there is no social component) so the analysis of plant genomes is of great interest and significance. Recent advances in molecular biology have given us a fascinating new view of plant genomes and the ways in which only a few of them have lent themselves to domestication. We will examine the remarkably fluid nature of plant genomes, with DNA constantly moving to and fro, both within and between species, sometimes to the extent that it becomes difficult even to define a particular plant species or genus. Unlike most animals, plants can also duplicate their genomes, often after hybridization with other species, and many of our most important crops are descended from such polyploid ancestors. The final two chapters of Part II deal specifically with the genetics of our major crops, and the ways in which their unusual genomic architecture, especially the clustering of certain genes in a few chromosomal regions, predisposed these plants to become domesticated by humans. One of the conclusions that may surprise some readers is that crop domestication in the Neolithic period almost certainly owed its success more to the structure of plant genomes than to the botanical skills of early protofarmers. Indeed, it is now widely accepted by geneticists that most or all of the ancient crop domestications were unconscious processes of plant–human coevolution, rather than deliberate strategies based on knowledge and foresight by the people involved.

In Part III, People and plants in prehistoric times: ten millennia of climatic and social change, the focus returns to humankind, and particularly the development of the early farming-based cultures that went on to create the dominant agrourban societies of Asia, Africa, Europe, and the Americas. The first two chapters describe the emergence of crops in various parts of the world over several millennia during the early to mid part of the Neolithic period. The decidedly mixed benefits of agriculture are discussed in the context of its sometimes-adverse effects on individual human health, especially compared to many of the better-nourished hunter–gatherers of the time. Despite often leading to a reduction in individual human fitness, farming was generally a highly adaptive strategy at the population level. In particular, farming enhanced the competitiveness of the growing agrarian societies compared to the smaller groups of hunter–gathers. We will also see how people have become modified genetically in response to farming, and how most of us carry relatively recent mutations that are directly attributable to our intimate associations with plant and animal domesticants.

The next three chapters of Part III deal in turn with the development of farming-based, agrourban cultures of varying size and complexity in the Near East, east and south Asia, Africa, Europe, and the Americas. Recent research shows how agrarian societies evolved independently in all of these regions, and also reveals many interesting similarities and differences between them. In particular, the speed of urbanization and development of complex, stratified social organizations varied considerably in different parts of the world, as did societal responses to vicissitudes such as climate change or resource depletion. One important point that emerges from these three chapters is the manner in which most (but by no means all) agrourban cultures have repeatedly and successfully modulated their size and complexity in response to environmental and social stresses. In particular, over the past twelve millennia, there have been many instances of retreat from complexity and often drastic population downsizing that sometimes involved considerable loss of knowledge and skills. However, such episodic setbacks were often, but not inevitably, followed by resumption of what used to be termed 'progress' towards increasing complexity, both in terms of social structures and technologies.

In Part IV, People and plants in historic times: globalization of agriculture and the rise of science, we move through the classical and medieval periods and the many ups and downs of technosocial evolution, particularly as related to agriculture. In Europe, the period after the Renaissance witnessed what I term a 'neonaissance' that involved more powerful paradigms for the discovery, dissemination, and exploitation of knowledge, with the rise of science and a vast suite of new technologies. In particular, during the post-Enlightenment era, there was a flowering of investigation into matters botanical and agronomic that underpinned a quantum leap in agricultural productivity. This was the era of 'imperial botany', with European explorer–entrepreneurs scouring the world for useful and profitable plants. Is also set the scene for the industrial revolution of the eighteenth and nineteenth centuries; the twentieth century globalization of

agriculture and technourban cultures; and the most recent population explosion that is only now beginning to level off.

Associated with these developments was the rise of a new and more evidence-based form of scientific plant breeding that by the twentieth century was benefiting from discoveries in genetics and physiology, and new technologies, from X-rays to tissue culture. Some of the subject matter in Chapters 14 and 16 overlaps with the more detailed discussions about the institutional context of modern plant breeding in my forthcoming book: *Plant Breeding and Biotechnology: Societal Context and the Future of Agriculture* (Murphy, 2007). Contemporary plant breeding is fast becoming a high-tech activity that uses the latest robotic and bioinformatic tools, often based on DNA and other sophisticated molecular marker methods. Modern scientifically-informed plant breeding has enabled food production to increase even faster than population growth. This has enabled the emergence of the impressive new megaeconomies of India and China, both with populations of over one billion people who, thanks to the 'Green Revolution' of the 1960s and 1970s, are now largely self-sufficient in crop production.

New methods of advanced plant breeding should enable us to keep pace with the predicted population growth over the next century, providing there is sufficient climatic and social stability to enable the research to bear fruit. Molecular tools may also enable us to domesticate some of the thousands of potentially useful plants that have hitherto proved genetically recalcitrant to all the best breeding efforts of our predecessors. In the final chapter, we finish with a brief retrospective and prospective glance at the broader context of plant–human interactions. Here, we will see how our new-found knowledge of genetics and human agrosocial development can do much to inform the choices that may be faced by our descendents. In particular, it gives us some ground for optimism for the ability of humanity to survive and prosper in the uncertain times that lie ahead, albeit perhaps with different societal models to those that currently prevail.

I am indebted to those who have inspired and helped me in various ways during writing of this book, especially the many colleagues with whom I had fruitful discussions. The award of a minisabbatical from the University of Glamorgan was of great assistance in ensuring the timely submission of the manuscript and in securing the services of three excellent graphic artists. David Massey drew Figures 3.2A and B, 4.2, 6.3A, B and C, 6.4A, B, C and D, 6.6A, 7.1, 8.1A and B, 8.3A, 10.3A, 10.5, 10.6, 10.8, 11.2B, 12.5, 13.1, 17.1, 17.2, 17.3; Anna Jones drew Figures 6.5A, B, C and D, 6.7A and B, 11.3A and B, and 12.2A and B; and Judith Hills drew Figures 2.1, 2.3, 3.1, 10.7, 10.10, 12.3, 12.5, 12.6. Special thanks to Steve Lee and the team at the University of Glamorgan Library for their support in obtaining the hundreds of additional texts and other references used in researching the book; and to all at Les Croupiers Running Club, Cardiff for helping me to maintain some vestige of sanity during the long months of deskbound writing. Finally, many thanks to Stefanie Gehrig, Ian Sherman, and the rest of the staff at OUP plus various anonymous referees for their advice, support, and encouragement during the gestation of this project.

Denis J. Murphy
Glamorgan, Wales
December 2006

Nomenclature and terminology

Common names	Botanical name (newer convention)	Botanical name (earlier convention)	Crop examples
Grasses, cereals	Poaceae	Gramineae	Rice, wheat, maize
Legumes, pulses	Fabaceae	Leguminoseae	Beans, pea, lentil
Solanaceous plants	Solanaceae	Solanaceae	Potato, tomato
Brassicas, crucifers	Brassicaceae	Cruciferae	Cabbage, rape
Cucurbits	Cucurbitaceae	Cucurbitaceae	Gourds, cucumber,
Spurges	Euphorbiaceae	Euphorbiaceae	Cassava, castorbean
Bindweed family	Convolvulaceae	Convolvulaceae	Sweet potatoes
None	Dioscoreaceae	Dioscoreaceae	Yams
Carrot family	Apiaceae	Umbellifereae	Parsley, coriander

Botanical names

Botanical names are sometimes troublesome for the layperson, but I can assure you that they can be even more vexatious for the plant scientist. This is because names of families, genera, and higher classifications are periodically altered, swapped, rearranged, and generally mixed up, much to everybody's confusion. In some cases, one group of experts might use one name while others use a different and seemingly unrelated name. This is most apparent in the case of family names where the more recent versions are widely used in the Americas but less frequently elsewhere. In this book, I have tried to use the most up-to-date versions of plant names, but in some cases this may cause confusion because many primary texts still use the older versions. The most important crops and their family names are shown above.

Measurements

The metric system is used throughout for all physical measurements except where quoting directly from historical sources. See Box 1.1 for an explanation of the various dating systems used here, and Box 1.2 for the chronological terms commonly used both here and in the geological and archaeological literature.

Initials and acronyms

A list of technical terms is given below. I have tried to forbear, as much as possible, from using unfamiliar initials and acronyms in the main text. Where this is impractical, I give the full version of each term in the text when it is first used. A list of such terms, and some explanation of their significance, is also given below.

Abbreviations and glossary

Abiotic stresses: non-living, environmental factors that may be harmful to growth or development of an organism: examples include drought, salinity, and mineral deficiency (see **Biotic stresses**).

BCE: Before Common Era, neutral dating term corresponding to BC, 'before Christ'.

Biotic stresses: living factors that may be harmful to an organism: examples include pathogens, pests, or competitors, often including members of the same species (see **Abiotic stresses**).

BP: Before Present—dating system used for the prehistorical period, where the 'present' is defined abitrarily as the year 1950 CE.

CE: Common Era—neutral dating term corresponding to AD, '*anno domini*'.

Chalcolithic: literally, 'copper stone', a transition period between the Neolithic and Bronze Ages where the first copper-based metal tools were used alongside stone implements. Early Chalcolithic cultures first arose in the Near East after *7000 BP*.

Corvée: system of conscripted labour, sometimes in lieu of tax and/or paid in-kind (e.g. with food), often used for agricultural work or for large construction projects and found in many societies throughout recorded history up to the present day.

Cultivar: cultivated variety of a crop—such varieties have normally been selected by breeding and are adapted for a particular agricultural use or climatic region.

Dansgaard-Oeschger event: one of at least 23 climatic episodes involving sudden warming followed by more gradual cooling that has occurred over the past 110,000 years (see **Heinrich event**).

Epigenetic: the transmission of information from a cell or multicellular organism to its descendants without that information being encoded in the DNA sequence of a gene. Epigenetic changes can be caused by differences in DNA methylation or in chromatic structure involving modification of histones.

FAO: Food and Agriculture Organization—a United Nations agency dedicated to improving agriculture and ending hunger across the world.

Genome: the genetic complement of an organism, including functional genes and an often-large amount of non-coding DNA. The principal genome of eukaryotes, such as plants and animals, resides in the nucleus but smaller genomes are also present in mitochondria and plastids.

Genotype: genetic constitution of an organism; see also **Phenotype**.

GM: genetically modified or genetically manipulated—a term normally used to describe an organism into which DNA, containing one or more genes, has been transferred from elsewhere. The transferred DNA is never itself actually from another organism, but may be an (exogenous) copy of DNA from a different organism. Alternatively the transferred DNA may be an extra copy of an (endogenous) gene from the same organism. Finally, the transferred DNA may be completely synthetic and hence of non-biological origin. An organism containing any of these categories of introduced gene is called **transgenic**.

Heinrich event: one of at least six abrupt and severe episodes of climatic change affecting large areas of the world during glacial periods over the past 60,000 years and having catastrophic consequences for many forms of flora and fauna (see **Dansgaard-Oeschger event**).

Hybrid: an organism resulting from a cross between parents of differing genotypes. Hybrids may be fertile or sterile, depending on qualitative and/or quantitative differences in the genomes of the two parents. Hybrids are most commonly formed by sexual cross-fertilization between compatible organisms, but cell fusion and tissue culture techniques now allow their production from less related organisms.

Inbreeding depression: a reduction in fitness and vigour of individuals as a result of increased homozygosity through inbreeding in a normally outbreeding population.

Input trait: a genetic character that affects how the crop is grown without changing the nature of the harvested product. For example herbicide tolerance and insect resistance are agronomically useful input traits in the context of crop management, but they do not normally alter seed quality or other so-called **output traits** that are related to the useful product of the crop.

Landrace: a genetically diverse and dynamic population of a given crop produced by traditional breeding. Landraces largely fell out of favour in commercial farming during the twentieth century and many have died out. Landraces are often seen as potentially useful sources of novel genetic variation and efforts are underway to conserve the survivors.

LTR: long terminal repeat—a common class of **retrotransposon**.

Neo-naissance: 'new birth'—period after the sixteenth century CE during which a new, scientifically based paradigm of knowledge production was invented in Europe. This period contrasts with the earlier postmedieval Renaissance, which was a 'rebirth' or rediscovery of pre-existing Classical and Oriental knowledge.

Output trait: a genetic character that alters the quality of the crop product itself, e.g. by altering its starch, protein, vitamin, or oil composition.

Paedomorphic trait: a juvenile character that becomes retained in the adult stage of an organism. Many domesticated animals carry such traits, as do humans who retain the flattened face, gracile features, and other attributes that are normally only found in juvenile stages of development in other primates.

PCR: Polymerase Chain Reaction—a technique for rapidly copying a particular piece of DNA in the test tube (rather than in living cells). PCR has made possible the detection of tiny amounts of specific DNA sequences in complex mixtures. It is now used for DNA fingerprinting in police work, in genetic testing, and in plant and animal breeding.

Phenotype: physical manifestation of the combined effects of the genotype and the environment for a given organism. Phenotypic traits include external appearance, composition, and behaviour.

Pleiotropic effect(s): multiple phenotypic effects of a single gene.

Quantitative genetics: the study of continuous traits (such as height or weight) and their underlying mechanisms.

Quantitative trait locus (QTL): DNA region associated with a particular trait, such as plant height. While QTLs are not necessarily genes themselves, they are closely linked to the genes that regulate the trait in question. QTLs normally regulate so-called complex or quantitative traits that vary continuously over a wide range. While a complex trait may be regulated by many QTLs, the majority of the variation in the trait can sometimes be traced to a few key genes.

Rachis: Structure holding cereal grains onto the stalk of the plant, which in wild plants normally becomes brittle as the ears mature. This enables the grains to break off from the plant, so they readily fall into the soil or are otherwise dispersed. Domesticated cereals have a non-brittle rachis trait, allowing them to retain grain on the stalk for easier harvesting by farmers.

Rainfed farming: also called dryland farming, this form of crop cultivation relies on rainfall rather than irrigation and is practiced on 80% of the global arable land area. Rainfed agriculture is only practical above the 200-mm isohyet and is only reliable in the longer term above the 300-mm isohyet.

Retrotransposons: the most abundant class of transposable elements (so-called 'jumping genes') in eukaryotes and especially common in plant genomes. Retrotransposons are particularly useful in phylogenetic and gene mapping studies and as DNA markers for advanced crop breeding.

Sedentism: settled lifestyle based on permanent or semipermanent habitations, rather then a wandering, nomadic existence. Most human groups were largely nomadic, although partial sedentism, perhaps to exploit seasonal resources, may have been commonplace well before permanent settlements were built. Although linked with the development of faming, sedentism was also practiced by certain non-farming cultures such as coastal fishing communities where nomadism was unnecessary.

Species: a group of organisms capable of interbreeding freely with each other but not with members of other species (this is a much simplified definition; the species concept is much more complex.). A species can also be defined as a taxonomic rank below a genus, consisting of similar individuals capable of exchanging genes or interbreeding.

TILLING: Targeting Induced Local Lesions IN Genomes—the directed identification of random mutations controlling a wide range of plant characters. A more sophisticated DNA-based version of mutagenesis breeding, TILLING does not involve transgenesis.

Transcription factor: DNA-binding protein often involved in the co-ordinated regulation of several genes. Mutations in genes encoding transcription factors are some of the most common mechanisms for radical phenotypic change in organisms, e.g. the transition from wild to domesticated crops.

Transgenic: an organism into which exogenous segments(s) of DNA, containing one or more genes, has been transferred from elsewhere (see **GM**).

Transgenesis: the process of creating a **transgenic** organism.

Transposon: sometimes called 'jumping genes', the most common class is the **retrotransposons**.

Wide crossing: in plant breeding this refers to a genetic cross where one parent is from outside the immediate gene pool of the other, e.g. a wild relative crossed with a modern crop cultivar.

Wild relative: plant or animal species taxonomically related to a crop or livestock species; a potential source of genes for breeding new crop or livestock varieties.

WHO: World Health Organization—a United Nations agency established in 1948 with a mission to improve human health around the world.

Younger Dryas Interval: period of sudden and profound climatic change involving widespread cooling and drying, from *12,800* to *11,600* BP. Although its effects on flora and fauna extended across the globe, they were most acute in Eurasia where they may have been instrumental in the genesis of agriculture.

PART I

People and plants: one hundred millennia of coevolution

All is flux, nothing stays still

Heraclitus, c. 540–480 BCE, from Diogenes Laertius,
Lives of the Ancient Philosophers

CHAPTER 1

Early human societies and their plants

Historians will have to face the fact that natural selection determined the evolution of cultures in the same manner as it did that of species

Konrad Lorenz, 1903–1989, *On Aggression*

Introduction

The development of agriculture is universally regarded as one of the defining moments in the evolution of humankind. Indeed, many accounts of human development still describe the so-called 'invention' of agriculture as if it were a sudden and singular transformative event.[1] The acquisition of the know-how and technology that enabled people to practice agriculture is conventionally portrayed as a dramatic and revolutionary change, which occurred about 11,000 years ago at the start of the Neolithic period (or 'New Stone Age').[2] We are told that this revolutionary event completely altered the diet, lifestyle, and structure of the human societies involved, most notably in the Near East. The epochal 'invention' of agriculture is then supposed to have led directly to urbanization and quantum leaps in technological and artistic development as part of a unidirectional and profoundly progressive process. This notion of a sudden agricultural revolution originated because of what appeared to be the almost overnight appearance and cultivation of new forms of several key plants, especially cereals and pulses, that had supposedly been deliberately 'domesticated' by people. Almost simultaneously, so it seemed, the new farming-based cultures began to build increasingly complex, permanent habitations that soon developed into elaborate urbanized cultures and, eventually, civilizations with imperial aspirations.

Moreover, it was also originally believed, and is still repeated in a surprisingly large number of textbooks, that agriculture was somehow 'invented' in the Near East and subsequently exported to Europe, Africa, and the Far East. The entire process of agricultural and societal development has also been decorated with Enlightenment and Victorian overtones of inevitability and progression, as if humanity was somehow 'destined' to tame plants and animals and to develop complex, technologically based societies. This 'revolutionary' thesis of the origins of agriculture is now being successfully challenged by manifold lines of evidence from a spectrum of scientific disciplines that includes archaeology, geology, climatology, genetics, and ecology.[3] It is now clear that several human cultures (possibly numbered in the dozens) independently developed distinctive systems of agriculture on at least four different continents.[4]

Over the past decade or so, detailed archaeological and genetic evidence has emerged supporting the view that widespread cultivation of crops evolved separately in various parts of Asia, Africa, Mesoamerica, and South America.[5] In contrast, in Europe, North America, and Australasia, crop cultivation occurred much later. In these latter three regions, crops and agronomic techniques were only secondarily acquired from the primary agricultural societies. These crops were then grown in places that were far from their initial centres of origin. In the comparatively few primary centres of crop cultivation, a relatively narrow range of locally available edible plants was domesticated as the major food staples. Wherever suitable species were available, it was the large-grained cereals that were the most favoured candidates for cultivation as staple crops. The most obvious examples are rice, wheat, and maize; these three plants were among the earliest domesticates and are still by far the most important crops grown across the world, supplying well over two-thirds of human calorific needs. The second most popular class of staple domesticants were the starchy tubers such as yams and potatoes, but these crops were not as versatile as cereals, especially as regards long-term storage, and this limited their more general use. The major class of supplementary crop is the pulses, or edible-seeded legumes, which provide useful proteins and nutrients lacking in cereals and tubers, as well as replenishing soil fertility with nitrogen compounds.

Domestication of these different crop species did not occur at the same time or in the same place.[6] Several overlapping, and sometimes lengthy,

primary domestication processes were in progress around the world over a period of at least eight millennia from about *13,000 BP* until *5000 BP* (see Box 1.1 for an explanation of the dating systems used here). In several cases, such as wheat and rice, a single plant species was domesticated completely independently on numerous occasions, by various unrelated human cultures living in different periods and in different regions of a continent. Moreover, it now appears that the systematic culti-vation of crops was preceded in most places by an extremely lengthy preagricultural phase of plant husbandry. During this period, many geograph-ically unconnected groups of humans started to col-lect, process, and even manage certain favoured plants for food use, while still relying on a nomadic hunter–gathering lifestyle to sustain the bulk of their livelihoods. In the Near East, this prefarming

---

Box 1.1 Dating systems

Dates in the text are presented in either *BP* or *BCE/CE* formats, in line with conventions in the primary literature. Dates relating to more ancient events and processes over archaeological and geological timescales are normally given as *BP*, or *Before Present*, where the present is arbitrarily defined as the year 1950. This dating system is followed in Parts I to III, which deal mostly with prehistoric periods ranging from several million years to about 4000 years ago. Here, dates expressed as *BP* are italicized in order to distinguish them further from dates within more recent historical periods.

Many of the *BP* dates quoted here are based on radiocarbon dating methods. These dates are always given in 'real' calendar years, rather than the potentially misleading (to the layperson) 'radiocarbon years' sometimes quoted in the primary literature. Because radioisotopes do not decay at a uniform rate, 'radiocarbon years' can vary significantly from 'real' calendar years. This is especially true for *BP* dates earlier than a few thousand years ago. For example, some radiocarbon-based chronologies place the end of the Younger Dryas Era at *10,000 BP* in so-called radiocarbon, or ^{14}C, years whereas the 'true' date is about 11,600 calendar years *BP*. Equally, the onset of the Younger Dryas Era and, possibly, of cereal cultivation, is often expressed as 11,000 ^{14}C years *BP*, although the 'true' date is more like 12,800 calendar years *BP*.

This practice can lead to confusion when comparing dates in the literature, especially in many secondary sources (including many popular books and the plethora of internet sites that cite human chronologies). Such sources frequently fail to state the type of dating method that is being used in a particular text so that a date like *10,000 BP* or 8000 BC can be ambiguous by a margin of as much as 1600 years. Hence, the admonition '*caveat lector*' when consulting such sources. In the present book, all radiocarbon dates have been adjusted, as far as possible, to true calendar years using a combination of correction formulae and by using other independent dating methods as a check. For a technical discussion of the vagaries of radiocarbon dating and conversion charts, see Stuiver and Becker (1993) and Stuiver *et al.* (1998). A simple online calibration chart from the present to as far back as *4500 BP* can be found at: http://www.sciencecourseware.org/VirtualDating/files/RC_5.5.html

In Part IV, which deals with events during the historic period, dates are generally given according to the modern convention as BCE (Before Common Era) or CE (Common Era). This corresponds to the former usage of BC (Before Christ) and AD (*anno domini*). In the later chapters that cover the post-Classical period, dates are usually given without a suffix when it is clear from the context that they relate to CE.

phase of informal plant management may have extended for many millennia and perhaps tens of millennia, from as long ago as *40,000* or *50,000* BP. It is also important to realize that agriculture is by no means the only successful and enduring option for the management and exploitation of plants. Indeed, numerous societies around the world opted over many millennia to remain wedded to a more flexible lifestyle of informal nurturing and collection of wild plants, rather than committing themselves to full-time agriculture.[7]

Why agriculture?

So, why did human societies, and especially those that had already been engaged in preagricultural plant cultivation for as much as ten millennia or more, not develop full-scale agriculture until so recently? These preagricultural people were certainly as intelligent as we are. They knew a great deal about the many different species of food plants that they utilized so effectively, including several species that were eventually to become our major crops. And yet, for some reason, these late-Palaeolithic people (see Box 1.2 for a discussion of the various chronologies used here) did not choose to exploit their preferred plants more intensively as their principal food source. It seems that people did not seriously contemplate alternatives to hunter gathering unless they had compelling reasons to do so. The reason is that hunter gathering is a very attractive lifestyle in terms of the effort expended and the nutritionally diversity of the resultant food. The major downside is that it normally entails a degree of nomadism, with all the attendant dislocation of regularly uprooting encampments and moving over often long distances before a new temporary base camp can be established. Such dislocation is especially difficult for nursing mothers and their relatively helpless infants, and can be a significant factor in the higher rates of both infant and maternal mortality in nomadic cultures.[8]

The issue of female and infant mortality in hunter–gatherer populations is still highly contentious and in particular the relevance of studies of recent societies to more ancient Neolithic and Palaeolithic cultures. One example is the assertion that systematic infanticide might have been used as a regular method for reducing the burden on mothers who needed to be both mobile and still maintain care of older dependent children.[9] It is difficult to know exactly how stressful regular migration would have been for Neolithic and Palaeolithic family groups as this would depend on such vagaries as the size of the group, the extent and difficulty of migratory journeys, and the climate. However, the stresses endured by women in hunter–gatherer groups might be minimized by the establishment of long-term base camps where small children could be left with carers, such as siblings and grandmothers, while their mothers foraged in the locality.[10] This highlights the importance of the unusually high postmenopausal longevity in humans that is the basis of the so-called 'grandmother hypothesis', as favoured by many evolutionists.[11] Although some authors have asserted that the 'grandmother effect' is a relatively recent, and therefore culturally explicable, phenomenon,[12] most anthropologists regard it as being a considerably more ancient, and hence evolutionarily determined, effect that dates back at least as far as the Mid-Palaeolithic Era.[13] Notwithstanding the stresses of dislocation and regular mobility, hunter–gathering can still provide an ample, balanced food supply for a lot less effort than farming.

Some idea of the efficiency of a hunter–gathering lifestyle comes from a well-known study of contemporary !Kung Bushmen from the Kalahari Desert. It has been estimated that these people only spend one-third of their time (or 2.3 days per week) in food gathering; for the rest of the week they are free to indulge in other pursuits.[14] Over the millennia, the !Kung have acquired an enormous amount of detailed botanical knowledge about each of the many dozens of different food plants that form a regular part of their diet. Some of these plants would be amenable to more systematic and intensive cultivation, should the people wish it. The !Kung are also well aware, from observation of their farming neighbours, of the methodology of crop cultivation. As the !Kung also know, parts of their home range might sometimes be suitable for cultivation of certain crops. However, and most importantly, the !Kung are also cognisant of the unfavourable logistics and the greater risks of relying solely on farming for their food supply.[15] These

Box 1.2 Geological and archaeological chronologies

Geological timescales

Geologists use a chronology based on Periods, such as the Jurassic (208–144 million years ago) and Cretaceous (144–65 million years ago). The most recent Periods are the Paleogene (65–23 million years ago) and Neogene (23–0 million years ago) The Neogene includes geological time up to the present day, covering what used to be called the later Tertiary and the Quaternary Periods (for a discussion of the latest geological nomenclature, see Gradstein *et al.*, 2004). The Neogene is divided into four Epochs: Miocene (23.03–5.332 million years ago), Pliocene (5.332–1.806 million years ago), Pleistocene (1.8 million–11,500 years ago), and Holocene (11,500 years ago to the present).

The vast majority of events described in this book occurred during the Pleistocene and Holocene Epochs. It was during the early Pleistocene Epoch, over one million years ago, that *Homo sapiens* emerged in Africa and subsequently spread across most of the world. This Epoch was characterised by dramatic climatic fluctuations, especially during the series of Ice Ages of the Late Pleistocene from 126,000 until 11,500 years ago. The Holocene Epoch, in which we are still living today, can be regarded as the latest interglacial interval (or interstadial) of the Pleistocene. The beginning of the Holocene coincides with the Neolithic era used by archaeologists to define the beginnings of agricultural societies.

Archaeological timescales

Archaeologists divide the prehistoric development of humans into a number of chronological stages. The Early Palaeolithic era is generally considered to have started with the emergence of the first members of the genus *Homo* about 2.5 million years ago. The Middle Palaeolithic era lasted from about *250,000 BP* until about *50,000 BP*, and was characterized by extensive use of chipped stone tools by human cultures around the world, including *H. erectus*, *H. ergaster*, *H. neanderthalis*, and *H. sapiens*.

Modern humans, capable of complex social and aesthetic behaviours, probably arose in Africa before *100,000 BP*. Around *50,000–40,000 BP*, at the onset of the Upper (or late) Palaeolithic, tools became smaller, more intricate, and much more diverse, and people created increasingly elaborate art forms. The final phase of the Palaeolithic (generally known as the Epipalaeolithic in the Near East), lasted from the end of the last major glaciation *c. 18,000 BP* until the end of the Younger Dryas *c. 11,600 BP*. This period marked the beginning of the long transition from hunter–gathering to farming in several regions of the world.

Finally, the Neolithic, or 'New Stone Age', began about *11,600 BP* with the introduction of superior grinding methods for the manufacture of stone tools, and the gradual adoption of more complex sedentary/agricultural lifestyles. In the Levant, the Neolithic is divided into a prepottery phase (actually two phases termed prepottery Neolithic A and B, or PPNA and PPNB) that lasted from *11,500* to *8,500 BP*, and the pottery Neolithic from *8,500* to *7,000 BP*. The Chalcolithic (Copper) Age lasted from *7000* to *4500 BP*, the Bronze Age from *4500* to *3200 BP*, and the Iron Age from *3200* to *2500 BP*. In some regions, such as Europe, the postglacial but prefarming period is known as the Mesolithic, which lasted in many areas until *5000 BP* or later.

Of course, these dates are approximate and overlap with each other to a great extent. Some cultures developed or acquired new technologies many centuries or even millennia before their contemporaries in different parts of the world. For example, as late as the mid-twentieth century, some isolated cultures in South America and Asia were still very successfully maintaining an essentially Palaeolithic-like lifestyle. Unfortunately, as with biological taxonomy (see Box 2.1), both primary and secondary geological and archaeological sources sometimes define their chronologies slightly differently to those described here. I have tried to follow the most consistent modern usages, but note that some literature sources may vary slightly.

---

sophisticated people are aware that farming in the Kalahari Desert does not bear comparison, in terms of an overall long-term cost/benefit analysis, with their current hunter–gatherer lifestyle.[16]

It seems likely that similar logic, whether or not it was consciously expressed as such, would have

prevailed in the remote past when our ancestors may have faced a choice between the more systematic exploitation of a few relatively abundant plants, or a more generalist hunter–gathering lifestyle. A key factor that probably tipped the balance in favour of the latter choice would have been an environment that was sufficiently productive of resources to sustain the sort of familiar hunter–gatherer lifestyle that had been pursed by most modern humans since they left Africa over 70,000 years ago. There was neither need nor motivation for these people to search for alternative means of generating biological resources for their sustenance. This does not mean that people did not constantly experiment with potential new food sources. Especially during lean periods during the constantly changing climates of the Palaeolithic, people would have sometimes been forced to rely more on larger fauna or perhaps to investigate any potentially edible plants, even small-seeded grasses.[17] In a few parts of the pre-Ice Age world, there was a periodic abundance of one rather special food source that would eventually become much more important to people, namely the starch-rich seeds of several pooid and panicoid grasses.

Some of these grassy species that grew in profusion throughout western Asia were those selfsame cereals that would eventually become domesticated as our most important staple crops. Useful pooid species included the wheats, barley, and rye; while exploitable panicoid species included many of the millet crops. In parts of the Near East, it is still possible for a modern forager to collect enough grain from wild cereals in a few hours to provide nourishment for an entire week.[18] This means that Palaeolithic people passing through such areas would have been highly rewarded if they stopped to gather any nutritious wild-growing plants that they came across, including cereal grains and fruits. However, at the same time, it would not have been particularly attractive to settle down in one place and try to grow such plants to the exclusion of other readily available foods. Such a strategy would be risky in its reliance on a few species, as well as involving a great deal of unnecessary, hard work. In order to understand why crops were ever domesticated at all, we must look more closely at the complex interactions between a host of interrelated factors, which gradually altered the cost/benefit equation away from the flexibility of the hunter–gatherer lifestyle and towards a less flexible, riskier, but ultimately more productive, sedentary/farming lifestyle.

The term 'productive' is applicable here in several senses. Farmers obtain far greater productivity than hunter–gatherers in terms of food calories per unit area of land. Farming can therefore sustain much greater populations, not all of whom need to be involved in food production. The greater numbers of people that could be supported in a farming-based society would give them an advantage in the case of conflict with groups of hunter–gatherers. The non-farmers would also be free to specialize in other pursuits such as tool making and building. Farming/sedentism is therefore immensely more productive in terms of technological innovation. Farming also engenders cultural changes that favour identification with larger groups than the family/clan, for example religious identities, allegiances with a city/state, specialized male fighting groups, etc. The existence of such organizations and social structures in turn enables urban/agrarian societies to operate effectively on a much larger scale than the relatively small groupings formed by clan-based hunter–gatherers.

Gradual transitions

The shift from exclusive hunter–gathering to farming probably occurred in a series of stages over several millennia. These stages would have established the necessary conditions for agriculture but would not have made it inevitable. The kinds of conditions needed for farming to begin include the availability of the 'right sort' of plants, that is plants that lent themselves to domestication due to their genetic make-up. People would also have needed to be very familiar with such plants; for example what they looked like, where they grew, when they set seed, what else ate them or competed with them, and so on. They would have needed the right technologies for harvesting and processing of the edible parts of the plants into easily digestible food. A degree of sedentism would also have been useful, but not necessarily essential. It has been suggested that some hunter–gatherer groups may have maintained a series of small gardens, which they visited periodically for tending and harvesting. This would have given such people the opportunity to familiarize themselves with the rudiments of plant cultivation and enabled them to experiment with strategies, such as tilling, sowing, and weeding, that would encourage better growth of their favoured plants. Such activities could readily occur within a peripatetic hunter–gatherer lifestyle without any kind of irrevocable commitment to full-time agriculture.[19]

However, even if all of the above conditions of incipient agriculture were in place, there would still be no need to make the change to more or less full-time farming, as long as there were plentiful and readily accessible sources of alternative food resources. Any prolonged threat to these alternative resources might have supplied the stimulus that pushed some communities towards a more serious investment of time and energy into the cultivation of just a few chosen plants. For example there may have been localized situations where many of the normal animal and plant resources became scarcer, possibly due to climatic changes.[20] Such events might have eliminated the more agreeable and more easily collected sources of food for a hunter–gatherer community that also happened to be well versed in preagricultural cultivation of domestication-friendly plants such as wild cereals. Hence, these people may have been forced into specializing in the cultivation of a few, relatively high-yielding food plants, simply because the alternative food collection strategies became too expensive and unproductive. Almost by default, they would have become the earliest farmers. But we must recall that the same people would have previously been growing very similar plants on an informal basis for a considerable time, and perhaps for many millennia.

There is increasing evidence from archaeological analysis, some of it very recent, that people were informally cultivating wild plants, including sowing their seeds into tilled soil, long before these plants evolved into the sorts of domesticated crops that we recognize today.[21] During this new type of manipulation by humans, the plants would have experienced a subtly different environment compared to their previous 'wild' condition. Some of the plants would adapt well and flourish in the new human-imposed conditions, while others would not.[22] Naturally, the human gatherers would have favoured those food plants that grew well and produced high yields under such conditions. This would have led to the gradual, unconscious selection of a number of genetic attributes in these favoured food plants, hence modifying the genetic profile of the species in that region and initiating the process of domestication. This kind of unintentional, preagricultural domestication would have altered some plant species more quickly and to a much greater extent than others. Those plants that became genetically altered in favourable ways for the human gatherers would have gradually (or, in a few cases, rapidly) evolved into our main crop species.[23] Far from a sudden 'agricultural revolution', therefore, it appears that there was a developmental continuum over tens of millennia during which some human groups and certain plants coevolved into a series of mutually beneficial associations. In different parts of the world, different plants became the favoured partners of human societies although, where they were available, cereals were invariably selected as the major staple crop.[24]

One remarkable aspect of early preagricultural human societies is that, right across the world, out of over 7000 plant species that were regularly used for food, only a tiny number of mainly grassy species were eventually selected and domesticated to serve as the primary dietary staples.[25] The importance of cereals to our ancestors is reflected in the word itself, which is derived from the name Ceres, who was the Roman goddess of plenty. Even today, cereals still supply 80% of our global food needs. In terms of dry matter per year, we produce 1530 million tonnes of cereals compared with about 400 million tonnes of all the other crops combined; including tubers, pulses, sugar cane, and the various fruits. It is especially noteworthy that, despite all the impressive developments in agriculture and breeding over the last twelve millennia, the dozen-or-so plant species that were originally chosen by early Neolithic farmers remain our most important dietary items to this day. This applies most particularly to the ancient crops from the grass family, including the cereals, wheat, rice, maize, barley, sorghum, millet, oats, and rye.[26] These plants still provide about 60 to 80% of the total protein and calorie intake of people across the world.[27] As with domesticated animals, therefore, only a tiny fraction of the potential riches of the plant kingdom has ever been domesticated by humankind.

These facts beg a number of important questions. Why did people focus on this extremely small group of plants when thousands of other, equally nutritious, species were also available? Was plant breeding ever a conscious and deliberate process on the part of the early agrarians, or did it all really just happen by chance? Is our repertoire of domesticated crops so small because these selected species are uniquely amenable to domestication? If so, what are the prospects for domesticating some of the thousands of other potentially useful plants that still represent one of the greatest untapped resources on the planet? In the coming chapters of this book we will examine these questions in detail and hopefully provide some of the often surprising answers now emerging from some very exciting areas of research, ranging from genetics and climatology to archaeology.[28]

Human beginnings

We will start our quest by looking at how modern humans arose as a distinct species and how their interaction with plants gradually became modified in the face of localized and global climatic changes which continually modified their physical and biological environments (see Figure 1.4 for a summary of the main processes). Humans originated in Africa, where several species of the genus *Homo* evolved over the past two million years and lived as omnivorous hunter–gatherers. As discussed in Box 1.3, recent archaeological evidence suggests that, from at least *100,000 BP*, and possibly earlier, there were groups of *Homo sapiens* in Africa and beyond that had many, and perhaps almost all, of the attributes and cognitive potential of modern people.[29] So-called 'modern' attributes are implied by findings of images in Middle Stone Age layers at the Blombos Cave in South Africa that have been dated to about *77,000 BP*.[30] The images predate the great migration of humans from Africa that gave rise to the modern populations of non-African people. The early evolution of complex behaviour in humans is also suggested by data from mortality profiles of the animals they hunted. The ability to select prime-age prey is indicative of a high level of technological and behavioural sophistication. It used to be thought that such behaviour only arose

after *50,000 BP*, but new studies of fossil assemblages in Africa and Eurasia show that it is much older, possibly dating from before *100,000 BP*.[31]

The prevailing view that cognitive modernity arose in Africa and that such people spread across the world during the post-*70,000 BP* migrations has recently been challenged.[32] In 2006, it was reported that shell beads dating from between *100,000* and *135,000 BP* had been apparently manufactured as items of symbolic display. Pierced shells of the marine gastropod, *Nassarius gibbosulus* were found at two widely separated sites in modern Israel and Algeria.[33] Both locations were inland, with the Algerian site being almost 200 kilometres from the sea, implying that the shells were valued sufficiently to merit long-distance transport and were possibly traded for other commodities. The findings demonstrate that aspects of cognitively modern behaviour were already developing in Africa and the Levant well before the advent of fully anatomically modern humans. This implies that the earliest *Homo sapiens*, who migrated from Africa well before *100,000 BP*, may have had some of the advanced cognitive attributes previously only ascribed to later forms of our species, such as the European Cro-Magnon cave painters after *40,000 BP*.[34]

Over the past two hundred millennia, as we now know from DNA evidence, there was a series of migrations from Africa that eventually reached each of the other inhabited continents, giving rise to all existing populations of our species, *Homo sapiens*.[35] One particular wave of African migrants, which left after *75,000* to *70,000 BP*, seems to have gradually supplanted existing groups of humans, including *Homo erectus*,[36] *Homo floresiensis*,[37] the Neanderthals,[38] and previous waves of *Homo sapiens*,[39] which had already spread across much of Eurasia.[40] Today, there remains just a single species of the genus *Homo*, most members of which are rather closely related in genetic terms. Genetic evidence, from analysis of Y-chromosome (representing the paternal lineage) and mitochondrial DNA (representing the maternal lineage), suggests that the vast majority of contemporary humans is descended from the relatively small groups of migrants that started to leave Africa some 70 millennia ago.[41] Those superficial differences that

---

Box 1.3 Cognitive modernity

Cognitive modernity is the suite of complex behaviours and potentials that is supposedly present in modern *Homo sapiens*, but absent in 'archaic' members of this and other species of the genus *Homo*. It is still often assumed that so-called 'cognitively modern' humans arose relatively recently, probably between *50,000* and *40,000 BP*, in a process epitomized by the growing complexity of Eurasian technological and cultural artefacts and the displacement of the Neanderthals between *40,000* and *28,000 BP* (e.g. Klein and Edgar, 2002). Probably the best-known examples of these 'advanced' artefacts are the Eurasian cave paintings dating from about *35,000 BP*. These abstract or depictional images are generally agreed to provide evidence for the types of cognitive abilities often considered integral to modern human behaviour. As described in the main text, this view has been challenged over the past decade following the discovery in Africa of much earlier human cultural artefacts, such as decorative jewellery and abstract representations that date back as far as *100,000 BP* (see Gabora, 2007, for a recent review).

One should also be cautious in attempting to define exactly what constitutes a 'modern' human. Such definitions are frequently used in a rather teleological manner to build and interpret behavioural models of the distant past. Of course, the definition of a 'modern' human also impinges on that elusive Holy Grail of philosophy: 'what it is to be human'. Here, one should beware of falling into tempting traps such as the essentialist perspective of humanity, or universalist definitions of what constitutes a modern human (Gamble, 2003). Such efforts often founder on the shoals of circular argumentation and progressivist, teleological, accounts of human evolution. In reality, the suite of attributes that we currently consider characteristic of modern humans is ever changing, especially as we continue to discover more about animal behaviour and human biology.

For example, as discussed in Box 1.4, it is now apparent that the Neanderthals may have shared many more attributes of cognitive modernity than previously believed, including complex speech and aesthetic senses. It is also apparent that some so-called 'advanced' human attributes can be latent in an individual and may only become overtly expressed within a particular physical and/or cultural context. People not subject to these conditions may appear to lack some attributes of cognitively modern humans, but still possess the potential to display such characters. A notorious example is the Victorian prejudice (still occasionally alive today) that many so-called 'primitive' peoples somehow lack the full range of cognitive attributes of more technological cultures. In reality, such people have all the latent potential of any other type of modern human, but it was not adaptive for such traits to be expressed in their particular culture. Such considerations make it especially challenging when deciding the limits of cognitive modernity in the sense discussed here. Perhaps it is better to accept that the attributes of so-called cognitive modernity are part of a complex suite of physical and mental changes that gradually arose over the past >150,000 years in anatomically modern versions of *Homo sapiens* and that some, but perhaps not all, of these characters may have been shared by other hominid species (McBrearty and Brooks, 2000).

It is probably as invidious to try to date the onset of human 'cognitive modernity' as it is to say when people first began to employ agriculture. Rather than being discrete and temporally defined events, both are arbitrary evolutionary processes with manifold causes, no predetermined trajectories, and no defined end-points. For example, David Harris succinctly describes 'An evolutionary continuum of people–plant interactions' (Harris, 1989, 2003). Several species of early African hunter–gathering hominids evolved complex social and cultural networks. They buried their dead and some of them produced representational art, as exemplified by shell jewellery, cave paintings, and bone sculptures. Could such people have developed agriculture over 80,000 years ago? The answer is quite possibly 'yes', at least in principle. But, as we will see in Box 3.2, in practice there were many additional prerequisites for agriculture, such as climatic stability and availability of suitable plant species, which were not in place until many tens of millennia later.

---

do exist between people around the world are due to the action of a tiny number of genes. Some of these genes can alter visually prominent features, such as skin pigmentation or eye shape, but otherwise we are a very homogeneous species indeed.

Because they are descended from relatively small groups of migrants, most non-Africans are genetically-speaking a rather uniform population.[42] In contrast, sub-Saharan Africans, being a much an older population, tend to be more genetically diverse.[43] This means that, notwithstanding their external appearance, the average Japanese person is likely to be much more closely related to an Icelander or Peruvian than the average Namibian is related to a typical Nigerian. Modern research makes it quite clear that there is no genetic basis for so-called 'racial' differences between people. There is no such thing as an Asiatic or an Aryan race; still less is there an English, Welsh, or French race in any genetically meaningful respect.[44] This means that concepts of 'purity' with regard to our ethnicity or genetic endowment[45] have absolutely no basis in terms of biology.[46] In contrast to the culturally convenient nineteenth century ideas of biologically determined racial identities, a more recent synthesis of knowledge across disciplines, including archaeology, climatology, geology, molecular genetics, linguistics, physical and social anthropology, and even parasitology, supports a much more inclusive view of human interrelatedness.[47]

Climate, migration, and food

Climatic change and small-scale migrations

Despite our surprisingly high degree of genetic interrelatedness, we humans are a particularly adaptable and culturally diverse species. This adaptability has been tested many times over the past hundred millennia, which has been, and potentially still is, a period of great variation and sudden change in the global climate.[48] The ever-changing local and global weather patterns have caused huge fluctuations in rainfall, temperature, and sea level, with dramatic consequences for the plant and animal life upon which emerging humanity depended. Thanks to evidence from ice-core samples, pollen records, fossil distributions, isotope abundances, and other sources, we now have a pretty fair understanding of the extent and consequences of climatic changes over the past few million years, and especially the last 150,000 years.[49] As shown in Figure 1.1, climatic oscillations increased markedly in amplitude about three million years ago, with the last one million years being an especially variable period. The last 450,000 years, which covers the emergence of hominids such as *Homo erectus* and *Homo sapiens*,

has been characterized by long spells of very cool conditions, punctuated by shorter periods of milder weather.[50]

Soon after anatomically modern groups of *Homo sapiens* appeared in Africa, there was a relatively warm period, called the Eemian interglacial, between *130,000* and *110,000* BP, and some populations emigrated to the Levant during this period.[51] After *110,000* BP, the global climate became cooler, although at first this may not have been so marked in much of Africa (Figures 1.1B and 1.2). The start of what many believe to be the last great human emigration from Africa after *75,000* BP[52] coincided with a glacial period, often called the Ice Age, when the world was much colder and drier than today.[53] Plant communities respond rapidly to relatively small climatic shifts, so the large climatic changes of the Upper Palaeolithic caused huge alterations in global vegetation patterns.[54] Thick ice sheets covered most of northern Europe and Canada, while further south lush forests were replaced by prairie-like grassland. From *75,000* to *12,000* BP, there was an extended period of particularly unstable climatic conditions covering the period when modern humans became dispersed across much of the world (Figures 1.2 and 1.3A). After *75,000* BP, *H. sapiens* populations in the Levant either died out or migrated, possibly due to competition from migrating Neanderthals retreating from the ice-bound continent of Europe. These Neanderthals became the sole human occupants of the Levant until the return of new groups of *H. sapiens* at around *45,000* BP.

During this key period of human development, the climate was much less stable than it has been during the relatively congenial Holocene Era that spans the past twelve millennia, and in which we are still living. Moreover, during the last 60,000 years, there have been at least 30 particularly severe climatic excursions that affected the entire global system. These excursions are referred to either as 'Heinrich events' or 'Dansgaard–Oeschger' events, and correspond respectively to sudden cooling and warming periods. Heinrich events are named after climatologist Hartmut Heinrich, who noted drastic fluctuations in parameters such as temperature, atmospheric CO_2 concentration, rainfall patterns, and sea level.[55] Although classical

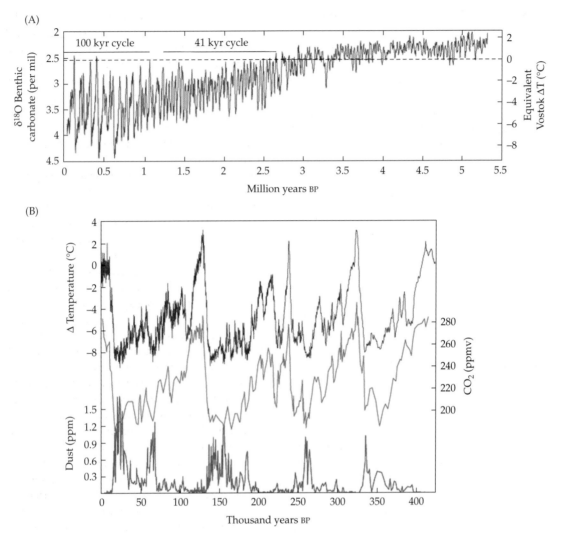

Figure 1.1 Climatic fluctuations over the past five million years. (A) Climate change over the last five million years showing the transition to much cooler and more variable conditions about three million years ago. Carbonate (per mil)—the units 'per mil' are parts per thousand difference from the isotope ratio of the reference standard. (B) Climate change over the last 450,000 years showing a series of brief warm spells interspersed with longer, cooler periods. Note that cooling tends to be gradual whereas rewarming is often very rapid. Both data sets are from Vostok ice and sediment cores in Antarctica. Figure 1.1A data from Lisiecki and Raymo (2005) as redrawn by RA Rhode, available online via Wikimedia Commons at: http://commons.wikimedia.org/wiki/Image:Five_Myr_Climate_Change.png. Figure 1.1B data from Petit *et al.* (1999) http://www.ngdc.noaa.gov/paleo/icecore/antarctica/vostok/vostok_data.html. Available online via Wikimedia Commons at: http://en.wikipedia.org/wiki/Image:Vostok-ice-core-petit.png

Heinrich events have only been described between about *60,000* and *17,000 BP*, it is likely that similar events have occurred before and since this period. Indeed, the Younger Dryas Interval of *12,800* to *11,600 BP*, which we will examine at length in Chapter 3, was probably a Heinrich-like event. Dansgaard–Oeschger events are named after the two geologists who first described them.[56] There were at least 23 Dansgaard–Oeschger warming events between *110,000* and *23,000 BP*, each involving an initial rapid increase in average temperature, normally over a few decades or less, followed by a much more gradual and extended period of cooling.[57] Therefore, although the Palaeolithic Era was

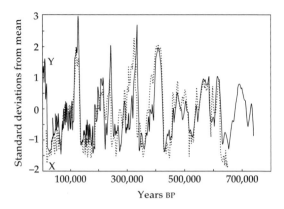

Figure 1.2 Correlation of atmospheric CO_2 levels (dotted line) with proxy temperature data (solid line) over the past 700,000 years. Three important conclusions can be derived from this figure: (1) average global temperatures are highly correlated with atmospheric CO_2 concentrations; (2) CO_2 levels have fluctuated greatly throughout the million-year history of *Homo sapiens*; and (3) CO_2 levels reached a low point during the depths of the most recent Ice Age, about 17,000 years ago (marked 'X' on graph), after which they rose rapidly as the world rewarmed up and vegetation recovered (marked 'Y' on graph. Image produced by Leland McInnes and available online via Wikimedia Commons at: http://en.wikipedia.org/wiki/Image:Co2-temperature-plot.png from original data of Jouzel *et al.* 2004, Siegenthaler *et al.* 2005, and Barnola
et al. 2005 from original data publicly available at NOAA (http://www.ncdc.noaa.gov).

appreciably cooler and drier than now, there were several sudden, dramatic oscillations leading to warmer periods of several centuries or more, plus spells of much wetter weather (Figure 1.3A).[58]

Research over the past decade, as summarized in Figures 1.1, 1.2, and 1.3, has led to a new paradigm of abrupt climatic changes, often over a timescale of a few decades or centuries, rather than over many millennia, as was the traditional view.[59] These sudden climatic events led in turn to often drastic changes in global geophysical and ecological conditions that affected life throughout the planet. Evidence from Greenland and Antarctic ice core data, and other sources, suggests that many of these drastic warming and cooling events happened very quickly indeed, sometimes within a single year.[60] Therefore, what was previously characterized as simply the 'Ice Age' is now known to have been a much more complex period with frequent and rapid climatic reversals. The ultimate causes of these climatic shifts are still controversial, but they

may well involve periodic fluctuations in solar activity and perturbations in the earth's orbit that lead to alterations in global climatic systems, such as oceanic circulation, glaciation, and rainfall patterns.[61] It is possible that the series of human migrations out of Africa during the Late Pleistocene was at least partially related to ecological disruption in their home areas and/or the opening up of new areas for colonization due to various forms of climatic change.[62]

It was during this particularly changeable period that new human migrants from Africa colonized much of the world (Figure 1.4).[63] By *67,000 BP* these people had reached the Pacific shores of Eastern Asia; Australia was probably settled by several waves of migrants from *60,000* to *40,000 BP*; and they had reached Europe by *40,000 BP*. This latter migration coincided with the demise of the indigenous Neanderthal species of humans, who may have been unable to compete technologically and/or reproductively with the new African immigrants (Box 1.4).[64] A final series of migrations took these dynamic people, via northern Asia, across Beringia into North America at about *25,000 BP*, ultimately settling throughout South America by *13,000 BP*.[65] Beringia was the 1600-kilometre-long land bridge linking America with Eurasia before its most recent inundation *c. 11,000* to *10,500 BP*. Beringia existed for many millennia prior to *35,000 BP*, covering a vast area from the Kolyma River in the Russian Far East to the Mackenzie River in the Northwest Territories of Canada. It was reformed during the period *24,000* to *11,000 BP*[66] and people were probably free to move between Eurasia and America until about *10,500 BP*.[67]

It is worth pointing out here that these transcontinental journeys were not necessarily epic treks of mass migration involving tens of thousands of people of the sort that occurred during the well-known *Völkerwanderung* at the end of the Western Roman Empire.[68] For example recent mitochondrial DNA data suggest that the number of founder members in the original group of African migrants, from whom most of today's five billion non-Africans are descended, may have been as low as 600 women.[69] While there may have been additional women in this group, the genetic evidence shows that none of them left any descendents that

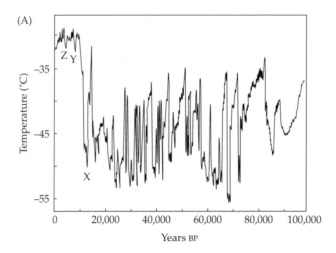

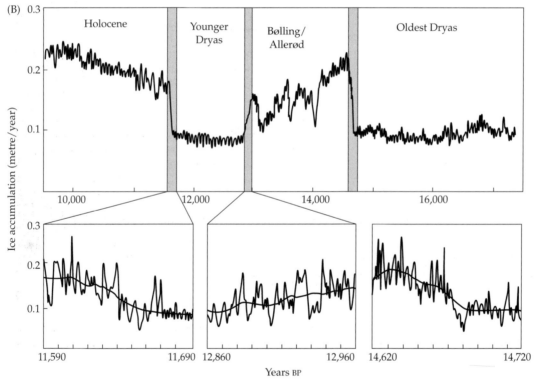

Figure 1.3 Climatic changes over the past 100,000 years. (A) Temperature record for central Greenland over the last 100,000 years. The large excursion at the Younger Dryas Interval (X), and the smaller temperature oscillations at about 8200 years (Y) and 4200 years ago (Z) are merely the most recent in a long sequence of such abrupt temperature fluctuations. Changes in materials from beyond Greenland trapped in the ice cores, including dust and methane, demonstrate that just as for the Younger Dryas and 8200 BP events, the earlier events shown in Figures 1.1 and 1.2 also affected much of the global climate. Graphs are based on Cuffey and Clow (1997) from original data of Grootes and Stuiver (1997). (B) Ice accumulation record for central Greenland over the last 15,000 years as a proxy measurement of temperature. Note the very sudden warming transitions between the relatively cold/arid conditions of the Oldest and Younger Dryas Intervals and the warm/moist conditions of the Bølling–Allerød and Holocene periods. These contrast with the more gradual cooling trend during the Bølling–Allerød period. As shown on the expanded lower scales, most of the warming at the end of the Younger Dryas occurred over as little as 20 years, between *c. 11,640* and *11,620 BP*, with an equally rapid rewarming at the end of the Oldest Dryas after *14,680 BP*. Modified from Alley *et al.* (1993).

are alive today. A similar genetic analysis of the descendents of the Amerind speakers who travelled across the Bering land bridge shows that the original ancestral founder group may have numbered fewer than 80 individuals.[70] It was this tiny group of people that gave rise to the most of the millions of North- and South-American Indians. Given the extremely small size of this founder population, it is possible that there were many other bands that had also attempted such journeys, and some of them may have even settled in parts of the Americas. However, few, if any, of

the descendents of these other groups appear to have survived to the present day.

The practical consequence of these very recent genetic findings is that we no longer need to think in terms of humans moving out to populate the world in a small series of epic mass migrations. The emerging paradigm is rather of many slow journeys by small bands of a few score people. Such journeys need not have been true migrations precipitated by some sort of dramatic crisis. A single band might have simply extended its foraging range because of local resource limitations or

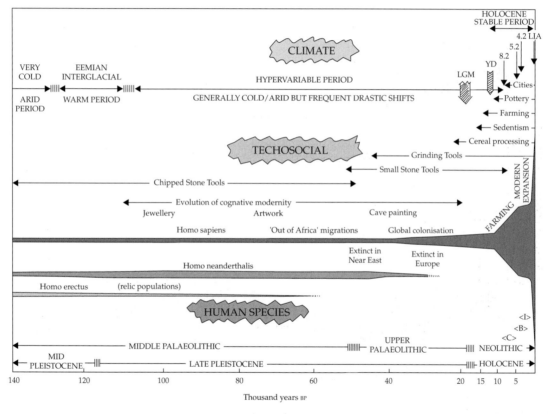

Figure 1.4 Technosocial and climatic contexts of human evolution. This period, which spans the Late Pleistocene and Holocene Eras, witnessed the most recent global migration of fully modern *Homo sapiens* from Africa and the demise of other species of *Homo*, including *H. erectus* and the Neanderthals. *Homo sapiens* developed complex technologies for the acquisition and manipulation of foods, ranging from cereal grains to caribou, as well as aesthetic sensibilities and skills that led to manufacture of jewellery and artwork. But humans of the Late Pleistocene were faced with a particularly variable climate that largely precluded the use of farming as an effective food-winning strategy. After *c. 12,000 BP*, the exceptionally stable, warm, and moist conditions of the Holocene stable period (albeit punctuated by several cooler, arid interludes, as arrowed) favoured the spread of several domestication-friendly plant species and their subsequent exploitation via agriculture in many parts of the world. C, Chalcolithic Age; B, Bronze Age; I, Iron Age; LGM, Last Glacial Maximum; YD, Younger Dryas Interval; 8.2, *8200 BP* cool/arid event; 5.2, *5200 BP* cool/arid event; 4.2, *4200 BP* cool/arid event; LIA, Little Ice Age.

Box 1.4 Could the Neanderthals have become farmers?

Although the Neanderthals probably died out soon after *30,000 BP*, is it conceivable that they could have developed agriculture, with all the technocultural consequences that this implies? This topic relates to other key issues such as cognitive modernity (Box 1.3), determinism (Box 3.1), and the prerequisites for agriculture (Box 3.2).

Were they clever enough?

As discussed in Box 1.3, it seems likely that a wide range of Middle Palaeolithic human populations possessed many attributes of cognitive modernity, including more complex forms of social structure, art, and tools. Can the Neanderthals be included in this group? Quite possibly. For example the discovery of a *H. sapiens*-like hyoid bone on a Neanderthal skeleton suggests that Neanderthals were fully capable of complex speech (Arensburg *et al.*, 1990; Bar-Yosef *et al.*, 1992). It is now believed that Neanderthals were also capable of sophisticated technocultural activities requiring advanced cognitive capacities, making them little different from modern human foragers (Henry, 2003; Zilhão *et al.*, 2006).

Were the right plants available?

The Neanderthals lived until about *38,000 BP* in the Near East and *28,000 BP* in southern Europe (Finlayson *et al.*, 2006; Jiménez-Espejo *et al.*, 2007; Finlayson and Carrión, 2007). While normally portrayed as hunters with a primarily animal diet, some Neanderthal populations living in the Levant at about *50,000 BP* enjoyed a surprisingly plant-rich diet (Henry, 2003). These people ate wild cereals, legumes, nuts, and fruits as supplements to their animal diet. Therefore, like other humans of the period, Neanderthals would have been familiar with such plants.

Was the physical environment suitable?

For much of the Neanderthal period, the Levantine climate was moister than today and game was sufficiently abundant to make plants a marginal dietary supplement. In Europe, the climate was much cooler and wild cereal stands were absent. In both cases, the environment militated against the need to exploit plants more intensively.

Did they have the right tools?

One of the interesting aspects of the eventual development of agriculture in the Near East is that it did not depend on the invention of a new suite of tools. People had been using sickles and grinding tools in non-agricultural contexts since at least *50,000 BP*, while flint adzes and hoes were developed as woodworking tools by Natufian hunter–gatherers many thousands of years before they were adapted for use in farming (Cauvin, 2000). Farming then proceeded successfully for about four millennia before the invention of the first agriculture-specific technologies, such as ploughs and animal traction. Neanderthals were able to use such complex tools but may have failed to invent technologies, such as food storage and improved clothing, quickly enough to adapt to the highly variable climate after *42,000 BP* (Figure 1.3) (Bar-Yosef, 2000; Henry, 2003).

Why did they die out?

The Neanderthals of the Near East were probably the only population of their species that knew enough about cereals to develop agriculture, but this outlying group died out or left the region by *38,000 BP*. The larger group of European Neanderthals persisted for another 10,000 years but eventually succumbed to higher mortality rates than *H. sapiens*, and probably also failed to innovate technically. By this time, 40% of Neanderthals died before adulthood and fewer than 10% survived beyond the age of 40 (Trinkaus and Thompson, 1987). Competing (but not necessarily warring) groups of better-equipped *H. sapiens* only needed a 1–2% lower mortality rate to have out-bred Neanderthals, resulting in their extinction in as little as 30 generations, or less than a single millennium (Zubrow, 1989).

Neanderthal farmers?

Some Neanderthals might have had sufficient intelligence and botanical knowledge to become farmers. Had they survived the various crises of *40,000–28,000 BP*, it is quite possible that a few groups of Neanderthals could have eventually become farmers eighteen millennia later, when agriculture eventually became an adaptive strategy of food production in many parts of the world. Whether such putative Neanderthal farmers would have been tolerated by neighbouring groups of *H. sapiens* is quite another matter. . . .

simply to follow a charismatic leader, to whom they would have been bound by strong social and/or kinship ties. Over a period of centuries the descendents of this small band might become separated by many hundreds of kilometres from neighbouring groups, as they continued to forage in search of an improved home range. The vast majority of such groups probably came to grief in various ways, leaving the few successful 'migrants' to become the genetic founders of populations that would eventually be numbered in the tens of millions. So, the emerging picture is that, from *75,000 BP*, relatively small groups of people across the world were gradually on the move. One of the major factors responsible for these minimigrations was probably the changeability of the climate, which in turn altered the availability of plants and animals upon which the people depended. For example a transition between glacial conditions to present day warmth could occur within a single decade or even less, that is well within the lifetime of many of the people who experienced these rapid shifts in weather patterns.[71]

The new human migrants from Africa proved to be extremely adaptable to the series of world-wide climatic fluctuations that would have repeatedly and drastically affected the local fauna and flora. This resilience in the face of climate change and its many consequences may have played a key part in the ability of the African immigrants to outcompete the many older, indigenous groups of humans across the world, including the remnant populations of *H. erectus*, which were still distributed throughout Southeast Asia, the Neanderthals, who were to be found throughout Europe, and the so-called archaic *H. sapiens*.[72] These older human communities were gradually marginalized, their populations declined, and they eventually became extinct (see Box 1.4).[73] Meanwhile, by about *50,000 BP*, the descendents of those rather more successful African migrants had spread as far as western Asia and the Mediterranean Basin where they were soon faced by a new set of challenges. In the remainder of this Chapter and the next we will focus mainly on the events in this region from about *50,000* until *15,000 BP*. The reason for concentrating on western Eurasia, rather than east Asia or Mesoamerica, is that there is far more evidence

available for events in the former region, which was eventually to be the site of wheat and barley domestication. We will come back to review matters in the other two regions, which were eventually to give rise to rice, maize, and squash crops, in Chapters 6, 8, and 11.

Moving down the food chain

During the Upper Palaeolithic Era (*c. 50,000– 11,500 BP*) human populations in the Mediterranean Basin and Near East gradually changed their hunting patterns. In particular, archaeological evidence has revealed that people began to hunt much smaller animals, switching from the likes of deer and gazelle to rabbits and birds.[74] This shift in prey preferences is unlikely to have been voluntary because the larger species would have been preferred in terms of the cost/benefit ratio of hunting them versus the amount of nutrition and other resources (such as skin, fur, and bone for tools or jewellery) obtained from them. The implication is that something was causing a decline in the numbers of larger prey animals. This selective population decline was probably due to a combination of environmental and biotic factors. Such factors would have included, but were certainly not limited to, climate change and over-hunting by humans. It seems, therefore, that the people in this region of Eurasia were gradually confronted with a shortage of larger prey species and so began to exploit smaller animals, such as birds, small mammals, and tortoises.[75]

Smaller prey animals would have been harder to catch and less rewarding than larger prey, and this may have resulted in shortages in the food supply. Earlier in the Upper Palaeolithic, human foragers had seldom bothered with such paltry and uneconomic prey. This was the first of several steps down the food chain that were made by these Palaeolithic people. As population pressures grew, and even the smaller prey animals became ever scarcer, the next step was to use plants of all kinds as an increasingly prominent dietary component. These dietary shifts would have occurred in localized areas where the previously preferred prey had become scarcer due to environmental and/or anthropogenic factors. Given that the Upper–Middle Palaeolithic was a particularly volatile climatic period, it is likely that human populations constantly had to adapt and modify their dietary and resource-gathering strategies.[76] The overall trend in western Eurasia was towards the hunting of smaller animal prey and an increased gathering of plant resources of all kinds.

Hominids had probably always been omnivorous to some extent, ever since their divergence from other anthropoid apes about four million years ago. For example, hominids developed thickened dental enamel and jaws, and larger, flatter teeth that allowed them to cope with a more varied diet than other apes.[77] Their dietary range was further enhanced by cultural innovations that favoured hunting, such as complex social networks, and the use of fire, tools, weapons, and other technologies.[78] By the early Upper Palaeolithic, many human populations exploited large protein-rich prey as a major component of their diet. In this respect, these people occupied the ecological niche of climax carnivores, such as wolves and the larger cats. But there was a crucial difference between people and true carnivores. The more successful climax carnivores, especially the large cats, have specialized to such a degree that they now find it very difficult to move away from this particular ecological niche, that is they are obligate carnivores. Their sharp canine teeth, superbly equipped as they are for ripping and tearing of relatively soft animal tissues, are poorly equipped to deal with any form of plant diet. Just try to imagine a lion or tiger trying to subsist on a diet of cereals and pulses. In contrast, humans are facultative carnivores who have retained a more generalist form of physiology and dentition.[79] So, fortunately for the future of *H. sapiens*, even during their time as specialist carnivores, they never lost their immense dietary flexibility. This meant that they were able to switch to alternative food sources whenever the need arose, as it did constantly during our ever-shifting climatic history.

The Middle-Upper Palaeolithic Era was marked by waves of expansion and migration from the Near Eastern end of the Mediterranean Basin towards the west.[80] By the end of the Palaeolithic, about *12,000* to *10,000 BP*, Mediterranean/Near Eastern humans had moved even further down the food chain, from being eaters of small animals to becoming mainly herbivores. A significant feature of this relatively rapid movement across trophic levels,[81] which is a highly unusual ecological phenomenon, is that lower trophic levels can support larger populations. Hence, there are more plants (in terms of biomass) than herbivores, and more herbivores than carnivores, while the climax carnivores at the top of the food chain have the smallest populations of all. By moving down several trophic levels, humans were able to increase their populations, albeit at the expense of higher energy expenditure in terms of food collection and processing. Their dietary flexibility gave humans a powerful tool, enabling them to adapt repeatedly to climatic changes and associated demographic changes in prey populations. It has also enabled them to migrate into a huge diversity of new ecological zones that lie well beyond their African homeland. It was their ecological malleability that gave people the capacity to build up their own populations, even as other species increased or declined in numbers during the ever-shifting conditions of the Palaeolithic period.[82] For example no other primates were able to move across from a diet based on forest fruits to steppe species such as cereals, or to leave Africa, in the way that humans have.[83]

Broadly similar shifts down the food chain towards such lower-ranking (both nutritionally and in terms of energy required to acquire them) food resources as wild grasses have recently been documented in late Palaeolithic northern China.[84] In this case it was wild millets that were exploited by human foragers as other more desirable food sources became scare due to cooling and aridification. These and similar developments elsewhere in the world during the late Palaeolithic set the scene for the much more extensive use of cereals, from *23,000 to 13,000 BP*, and led to the first experiments in plant cultivation. This special ecological flexibility that modern humans possess is largely due to a physiological and behavioural ability to adapt their diet and lifestyle according to what is available at the time. Although we are unable to digest certain complex organic polymers such as cellulose, lignin, or chitin,[85] we are still able to eat almost anything else, from the tiny seeds and large fruits of plants to the flesh of all animals from fish to mammoths. Our flexible genetics has also allowed some modern human populations to develop an ability to use milk if it is available in abundance, but not to develop this ability if it is not required. Technology and custom have also played important roles in food exploitation. For example many seeds and tubers are poisonous but can be rendered safe by the appropriate treatment, such as prolonged soaking in water and/or extensive cooking. Such manipulations can also alter the taste, nutritional quality, and even storage potential of a foodstuff.[86] Unfortunately, it is often difficult to assess if, and to what extent, a given group of people used such methods to improve their food, so the mere presence of seed remains at a site will not necessarily give the full picture of how effectively the seed was exploited.

The dietary resilience of many late-Palaeolithic populations was called upon when the world entered what is called the 'Last Glacial Maximum', from *25,000* to *15,000 BP*.[87] As its name implies, the Last Glacial Maximum was a full-blown ice age with extensive snow cover for much of the year in temperate regions, coupled with a drier and more arid climate with appreciably lower sea levels than today. Obviously, such a drastic climatic change had an enormous impact on the type and distribution of animals and plants throughout the world. In turn, this meant that human populations in many parts of the world could either try to adapt to the new conditions, migrate away from the worst affected areas, or face the oblivion that was the fate of many other animal and plant species. As in previous ice ages, many temperate and subtropical forests died out and were replaced by grasses, including members of the cereal family. Across vast regions of the world, only a few relict woodlands survived as isolated refugia, surrounded by huge expanses of treeless, dry grassland. In some areas, these prairie-like ecosystems supported large populations of grassy plants that had somewhat larger-than-average starchy seeds. These plants were to change the course of human development: they were, of course, what we now refer to as the cereals.[88]

CHAPTER 2

Plant management and agriculture

Man, despite his artistic pretensions, his sophistication and many accomplishments, owes the fact of his existence to a six-inch layer of topsoil and the fact that it rains.

Anonymous

Introduction

The behavioural and dietary flexibility of modern humans enabled them to alter their hunting and foraging strategies on repeated occasions during the climatically turbulent millennia of the Upper Pleistocene. Part of the human response to constantly changing faunal and floral distributions in this period was to spread across much of Africa, Eurasia, and Australasia.[89] In some of these regions, people encountered a relatively new type of food resource, the grasses, which required considerable ingenuity to process into an easily edible form. Presumably, the initial stimulus to exploit the starchy seeds of wild grasses was a combination of an overall dearth of more convenient resources, such as animals or fruits, and the relative abundance of grasses during their early-summer ripening season. These people were already proficient toolmakers and had been using grinding implements since at least *40,000 BP*,[90] and possibly as long ago as *200,000 BP*.[91] Therefore the means to exploit this versatile and easily stored food resource were already at hand, at least potentially. We will now look at how people started to gather and harvest wild cereals at least ten millennia before they cultivated them, and how such activities established new ecological conditions that favoured the evolution of so-called 'domesticated' varieties of several cereal species.

The rise of cereals after *25,000 BP*

The onset of cooler and drier climatic conditions after *25,000 BP* favoured the spread of many grass species throughout Southwest Asia (Figure 2.1). It is here that we get the first glimpse of a kind of prepastoral use of some of the plant species that were to become the ancestors of many of today's major crops, including barley and wheat. Some of the initial insights into the early use of cereals by hunter–gatherers came from excavations supervised by US archaeologist, Robert Braidwood, in the 1950s and 1960s. Braidwood's team showed that Near Eastern peoples were collecting and processing wild cereals by at least *15,000 BP*. They also found that in *9000 BP* people in the farming village of Jarmo were still using exactly the same seed-processing technologies employed by their hunter–gatherer ancestors many millennia previously.[92] In other words, these technologies were already mature well before the people actually used them in a farming context.[93] It is clear that the Palaeolithic hunter–gatherers in this region, centred on Anatolia and Syria, did not actually cultivate the cereal plants that grew in such profusion every summer. Rather, they collected the grain from existing stands of wild cereal plants that they found during their continual forays for food plants. As cereals became more common, as a result of the changing climate, the people in this part of the Near East would have become more familiar with these plants.[94]

Eventually, these people would have recognized that the seeds of the wild cereals were edible and would have started to collect cereal grains wherever they found them.[95] However, a cereal grain is

Figure 2.1 Beginnings of semisedentism and cereal harvesting in the late Palaeolithic Levant, *c. 23,000 BP*. During some of the warmer interludes of the late Palaeolithic, much of the Levant was populated by a biologically rich mixture of woodland and grassland. This ecosystem supported semisedentary communities of human hunter–gatherers subsisting on wild plants (including cereals), fish, and small game. As they settled in an area to exploit its seasonal resources, such people constructed temporary shelters in the form of simple huts of branches and reeds, as shown in this example from *23,000 BP*, based on excavations at Ohalo near the Sea of Galilee (e.g. see Nadel *et al.*, 2004).

not as easy or pleasant to chew on as a sweet, juicy fruit such as a pear or a tasty nut such as an almond. Pear, *Pyrus* spp., and almond trees, *Prunus dulcis* (*syn. Amygdalus communis*), were relatively common in the Near East after *25,000 BP* and would have been much more attractive food sources than raw cereal grains. The breakthrough that made it

not just feasible but positively beneficial for people to start exploiting cereals on a larger scale was the discovery that cereal grains could be ground and processed to render them more edible. So when did people devise methods to process cereal grains to more the palatable foodstuffs that we are familiar with today, such as breads and cakes? The answer was only found recently, and it now appears that our ancestors were engaged in food technology a lot earlier that anybody had previously suspected.

In 2004, archaeologists came across surprising evidence that people living in an encampment at Ohalo on the shore of the Sea of Galilee were using stone grinding tools to process seeds of wheat, barley, and other cereals into flour as long ago as 23,000 BP.[96] An oven-like hearth found at the site suggests that the flour dough was also baked into seed cakes, as is still done today by people in the region. At the same site, there was a profusion of seed remains, including wild forms of wheat and barley plus simple huts that served both as shelters and as sites for cereal processing.[97] This shows that relatively sophisticated processing of cereals into foodstuffs was underway more than twice as long ago as the earliest firm evidence for the beginning of agriculture, which dates from around 13,000 to 11,000 BP. The Ohalo fishing and hunter–gathering community of 23,000 BP used a highly varied mixture of grass seeds, including many small-grain species as well as larger-grained cereals. The switch to the predominant use of cereal grains in their diet was a gradual process that took place over as much as 15,000 years.[98] The Ohalo discoveries show that this Levantine hunter–gathering culture, and possibly others in the region, was already familiar with the collection and manipulation of grasses, including cereals, for the manufacture of foods more than ten millennia before people grew plants in any systematic way as crops.

Similar grinding stones have been found at much older African and Asian sites, some dating from as long ago as 200,000 to 50,000 BP. It was presumed that these older stones were used primarily to grind plant and animal materials, or minerals, to make pigments, rather than for the preparation of foodstuffs. However, the new findings from the Sea of Galilee raise the intriguing possibility that some human groups may have been using grinding stones to process cereal grains, and maybe other types of edible plant, as early as the Middle Palaeolithic Era (i.e. before 50,000 BP). But why is it such an advantage to grind cereal grains before eating them? The main reason is that grinding breaks down the hard, fibrous cereal grain to release the

easily digestible starch granules contained within. This serves two purposes. Firstly, it enabled people to save enormously on the wear and tear of their teeth, compared to eating raw, unprocessed grains. Unlike the teeth of grazing animals, human teeth do not continue to grow after childhood. Tooth wear due to a diet enriched in high-fibre, raw plant products can result in the substantial erosion of molars by early adulthood. People with worn or absent teeth faced starvation, unless they could find alternative types of food that did not require chewing. Alternatively, they could try to find another way of grinding the fibrous plant material before eating it. Perhaps this was one of the incentives that led to the use of stone grinding tools for seed processing.

The development of grinding technology would have been socially advantageous to a human group. Not only would people who could grind or mill plant products need to hunt less frequently, they would also tend keep their teeth for much longer, despite subsisting on a largely plant-based diet. This might have also been a factor in enabling older, more experienced individuals to live longer, despite the ultimate loss of their teeth. Such people could then earn their keep either as 'grandmother' child carers or by acting as media for the innovation and transmission of oral culture. The latter role was a key adaptation in preliterate societies, particularly in relation to strategies for food acquisition and technology in an era of considerable climatic flux. The remembered knowledge of how their grandparents dealt with the last arid period, including alternative food acquisition strategies, would have enabled such surviving elders to greatly enhance the ability of their clan to deal with such contingencies. Unfortunately, as we will see later (Chapter 9), grinding seeds to make flour could be a mixed blessing. Depending on the type of stone used, the prolonged and laborious process of grinding cereal grains could produce small chips of stone that would get into the flour. People eating the products of such flour every day would be repeatedly exposed to the stone chips as they chewed their food, and eventually their teeth might become chipped and worn. As discussed in Chapter 10, this problem was partially alleviated many millennia later by the invention of pottery, which enabled a porridge to be made from grains mixed with water and boiled.

The second, and more immediate, reason for grinding cereal grains is that it enables us to produce a much more attractive, sweeter tasting, more nutritious, and calorie-rich foodstuff. Rather than a hard, dry, indigestible, tooth-destroying cereal grain, people could enjoy foods such as seed cakes, biscuits, and all the various forms of bread that we still relish so much today. Cereal grains that have been ground and processed into flour can be much more easily digested due to the higher surface area that is available for gastric enzymes. This means that, not only the plentiful starches, but also the grain proteins and the much less abundant micronutrients, are more easily assimilated from processed cereals. In the cold, dry climate of the Last Glacial Maximum, plants of the grass family, such as cereals, would have been a more reliable source of food than woodland plants (e.g. 'nuts and berries'). Many of these woodland plants would have had died out as the weather worsened, and edible animals would have also become increasingly unavailable as they migrated to warmer climes, leaving cereals as one of the few remaining options for the people who chose, or were obliged, to remain in this area of the Near East.

The most common cereal found at the Sea of Galilee site at Ohalo was not wheat but barley. This is because barley is an especially resilient cereal with a larger geographical range than most types of wheat or rye (Figure 2.2). This tough plant can grow in the mild summers found in northern temperate regions today, and which were prevalent across much of the Near East twenty-odd millennia ago. Moreover, barley is relatively tolerant of the arid, salty soils produced as the climate cooled suddenly down after *13,000 BP*, when it is believed that the first attempts were made at its systematic cultivation. The type of barley found in the original Ohalo site from *23,000 BP* is so closely related to the modern crop that it has been classified as part of the same species, *Hordeum vulgare*, albeit as an undomesticated genotype. This Levantine wild barley is likely to have been the major progenitor of all the modern domesticated varieties of barley.[99] In contrast, the types of wild wheat found at the same site, *Triticum monococcum* (einkorn) and *Triticum*

turgidum (emmer), are from the same genus as modern breadwheat, *Triticum aestivum*, but are classified as different species (see Box 2.1 for a discussion of cereal nomenclature). The wild emmer wheat found at Ohalo is the progenitor of the modern glutinous durum wheats, from which semolina and all of the many and varied forms of contemporary pasta are made.

Although people in some parts of western Asia relied on cereal grains for at least some of their diet, there is no evidence that there was any organized effort to cultivate these plants as crops during the Last Glacial Maximum of *25,000* to *15,500 BP*. It is likely that the wild cereals grew in such profusion that grain could be gathered quite readily, and hence used to supplement an existing diet. Many types of wild cereals would have been collected at this time. However, wild barley and wild emmer wheat would have been especially suitable for exploitation by human groups. Unlike other the other types of wild cereal growing in these regions, wild barley and emmer produce large, durable seeds that would have been available for several months in early summer and could be collected by hand.[100] Another advantage of these grains is that, unlike many fruits, any surplus could be stored until hungry periods, such as winter. The accidental dropping of some seeds around storage and living sites would have ensured a more reliable supply of cereals as they germinated and matured during the following season. A third type of wild cereal, wild einkorn wheat or *Triticum monococcum boeoticum*, was especially common in the northern Levant and across the Anatolian plateau of modern Turkey (Figure 2.2).

The potential of einkorn wheat as a food source was shown by an experiment conducted in Turkey in the 1960s. An American geneticist, called Jack Harlan, demonstrated that a small family equipped with a typical Palaeolithic stone sickle could gather enough wild einkorn in only 3 weeks of hand harvesting to last them a full year.[101] Harlan's experiment involved harvesting wild stands of einkorn wheat that were growing in one of its putative centres of origin in the highlands of eastern Turkey. Although this type of einkorn is in the same genus as modern breadwheat, it was not expected to yield nearly as much grain and the quality was

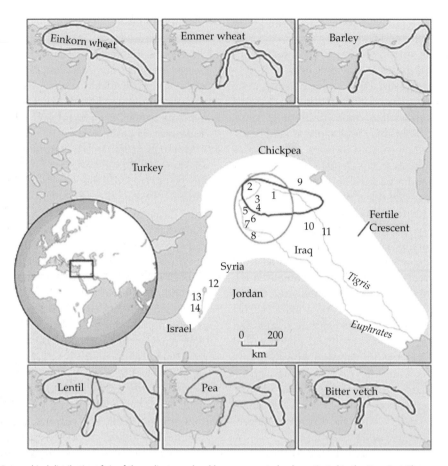

Figure 2.2 Geographical distribution of six of the earliest cereal and legume crops to be domesticated in the Near East. The maps show the distribution of the original wild ancestors of these six crops in the early Neolithic, immediately prior to their domestication. Note the overlap of several crops in the 'core domestication area' of the Levant and Upper Tigris/Euphrates Valleys. The upper row shows the distribution of the cereal crops: einkorn wheat, emmer wheat, and barley. The lower row shows the distribution of the legumes: lentil, pea, and bitter vetch. The larger map shows the distribution of Neolithic sites: (1) Cayönü, (2) Cafer Hüyük, (3) Nevali Çori, (4) Göbekli Tepe, (5) Djade, (6) Jerf el-Ahmar, (7) Mureybet, (8) Abu Hureyra, (9) Hallan Çemi Tepesi, (10) Qermez Dere, (11) Milefaat, (12) Aswad, (13) Yiftahíel, and (14) Jericho. Adapted from Lev-Yadun *et al.* (2000). Botanical data compiled from Heun *et al.* (1997); Zohary and Hopf (1993); Ladizinsky (1999); Zohary (1996).

expected to be lower. Surprisingly, Harlan's group was able to use their primitive sickle to harvest almost a kilogram of grain per hour. Moreover, the wild grain had a high protein content and was amenable to boiling or steaming to make a nutritious porridge. Harlan's experiment also showed that the use of harvesting tools, such as sickles, does not necessarily mean that crop cultivation is taking place.[102]

It is quite possible that, many centuries before they began to grow crops for themselves, people had already developed harvesting tools to facilitate the gathering of seeds from stands of wild plants.

Harlan's study demonstrates that wild cereals could have been a significant food resource for nomadic hunter–gatherers well before any such human groups adopted a sedentary lifestyle. We should also remember, however, that not all human groups were necessarily nomadic hunter–gatherers in the Palaeolithic Era. Indeed, only terrestrial hunter–gatherers are really obliged by the mobile nature of much of their food supply to be nomadic. Riverine (riverside), lacustrine (lakeside), and littoral (coastal) communities, who are able to subsist largely on aquatic resources that are available for much of the year, often adopt sedentary or

Box 2.1 Cereal nomenclature

Scientific nomenclature is often bedevilled with inconsistencies and disputes about the best system to adopt. This problem is growing in several fields, such as taxonomy, as new genetic and other data suggest that some current usages might be misleading. However, by creating a new system of nomenclature, it becomes increasingly difficult for non-specialists (and sometimes specialists too) to refer to the older literature where one or more alternative systems have been used instead. I had not realized just how acute this problem was until I started to research the genetics of cereals, and especially to read some of the older literature. It soon became evident that the systematic nomenclature of the cereals, and especially the various types of wheat, was in a particularly convoluted and inconsistent state of flux with one species sometimes being referred to with as many as four alternative formal Linnean names by different authors. To some extent, similar problems are also encountered with non-cereal crops, as recognized in the new system for intraspecific classification called the International Code of Nomenclature for Cultivated Plants (Trehane *et al.*, 1995), but the situation in cereals is especially problematic.

One of the major international repositories of cereal genomic information, called GrainGenes, contains the following somewhat rueful appeal. 'Although GrainGenes does not endorse any particular taxonomic treatment, we are very interested in the development of synonymy tables to help de-Babelize the various existing classifications' (see the GrainGenes website (http://wheat.pw.usda.gov/GG2/index.shtml) for details). Probably only an American could come up with that wonderfully evocative term 'de-Babelize', which just about sums up the whole sorry situation, and the frustration of those affected. In August 1998, participants at the IWGS (International Wheat Genetics Symposium) workshop on taxonomy drew attention to the following problems: (1) the large number of different classifications currently in use; (2) the lack of uniformity in the nomenclature of wheat species; (3) the failure of researchers to consistently follow a given classification; and (4) misunderstandings of species concepts, which have caused serious errors in the selection of germplasm.

These wheat scientists agreed that the confusing condition of wheat taxonomy is particularly difficult for new researchers in the field—never mind the interested outsider, no matter how well informed. The situation is so bad that it has been necessary for the Wheat Genetic Resources Center (Kansas State University) to publish a list of 'unaccepted' names of wheat species that are regarded as 'illegitimate, not validly published or ambiguous'. Another useful account of the problem is given in Morrison (1998). In this book, I have attempted to use the most recent and/or the most commonly used names as much as possible and therefore follow the system of Van Slageren (1994), as also employed by Waines and Barnhart (1992); Hancock (2004); and Yen *et al.* (2005). See also the section on 'Wheat' in Chapter 6, where some of the basic taxonomy is described further. A useful website is that of the Kansas State University Wheat Genetic Resources Center: http://www.k-state.edu/wgrc/Taxonomy/taxintro.html. However, as with the warning about radiocarbon chronologies in Box 1.1, readers should be cautious about cereal names in much of the literature, whether primary and secondary.

---

semisedentary lifestyles. In some cases, as at Ohalo, lacustrine communities also dabbled in cereal processing, which may have reinforced their tendency to maintain a semisedentary lifestyle. In general, it seems that experience with and the availability of potential plant domesticants was probably far more important than factors such as sedentism and population pressure in setting the scene for the development of agriculture.

We should also remember that the Ohalo people, who were grinding barley and wheat at around *23,000 BP*, did not rely on cereals to anything like the extent of later farming cultures. Firstly, the plant portion of their diet merely supplemented a plentiful supply of fish and game. Secondly, in addition to the relatively large-grained cereals, these people collected and processed a wide range of small-grained grasses that were almost of equal importance to the cereals in their dietary starch intake. The collection and processing of small-grained, non-cereal grasses has continued in numerous human cultures to the present day. For example, in Ethiopia, the minute grains of tef, *Eragrostis tef*; and in Australia, the small-grained grasses, *Panicum australiense* and *Fimbristylis oxystchya*, are still important dietary staples.[103]

People of the Ohalo culture augmented their cereal/grass diet with a plethora of other plant foods that included almonds, pistachios, acorns, wild olives, wild figs, and wild grapes, to mention only a few.[104] This varied and nutritionally balanced diet of fish and plants was supplemented still further by small game, such as hare and partridge. As we will see in the next chapter, it was not so much the inherent attraction of cereals that led to their eventual adoption as the primary staple. Rather it was the disappearance of most of the alternative edible plant and animal resources from the ecosystem inhabited by people such as the Ohalo culture and their neighbours in the Near East. But before this happened, there was a pleasant interlude of milder weather during which the importance of cereals actually declined in the region. This climatic amelioration effectively postponed the need to develop agricultural systems for a few more millennia.

A warm interlude after *15,500 BP*

After a 10,000-year ice age, the global climate changed yet again at about *15,500 BP*, with a rapid warming and deglaciation that was especially marked in the northern hemisphere.[105] This may have lessened any reliance on cereals for people in the Near East because, as forests became re-established, other more attractive edible plants and animals were available for exploitation. By *14,000 BP*, the climate in almost all parts of the word was at least as warm and moist as today, and in some areas it was even warmer. In addition to this warming trend, two other factors may have been even more important in favouring the more prolific growth of vegetation of all kinds. Firstly, the climate was much less arid than it had been during the Ice Age. As discussed in Box 2.2, aridity is a far greater threat to most plants than extremes of temperature. The second factor favouring a resurgence of plant growth was the huge increase in atmospheric CO_2 concentration (see Figure 1.2) after the Ice Age.[106] By about *15,000 BP*, the atmospheric CO_2 concentration had increased by almost 40% from less than 200 ppm to over 275 ppm.[107] For many centuries after *15,500 BP*, there was a steady northward progression of plants and animals

to recolonize the formerly icebound temperate regions of the north.

This would have been a time of relative plenty for many human populations who, thanks to strong group co-operation and improved technologies, were now emerging as some of the most effective predators on the planet. The ready availability of more easily processed and digested foodstuffs, such as meat and fruits, during this halcyon period would probably have led to a greatly reduced necessity to exploit cereals, and it would certainly have removed any serious incentive to try to cultivate such plants. Even during these climatically benign times, however, the use of cereals as a seasonal foodstuff continued in some areas. At about *13,000 BP*, the climate was relatively wet and mild in upland parts of south-western Asia, from the inland Levant up to southern Anatolia.[108] Such conditions favoured the growth of dense stands of large-grained wild grasses (Figure 2.1). Many human groups across this part of the Near East would have become accustomed to harvesting these grains, possibly coming to rely upon them for much of their food supply at certain times of the year. It is likely that, as these people became more familiar with the wild cereals, they began to assist their growth by practices such as controlled burning, clearing away competing vegetation (weeding), and even sowing or planting seeds. We think that this may be the case because of evidence that other hunter–gatherer societies have repeatedly adopted similar practices for the exploitation of their most favoured plant resources. Such practices, termed non-agricultural plant management (see below), enabled societies to become intimately familiar with many aspects of plant husbandry without necessarily making a commitment to the formal cultivation of crops.

The Kebaran hunter–gatherer culture

The Kebaran culture occupied the general area of the Levant and Sinai from about *20,000* to *15,500 BP*, and was the immediate precursor of the Natufian culture, which lived in the same area and went on to domesticate wheat and barley after the Younger Dryas Interval. The Kebarans are named after the site at Mugharet Kebara, near the Mediterranean

---

Box 2.2 Aridity and agriculture

By far the most important climatic prerequisite for successful agriculture is a reliable supply of water. Availability of water is a much more serious limitation on plant growth than temperature. For example many of our staple crops, including wheat, barley, maize, beans, and brassicas, can flourish equally well in the fierce heat of West Australia and the Punjabi Plains or in the cool damp climate of the Scottish Borders and Alpine foothills. But even these versatile crops will falter in the absence of a steady water supply. Indeed, if global warming turns out to be a long-term reality, it will not be high temperatures that affect farming so much as disruption to rainfall patterns. It is also important to recall that all the major climatic episodes of the Holocene Era affecting agricultural and societal development, from the Younger Dryas to the postmedieval 'Little Ice Age', have involved a cooler rather than a warmer climate. Moreover, in each case, the cold was not the real problem; rather it was a reduction in rainfall, sometimes by as little as 10–20%, that often resulted in widespread collapse of agricultural systems and the complex societies that they underpinned.

For at least 8000 years, farming cultures have adopted either of two strategies to ensure their crops are watered. The original strategy was rain-fed agriculture, which relies on an adequate supply of rainfall during the growing season of the crop. The absolute limit of rain-fed agriculture is the 200-mm isohyet (an isohyet is a line joining points of equal precipitation on a map, equivalent to an isotherm [temperature] or isobar [pressure]). Even this amount of precipitation may not be reliable in the long term due to annual fluctuations that may limit crop production for several years running and hence force people to use up surpluses generated in better years. A more reliable limit for dependable farming over a longer period (of centuries) is the 300-mm isohyet (Oates and Oates, 1976).

In the case of the Near East (see Figures 10.1 and 10.2), the millennia-long evolution of agriculture had its origins in the valleys of the Levant and the upper reaches of the Tigris/Euphrates at the foothills of the Taurus and

Zagros Ranges in those regions between the 200- and 300-mm isohyets and in adjacent wetter areas, with plentiful supplies of water, grazing, game, fruit, nuts, and wild cereals. Gradually, the early rain-fed form of agriculture spread to slightly dryer regions of Northern Mesopotamia that lie between the 200- and 300-mm isohyets. As discussed in Chapter 10, farming and societal development in these regions were repeatedly disrupted by interruptions in rainfall during the early to mid-Holocene, although some of these events may have also acted as stimuli for increased societal complexity (Brooks, 2006).

The second farming strategy is to bring water to the crop via irrigation systems, such as canals or ditches. Irrigation was probably first used in the Near East by the Samarrans as they migrated along the Tigris/Euphrates Basin into more arid regions below the 200-mm isohyet and eventually founded the enduring civilization of the Sumerians (Chapter 10). Although the flow of these rivers was seasonal, there was usually sufficient springtime water to supply a vast area of adjacent fertile alluvial soil during the period of maximum crop growth. The main downside of such irrigation agriculture was its enormous demands on organized human labour, requiring complex management skills and a high degree of social cohesion for its success. Poor management can result in long-term soil damage due to salinization but well-managed irrigation farming can be sustainable in the long term. After the drought of *4200 BP*, irrigation farming enabled the Sumerians to maintain their urban culture for several centuries after the collapse of many rain-fed farming cultures in the North.

Today, only about 20% of farmland is irrigated, but this provides 40% of our global food supply. It is estimated that by 2050, as much as two-thirds of the global population will live in water-scarce areas (Wallace, 2000). As discussed in Chapter 17, one of the major future challenges will be to maintain the supply of water to agriculture in the face of depleted groundwater supplies, increasing salinization, alterations in river flow, and changing rainfall patterns across the world.

---

coast, where many of the initial excavations of their artefacts occurred.[109] A common feature of the Kebaran hunter–gatherer culture is the use of small geometric microlithic tools. The Kebarans were highly mobile hunter–gatherers living in relatively small bands that were well adapted to the

changeable climatic conditions that prevailed during and immediately after the Last Glacial Maximum.[110] The Kebarans lived in the same general region as the Ohalo people who, as we saw above, were busily grinding and baking barley and wheat at around *23,000 BP*. During the intervening millennia, human populations in the Near East declined as the cooler, dryer conditions of the Ice Age took hold. It is possible that there was some cultural continuity between the Kebarans and previous human populations in the region, but it is just as likely that cereal-processing technologies were lost as the plants themselves died out and people searched for other types of food.

Later in the Kebaran period, the people made and used stone grinding tools, including pestles and mortars, as well as sickles suitable for cutting plants, although the latter were rather rare. It is uncertain whether any of these tools were used in the processing of grain for food until plants became more important again in the late Kebaran period and the transition into the Natufian era. Following the end of the Last Glacial Maximum at about *20,000* to *18,000 BP*, the Kebarans who returned to the Levant consumed a high proportion of plant matter in their diet, but there is no evidence of a return to large-scale cereal processing or semi-sedentism as seen at Ohalo five millennia earlier. At this stage, from *20,000* to *15,500 BP*, the Kebarans were largely restricted to the Levantine coastal strip and a few isolated inland oases by a climate that was still relatively cool and dry.[111] By *17,000 BP*, the so-called Geometric Kebaran culture, using characteristic geometrically shaped tools, had developed in the Levant. Later Kebaran groups seem to have progressively reduced their intake of plant-derived foodstuffs and increased their geographical range. This is correlated with the climatic amelioration after *15,500 BP*, during which the relatively arid semidesert of the Levantine interior would have given way to a much more varied habitat of mixed woodland/steppeland, rich in both game and edible plants. Animals such as fallow deer, gazelle, and wild boar were hunted in the woodlands of the Central Levant, while gazelle, ibex, and hare were common in the steppes beyond.

Slightly further afield, in the Taurus and Zagros Mountains that mark the traditional extremities of the 'Fertile Crescent', wild goats, sheep, and aurochs, progenitors of the future domesticants, were commonplace game species. In lacustrine and marine regions of their home range, Kebarans also exploited fish and all manner of invertebrate seafood, although in many places these areas have

now been inundated and destroyed as archaeological sites by rising sea levels.[112] Many Kebaran groups turned increasingly to hunting the newly-prolific game, and their dependence on plants became dramatically reduced. Other groups adopted a different strategy by reducing their mobility in order to exploit more intensively all types of local resources, both plant and animal. With an increasingly sedentary lifestyle, these latter groups were also able to develop heavier and bulkier plant processing tools and technologies (including large grinding stones, kilns, and baking ovens) than their more mobile hunter–gatherer neighbours. One of the better known such groups of this period, at *14,500* to *13,000 BP*, is the so-called Natufian culture that lived around the Levant and its immediate environs.[113]

The early Natufians and sedentism

The Natufians were one of several Levantine cultures of this period, all of which developed from the Geometric Kebaran culture.[114] They originally occupied the entire Jordan valley and beyond, as far as the Mediterranean coast from present-day Jaffa to Tyre. The culture is named after Wadi en-Natuf near modern Ramallah, where the first finds were made by Dorothy Garrod in 1928.[115] A subsequent expansion of their range during the Younger Dryas Era took the Natufian culture to the north along an inland axis straddling the eastern face of the Anti-Lebanon Mountains as far as the Aleppo Plain of Syria and the southern flanks of the Taurus Mountains in Turkey. The earliest Natufians were mainly hunter–gatherers but they were also familiar with wild cereals such as emmer wheat and barley, which they ground into flour and baked (Figure 2.3). Archaeological evidence suggests that the Natufians were one of the first human cultures to adopt a predominantly sedentary lifestyle based in semipermanent villages, and that this occurred well before the development of agriculture.[116]

The impetus for this form of village-based sedentism may have been a brief cold spell at about *14,500 BP* that was immediately followed by an increase in precipitation and an expansion of woodland and parkland within the home rage of the Natufians.[117] This newly bountiful ecosystem

Figure 2.3 Semisedentary Natufian foragers collecting wild cereals. The Natufian culture of the Levant spanned the Palaeolithic/Neolithic transition from about *15,000* to *11,000* BP, during which the first crops were domesticated in the region. The Natufians were semisedentary hunter–gatherers who built some of the first true villages. In this artist's representation, a mixed band of Natufian foragers is collecting wild cereals with flint-bladed sickles and carrying the grain to the nearby village where it was processed into flour using stone pestles and mortars. These tools and the practice of sedentary village life, which are normally associated with farming communities, were invented by hunter–gatherers such as the Natufians many millennia before the beginnings of agriculture.

provided a profusion of faunal and floral resources within a relatively small area, hence reducing the need for a highly mobile lifestyle. A more static population would be better able to manage and exploit these plentiful resources, possibly in analogous ways to the Kumeyaay and the other more recent human cultures that are discussed in the next section (see below). Establishment of semipermanent or even permanent dwelling sites would also have enabled the Natufians to defend their valuable food resources against interlopers. The newly sedentary Natufians had access to a rich woodland flora that was dominated by oak and pistachio trees but which also included a prolific undergrowth of grasses with high frequencies of cereals. In addition to dwelling places, their small settlements contained storage sites, for collected food such as grain, and burial sites. The burial sites show a degree of social differentiation that was absent from previous societies, with a few, presumably privileged, people being interred with valuable grave goods such as seashells and bowls.[118]

A predominantly plant-based diet in these early villages is suggested by the number of tools for plant acquisition and processing, such as sickles, mortars, bowls, and pestles. Edible plants were supplemented by seasonal game in some areas and by aquatic food in riverine and lacustrine areas, for example waterfowl along the Jordan Valley and freshwater fish in lakes such as Hula and Jordan. The relatively benign climatic conditions that favoured the establishment of quasi- or fully sedentary settlements by the Natufians lasted for almost two millennia. This enabled the evolution of a robust and distinctive culture, with its own characteristic decorative artefacts and styles of construction. But, quite suddenly at about *12,800 BP*, the Natufians were confronted with a climatic disaster that almost eradicated them. Many Natufians did not survive as they made abortive attempts to flee elsewhere. It was the people who stayed put in their settlements, who not only survived the disaster, but went on to flourish. The reason for the success of these particular Natufians was that they found a new way of managing their increasingly scarce and restricted plant resources. They discovered what we now know as 'agriculture'.[119] We will consider how these momentous events unfolded in the next Chapter. Meanwhile, for the remainder of the present Chapter, we will examine how human societies like the Natufians might have managed their plant resources in the absence of formal agriculture.

Non-agricultural plant management

In many books and scholarly articles, the title of this section would be '*pre*-agricultural plant management', with the implication that such practices are considered as preludes to formal agriculture. However, we will see that this is a misleading interpretation based on a combination of lack of firm evidence for non-agricultural plant management and a tendency, that is still surprisingly common, to assume that agriculture was somehow both progressive and perhaps even inevitable. For our purposes, we can define non-agricultural plant management as follows: 'the manipulation of plant development and distribution for the purpose of human exploitation without the practice of formal cultivation'. As discussed below, non-agricultural plant management might involve

techniques such as transplantation, controlled burning, and sowing of gathered wild seed. However, in the case of these particular plant species, such management practices did not give rise to the cascade of genetic changes that gave rise to domesticated varieties. Hence, the managed plants remained as wild forms that were favoured by humans, but never became as dependent on them as did fully domesticated crop species. As we will see in Part II, the reasons why many plants never became crops are largely related to the organization of their genomes.

One of the greatest challenges that bedevils the study of many non-agricultural or non-sedentary cultures is the lack of visible traces that they generally leave behind. In the case of grain farmers, we have readily identifiable remains in the form of domesticated seeds, processing tools, and even traces of old field patterns or irrigation systems. Sedentary cultures leave durable evidence of their habitations, as well as tools and other artefacts. However, a more mobile culture, with temporary seasonal camps, that managed a large area of plant resources without actually cultivating or domesticating these species, might leave no trace at all for future generations to find. It is possible that many of the prevailing ideas about the supposed advantages of agriculture have been skewed by this dearth of evidence for alternative lifestyles that may often have been just as viable as farming.

Our views of non-agricultural plant management are gradually being modified as we discover some of the surprisingly sophisticated practices that were commonplace across the world until recently. In fact, there are many well-documented cases of various types of non-farming husbandry that were still in widespread use until the twentieth century and, in some of the more remote areas, these practices have only died out in the last few decades. For example the Bagundji people of southeast Australia used repeated firing of grassland to increase seed production of Mitchell grass, *Astrelba pectinata*. In other parts of Australia, people would dig up and collect the edible tubers of wild yams, (*Dioscorea* spp.) or bush potatoes (*Ipomoea costata*), as well as grain-bearing plants such as wild rice (*Oryza rufipogon*).[120] These hunter–gatherers would replace the stems attached to the tops of the tubers in the ground to ensure that more tubers would be propagated for harvesting in future years.

Meanwhile, across the world in the lower Colorado River Basin of North America, the Cocopa people actively planted seeds, but not in a formal agricultural context. The Cocopa supplemented their diets of game by sowing two species of panic grass, *Panicum* spp. (a type of wild millet), on the floodplain of the Colorado River after the waters receded. Further west, in California, the Miwok used burning, sowing, and harrowing to favour the growth of six wild species of grass seed, including the splendidly named 'farewell-to-spring', *Clarkia purpurea* ssp. *viminea*, and mule ears, *Wyethia helenoides*.[121] Other Californian tribes sowed seeds of wild herbaceous plants as well as grasses, while tribes in the Great Plains used fire and sowing to grow productive stands of Indian rice grass, *Achnatherum hymenoides*. These people, and many others, were relatively mobile hunter–gatherers, who were also capable of actively managing and exploiting plant resources on a wide scale and over a period of many millennia, without any recourse to formal agriculture.[122]

The remarkable Kumeyaay people

In some cases, the non-agricultural exploitation of plant resources reached a very high level of sophistication that involved a particularly impressive degree of botanical knowledge. One example of such a culture is the Kumeyaay people of southern California.[123] The Kumeyaay home range once extended throughout modern San Diego County and southwards into the northern part of the Mexican State of Baja California. The region has a Mediterranean climate, with relatively sparse summer rainfall and a wet season in winter. The Kumeyaay were essentially semisedentary hunter–gatherers who manipulated their floral landscape to an extent that now seems extraordinary, not only in its breadth and complexity, but also in its adaptability throughout the periodic severe droughts that still affect the region. Moreover, and in contrast to many farming-based cultures, the Kumeyaay successfully maintained and modified their non-agricultural lifestyle during at least a

millennium of constant climatic and social change, and probably much longer.[124]

The normally arid, semidesert environment of the Kumeyaay home range is especially problematic for a would-be plant exploiter. For a start, the area is not naturally rich in edible plants. Furthermore, the occurrence of any given plant species (or edible portion thereof) is often acutely seasonal and can be disrupted by over-long summer droughts and excessive winter floods. However, by transplanting various useful species across the range of habitat types that existed in their territory, the Kumeyaay were able to achieve a notably more diverse resource base of flora than would otherwise be found. From the coastal sandbars and dune systems, through valleys and foothills, to the arid deserts of the high mountains, the Kumeyaay experimented with a host of potential food and medicinal plants. These practices of habitat dispersal and multiple sourcing also provided a more predictable availability of plants throughout the year. By utilizing a wide range of plants and locations, the Kumeyaay buffered themselves against the regular, but unpredictable, climatic vagaries that might wipe out all the plants in a particular area or decimate a single species throughout their home range. In contrast, as later farming cultures have found to their cost, attempts to move to an agricultural lifestyle in such areas have been dogged by repeated crop failures due to the unpredictable climate and the over-reliance on a small number of food staples or, even worse, on a single key crop.

Although they deliberately moved and replanted certain species that were useful to them, the Kumeyaay did not actually cultivate any plants. A few examples will give a flavour of the extent of the impressive botanical activities and achievements of these resourceful people. The Kumeyaay created groves of wild oak and pine in the areas of their home range at higher altitude. These trees were then harvested for their edible nuts. They established desert palm and mesquite along the coast. They planted agave, yucca, and wild grapes in appropriate microhabitats in various parts of their range. And they planted cacti, which were used as emergency sources of water, as close as possible to their villages and campsites. In addition to these transplantations, the Kumeyaay managed their floral environment by the systematic burning of tree groves to increase fruit yield; they used the controlled burning of chaparral grassland to improve forage for the (non-domesticated) deer that they hunted; and they resowed a proportion of the edible grain from wild grasses that they had harvested.[125]

The Kumeyaay people lived in this manner for centuries, perhaps millennia. During this period, what appeared to the uneducated eye to be a barren, arid, and hostile landscape was actually a bountiful area that supported tens of thousands of people with a unique series of botanical and resource management skills. Later 'sophisticated' European travellers would starve or die of thirst within a few metres of abundant sustenance, had they only known what plants to look for and how to process them for eating or how to extract water from them. The non-agricultural, hunter–gatherer lifestyle of the Kumeyaay was eventually dealt a mortal blow by the arrival of a Spanish–Mexican expedition led by Fray Junipero Serra. This well-meaning, but narrow-minded, cleric arrived in the region in 1769 and established a series of Missions, including the large Mission and Presidio at San Diego at the core of the Kumeyaay home range. Within decades, their population had collapsed as the people were severed from their livelihoods and forcibly settled in guarded Missions where they were obliged to raise and subsist on unsuitable, and often unsuccessful, crops such as maize.[126] A few decades later, North American settlers expropriated most of the remaining Kumeyaay land, following the seizure of Upper California from Mexico in 1848.[127] Today, only a few scattered bands of Kumeyaay remain on inadequately sized reservations and their unique lifestyle has completely vanished.[128]

Plant management does not necessarily lead to agriculture

To the uncomprehending European incomers, people such as the Kumeyaay seemed to be living a 'simple' life of gathering plant products that grew 'naturally' in the region. In fact, the Kumeyaay and other Amerindian cultures created huge and highly complex botanical gardens, which they carefully maintained, adapted, and exploited for their own use. Similar modes of seemingly basic, but in reality tremendously sophisticated, strategies of floral resource management have been found in other cultures in California. For example the coastal people of Central California were hunter–gatherers who also relied greatly on the seasonal abundance of acorns.[129] These fruits could be collected and stored for consumption during the winter period, when game was relatively scarce. Although some oaks were more favoured than others for their type of acorn, all oaks were equally encouraged to grow in the coastal woodland.[130] This was because acorn production by one tree or even one oak species was highly variable from year to year, but taken together the sum of all the oaks tended to have a similar annual productivity.[131] By spreading their plant resource base to include less desirable species, these coastal communities greatly reduced the risks that would accompany reliance on a single staple food resource. Similarly, the Nomlaki people of the Upper Sacramento Valley sampled an especially diverse flora in the mixed chaparral/oak and conifer/oak woodland in the highlands at 1300 metres during the warmer months, before moving to long-term residential sites in the lowlands in the winter.[132] As we will see in later chapters, many farming cultures across the world would repeatedly fall into the trap of relying on monocultures, and were regularly blighted by famine when their single crop staple failed.

One of the most dramatic examples of botanical resourcefulness in a seemingly hostile environment comes from published ethnographies of the Paiute culture in North America.[133] These people lived in the Owens Valley and Mono Basin areas immediately to the east of the Californian Sierra Nevada. The Paiute lived in a land that had been described as follows by early white explorers: 'The country on this side is much inferior to that on the opposite side (of the Sierra Nevada)—the soil being thin and rather sandy, producing but little grass, which was very discouraging to our stock.'[134] In fact, the countryside had for millennia supported several enterprising cultures, including the Paiute, who both nurtured and exploited all of the available plant resources. Written records describe how the Paiute propagated wild hyacinth (*Camassia quamash*), nut-grass (*Cyperus rotundus*), and spike rush (*Eleocharis palustris*), all of which are root crops that grew abundantly in seasonal water meadows bordering the Owens River. Higher up in the Sierra foothills, were extensive pine forests dominated by several variants of pinyon pines, especially the single-leaf pinyon, *Pinus monophylla*. Every autumn, soon after their beloved vitamin-rich rosehips turned red and ripe for gathering, entire families of Paiute would trek up to these forests and harvest the nutritious pine nuts as a winter food.[135] With a protein content of over 30% by weight, which is higher than any other nut or seed, pine nuts are a greatly prized foodstuff that we still use today as the basis of pesto sauce.[136]

In the spring, Paiute men dammed tributary creeks in the hills near their low-altitude winter camps, and dug a series of irrigation ditches up to 6 km long to the meadows in the Owens Valley, thus creating many hectares of new habitat for their edible plants. Although the Paiute did not deliberately sow seeds, their activities resulted in a considerable expansion of the habitat of certain naturally occurring plants, which in turn increased the yield and productivity of these important food sources. The Paiute dismantled their dams every year, so without the written records of eyewitnesses, their work would have been invisible to archaeologists. As with many similar examples of non-agricultural plant management, the Paiute culture collapsed abruptly in the late-nineteenth century, as invading miners and prospectors cut down the stands of pinyon pines that were both one of the key food resources of the people and the keystone of their entire semimontane ecosystem.[137] To make matters worse, the cattle introduced by the newcomers to feed their mining camps roamed across the Paiute lands, eating precious stands of wild hyacinth and nutgrass. Reduced to near starvation by the 1860s, desperate bands of Paiute started to raid the cattle that were devouring their food supply. Despite some early successes, the Paiute resistors were doomed as the US Army took the field against them in force. Following their military defeat, most of the survivors were deported in 1863 to the San Sebastian Indian Reservation, near Fort Tejon, just north of Los Angeles.[138]

The achievements of such peoples should stand alongside the more formal categories of agriculture with regard to the ingenious and sustainable long-term exploitation of plant resources. It is likely that there were hundreds, perhaps thousands, of human cultures existing according to this kind of highly adaptable, mixed plant husbandry/hunter–gatherer lifestyle over the past 50,000 years. Unlike farming, such practices leave few traces, which means that their importance to the development of plant exploitation has almost certainly been seriously underestimated. We can regard these activities as a kind of 'quasiagriculture', whereby people gradually learned more about how to manipulate those potentially useful plants that grew in their home range without formally growing them as crops. Such knowledge included methods to promote plant growth, how best to harvest the seeds, and how in general to manipulate such plants for their own benefit. This sort of systematic gathering and management of wild plants was not only done for the production of food. Many plants had other uses, such as in the manufacture of clothing, basketry, cordage, medicines, weapons, utensils, tools, and of cultural artefacts such as musical instruments and toys.[139] It must be stressed, however, that the use of the term 'quasiagriculture' is not meant to imply that this sort of lifestyle was merely a stage on the way towards the evolution of farming.

As we saw in the case of the Kumeyaay Indians (see above), such elaborate exploitation of plants did not necessarily lead to the development of formal agriculture. These people, and many other comparable cultures in other locations, remained as very successful non-agricultural hunter–gatherers. In the majority of cases, they maintained stable cultures for centuries and millennia, until their sudden demise following the abrupt disruption of their habitats and social organizations by technologically well-equipped, disease-ridden, and highly aggressive modern European invaders (see Chapter 9 for a discussion of the adaptive advantages of disease tolerance in agricultural societies).

In the context of the late Palaeolithic Era, we can imagine that at least some human cultures managed their floral landscapes in an analogous manner to these later Australian and American cultures.

During the improved climatic conditions of the immediate preagricultural period of *15,500* to *13,000* BP, human populations across the world expanded. Such groups would tend to become increasingly territorial as they came into contact with neighbouring groups, who would often be their most threatening competitors. For example the group controlling a region that was enriched in wild, large-grained cereals would not need to leave the area to forage as frequently as groups in cereal-poor regions. The people in such a cereal-rich region would therefore be more likely to stay put, so as to manage and defend this valuable resource against incursions from competitors, whether animal or human. Eventually this may have led to a shift from a primarily nomadic lifestyle, which largely precludes organized agriculture, to an increasingly sedentary mode based on semipermanent habitations.

Finally, in considering the development of both foraging and agriculture as at least partially biologically-driven processes of coevolution (Box 2.3), it is instructive to note that the seemingly well-organized exploitation of plants in the absence of either domestication or formal agriculture is by no means a solely human attribute. In parts of the Amazonian rainforest there are what appear to be extensive monocultures of a single species of tree from the madder family, *Duroia hirsuta*. These stands of *D. hirsuta* can be several hundred metres wide and are virtually devoid of other plants. Local legends tell of evil forest spirits that cultivate these so-called 'devil's gardens'. The reality is perhaps even more remarkable in that these tree 'plantations' are effectively being cultivated by lemon ants, *Myrmelachista schumanni*, a species that constructs its nests only in this particular tree.[140] Lemon ants attack and eventually kill all other plants by injecting their leaves with formic acid, but will tolerate the growth of saplings of their preferred host tree, *D. hirsuta*. As a result, the ants create large monocultures of their preferred 'crop' plant, without recourse to domestication or formal agriculture. The *D. hirsuta* monocultures provide abundant and secure nest sites for ant colonies that can live for as much as 800 years. Obviously the ants are not conscious agents in this process, but they are effectively acting as plant managers and exploiters on a large scale that has been sustained successfully for many millions of years. One wonders whether agriculture as practiced by people will last for even one percent of this time.

Box 2.3 Agriculture as a coevolutionary process

One of the original, and still clearest, exponents of agriculture as a coevolutionary series of interactions between people and plants was David Rindos (Rindos, 1980; 1984). This perspective is both stimulating and revealing, but should not be regarded as the only useful way of looking at what most people still regard as a form of human-invented technology. The hypothesis is satisfyingly parsimonious in eschewing human intentionality, and in placing agriculture alongside the many other examples of adaptive coevolutionary associations that occur throughout the biological world.

One of the most important predictions of the hypothesis is that this mutually beneficial process would have led to adaptive changes in both partners in each of the many domestication dyads that are involved in agriculture. Hence, most crops have dispensed with their ability to shed seed freely, but benefit from the vastly more efficient propagation mechanism provided by farmers. On the other hand, farming societies have adapted to crops by creating new social structures that have locked them into an ever increasing dependence on these crops. More recently, it has become clear that people have also adapted genetically to their new association with crops (see Box 9.1). Some of these genetic adaptations, such as craniofacial reduction and lactose tolerance, have reduced our ability to survive in the absence of plant and animal domesticants, and therefore tend to tighten our dependence on crops in the same way (although not to the same extent) as crops now depend on us for their reproduction. One should not go too far down this road, however, and coevolution does not imply a more or less equal symbiotic association between crops and people.

continues

Box 2.3 *continued*

Clearly, people are the dominant partners in this venture. But it is nevertheless salutary to remind ourselves that we have been genetically modified too as part of our profitable, if not always healthy, relationship with domesticated plants.

An interesting alternative perspective about people/plant relationships is provided by behavioural ecology, which emphasizes the active manipulation of the environment by human groups (Kennett and Winterhalder, 2006). This viewpoint stresses the increasingly sophisticated management of plant by foragers, which in some cases (as modulated by the environment and the nature of the plant resources) led to the adoption of agriculture when its marginal returns exceeded those of foraging (Pearsall, 2006). It is also useful to recall that agriculture is not necessarily an either/or alternative to hunter–gathering. Hence, in many societies the two forms of resource exploitation were practiced simultaneously, with their relative importance at any particular time depending on their relative efficiency under the prevailing circumstances.

As with many complex processes, the study of the origins of agriculture will benefit in future from a more broadminded and multifaceted approach that embraces evolutionary, ecological, economic, cultural, and technological perspectives. One very good reason for adopting such a holistic approach is that agriculture involves all of the above elements, and possibly more. Nevertheless, the coevolutionary perspective, as is apparent from recent genetic studies, can establish useful limits to the potential for agricultural development. Hence, notwithstanding the cognitive abilities and impressive botanical knowledge of the indigenous people of Australia, the absence of domestication-friendly genotypes of food plants rendered farming impossible there for almost 50,000-years. In Mesoamerica, the early domestication of a modest yielding form of maize only allowed for relatively small-scale farming, supplemented by foraging, for several thousand years, before the eventual evolution of larger cobs suddenly made it possible to switch to intensive farming and led to the development of city states and empires. In these and many other cases, the trajectory of agricultural development was clearly modulated, to a large degree, by biological factors residing in the genomes of those plants selected for exploitation by different human societies.

CHAPTER 3

How some people became farmers

And the days passed. And the years.
And Death came and swept them from their refuge; all of that race disappeared with all of its tales and all of its history.
But all things came back to life in that place. Other trees stood tall and other men bent to the ground. Newborn litters roiled in the caves; the tapestry never unravelled.

Wenceslao Fernández Florez, 1943,
El Bosque Animado

Introduction

Throughout the Palaeolithic Era, many groups of people lived in close contact with a wide variety of plants that they exploited not only for food but also for a host of non-edible purposes. At various times during these tens of millennia, it is likely that particular groups of people were forced by various environmental circumstances into a temporary reliance on more restricted groups of plants. However, given the ever-changing climatic conditions, such episodes of dependence on a few types of plant would have been ephemeral. A significant new factor, as the Pleistocene gave way to the Holocene Era, was a relatively stable climatic period in some regions that favoured longer-term exploitation of certain plants such as cereals and tubers. This long-term climatic stability also allowed sufficient time for a few of the favoured plants to adapt to the new human-imposed conditions of floral management by developing traits that tended to facilitate coevolution of the mutually beneficial interdependent association between people and plants that we now call agriculture.

As we will see, agriculture probably could have (and maybe did) evolve before the Holocene, but the climatic conditions were far too variable to support its continued existence for more than a millennium or two before the next cold, hot, or arid episode made it impossible to continue. As we will discuss in Part III, the Holocene itself has been far from free of climatic changes, some of which have had profound effects on agriculture and human societies, but these have been on a much less drastic scale than the dramatic climatic events of the Pleistocene. The respective roles of climatic, cultural, and genetic factors in influencing the development of agriculture are discussed in Box 3.1.

A cold, dry shock—the Younger Dryas Interval

In the previous chapter, we were introduced to a Near Eastern cultural group called the Natufians. During their early period of development, between about *14,500* to *13,000* BP, these people inhabited a relatively benign, postglacial environment, rather like a slightly moister version of today's Mediterranean biomes, which was rich in plant and animal resources. But the good times for the Natufians, and for numerous other human cultures around much of the world, finished abruptly at about *12,800* BP with the onset of a new, short but sharp mini-Ice Age, known as the Younger Dryas Interval.[141] The Younger Dryas Interval was so named by the Scandinavian palaeobotanist, K. Jansen, who noticed unusual accumulations of the arctic-alpine herb, *Dryas octopetala*, at two strata in organic sediments. These accumulations suggested that the otherwise mild climate had undergone a return to relatively frigid conditions on two occasions that Janson termed the Oldest Dryas at pre-*14,700* BP and the Younger Dryas from *12,800* to *11,600* BP.[142] This latter climatic shock profoundly affected many human societies, and is generally regarded as one of the key factors that led to the

Box 3.1 Genetic, environmental, or cultural determinism?

For much of the twentieth century, the study of human development was been plagued by divisions and misunderstandings between specialists from different fields. One of the most heartening aspects of many recent advances in this area has been the emergence of a genuine multidisciplinary approach and a more open-minded willingness to synthesize knowledge from a variety of sources. More rigorous archaeological methods that systematically contextualize artefacts, rather than simply listing and describing them, have greatly enriched the understanding of past human activities. Gradually, such methods are being applied in sites across the world, and are revolutionizing our ideas of societal development in places as far apart as China and the Andes. The use of molecular genetics to track population movements has completely altered our views of human evolution and is enhancing understanding of how and when agriculture was disseminated from its centres of origin. Botanical methods, including pollen and phytolith analysis, have recently pushed back the date for wild cereal exploitation by over 12,000 years. Physical techniques, such as analysis of core samples and isotopic ratios, can give us detailed data on climatic conditions and enable the more accurate dating of artefacts from millions of years ago. New computational and linguistic methods are enhancing the analysis of ancient texts. Techniques from disciplines as diverse as population biology, economics, sociology, risk analysis, statistics, and biochemistry are all making important impacts on our understanding of human development.

Unfortunately, one of the residues of the now-outdated compartmentalized approach to human studies is the lingering controversy about various forms of determinism, and especially so-called genetic, environmental, or cultural determinism. We seemingly live in an age where, although it is apparent that the world is a very complex place, many people still use simplistic labels to describe intricate networks in terms of just one of their properties. It was just such a tendency that produced the false dichotomy of 'nature versus nurture', when obviously both genetics and environment contribute in a variable ratio, depending on the trait(s) in question, to its/their manifestation as part of

the human phenotype. In much the same vein are those more recent controversies about the extent to which humanity, and especially its agrosocial development, has been influenced, or even determined, by cultural, genetic, or environmental factors.

In this book, I present a great deal of evidence showing the importance of genome organization in facilitating crop domestication; and the impact of climatic events on processes ranging from cereal farming in the Sahara to the fall of the Akkadian Empire. But I also show instances where social factors have overridden climatic evens, such as the failure of some, but not all, medieval Welsh farmers to recolonize upland areas when the climate improved—this failure was due to their displacement by English incomers who wished to preserve the uplands as a pristine habitat for hunting. Recent DNA evidence shows that many earlier human migrations, such as the leaving of Africa or the colonization of the Americas, may have been very small scale, non goal-orientated affairs involving only a few hundred people gradually moving over small distances at a time. On the other hand, we also know of larger epic mass-migrations, such as the late Roman *Völkerwanderung*, where hundreds of thousands of people, driven mainly by social pressures, moved in a very organized fashion over large distances to set up new permanent settlements elsewhere. Hence, social, genetic, and climatic factors can exert variable, and largely unpredictable, influences on apparently similar processes in different places and times.

The take-home lesson is that the story of human development and our interactions with the biological (plants, animals, microbes, etc.) and physical (climate, soil, water, etc.) environments is both complex and contingent. These processes are influenced, but not predetermined, by manifold factors that include genetics, environment, and society. While we are by no means slaves to such processes, we cannot ignore their potential to affect us. Therefore, we should continue to study the totality of influences on past societies in order to understand some of the options that may be available to confront the many future challenges to humanity.

development of the first examples of organized agriculture in western Eurasia. The main cooling event took less than 100 years, during which forests that had recolonized the northern regions of the world during the previous warm spell rapidly

died back, together with much of their associated animal life.

During the subsequent cold, dry spell, average temperatures in the most highly affected regions of the world would have been as much as 5 to 10°C below current values. However, the more serious climatic change was not so much the increasingly severe cold, but rather the extreme dryness. This fall in overall annual precipitation, and the displacement of seasonal rains such as the Asian monsoons, led to enormous changes in vegetation patterns, especially in the northern hemisphere.[143] The cold/dry period lasted for a full 1000 years, or more than 40 generations of people, who had to adapt to their newly hostile climate. And then, around *11,600 BP*, the Younger Dryas ended even more suddenly than it had started. This new and even more abrupt climatic transition involved an exceptionally rapid global warming, with an 8 to 10°C average temperature rise in just over a decade. There was also a doubling of average precipitation values in some areas, most of which occurred in a single year.[144] The large magnitude and sudden onset of these climatic changes is shown graphically in Figure 1.3. To put this into a contemporary context, the magnitude and rapidity of the post Younger Dryas climatic change far exceeds even the direst predictions of the various models of putative anthropogenic climate change, which have received so much attention over the past decade.

Biological and human consequences

The Younger Dryas climatic changes seem to have been more rapid than most of the previous entries into and exits from the various Ice Ages from the more distant past, in which the transitions from cold to warm climates, and vice versa, typically occurred over decades or centuries. The latest data from ice core and sediment core samples from around the world show that within the relatively recent past, some of our ancestors would have been subjected to sudden and serious climatic changes, many of which occurred within a very brief period, and certainly within a single human lifetime. It is hardly surprising that human groups in the affected regions tried to modify their lifestyles and dietary habits in order to adapt to these rapid and unexpected changes. These people were faced with unprecedented challenges and would have responded by using their existing knowledge to adapt to the drastically modified new world into which they had been plunged at such short notice.

What is perhaps surprising is the considerable measure of success that was achieved in making such adaptations, albeit at a sometimes considerable cost to the health of the population. Even more surprising is the unexpectedly profound consequences that the post-Younger Dryas adaptations would have for our subsequent development as a species.

The vulnerability of ecosystems to the kind of rapid climate changes exemplified by the Younger Dryas Interval is shown by the often-dramatic changes in species diversity that can occur at a local level. For example in southern New England, cool-adapted trees such as spruce, fir, and paper birch experienced local extinctions within a period of 50 years during the rapid warming phase that marked the end of the Younger Dryas.[145] In North America as a whole, at this time, there was a massive series of extinctions of many large mammals including horses, mastodons, mammoths, and sabre-toothed tigers. The loss of animal species during this period was greater than at any other extinction event over the preceding 20 million years.[146] While this series of large-animal extinctions is mainly linked to climatic changes,[147] there is also evidence that the impact of human hunter–gatherers may have accelerated the process.[148] Human activities have also been implicated in earlier megafaunal extinctions, such as the events in Australia at *50,000* to *45,000 BP* that apparently led to the demise of all of the large mammals on the continent, although the extent of human culpability in either of these extinction episodes remains controversial.[149]

In addition to the dramatic events of the Younger Dryas Interval, there have been numerous more recent examples of the drastic effects of rapid climatic change on human societies. In many cases, sudden climatic changes are associated with the precipitate collapse of previously successful human societies, both agricultural and non-agricultural.[150] These include the demise of cultures such as: the north Mesopotamian civilization at about *4200 BP*; the sedentary, lacustrine (lake-dwelling) and farming people of the Sahara during the African Humid Period at *5200 BP*; and the early medieval Mayans of the Yucatán Peninsula soon after *1200 BP*. We will consider these examples in much greater detail in

Chapters 10 to 12. For now, we can make the point that such accounts of extreme societal disruption merit further attention because they demonstrate the fragility of at least some historical human cultures in the face of sudden climate change. They also reinforce our impression of the resilience of those early Neolithic protoagriculturalists of the Near East, who surmounted even worse challenges and survived by turning into farmers at around the time of the Younger Dryas Interval.

The periods of abrupt climatic transition into and out of the Younger Dryas Interval, and the exceptionally cold, dry conditions of the Interval itself, would have placed significant stresses on relatively sedentary human groups like the Natufians. These non-nomads were largely dependent on those plants and animals that were present in their immediate vicinity. They were therefore especially sensitive to effects on these resources of the rapidly changing climate. It is therefore of interest that the period during and immediately after the Younger Dryas Interval is marked by the first good evidence for the use of systematic cultivation and selection of crops by any human group. It is likely that this series of large climatic shocks was a major factor in the emergence of agriculture, but other factors were also important. As we have seen above, modern humans had experienced great climatic change before. Indeed, data from ice-core records show that, over the past 110,000 years, there have been no fewer than 23 climatic events of comparable magnitude (albeit not as sudden) to the Younger Dryas.[151] Despite this, we have no evidence of the emergence of any sort of systematic agriculture until about 12,000 BP. So what were the factors that led to the emergence of agriculture at this particular time and in this particular place?

One of the differences between the Younger Dryas Interval and previous climatic events might be that, by this time, many groups of people in southwest Asia would have had several millennia-worth of experience of ever more intensive protoagricultural methods (Box 3.2). Cultures such as the Natufians would have built a rich store of knowledge about the prolific, large-grained cereals that they increasingly depended on. Their knowledge would have included agronomically-relevant facts such as: the best locations and soil types for growth

of dense stands of healthy cereals; potential plant and animal enemies and how to deter them; the optimal time to gather grain from the plants; methods for storing and protecting harvested grain; and so on. This biological expertise was combined with technological innovations including: wooden and stone tools for harvesting and winnowing grain; implements for grinding the grain; and methods for processing the flour to make various types of food. It is likely that, by the time the Younger Dryas began, at around *13,000 BP*, groups such as the Natufians already had over 10,000 years of knowledge gleaned from their sophisticated non-agricultural cereal husbandry.

A stimulus towards sedentism?

A parallel development that probably occurred after the development of preagricultural cereal husbandry, but before that of agriculture, was the increasing tendency towards sedentism that is found in many populations in the Near East at this time. Such sedentism took the form of groups of people who tended increasingly to stay in one relatively restricted area, sometimes in permanent dwellings, for an extended period. Dozens of Natufian settlements, dating from as early as *14,500* to *12,800 BP* (i.e. during the warm spell immediately prior to the onset of the Younger Dryas Interval) have been found throughout the Levant.[152] The Natufian settlements can be thought of as base camps to which part or all of the group would return after periodic forays. In some cases, the settlements would have acted as early villages, with a more-or-less permanent residual population of less mobile individuals, supplemented by a transient population of more active foragers. Sedentism would have been favoured by the milder and relatively stable climatic conditions of this period. Although sedentism involves additional initial costs in the construction of more durable habitations, these costs are more than offset by the energy savings from not having to repeatedly move an entire community campsite, plus all the human occupants whether ambulatory or not, and relocate them elsewhere. As with much later semisedentary communities such as the Kumeyaay (see Chapter 2), the Natufians would have become much more

Box 3.2 Prerequisites for the evolution of agriculture

Intelligent hominids have been eating plant products for millions of years, but agriculture only dates back about 12 millennia. What were the factors that apparently precluded the evolution of agriculture for over 99% of human evolution, but then facilitated its appearance throughout the world during the Holocene era? Agriculture was not possible without: (i) the right sorts of plants and people to set up the process; (ii) the right environmental and cultural conditions to sustain the process; and (iii) the right stimuli to push people away from tried and tested hunter–gathering lifestyles towards this new, and untested, means of subsistence.

(i) The right plants and people

Domestication-ready plant species: Agriculture could not have developed without the ready availability of starch-rich edible plants of moderately high yield with appropriate genomic architecture, such as the cereals, legumes, and tubers. Such plant species existed alongside hominids for several million years and were often exploited as seasonal foods. During this period, domestication-friendly mutations would have occurred regularly, but in the absence of human selection such variants would have been rapidly eliminated from wild populations. Hence, plant material potentially suitable for farming was available long before the arrival of *Homo sapiens*, but could only persist with the assistance of a human coevolutionary partner. Such plants were also very limited in their geographical distribution and ease of identification and selection by humans.

Human cognitive capacity: We have seen that cognitively modern humans have possibly been around since before *100,000 BP* (Box 1.3). By *30,000 BP*, people were producing very sophisticated artwork and probably had the capacity for the kinds of insights and forward planning required for farming. Furthermore, much of the technology used by early farmers, including sickles and grinders, had already been in use for other purposes many millennia before crops were grown.

(ii) The right environmental and cultural conditions

Climatic conditions: The right climatic conditions for farming are twofold; first you need an adverse period to diminish returns from hunter–gathering, and second a prolonged favourable period to enable the fragile seedling of agriculture to take root. Such conditions were provided in some regions by the Younger Dryas episode at the Pleistocene/Holocene transition followed by an unprecedentedly long period of *relative* stability that persists to this day.

Cultural conditions: Farming is a unique method of food generation, representing a paradigm shift from nomadism. A shift to farming might have entailed a high degree of cultural flexibility to circumvent prohibitions on land ownership by individuals or small groups. In some prefarming cultures such as the Natufians, this cultural shift may have begun earlier as they became semisedentary. Such cultural flexibility became increasingly adaptive as the returns from farming at the societal level far exceeded those of hunter–gathering. Farming societies then rapidly evolved new cultural forms and ideologies, such as religion, inequality, and kingship, as urbanized cultures became more powerful than smaller dispersed units.

(iii) The right stimuli

The conjunction of these prerequisites for agriculture did not occur in the Pleistocene, mainly because of climatic instability and the rarity of the right sorts of food plant. The Younger Dryas supplied the appropriate carrot and stick, where the stick was the steady decline in availability of the majority of traditional food resources, and the carrot was the presence of high-yielding protodomesticants that could be stored for months or even years. The result was a gradual switch to farming by several societies in Asia and Africa soon after *11,000 BP*.

familiar with their home ranges, and therefore better able to both exploit and defend them.

There was probably no sudden transition from a mobile to a more sedentary lifestyle. Rather, there is likely to have been a fluid balance between nomadism and sedentism within and between human groups, depending on the external circumstances. Hence, less-mobile individuals such as the old, the sick, nursing mothers, and young children would preferentially be more sedentary, and therefore available for the kinds of plant husbandry close to the village that eventually led to the cultivation of crops. More mobile individuals, including older children and most of the healthy adults, would be available for external activities such as foraging and hunting. Hunting for animals seems to have been a largely male preserve in early human cultures across the world, while foraging and collecting tended to be a more female-dominated (perhaps accompanied by children) pursuit. Hunting and collecting forays may have taken the mobile groups away from their home base for extended periods if local game and floral resources were scarce. In this kind of flexible social structure, the ratio between the foragers and the stay-at-homes could be varied according to resource availability. If the village were situated close to sites of wild-cereal stands, more foragers might be recruited to join the stay-at-homes, especially when required for plant harvesting and processing. As cultivation in the vicinity of the village became more time consuming and more productive, and the more distant foraging and hunting became less rewarding, the majority of the group would eventually tend to become stay-at-homes. However, it is likely that there was always an element of opportunistic hunting and foraging, especially when the seasonal demands of crop cultivation were reduced and/or when external resources became more abundant.

The early Natufians lived in stone and wood, semisubterranean dwellings, sometimes called pit-houses (Figure 3.1). Some of the structures were used as tool sheds, both for manufacture and storage. As well as utilitarian artefacts such as sickles, pestles, and axes, there were many objects with decorative or ornamental functions, such as earrings, necklaces, and bracelets. Village sites such as Mureybit and Abu Hureyra on the Euphrates in Syria, Hayonim in Israel, Wadi Hammeh in Jordan, and a little later Qermez Dere, Nemrik 9, and M'lefaat in northern Iraq have round architecture, large hearths, and grinding stones for seeds. The largest settlements, some of which extend over more than 10 hectares, were all located in the core region of the Natufian home range in a woodland belt dominated by a canopy of oak and pistachio (terebinth) trees. The forest would have also supported an undergrowth of grasses, including a profusion of edible cereals. On these sites the archaeobotanical evidence indicates that wild cereals were exploited together with a number of edible fruits and pulses. At sites where plant remains were not recovered, indirect evidence for the use of grasses comes from glossed flint tools, indicating the harvesting of plants with the sort of high silica content that is characteristic of the grass family but absent from most other food plants.[153] Before the Younger Dryas Interval, large and small animals that could be hunted to supplement a plant-based diet were relatively plentiful in this area; these animals would have been readily accessible, even to the ever more sedentary Natufians, via relatively brief hunting forays into the surrounding wooded countryside.

As we have just seen, the intrusive management of plant resources, and especially cereals, by various groups of humans had been going on for many millennia before the Younger Dryas. These actions had already resulted in some local changes in the genetics of the wild plant populations. The kind of preagricultural selection seen in many hunter–gatherer societies would have produced what is termed an 'incipient state of domestication' in the plants.[154] It should be stressed that such activities would have been predominantly or exclusively non-intentional and can be regarded as a kind of coevolutionary process between the plants and humans, just as the early stages of animal–human commensalism can be so described.[155] The kinds of genetically regulated characters that could have been inadvertently selected by preagricultural hunter–gatherers include rapid and uniform seed germination, seed colour, synchronization of flowering and maturation, adaptation to disturbed soil conditions, and some degree of seed retention on the parent plant.

Rapid and uniform seed germination involves a loss of the dormancy period whereby seeds of many wild plants normally enter a quiescent, or dormant, stage that delays their germination until triggered by an environmental change, such as increasing average temperature or day length in spring. Seeds that had lost this dormancy characteristic could be sown in prepared ground for immediate germination. This would be a great advantage for the human users of the plants, especially if they were sowing their seed in the autumn. Adaptation to deliberately disturbed soil

Figure 3.1 Near Eastern pit dwellings during the transition to farming. Artist's impression of life around a cluster of semisubterranean pit dwellings during the early Neolithic period, just before the transition to farming. At this stage, semisedentary hunter–gatherer societies in the Near East occupied simple pit-houses. In the Levant, the Natufians constructed their buildings using a foundation wall of sun-dried mud bricks over which a framework of wood and thatch was laid. The interior of such dwellings was dug out to create one or more shallow pits, both to create more living space and to protect their vital stores of grain. It was the ability of people to store their harvested grain for long periods that enabled them to become increasingly sedentary. With the gradual spread of farming after *12,000 BP*, grain storage became even more important in the increasingly large villages and towns, such as Abu Hureyra and Jericho, that sprung up across the region. These larger communities began as clusters of pit dwellings similar to those shown here, but were eventually constructed on more regular lines using highly durable stone buildings, some of which still survive today.

conditions would also be a useful seed character under such conditions. Synchronization of flowering and maturation and seed retention on the parent plant would facilitate more efficient harvesting. The latter characters could readily be selected unconsciously because people would tend to collect more seed from synchronously flowering plants that retained their seeds in readily accessible clusters. Some of these seeds would be sown or accidentally dropped to grow into the next generation, thereby favouring plants expressing the new characters, without the need for any intentional

human selection. Note that both parties in this process, the humans and the plants, are acting as classical biological agents of Darwinian selection and that their interactions are reciprocal.[156] The humans modify their behaviour (mainly via cognitive and cultural mechanisms) to maximize their ability to interact with (manage and collect) the plants, while the structure and function of the plants becomes modified due to selection in favour of mutations that maximize their ability to grow and reproduce in the new environment caused by the human activity.

There have been claims that some of the early Natufians may have systematically cultivated cereals, including tilling the soil, prior to their domestication as true crops. Although some of the earlier evidence for these claims has been questioned, there is now growing support for such a proposal.[157] Meanwhile, there is still something of a 'chicken and egg' controversy about whether sedentism preceded agriculture or vice versa.[158] Was the establishment of more permanent human settlements driven by the need to remain in one place, in order to exploit the increasing important cereal crops?[159] Or did the settlements precede the husbandry, and was it the fact of their existence that provided the stimulus to develop a productive food resource in their immediate vicinity?[160] As is so often the case, the true situation may involve a combination of these two alternatives.[161] For example one can imagine that groups of humans who were increasingly familiar with the preagricultural exploitation of wild cereals would have had less need to forage. These cereal-specialists might have gained an advantage over other groups by constructing readily defensible, semipermanent habitations that gave them much more effective control of the local stands of wild cereals.

A more sedentary lifestyle would have also been more conducive to the development of the gamut of technologies, ranging from improved tools to better storage facilities, which would facilitate even more efficient exploitation of wild and, eventually, of domesticated cereals. Therefore, sedentism and agriculture would tend to act synergistically, each feeding off the other and accelerating the development of increasingly permanent and ever more elaborate settlement-based agricultural societies. The concept of 'husbandry/settlement synergy' discussed here is a more parsimonious and hence more scientifically satisfactory hypothesis.[162] Such a concept is also in line with the view taken here that the evolution of human–plant interactions should be regarded in terms of an extended developmental continuum, rather than as a sudden and revolutionary change to full-blown agriculture.[163] This gradualist view has implications for the genetic processes involved in crop domestication, as discussed later. By the same token, it is dangerous

to overemphasize the importance of climatic change as the catalyst for the emergence of agriculture in all societies. As outlined below, there seems to be good evidence that climate change was indeed of great significance in the adoption of agriculture by some of the Near Eastern societies. However, it is far from clear that similar climatic causes can be invoked for many of the independent and less well-documented development of agriculture in other locations, such as east Asia or the Americas.

For this reason, one should be cautious about 'climate change' hypotheses relating to early agriculture in general, for example the suggestion that agriculture may have been impossible in the Pleistocene, mainly for climatic reasons, but was more or less 'compulsory' in the Holocene.[164] As we now know, there was extensive collection and processing of wild cereals even at the height of the Last Glacial Maximum, about *23,000 BP*.[165] Several of the subsequent warm, moist interstadial eras lasted for as much as 2000 years, which might have been sufficient for the establishment of agriculture for cultures already familiar with the wild versions of the candidate crops. After all, the uptake of agriculture in the Near East following the Younger Dryas Interval was followed after only about 3000 years by a sudden cool dry spell at *8200 BP*. Despite much hardship, this climatic shift did not kill off the incipient agricultural societies, although later climate shifts did contribute to the demise of many cultures, as we will explore later. So, if it did develop in the Pleistocene, why did agriculture not persist and spread as it did in the Holocene? We do not know the answer to this question but, rather than asserting that agriculture was impossible during this period for mainly climatic reasons, it may be better to explore some of the other prerequisites that may not have been in place at that time.[166] These include the degree of sedentism, the nature of human social organization, the genetic status of the putative crops, and the availability of alternative food resources (see Boxes 1.4 and 3.2). Once these pre- or corequisites were in place, full-scale agriculture could probably develop fairly rapidly, that is within one-to-two millennia, as it indeed did in some localities.

The human response

The later Natufians

The Younger Dryas Interval of 12,800 to 11,600 BP imposed considerable stresses on many human groups, but the magnitude of the associated climatic changes was especially severe in the northern temperate regions of Eurasia and America. This period coincided with the local extinction of several species of medium-sized mammals, such as gazelle, *Gazella* spp., aurochs, *Bos taurus primigenius*, onager, *Equus hemionus*, and wild boar, *Sus scrofa*, that were hunted by southwest Asian populations, including the Natufians. The response of some Natufian groups to these events was to increase their mobility, expand their home range, and hence decrease their population density.[167] Natufian groups expanded to the northern Levant and into the area of the anti-Lebanon mountains and the southern Anatolian plateau, an area that included some of the best habitats for cereals under the new, colder, drier climatic regime of the Younger Dryas. In contrast, an attempted southerly expansion by some of the Late Natufians to the even drier, but somewhat warmer, regions of the Sinai and Negev was apparently unsuccessful. Despite a change from sedentism to more mobile hunter–gathering, and the development of useful technical innovations such as the invention of the Harif arrowhead, these southern groups disappeared from the archaeological record within a few hundred years.[168]

Unlike the unfortunate southern migrants, the more northerly, cereal-utilizing Natufian groups survived, and even thrived, despite the deterioration in their climate. As many forests receded across the Levant, the kinds of cereals with which the Natufians were already familiar became more common and prolific. At the same time, animal game was less accessible and the decline of the forests meant that there were fewer alternative plant resources. At this stage, the cereal-exploiting Natufians were still able to rely on small-game hunting to supplement their cereal diet and recent evidence suggests that late Natufian groups did not suffer the dietary deficiencies that are found in the later more specialist agrarian cultures of the mid-Neolithic period.[169] During the long millennium of the Younger Dryas, the steady disappearance of

alternative food sources would have driven the Natufians to an increasing reliance on wild cereals, especially barley and wheat. As the cold, dry period progressed, even these wild cereals would have been affected. Only those groups that were able to protect and nurture their vital stands of wild cereals would have survived.

The Natufians would have been compelled to spend more time on cereal husbandry—by now cereals would have been almost (but not quite) the 'only show in town', as regards a reliable food supply. As the wild stands of cereals diminished due to drought, the people would have selected sites where some moisture might still be available. Thus they would have planted the first fields and these would be vigorously nurtured and protected from all competitors and intruders. Such competitors and intruders would have included any competing plants, that is weeds;[170] herbivorous animals, such as rodents that might feed on the crop; and other humans who might steal the grain. There is little doubt that some of the genetic changes associated with domestication were already underway during the previous ten millennia of non-agricultural cereal husbandry. For example, there was selection for larger seeds as both the human themselves, and their methods of grain gathering, favoured increased grain size. However, the entirely new circumstances in which the cereals were now grown would have immensely increased selective pressures in favour of those traits that are regarded as characterizing domesticated, as opposed to wild, crop plants. We will consider these genetic changes in some detail in the next two chapters.

One recent finding that has taken many investigators by surprise is the apparent discovery of domesticated figs, *Ficus carica*, in the Jordan Valley, dating from 11,400 BP.[171] Several fig fruits were found at a Natufian site near the village of Gilgal (near Netiv Hagdud), which were parthenocarpic, similar to modern cultivated figs. Fig trees that produce such fruits are effectively sterile unless humans plant cuttings to enable the plants to propagate vegetatively. The same storage site at Gilgal contained wild barley, wild oat, and a type of acorn (*Quercus ithaburensis*). This implies that Natufian populations may have 'domesticated' figs, and were already using these fruits, plus acorns from returning oak trees, to supplement their cereal-based diet only one or two centuries after the end of the Younger Dryas. Meanwhile, the Natufians not only evolved new relationships with plants, they also developed a new and unexpected type of association with an animal that would become an abiding companion to people across the world, often called 'man's best friend'—the dog.

Domestication of canids

The Natufians were among the first humans that are definitely known to have kept and valued dogs, which they had presumably derived from carefully selected, tamed wolf cubs, as long ago as *12,000 BP*.[172] It is interesting that some of the earliest archaeological evidence of domesticated animals comes from the same period in which we see the beginnings of agriculture. As we will see in the following chapters, modern genetic research implies that most instances of early plant domestication were largely non-intentional, but how did our ancestors come to domesticate animals, and in particular a wild, pack-living, canid species such as the wolf? Recent work from a group in Russia has shed surprising light on the probable process of canid (i.e. the dog family) domestication. In a series of groundbreaking studies, Russian geneticist Dmitri Belyaev and his group have shown that it is possible to 'domesticate' individuals from certain wild species of canid in only a few decades, simply by only keeping and breeding from those animals that display a particular trait that we can describe most plainly as 'friendliness'.

Belyaev began this pioneering study in 1959 by attempting to domesticate the Siberian silver fox, which is a conspecific variant of the European red fox, *Vulpes vulpes*. Foxes had never been domesticated previously and are normally both fearful and aggressive when confronted by humans. Belyaev decided to select wild foxes for further breeding, simply according to one behavioural attribute, namely their friendliness to humans. No other criteria were used and the selection was done just once for each animal, when the fox kits were still young. A person gradually put their hand into a kit's cage and observed the consequences. Most wild kits reacted with the usual mixture of fear and

aggression or ignored the hand. But a few kits approached it in an inquisitive and non-threatening manner. Only these relatively 'friendly' (to humans) foxes were selected for further breeding and they were only bred with other 'friendly' foxes. Within a few generations, there were dramatic changes, not only in the behaviour of selected foxes, but also in their physiology and external appearance. In particular, the adult domesticated foxes tended to retain several traits normally only found in juveniles, including whining, barking, and submissiveness, as do modern domestic dogs.

This phenomenon is called paedomorphosis and turns out to be common in many domesticated animals. It also happens to be a very human trait. Modern human adults exhibit many 'juvenile' traits that are seen in young apes but not in adult apes.[173] It is possible that paedomorphic traits in domesticated animals, in addition to making them more placid and friendly, also make them seem more appealing (i.e. 'cuter'), leading to the extension of affection by humans to such creatures. Belyaev's friendly, paedomorphic foxes had several other physiological changes, such as lower levels of stress-related hormones, including powerful mood enhancers such as serotonin, which may well be related to their greater placidity.[174] The foxes also showed morphological changes, such as floppy ears, curled tails, and mottled coats, as found in many breeds of contemporary domestic dog. In a decade or so, the selected silver foxes were to all intents and purposes as domesticated and as suitable to be human companions as any dog could be. This pioneering Russian study shows that simply selecting for one trait like friendliness can very quickly lead to a host of complex and profound genetic, behavioural, and biochemical changes. The suite of changes caused by this type of selection, especially paedomorphosis, appear to be similar in very different domesticated animals, including dogs, cats, sheep, and mink.[175] The fact that selection for a single trait like friendliness results in selection for additional traits suggests that all of these traits are genetically linked in the genomes of the animals concerned. As we will see in Part II, it was exactly this sort of genetic linkage of useful traits that was the cornerstone of the successful Neolithic domestications of both animals and plants.

Taming the silver foxes only took a few decades, and if the Neolithic wolf/dog transition occurred over a similar timescale, the process could be accomplished quite readily by a single person over just part of their adult life. Perhaps it is no coincidence that the first archaeological records of domesticated dogs date from the most severe phase of the Younger Dryas. The close bond between the Natufian people and their newly tamed dogs is evident in burials at Ain Mallaha in the Jordan valley.[176] In one grave at this site, a skeleton of an elderly person was interred with a puppy cradled in their left hand.[177] It was to be several thousand years before any other animals were domesticated by people. Finally, lest we regard the ability to domesticate other species as a uniquely human attribute, there are well-documented cases of both animal and fungal 'domestication' by ants that predate the human efforts by tens of millions of years.[178] It is most unlikely that we would regard the spectacularly successful domestication efforts of attine ants as proceeding from conscious actions that were informed by foresight as to the consequences.

In much the same vein, we need not assume that people domesticated wolves into dogs as part of a deliberate, long-term stratagem. As with the plants that eventually became domesticated as crops, our ancestors lived in close association with groups of scavenging wolves for an extended period. During this time, a coevolutionary process would have developed that favoured genetic changes in the wolves. For example only those animals that did not pose a threat, that is were 'friendly', would have been tolerated near a human settlement. Eventually the animals would have become positively useful, for example by providing an early warning of potential trespassers, and assisting in the defence of what was now their home territory as well. The 'friendlier' and less fearful wolf cubs may have sought to play with the children of the settlement and the 'cutest' may have been adopted by individual families. Adopted cubs and their progeny would have much enhanced survival prospects and gradually came to dominate the local canid population. And thus the slide down the slippery genetic slope from wolf to poodle would have been well underway.[179]

Early Abu Hureyra cultures 14,000 to 11,000 BP

Although they are one of the best studied cultures of the Palaeolithic/Neolithic transition period, the Natufians of the Levant were by no means the only south-west Asian culture that had extensive experience of working with wild cereals. Nor were they the only people who were subject to the climatic and ecological rigours of the Younger Dryas Interval. Another especially well-researched archaeological site of this period is Abu Hureyra in the middle Euphrates region of modern Syria in the northernmost part of the Levantine Natufian cultural zone.[180] Studies at the Abu Hureyra site have revealed several interesting characteristics that shed light on the beginnings of crop domestication in this part of the world. During the favourable climatic interval between the end of the Last Glacial Maximum and the beginning of the Younger Dryas, hunter–gatherers regularly foraged in this area, which would have consisted of a riverine zone with prolific stands of wild cereals, bounded by wooded parkland merging into dense oak forests.[181] These are broadly similar floral and faunal assemblies to those found in other pre-Younger Dryas, Natufian sites to the south-west. From about 13,500 BP, the Abu Hureyra site was occupied on a more permanent basis by semisedentary hunter–gatherers. The main village at Abu Hureyra consisted of a series of relatively simple semisubterranean pit dwellings and was inhabited by a total of 100 to 200 people.[182] While they were almost entirely self-sufficient, the people of Abu Hureyra used many artefacts similar to those found in contiguous Natufian areas, and can be considered as outliers of the broader Natufian cultural groups that had migrated from the south and with whom they must have been in regular contact.[183]

The first few centuries of the human occupation of Abu Hureyra were times of relative plenty, with abundant stands of almond- and oak-dominated woodland yielding nuts and acorns, and wide swathes of cereal-rich grassland yielding a rich harvest of edible grains. Nowadays, the nearest almond and pistachio woodland is in the highlands, more than 90 km away, but the climatic conditions were distinctly kinder during the first half-millennium of human occupation at Abu

Hureyra. Archaeological evidence shows that people collected and processed edible seeds and fruits from more than a hundred plant species, doubtless supplemented by a wide range of edible roots and leafy foods; plus small and large game, such as gazelle, when available. During this period of relative plenty, sedentism was a successful adaptive strategy for this hunter–gatherer community. As well as enabling the people to exploit their home range more efficiently, sedentism allowed them to develop devices such as larger grinding stones and other heavy tools that facilitated the processing of wild seeds, but would be too heavy to be carried around by a mobile hunter–gathering community.[184] However, one should also note that this adoption of an increasingly specialized sedentary lifestyle did not in any way commit the population to becoming farmers. These people lived successfully as sedentary hunter–gatherers for about 700 years, and there is no reason why they could not have continued in this vein for many millennia to come, were it not for a wholly contingent set of external circumstances that arose during the Younger Dryas Interval.

Archaeological studies at Abu Hureyra show that, soon after 12,800 BP, and after hundreds of years during which they gathered an impressively diverse collection of wild fruits and seeds, the people gradually stopped using these resources altogether over a period of a few decades. This period coincided with the first floral changes of the Younger Dryas. The increasingly arid climate spelt the demise of nearly all the forest vegetation for hundreds of kilometres around the village of Abu Hureyra. Among the first plants that died out were the trees and shrubs, which had hitherto produced the vast range of edible fruits and berries gathered by the people as a major part of their diet. In particular, loss of the calorie-rich acorns that were collected during the autumn for winter sustenance would have been an especially grievous blow. The next food plants to go were the wild lentils and other large-seeded legumes. People focused ever more intensively on collecting the remaining wild cereals, such as feather grasses, *Stipa* spp., as well as the more familiar, larger-grained species such as wild wheats and rye, which at that time grew abundantly in the area.[185] Eventually, even the relatively

drought-tolerant wild grasses, including cereals, began to decline and, by about 11,600 BP, the seed species found around the Abu Hureyra site became dominated by classic arid-zone weeds such as gromwells.

Over this period, the appearance of the countryside would have changed dramatically from a moist, species-rich woodland/grassland to a relatively featureless, arid, treeless steppe containing just a few specialized drought-tolerant plants and even fewer animals. Unlike some of the Natufians, the Abu Hureyra people did not migrate (or, rather, there is no record of any successful attempt at migration). It is likely that successful migration was precluded as the entire surrounding region would have been equally affected by the climatic changes, and may have been occupied anyway by other people who were suffering similar privations and would not welcome any interlopers. Whatever the reasons, most of the Abu Hureyra folk stayed put in the vicinity of their small village. The diminishing options for obtaining food, and especially the scarcity of edible plants, seem to have led these people to redouble their efforts to somehow assist the growth of the large-grained wild cereals, some of which were still available and which they had been using for millennia as a productive and nutritious source of flour. Like many other south-west Asian cultures, the Abu Hureyra people would have had a vast store of knowledge about this important food resource.

There is evidence of a gradual shift from the prepastoral use of cereals to a more organized and deliberate cultivation over a few centuries. Despite the increasing aridity that should have drastically reduced the numbers of wild cereals, wild-type wheat and barley grains were still present at Abu Hureyra well after 12,800 BP. By about 12,000 BP, seeds of drought-intolerant weeds characteristic of rain-fed, arable cultivation had appeared. The implication is that the people were now growing cereals in locations, such as breaks in slopes and shallow wadi bottoms, where the scarce rainfall could be better retained by the soil. Even today, similar locations can support crops such as cotton that would normally require irrigation. To grow wild cereals in such places, would have involved vigorous clearing and weeding of the dense scrub that would normally have out-competed the cereals in the absence of human intervention. During this period, the cereal crops (as we can now call them) would have experienced a drastic change in their environment, not only due to the climate but also due to the huge range of new conditions imposed by the human cultivators. These included different germination times, new soil types, selective harvesting, seed storage and, possible visual selection before resowing. Within a century of these developments, a new type of cereal grain had appeared at Abu Hureyra.

After about *12,000 BP*, at the Abu Hureyra site, there is the earliest evidence of a putative domesticated cereal in south-west Asia. Surprisingly, this first domesticate is not wheat, but another some-what less-common cereal, namely rye, *Secale cereale*. Like all domesticated cereals, the new type of rye found at Abu Hureyra has a much larger seed than is normally found in wild populations of rye (Figure 3.2). These larger rye seeds also show evidence of threshing, implying that some force was needed to remove them from the plant. In contrast, wild-type seeds tend to be shed from the plant more easily (and are therefore lost), so vigorous threshing is not required. Similar sorts of large-gained, domestic-type rye grains are found at all of the subsequent strata at Abu Hureyra dating from the period between *10,500* and *8000 BP*. During this time, agriculture and urbanization became thoroughly established throughout the entire Mesopotamian/Levantine region. Domesticated rye continued to be cultivated as a major crop for at least 2000 years at Abu Hureyra but was then supplanted by emmer and einkorn wheat and barley. After this, rye occurs either as a minor crop or as a weed in other grain crops. However, as we will see in Chapter 6, rye eventually made a comeback as agriculture spread into the cooler climates of Europe, where it often had a competitive advantage over the other major cereal crops.

The discovery of rye as one of the earliest domesticants in the Near East begs the question; why did the Abu Hureyra villagers choose to cultivate rye when wheat and barley also grew in this region and have larger and more nutritious grains? We can only make educated guesses at present. Probably the best hypothesis is that there were temporary, local conditions that favoured the development of rye cultivation at this particular site. Rye is more easily threshed and prepared for eating than wheat or barley. Rye starch also releases its sugars more slowly upon digestion, leading to a lower insulin response,[186] and thereby acting as a longer lasting, more sustaining food.[187] On the negative side, however, rye suffers from the acid taste that develops in its products after cooking, and, given the choice, human populations almost universally prefer wheat products, due their superior taste and more rapid digestibility. Rye is also slower to mature than wheat, which would have made it particularly vulnerable to an early onset of the annual summer drought in this region.[188] The crop is also taller

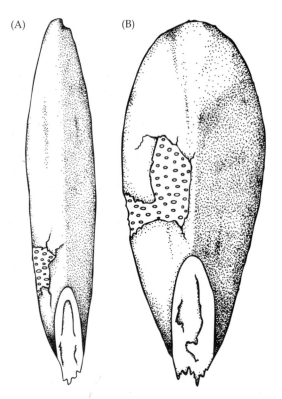

Figure 3.2 Rye, the first domesticated cereal crop? (A) Wild rye grain and (B) cultivated grain. Both grains were found at the early Neolithic village of Abu Hureyra in Syria, which was one of the earliest sites of plant cultivation. The cultivated grain, which dates from the middle of the Younger Dryas cool/arid period at about *11,800 BP*, is significantly larger and richer in starch than the wild grain. Redrawn from Hillman *et al.* (2001, Fig 4).

than wheat, which makes it more prone to lodging (toppling over) in windy or rainy conditions. During wet spells, rye is also susceptible to the potentially deadly ergot fungus and its grains are readily attacked by rodent pests.[189]

Rye has a further genetic characteristic that would have impeded its agronomic performance, namely its propensity to interbreed with wild relatives. Modern domesticated rye is a strongly outbreeding species, and its pollen can fertilize other rye plants as much as 1 km away.[190] This means that domesticated rye can readily interbreed with neighbouring wild rye plants to produce seed in which the advantageous characters, such as large grain size, would become progressively eroded. This implies that the early, predomesticated rye was either already self-fertile or rapidly became so, possibly due to a temporary breakdown in its ability to outbreed. It has been hypothesized that a few decades of warm summers might have sufficed to change the early domestic-type rye into a self-fertile plant that would not so readily interbreed with any nearby wild rye.[191] The fairly sudden appearance of domestic-type rye, so soon after people started to cultivate the wild form, has been taken as evidence of the possible application of intensive and conscious selection by the human population at Abu Hureyra.[192] When the newly domesticated rye plants eventually reverted to their original outbreeding phenotype, interbreeding with wild relatives could be minimized by human interventions, such as vigorous weeding out of wild plants and elimination of any low-performing hybrids that were identified. However, this is not necessarily the case. As with wheat and barley, we now know that many of the most important domestication-related traits in the rye genome are determined by only a few genes. This would have made it much easier for agronomically suited varieties to arise via unconscious selection in as little as a few decades, under the known conditions that prevailed at settlements such as Abu Hureyra in the immediate aftermath of the Younger Dryas.[193]

Rye cultivation at Abu Hureyra is the earliest proven example of agriculture anywhere in the world. The key preconditions for the relatively rapid domestication of this cereal can be summarized as follows:

(1) a genetic predisposition that enabled certain wild plants to respond to cultivation by the speedy development of domestication-friendly traits;
(2) an environmental shift (normally climatic) that denied the local human population access to existing alternative edible staples, both plant and animal;
(3) sufficient environmental stability to enable continued cultivation of the domesticated crops in the longer term;
(4) the ability and willingness of the human population to exploit the new resource by engaging in possible taboo-breaking activities such as tilling and harrowing the soil.[194]

Some, but not all, of these preconditions could have been met before the end of the Last Glacial Maximum. Humans might well have had the cognitive capacity for the sorts of complex tasks requiring foresight and planning that are involved in agriculture since as early as *100,000–80,000 BP*. However, alternative resource-exploitation strategies were available that may have been more efficient and straightforward. For much of the Pleistocene, the climate was too variable to allow agriculture to succeed as a viable lifestyle in the long term. And finally, those few plant species that were genetically predisposed to domestication were only patchily distributed across the world, and were mostly located well away from the African centres of human origin and initial expansion. It is quite possible that there were limited experiments with plant cultivation before the Younger Dryas, as is suggested by preliminary data from China (see Chapter 11). But these isolated examples of early farming all appear to have ended in abandonment or failure until a few groups of Near Eastern people, such as those at Abu Hureyra, were driven to repeat the experiments, this time with more enduring success.

Plant domestication and acquisition of agriculture are reversible processes

It should now be evident that there was nothing inevitable about the development of agriculture and the eventual displacement of hunter–gathering in much of the world. Farming is simply an alternative and sometimes more adaptive method for the exploitation of environmental resources such as edible plants. Agriculture has both advantages and disadvantages compared to hunter–gathering, and only seems to have been adopted when the latter strategy was rendered more costly and less rewarding by factors such as climate change. Equally, as we will see with some midwestern Amerindian cultures (Chapters 8 and 12), when the cost/benefit analysis proved unfavourable, and when viable alternative methods of resource exploitation were available, people have not hesitated to eschew agriculture altogether and return to hunter–gathering.[195] There was a great deal of trial-and-error during crop domestication and several abortive attempts to cultivate edible plants. In ancient Mesoamerica, foxtail seeds, *Setaria* spp., suddenly increased in size as they were cultivated, but the crop was abandoned when maize was introduced and wild foxtail populations soon reverted to the small-seeded form.[196] In ancient China, the common mallow, *Malva sylvestris*, was originally the most important green vegetable, but is was subsequently replaced by Chinese cabbage *Brassica chinensis* and common mallow reverted to the wild type. In the Near East, lucerne, *Medicago sativa*, was an abundant crop in some of the earliest agricultural sites, but it then disappeared as other legumes such as pea and lentil were adopted instead.

It was not just individual crops that were regularly tried and rejected; the entire agricultural lifestyle was regularly abandoned by communities. Indeed, both the acquisition and subsequent rejection of agriculture are becoming increasingly recognized as adaptive strategies to local conditions that may have occurred repeatedly over the past ten millennia. For example, in a recent study of the Mlabri, a modern hunter–gatherer group from northern Thailand, it was found that these people had previously been farmers, but had abandoned agriculture about 500 years ago.[197] This raises the interesting question as to how many of the diminishing band of contemporary hunter–gatherer cultures are in fact the descendents of farmers who have only secondarily readopted hunter–gathering as a more useful lifestyle, perhaps after suffering from crop failures, dietary deficiencies, or climatic changes. Therefore, the process of what may be termed the 'agriculturalization' of human societies was not necessarily irreversible, at least on a local level. Hunter–gatherer cultures across the world, from midwestern Amerindians to !Kung in the African Kalahari, have adopted and subsequently discarded agriculture, possibly on several occasions over their history, in response to factors such as game abundance, climatic change, and so on. However, it is also true that these were relatively isolated groups who were remote from major centres of population or mass agriculture.

In contrast, in the principal centres of crop diversity, such as the Near East, Mexico, China, and India, agriculture soon came to dominate the available landscape. As agriculture developed and spread, human populations increased and spread out; towns grew up; animals were domesticated; crops were improved to produce higher yields; and new crops were introduced from other regions. This set up a kind of positive-feedback loop that made it gradually more and more difficult to reverse the process of agriculturalization on anything but an extremely localized level. Although these events have often been interpreted as showing that agriculture was inevitable and 'progressive', such is not the case. As with all evolutionary processes, agriculture arose in several parts of the world due to very particular local circumstances that include genetic happenstance (as in the teosinte/maize transformation that is described in Chapters 5 and 6) and the vagaries of an ever-changing climate.

The eventual global triumph of agrarian-based societies is a relatively recent phenomenon that is based largely on their overwhelming numerical and technological advantages over other types of human culture. It certainly does not mean that agriculture is necessarily here to stay. That depends very much on whether it turns out to be a food acquisition strategy that is sustainable in the very long term, for example during some of the periods of more drastic climate change, like the Younger Dryas, that will surely recur at some time in the future. Meanwhile, it is worth reminding ourselves about the relative fragility of our agrarian systems and we will do this in Part III by considering some examples of more localized climate-induced societal collapse during the last ten millennia.

Before we continue with the human story of agriculture in Part III, however, we will pause to examine the process from the perspective of the plants that eventually became our major crops in Part II. In the next four chapters, we will switch to a more biological viewpoint as we delve into plant genetics to see how a series of accidents of genome organization has largely determined the nature of our crop species and hence the very food that we eat today.

Crops and genetics: 90 million years of plant evolution

The history and origin of human civilizations and agriculture are, no doubt, much older than what any ancient documentation . . . reveals to us. A more intimate knowledge of cultivated plants . . . helps us attribute their origin to very remote epochs, where 5000 to 10,000 years represent but a short moment.

Nikolai Ivanovich Vavilov, 1924, *Origin and Geography of Cultivated Plants*

CHAPTER 4

Plant genomes

E pluribus unum

Virgil (attrib), 70–19 BCE, *Moretum*

Introduction

In Part II, we will explore the weird and wonderful world of plant genetics. We will particularly focus on the genetic attributes that made possible the largely accidental domestication of the major crops by several early Neolithic cultures. Several fundamental topics will be addressed, such as how the peculiar and inconstant genetic constitutions of plants set the scene for the emergence of crops. We will also ask: why is it, after more than ten millennia of agriculture, that people around the world still cultivate so few plants as their major staple crops? We will begin to answer this question by looking at the often surprising results of recent research into plant genome organization. These discoveries in plant genetics are providing powerful insights that are enabling us to elucidate the biological mechanisms involved in crop domestication. The new findings also reinforce the hypothesis that the first domestications, at least in their early stages, were largely non-intentional processes on the part of human farmers.

It is estimated that there are at least 400,000 plant species, many tens of thousands of which are edible or useful in some other way. In principle, each of these tens of thousands of plants should potentially be suitable to cultivate as crops.[198] Despite this seeming plenitude of botanical wealth and many millennia of experience of domestication and breeding, we still only cultivate about 150 species of food crops.[199] Even more remarkably, the vast majority of the world food supply comes from fewer than 20 major crops.[200] Our dependence on such a narrow range of crops is not simply due to a lack of effort to utilize other species. Indeed, people around the world have tried, and failed, on repeated occasions to domesticate nutritious food plants that turned out to be recalcitrant to such cultivation. Several examples of such intractable species are found in the genus *Zizania*, which is closely related to Asian rice (see Chapter 6). So, we come back to the main question: why do we grow so few crops? Is it because only a few species possess those unusual and special characteristics that make them relatively amenable to domestication? If this is the case, what are these mysterious properties and how, if at all, did the early farmers learn to manipulate them? Or was domestication simply an accident of evolution—a series of contingent events and processes that led to the coevolution of what eventually turned out to be a very successful symbiotic partnership between humans and crop plants?

Darwin, de Candolle, and Vavilov

The questions posed above have preoccupied scientists and breeders for centuries. No less an authority than Charles Darwin devoted an entire book to the subject of plant and animal domestication.[201] For example he discussed a lengthy experiment to convert the wild English oat, *Avena fatua*, into a useful crop. The wild oat is normally a rather troublesome weedy species that bedevils cereal growers, but the experiment succeeded in producing a new, agronomically useful form that was almost identical to the cultivated oat, *Avena sativa*.[202] One of the most influential early figures in the study of crop domestication was the Swiss botanist, Alphonse de Candolle, a contemporary of, and correspondent with, Darwin. De Candolle recognized that the key to understanding the domestication of crops was to determine their places of origin. He also realized that crop domestication had occurred relatively recently in terms of geological time, and probably after the last major Ice Age, which ended about *15,000 BP*. His major arguments were set out in the treatise entitled: *L'Origine des plantes cultivées*.[203] De Candolle was a rare interdisciplinary scientist and scholar; a botanist by training who also recognized the value for his work of other fields, such as linguistics, historical texts, and archaeology.

Following in the footsteps of De Candolle, who established the study of crop origins and genetics as a rigorous academic discipline, was the towering figure of Nikolai Ivanovich Vavilov. This Russian geneticist has been described as the 'Darwin of the twentieth century'.[204] Such an epithet may seem exaggerated to those who are unaware of Vavilov's life and times. However, he is increasing being recognized as one of the foremost scientists of the twentieth century, not only for his contribution to biology but also for his heroism in the face of appalling adversity during the Stalinist purges of the late 1930s (Box 4.1).[205] In 1956, his belated rehabilitation in the USSR began with the republication of his works.[206] Despite his tragic fate, which has echoes of the persecution of Galileo in the seventeenth century, Vavilov and his successors made many enduring contributions to the understanding of crop domestication.[207] The most important of these was the demonstration that the major crops come from a few localized regions, dubbed 'Centres of Origin'. We will now look in detail at this concept and its implications for the mechanism of plant domestication.

Origin and domestication of the major crops

Prior to Vavilov's discoveries in the 1920s and 1930s, most people believed that agriculture arose in the 'Fertile Crescent' area of the Near East. Little was known about events in other parts of the world, and their possible role in the crop development. Vavilov noted that some parts of the world were relatively rich in crop species that had been grown by local farmers for many millennia. He also noticed that such crop-rich regions tended to have many wild relatives of the cultivated crops. In contrast, there were large areas of the world where people only grew a few staple crops and no wild relatives were present. Vavilov called the crop-rich areas 'Centres of Diversity'. For example he noted the immense diversity of maize and squash varieties in Mesoamerica, as well as long-held local traditions about their use and ancient myths about their provenance.[208] Mesoamerica is also the unique location of many wild relatives of these crops.[209] Vavilov concluded that the 'Centres of Diversity' were also likely to be the places where domestication of such crops had begun. In other words, these 'Centres of Diversity' were also the 'Centres of Origin' for crops. He listed six principal 'Centres of Diversity' throughout the world; namely the Andes, Mesoamerica, Mediterranean/Near East/Central Asia, China, India, and Ethiopia.[210] Vavilov's ideas have since been modified and extended by others, most notably by US geneticist, Jack Harlan.[211] One notable omission from the above list is non-Ethiopian Africa, now known to be an important centre of origin for crops such as sorghum and yams (see Chapter 12). A modern view of the major centres of crop origin and diversity is shown in Figure 4.1.[212]

The Centres of Origin concept is significant in two ways. First, it shows that crop domestication happened independently in different areas of the world. Second, it demonstrates that such domestications were relatively rare events—hence the small number of primary centres of diversity. People outside these primary centres were, and in a few cases still are, constantly experimenting with non-agricultural uses of food plants. However, it seems that the types of plants available to such cultures were often simply not amenable to domestication. For example hunter–gatherers in southern Africa, Australia, and California have been living off wild grasses for many millennia.[213] During this time, they have employed techniques such as burning, sowing, and harrowing to encourage growth and improve the yield of their edible grasses. And yet, they never managed produce any domesticated versions of these grasses, while other cultures appear to have more or less stumbled into domestication of other grass species, namely the cereals, with relative ease.

Box 4.1 Nikolai Ivanovich Vavilov, the doyen of modern crop genetics

Born in 1887, Vavilov was a polymath with interests in botany, genetics, agronomy, and geography. He also possessed the impressive organizational talents that led him to become one of the most senior academicians in the USSR. He made several lengthy and productive visits to Asia and the Americas, and was particularly inspired by William Bateson (one of the British rediscoverers of Mendel's work, who coined the term 'genetics') in the UK. During the 1920s, and despite occupying the onerous administrative positions of Director at the Institute of Genetics and President of the Lenin Academy of Agricultural Science in Moscow, Vavilov personally organized and/or participated in over 100 expeditions to 64 countries across the world, in order to investigate the origins of crops. On one expedition to North America, Vavilov and his team collected several thousand plant samples, which they took back for preservation in the USSR.

Eventually, Vavilov amassed a seed bank of incalculable value that today numbers 380,000 genotypes of 2500 species. He published dozens of papers and books on his research (Vavilov, 1926, 1935, 1992); he founded over 400 research institutes across the USSR; and he transformed our thinking of about the origins of agriculture. Vavilov's motivation for such Stakhanovite exertions was a judicious combination of scientific curiosity and a genuine desire to improve agricultural production in his native land. By 1930, Vavilov was at the height of his fame, the recipient of the Lenin Prize with an international reputation as an innovative researcher in plant genetics. At the same time, he was a deeply practical scientist who vigorously applied his knowledge of modern genetics in the difficult effort to increase crop yields, especially in the then famine-prone farming regions of the southern USSR, once dubbed the 'breadbasket of Europe'. Suddenly, in the mid-1930s, and to the great consternation of biologists in the USSR and overseas, this eminent scientist dramatically fell from official grace and his reputation was maliciously undermined by the pseudoscientific machinations of a cabal led by the notorious Trofim Denisovich Lysenko (Sheehan, 1993).

Lysenko was a promising exstudent of Vavilov, who initially published some useful studies on crop physiology but then developed an extraordinary version of the discredited Lamarckian theory of evolution, whereby acquired characteristics can supposedly be inherited. He also attacked Mendelian genetics and its supporters, including his erstwhile mentor, Vavilov. Lysenko's heterodox views suited the prevailing Soviet ideology that regarded (potentially controllable) environmental influences as more important in biology (and society) than (uncontrollable) genetic factors. As a loyal *vydvizhenets*, Lysenko soon became a favourite of Stalin. The term '*vydvizhenets*' literally means 'pushed up' and was applied to people of modest backgrounds and often little education who were promoted to senior positions in the Stalinist era. As the son of a Ukrainian peasant, Lysenko did not learn to read or write until the age of 13 and was always insecure about his knowledge of science in general and biology in particular (Hossfeld and Olsson, 2002; Roll-Hansen, 2004). With Stalin's connivance, Lysenko eventually replaced Vavilov in all of his major posts and was free to apply his flawed theories to crop production—with predictably disastrous consequence for food output in the USSR. Lysenko spent most of the late 1930s denouncing his erstwhile teacher in ever more vituperative terms. For several years the hapless Vavilov was repeatedly harassed and persecuted for his Galileo-like adherence to Mendelian genetics. In the face of ever more hysterical accusations of such capital crimes as sabotage, espionage, and terrorism, he steadfastly stuck to his principles, famously declaring in 1939:

We shall go the pyre, we shall burn, but we shall not retreat from our convictions. I tell you, in all frankness, that I believed and still believe and insist on what I think is right. . . . This is a fact, and to retreat from it simply because some occupying high posts desire it is impossible.

Alas, this was no mere empty rhetoric. Within a year, Vavilov had been arrested on a trumped up charge of agricultural sabotage, plus a string of other equally false allegations. Following an often-brutal 11-month interrogation, he was subjected to a show trial, and in 1941 was sentenced to death (later commuted to life imprisonment). For a further year, Vavilov endured a miserable period of extreme privation, first in a concentration camp at Saratov on the Volga, and finally in the bleak Magadan forced-labour camp in eastern Siberia. On 26 Jan 1943, this giant of science who contributed so much to agriculture, died of scurvy and dystrophy, caused by prolonged malnutrition. It is supremely ironic that a man who had done so much to improve the supply of food to his country was effectively beaten and starved to death by agents of that same state. If any scientist deserves a posthumous Nobel Prize, surely it is Nikolai Ivanovich Vavilov.

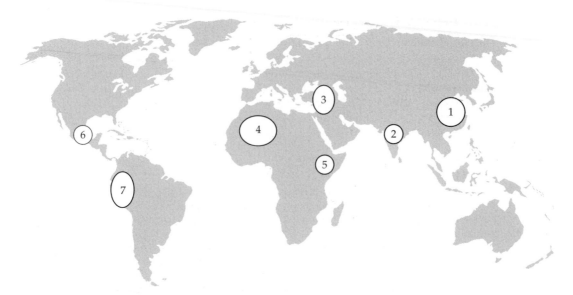

Figure 4.1 Centres of origin of the major crops. The concept of areas of crop origin was elaborated in the 1920s by the Soviet botanist Nikolai Vavilov. He identified several areas that he considered to be the original homes of the world's most economically important crops. The areas shown below relate to the major crops discussed in the text. (1) China—rice, millets (*Panicum* spp.), hemp (*Cannabis sativa*), mulberry (*Morus alba*), onion (*Allium chinense*), tea (*Camellia sinensis*), soybean (*Glycine max*), sugar cane (*Saccharum sinense*). (2) India—rice, banana, breadfruit (*Artocarpus communis*), coconut (*Cocos nucifera*), mango (*Mangifera indica*), orange (*Citrus sinensis*). (3) Western Eurasia—barley, wheat, rye, fig, lentil (*Lens esculenta*), flax/linseed (*Linum usitatissimum*), cabbage (*Brassica oleracea*), olive (*Olea europea*). (4) Sahara/West Africa—pearl millet. (5) Ethiopia—coffee (*Coffea* spp.), castor bean (*Ricinus communis*), cowpea (*Vigna sinensis*), sesame (*Sesamum indicum*). (6) Mesoamerica—squash (*Cucurbita pepo*), avocado (*Persea americana*), common bean (*Phaseolus vulgaris*), cocoa (*Theobroma cacao*), maize, cotton (*Gossypium hirsutum*), pepper (*Capsicum annuum*). (7) Andes—manioc (*Manihot utilissima*), peanut (*Arachis hypogea*), tobacco (*Nicotiana tabacum*), potato (*Solanum tuberosum*), tomato (*Lycopersicon esculentum*).

This implies that it is not necessarily the activity of the humans that is the primary determinant of crop domestication, but rather the availability of the 'right sort' of plant. In other words, crop domestication as an historical process was determined as much (or perhaps more) by a combination of plant genetics and environmental factors, such as climatic change, rather than being the exclusive product of conscious human intervention. Of course, domestication also requires human participation, but possibly as unconscious partners in a coevolutionary process, instead of acting as conscious protobreeders of crop plants (see Box 2.3).[214] Therefore, if edible plants with the genetic potential for domestication happened to be present in an area, then the chances for the development of agriculture would have been increased. On the other hand, if no edible plants in a region possessed such genetic attributes, no agriculture based on indigenous crops could have developed, no matter how clever or resourceful the local human population. The only way that such people, including most Europeans and North Americans, could develop agriculture would be to import the technology, including already-domesticated crops, from elsewhere.

An analogous argument has been used to explain the remarkably small repertoire of domesticated herbivorous mammals.[215] For example the majority of our domesticated animals originated in Eurasia (cattle, sheep, goats, horses, and pigs), plus a few in the Andes (llamas, alpacas), but very few examples can be found elsewhere in the world. In particular, over the whole of sub-Saharan Africa, with its massive diversity of animals of all kinds, not a single species of native herbivorous mammal was ever domesticated.[216] This situation is all the more remarkable because humans have lived in Africa for much longer than the other continents and would have been very familiar with the native

Table 4.1 Some of the key domestication-related traits in crop plants

Trait	Wild plant	Domesticated crop
Height	Tall	Short or dwarf
Growth habit	Branched and bushy	Unbranched and compact
Ripening	Asynchronous	Synchronous
Seed dormancy	Present	Absent
Seed shattering	Shattering heads	Non-shattering heads
Seed size	Small	Large
Ease of dispersal	Highly dispersible	Loss of dispersal
Threshing	Hard	Easy
Reproduction	Outbreeding	Self-fertilizing
Germination	Asynchronous	Synchronous
Hairs and/or spines	Present	Absent or reduced
Toxins	Present	Absent or reduced

Any given crop will not necessarily carry all of these traits and their relative importance will vary considerably according to the crop type and farming system. Hence, grain crops invariably have much larger seeds than their wild relatives, but this may not apply to root crops such as potatoes, where tuber size and the presence of toxins are much more important traits than seed size.

fauna. Like other cultures, African people have had powerful incentives to attempt the domestication of their native fauna; they have also had tens of millennia to achieve this, but never succeeded. It seems clear, therefore, that, as with crop plants, the genetic endowment of an animal is one of the major factors contributing to its domesticability by humans.

The second point to come out of the Centres of Origin concept is that some very different types of plant have been domesticated in each of the centres. In addition to the 'big three' cereals (rice, wheat, and maize) there were potatoes, squash, and the various pulses, such as lentils, peas, and beans. These plants are members of widely diverse families, as different in genetic terms as are reptiles, birds, and mammals. Yet these very different types of plant have each responded in broadly similar ways during their journey to domestication. In contrast, there are close relatives of readily-domesticated cereals, pulses, and root crops that have never been domesticated. We now have genetic evidence that despite their diverse origins, the various successful domesticants generally display remarkably similar adaptations to even the most basic forms of informal plant management and cultivation. Such adaptations in turn make the plants progressively better able to respond to

ever-more intensive forms of cultivation. A list of some of the most important domestication-related crop traits is given in Table 4.1. One genetic feature shared by the most successful crop species is that the control of this suite of genetic traits, often collectively called the 'domestication syndrome', resides in a very small number of genes, as we will discuss in more detail later in Chapter 5.

Once again, these findings lead to the conclusion that the potential of a plant to be cultivated (or not) as a crop may reside more in its genetic endowment, than its nutritional or other qualities. In other words, our crops may have been selected by early farmers because they happened to exhibit a relatively simple, and hence easily manipulated, genetic control of domestication traits. It is these traits that mark out a plant as a potential cultivated crop, rather than simply being a species that is only suitable for gathering. In the next two chapters, we will test this hypothesis further by examining the genetics of the major crops. We will pay particular attention to any unusual characteristics that may give some clue as to how and why these particular species were selected for domestication. First of all, however, we will look at recent findings from molecular genetics that concern some remarkable genetic attributes of plants in general, and crop species in particular.

These findings are beginning to give us a much better picture of the rather odd and fluid structure that constitutes a plant genome. This genome is often polyploid rather than diploid; it usually contains vast amounts of 'extra' DNA that does not encode proteins and which seems to have originated from non-plant sources; and, finally, the plant genome is in a constant state of flux, with genes and other fragments of DNA continually entering and leaving genomes at rates that vary greatly from one species to another. These findings are beginning to challenge our notion of what constitutes a genome, or indeed what we mean by a biological species. All of this new knowledge about plant genetics is therefore particularly germane to any discussion on the genetic manipulation of plants, whether by a would-be Neolithic farmer or a modern corporate biotechnologist.

Polyploidy and crops

What is polyploidy?

Polyploidy is one of the key phenomena in plant genetics that is responsible for the creation of new genetic variation.[217] It is especially important in the evolution and domestication of many major crop plants.[218] Polyploidy simply means the presence of more than two sets of chromosomes in the genome of an organism. Humans, and most other large animals, normally have two sets of chromosomes in each cell, one set inherited from each parent. This means that we humans, and most animals, are diploid (diploid means 'two-fold'). In contrast, it is likely that well over half of all species of flowering plants possess more than two sets of chromosomes, and are therefore polyploid (polyploid means 'many-fold').[219] Crop plants are especially likely to be polyploids.[220] For example durum wheat is a tetraploid (four sets of chromosomes), while spelt, breadwheat, and oats are all hexaploid (six sets of chromosomes) (see Figure 4.2). Even rice, which has the smallest genome of any major crop, is now believed to be an ancient polyploid species.[221]

It used to be thought that polyploidy was an unusual and aberrant condition, especially in animals,[222] and that it was not in any way an adaptive trait.[223] However, it turns out, from some very

(A)   (B)   (C)

Figure 4.2 Polyploidy—the effects of genome multiplication in wheat. Polyploidy can result in greatly increased fruit and grain size, and many of our major crop families contain polyploid genomes. Here we see the result of hybridization between the diploid wild grass, *Aegilops tauschii* (A) with the tetraploid crop, emmer wheat, *Triticum dicoccoides* (C), to form a new and more productive form of wheat called spelt (B). Spelt wheat is a hexaploid hybrid species, now called *Triticum aestivum*. Spelt went on to develop into the most commonly grown form of modern wheat, namely breadwheat. Note the greatly increased size of ears (containing the grains) in spelt wheat compared to the two parental species.

recent research, that polyploid animals are rather more common than we have hitherto suspected.[224] Genome doubling seems to have been a significant aspect of the evolution of all complex animals, including the vertebrates, so even humans are descended from ancient polyploids that subsequently adopted a pseudodiploid genome organization. In 1970, Susumu Ohno proposed that two rounds of polyploidy occurred early in vertebrate evolution.[225] This model has been supported by more recent evidence using molecular genetic analyses.[226] Many simpler animals such as amphibians are now known to be polyploid, and a few years ago the first example of a polyploid mammal (a rodent) was reported, although the latter is still very much an exceptional case.[227]

Not even humans are exempt from the possibility of polyploidy. The vast majority of us are functionally diploid because we arise from the fusion of a haploid sperm with a haploid egg, each of which carries a single set of chromosomes. However, among the billions of haploid sperm that a man produces in his lifetime, several thousand diploid sperm might be produced due to a rare failure in the process of meiosis during spermatogenesis. If one of these abnormal diploid sperm, or two normal sperm, should fertilize a normal haploid egg, the result is a triploid zygotic cell. Although triploid zygotes occasionally divide to become embryos and then foetuses, they rarely survive in the womb. In the extremely rare case of the birth of a triploid child, it will hardly ever survive to adulthood. Triploid individuals are generally non-viable and/or sterile because they contain an odd number of groups of chromosomes. During cell division, the chromosomes normally line up in pairs, which is fine for diploids (one pair) and tetraploids (two pairs), but not so good for triploids ($1\frac{1}{2}$ pairs). As a result, there tend to be often-fatal abnormalities during cell division in triploids.[228] Humans can also occasionally produce tetraploid offspring. If a woman were to produce some eggs that were diploid, instead of the normal haploid state, and if one of these were fertilized by a diploid sperm, it is possible that viable tetraploid human progeny might result. However, this is extremely rare and such unfortunate individuals invariably suffer from severe abnormalities, both mental and physical.[229]

Triploid genomes also lead to sterility in plants and are therefore rarely found. There are, however, two interesting exceptions to this rule, both of which relate to edible crops that are important in the human diet. The first example concerns what is termed somatic triploidy, which is when a single tissue is triploid, rather than the entire plant. All flowering plants produce a nutritive tissue called the endosperm, which helps to sustain the growth of the young embryo during seed development in an analogous manner to the mammalian placenta. The endosperm is triploid due to the fusion of a diploid maternal cell with a haploid nucleus from the pollen during fertilization. In some plants the endosperm persists to form a store of starch, protein, and/or oil that can occupy most of the volume of the mature seed. The grains of our cereal crops are mostly made up of a starchy endosperm, and it is this triploid tissue that is by far the major source of calories for human societies around the world. Hence, whenever you eat a slice of bread, a forkful of pasta, or a pinch of rice, you are eating the product of a triploid genome. The second example of triploidy in a crop is the commercial banana, *Musa acuminata*. Virtually all of the commercially traded bananas in the world consist of a single triploid clonal variety, called Cavendish.[230] This means that all commercial bananas, that is the sort that one might buy in a typical Western store or supermarket, are not just extremely inbred; they are genetically identical to each other. Because these bananas are triploid, they are also sterile and must be propagated vegetatively, which makes life rather difficult for their breeders.[231]

Polyploidy, especially tetraploidy and hexaploidy, is much commoner in plants than in animals, and is considerably more benign in its effects. It has been especially important in the speciation and evolution of flowering plants (or angiosperms), from which all of our most important crops are derived. It is estimated that at least 50 to 70% of extant angiosperm species have undergone one or more episodes of chromosome doubling during their evolutionary history.[232] A recent survey of New Zealand grasses found that a mere 10% of the 55 species analysed was diploid—the remainder having various degrees of polyploidy.[233] There is no reason to believe that New Zealand flora are unusual in this regard, and as further data are accumulated it seems likely that the polyploid state will prove to be the norm for the vast majority of plant genomes. Indeed, a report in 2006 suggested that all flowering plants, except for the very primitive genus *Amborella*, are probably descended from polyploid ancestors.[234] Interestingly, polyploidy does not automatically mean a larger-sized genome. If anything the reverse may be true, with good evidence that genome size actually decreases with polyploidy.[235]

This may seem paradoxical, but it might be the case that, although polyploid species have more genes than diploids, the polyploids are better able to prevent the colonization of their genomes by the vast amount of so-called 'extra' DNA, often of ancient viral origin, that is found in many plant and animal species. This is certainly the case for the multiple-polyploid members of the brassica family, which have several different sets of chromosomes in their genomes but relatively little 'extra' DNA (as discussed towards the end of Chapter 7). The genomes of even quite closely related grass species often vary enormously in size due either to polyploidy and/or the presence of 'extra' DNA.[236] We will return to look at the fascinating topic of 'extra' DNA in some detail in Chapter 5. Meanwhile, there is an emerging consensus on the virtual ubiquity of polyploidy in plants, and appreciation of its evolutionary importance. Looking specifically at crop species, we see that polyploidy has been one of the main driving forces of genetic variation, and that it has played an especially important role in the process of crop domestication.[237] There are two principal categories of polyploid organisms, both of which have played roles in crop evolution and domestication. The first group is the autopolyploids, which come from a single ancestor via a duplication of an entire single genome.[238] The second group is the allopolyploids, which derive their several genomes from hybridization between two or more different species.[239]

Autopolyploidy and allopolyploidy

Autopolyploids tend to remain similar to their diploid ancestors over many generations. This is because they contain almost exactly the same genome as their parental species, except that it has become duplicated. Compared with its diploid parent(s), an autotetraploid organism has an additional copy of each of the genes in its genome. As new autotetraploid individuals reproduce and are subject to mutation and evolutionary selection, many of the genes in the 'extra' genome will be lost or will diverge in structure and/or function from equivalent copies in the other genome.[240] As a result, the descendents of a newly formed autotetraploid organism will increasingly differ from its diploid cousins, but this will occur over many generations, and possibly much longer.[241] Therefore, autotetraploidy will result in new variants, and

eventually perhaps in new species, but normally over a long timescale. Of course, the relatively long timescale of phenotypic divergence in autopolyploids is only true in relative terms. In terms of evolutionary changes that span millions of years, both autopolyploid and allopolyploid divergences are actually quite rapid processes and both of them are major drivers of plant speciation.

In contrast to autopolyploids, allotetraploids are much more likely to give rise to radically different species from their parents over a short timescale, possibly as brief as a few generations or even immediately after they are formed. This is because allotetraploids are hybrids with a complete set of genomes from two dissimilar parents of different species. This dramatic reshuffling of its genetic endowment means that an allotetraploid organism will automatically constitute a new species, carrying a mixture of characteristics from each parental species. Sometimes, equivalent genes derived from the respective parental species, which encode the same protein (such genes are called homeologues), will 'specialize' almost immediately after polyploidization. For example one of the two homeologous genes might be silenced in one set of tissues and organs, while the other is silenced in different set of organs during plant development.[242] In newly formed allopolyploids, there is also an element of 'cross-talk' between the two parental genomes that now coexist in the new hybrid plant.[243]

Indeed, there are several important genetic processes that occur in both auto- and allopolyploids that are above the organizational level of their duplicated (or homeologous) genes.[244] These include the sort of intergenomic 'cross-talk' mentioned above, but also such phenomena as saltational variation, intergenomic invasion, and cytonuclear stabilization.[245] One of the most important problems facing a newly formed polyploid is the pairing of homeologous, rather than homologous, chromosomes at meiosis. Such inappropriate chromosomal pairing can lead to sterility and at least two mechanisms to avoid it have developed in successful allopolyploids. The first mechanism is the selective elimination of large amounts of non-coding DNA from different parts of the genome such that homeologous chromosomes no longer resemble one another enough to form such illegitimate pairings.[246] Remarkably, tens of millions of DNA base pairs, and as much as 15% of the entire genome, can be rapidly lost in this manner.[247] The second mechanism involves more direct suppression of homeologous pairing, such as the system regulated by the *Ph1* locus in polyploid species of wheat (see below).[248]

Evolutionary significance of polyploidy and hybridization

In addition to its role in speciation, polyploidy is of considerable adaptive significance for plants.[249] For a start, polyploid individuals tend to have larger average cell sizes and often produce larger adult forms. This may or may not be useful in an open ecosystem but, in the context of incipient domestication, a larger plant might mean larger fruits or seeds, which would be of great interest to a hunter–gatherer or aspiring farmer.[250] Therefore, one can immediately see that larger polyploid plants would tend to be selected over smaller diploids by foragers. Sometimes, for example when there was a particular abundance of seed, the forager might have planted (or perhaps accidentally dropped) some of the seeds, instead of eating them all at once. Larger fruits or seeds from larger plants, that are also more likely to be polyploid, would be more likely to be chosen for collection and dissemination in this way. Eventually, such selection would have favoured dispersal and reproduction of the new, larger polyploids over their smaller diploid relatives. As time went by, the new polyploid forms would have gradually become the main variety of that particular food plant in those regions where foraging humans were active.

In this example, we see that polyploidy has created a favourable variation (e.g. large seed size) and humans have, wittingly or unwittingly, acted as agents in a process of selection. The result is the evolution of new varieties of plant that are more closely adapted to growth in association with foraging or farming humans. In addition to increased size, polyploids can have other advantages, including improved resistance to insect pests. For example autotetraploid forms of the saxifrage, *Heuchera grossulariifolia*, are 12-fold less likely than diploids to be attacked by some lepidopteran pests,

although they may be more susceptible to attack by other insects.[251] This was one of the first studies on the ecological consequences of polyploidy and demonstrates that it may play an important, and hitherto unrecognized, role in modulating plant–herbivore interactions in terrestrial communities. These results suggest one reason why polyploidy arises so frequently in plants, even in the absence of human intervention.[252]

Not all instances of hybridization between two different plant species results in the formation of an allopolyploid. If the two hybridizing parent species are sufficiently related to each other, their sets of haploid chromosomes can pair successfully, which means that pollen from one species can fertilize the eggs of the other. In this case, the hybrid progeny of two diploid parents will be a diploid that has attributes of both parental species. This type of interspecific hybridization between sexually compatible species is much more common in plants than in animals. It can also be an effective mechanism of evolutionary change and speciation in cultivated and wild species alike. This point is well illustrated by a recent example of a new hybrid created between a crop and a wild plant that was so well adapted that it then caused the localized extinction of both parental species.[253] The parents were the cultivated radish, *Raphanus sativus*, and the wild radish, *Raphanus raphanistrum*.

These two species of the radish family have long coexisted in their native European habitats with only occasional hybridization that almost invariably results in unfit progeny. Both species were introduced into California in the nineteenth century and now grow wild throughout the West Coast from Baja California to Oregon. In the past few decades, a new intermediate hybrid form has been observed across this region. The hybrid has several advantageous traits compared to its parental species. It has three to four-fold larger fruits than either parent; its fruits are tougher and can better resist attack from local avian herbivores; and the plants have slender roots that are much less susceptible to disease. The new genetic combination present in the hybrid has also converted this form of radish into a much more aggressive colonizer of new habitats.[254] The hybrid has been so invasive that it has now completely displaced both parents in the area where it originated and may spread to other localities. Similar displacements have occurred with other plant species. For example in several parts of the US Midwest, allopolyploid hybrid versions of the amaranths, *Amaranthus tamariscinus* and *A. tuberculatus*, have completely replaced the two parental species, and the same is probably the case for the stripeseed species, *Piriqueta caroliniana* and *P. viridis*, in central Florida.[255]

Polyploidy and agriculture

Some crops, such as breadwheat and oats, have adopted a more radical form of polyploidy than mere tetraploidy. In these cases, an allotetraploid species has hybridized with a diploid species to create a new hexaploid species that now contains six sets of chromosomes.[256] Surprisingly, such genetic 'monsters' are sometimes fertile and successful plants. In the case of breadwheat and oats, the new hexaploid species that arose after domestication were just as fertile and vigorous as their diploid parents. As a bonus, these new hexaploid cereals also tended to be higher yielding; they produced better quality grain; and they had a greater tolerance to cold and drought than their parental species. For these reasons, cereal polyploids, which probably arose several times via spontaneous hybridization events, were selected by early farmers and became the favoured crop varieties. However, polyploids do not always make better crops. Barley is a diploid plant and polyploid versions, whether spontaneous or man-made, do not have an increased performance as crops so all modern varieties of barley remain diploid.[257] Therefore, although polyploidy is often a considerable advantage for a crop, this is not always the case and there do not seem to be any universal rules that apply to all species.

The significance of allopolyploidy for one of our major crops can be seen from several recent studies of wheat genetics.[258] These show that allopolyploid formation triggers two types of genetic change: a radical series of cardinal genetic and epigenetic alterations;[259] and a sporadic set of slower evolutionary changes that are not possible in diploid plants.[260] One of the key factors that enables allopolyploid crops, such as the wheats, to successfully propagate is the presence of a mechanism to ensure correct pairing of their different sets of chromosomes, so that they behave like diploids at meiosis. In the absence of such a mechanism, pairing between homeologous chromosomes (i.e. from the different parental genomes) can lead to sterility or other gross abnormalities. In tetraploid (durum) and hexaploid (bread) forms of domesticated wheat, correct pairing of homologous (rather than homeologous) chromosomes is controlled by a chromosomal region called *Ph1*.[261] This region is not active in diploid wheats, suggesting that it only arose after polyploidization.[262] Characterization of the *Ph1* locus in wheat has required more than 50 years of sustained effort from the pioneering genetic work of Riley *et al.* to the recent molecular and cytogenetic analyses of Moore *et al.*[263] Despite these advances, the precise mechanism by which *Ph1* is able to prevent pairing of homeologous chromosomes is still unresolved.[264]

To summarize, it now appears that polyploidy may be almost ubiquitous in plants and that it has been a major mechanism for evolutionary changes such as speciation. Polyploidy has also played an important role in the adaptation of some plants to cultivation, and hence in the process of crop domestication. What used to be thought of as a genetic aberration is now known to have played a role in such momentous processes as the development of the various forms of wheat as some of the most important of the world's food crops. In this chapter, we have seen how plants are much more promiscuous in their breeding practices than animals, readily forming fertile interspecific hybrids. In the next chapter, we will look in more detail at the phenomenon of this and other forms of genomic fluidity and their implications for crop domestication.

CHAPTER 5

Fluid genomes, uncertain species, and the genetics of crop domestication

Nec species sua cuique manet, rerumque novatrix ex aliis alias reparat natura figuras

[No species of thing keeps its own form, and renewing Nature reforms one shape from another]

Ovid, 43 BCE–17 CE, *Metamorphoses XV*, 254

Introduction

Most people, including a surprising number of scientists, are still under the impression that genomes and biological species are relatively fixed entities. For example it is frequently asserted that the addition of a tiny amount of so-called 'foreign' DNA (as in transgenic organisms or GMOs) is undesirable, both in principle and in practice, and there have been many dire warnings of the potential, deleterious consequences of such genetic manipulation. However, research over the past decade has revealed that the genomes of all organisms, from the simplest bacteria to the most complex multicellular plants and animals (including humans), are extremely dynamic and inconstant entities. Even the once-hallowed concept of the biological species is becoming increasingly difficult to define with any satisfactory degree of precision. The latter point applies especially to plants, and it is precisely this genomic fluidity and promiscuity of some plants that has underpinned the development of domesticated crops, hence making agriculture possible.

Fluid genomes, 'extra' DNA, and mobile genes in plants

The fluid genome

Most genomes, from bacteria to humans, are in a constant state of flux, both in terms of their size and DNA composition. For example the genome size of

many relatively closely related plants can vary considerably, even though they may have similar numbers of chromosomes. Rice and maize are both diploid members of the grasses, or Gramineae, and contain 24 and 20 chromosomes respectively.[265] They also have about the same number of genes in their genomes—currently estimated at around 40,000. However, whereas the rice genome contains only 400 Mb (megabases[266]) of DNA, the maize genome contains over six-fold more DNA, totalling 2500 Mb.

Even more remarkably, the genome of a single plant species, such as maize, can vary greatly in size in different geographical locations. Depending on their climatic adaptations, some varieties of maize have been found to have twice the genome size of other seemingly indistinguishable varieties.[267] Finally, the genome size of a species can vary greatly over time; for example the rice genome has more than doubled in size and then contracted again to lose two-thirds of this additional DNA.[268] The reason for these huge variations in genome size, both within and between species, is that the genomes of many, but not all, plants harbour large quantities of what we can term 'extra' DNA, that is DNA that does not encode functional genes. In fact, more than 80% of the genome of maize, and the various species of wheat and barley, consists of such 'extra' DNA.[269] Cereals and other monocots are unique in the plant world in the diversity and often-massive size of their genomes, which range from 400 Mb in rice to 123,000 Mb in the fritillary, *Fritilaria assyriaca*.[270]

The range of genome sizes found in higher plants as a whole is far greater than in animals, extending 1000-fold from a mere 125 Mb in the thale cress, *Arabidopsis thaliana*, to 123,000 Mb in the fritillary.[271]

Despite the fact that it has 1000-times the amount of DNA of *Arabidopsis*, the fritillary is an extremely modest-sized plant of the lily family, standing no more than 15 to 30 cm tall at maturity. What the fritillary genome does with its additional 99.9% of DNA compared to *Arabidopsis* is still a mystery. It has been proposed that all flowering plants originally had small genomes and, although many species still have fairly modest-sized genomes, a sizeable minority of species (most notably in the monocots) has acquired as much as 1000-fold more DNA than most of the rest of their plant brethren. There is recent evidence that very large genome sizes in some plants can be selectively disadvantageous in the long term, although this has not been a sufficient constraint to prevent the extreme diversity and continual flux in genome size that we see in many plants today.[272]

What is 'extra' DNA?

The majority of the 'extra' DNA in plants is made up of highly repetitive regions called LTR (long terminal repeat) retrotransposons, or retroelements. These small sections of DNA can duplicate themselves without excision from the chromosome, resulting in a steady process of multiplication within the genome. This means that, as time goes by, more and more of the LTR DNA accumulates in the genome. But some plant genomes do not just passively gain additional DNA—they can also lose it, albeit with varying degrees of facility in different species. One study has shown that the rice genome, which currently contains 400 Mb, was originally much larger but has lost about 200 Mb of DNA over the past eight million years (a relatively brief period in terms of plant genome evolution).[273] It may be that rice is just better at removing its exogenous DNA than most other plants and animals. Alternatively, the 'extra' DNA might be performing a useful function in those organisms in which it has not been removed, as we will see below.[274] In contrast, the maize genome has doubled in size over the past three million years, since its divergence from the related cereal, sorghum, *Sorghum bicolor*, and is still increasing to this day.[275]

It is not just plants that contain a lot of repetitive DNA. In the genomes of most mammals, including humans, from 50 to 98% of the total DNA does not consist of protein-encoding or regulatory sequences and is of unknown function.[276] And as with rice, it seems that large tracts of this DNA can sometimes be removed from mammalian genomes without any apparent effects, either for good or ill.[277] An obvious question of great interest to molecular geneticists is: where does this 'extra' DNA come from and what, if anything, is its function? We are not completely sure about the answer to these questions, but there are some intriguing clues from the structure of some of the repetitive DNA sequences, and especially LTR retrotransposons.[278] Some retrotransposons closely resemble a class of viruses called retroviruses,[279] and may indeed be derived from such viruses.[280] Instead of replicating within host cells and causing disease, some viruses integrate their DNA into the genome of their host and are then duplicated, along with the host DNA, during cell division.[281] Many familiar viruses that infect humans, including papovaviruses (causing warts and some cancers) and the human immunodeficiency virus (HIV, which causes to AIDS), can integrate their DNA into the human genome, effectively acting as genetic engineers. In the case of retroviruses, if viral DNA remains in the host genome, it can develop into a retroelement, or retrotransposon, that can then amplify itself at will. To quote from a recent study on *Dynamic DNA and genome evolution*: '. . . the majority of that repetitive DNA consists of retrotransposons and their derivatives. The retrotransposon life cycle resembles the intracellular phase of retroviruses and is replicative. The integration back into the genome of retrotransposon daughter copies has the potential to be highly mutagenic and disruptive to the genome and can lead to retrotransposons occupying major fractions of genomes. This appears to be a major factor in explaining the wide variation in genome size within many groups of plants and other eukaryotes.'[282]

It is important to appreciate that not all exogenous ('foreign') DNA is useless or parasitic, as was first thought when it was erroneously called 'junk' DNA.[283] In some cases, this 'extra' DNA has now acquired various functions in its plant or animal hosts. For example it can function as part of the process of DNA repair following the sort of double-strand breakage that can occur as a result of oxidative stress or other environmental insults to an organism.[284] It also appears that retrotransposons are much more active in plants of the grass group (especially cereals) compared to the broadleaf plants (such as beans and potatoes).[285] Transposons and other forms of repetitive DNA are implicated in some of the events that follow hybridization between two plant species and the formation of a new allotetraploid species. As we have already seen, and as we will also see below, such hybridization events have been of crucial importance in the evolution of many of our staple crops, most notably the wheat family, but also in many others including brassicas, oats, and cotton.[286] It seems that repetitive DNA elements are often exchanged between the two genomes in a new hybrid plant, sometimes silencing genes in one genome and sometimes overwriting them to generate identical sequences in otherwise different regions of the two genomes.

One recent report suggests that the activity of mobile DNA elements has been one of the major factors contributing to the evolution of the domesticated forms of wheat.[287] Breadwheat is classified according to the hardness of its grain. Hard-textured grains require more grinding than soft-textured grains, in order to reduce the rather gritty endosperm to a softer, more powdery flour. During this milling process more starch granules become physically damaged in endosperm from hard than from soft wheats. Since damaged starch granules absorb more water and enable more gas production by yeast than undamaged granules, flour from hard wheats is preferred for yeast-leavened breadmaking because it produces soft, fluffy, light breads. In contrast, flour from soft wheats is preferred for manufacturing heavier and denser products such as cookies, biscuits, and cakes. The hard-starch trait in the wheat plant is controlled by the *Hardness* (*Ha*) locus and is a classical example of a trait whose variation arose from gene loss after polyploidization. In some varieties of polyploid wheat, the *Ha* locus was disrupted by the insertion of non-coding, mobile DNA elements, leading to a soft grain trait, while in others the locus remained intact, leading to a hard grain phenotype. Therefore, we can see that repetitive DNA can sometimes play useful roles in

the processes of hybridization, polyploidization, and ultimately the enhancement of genetic variation and speciation in a major crop plant like wheat. Indeed, the action of mobile DNA is the main reason that we can enjoy the light, 'cotton wool' textured bread that is so popular with many consumers (although not perhaps with dietary advisors).

To summarize, it seems that the massive amounts of (originally) exogenous DNA that accumulate in many genomes are sometimes undesirable, and are gradually removed from the host organism (e.g. rice). In general, there seems to be an attempt by most organisms to rid their genomes of non-coding DNA; or to put it another (less teleological) way, large amounts of 'extra' DNA appear to be maladaptive in many cases and in some cases can result in the extinction of a plant species.[288] However, many other species, including some of the major crops, do not or cannot remove the vast bulk of their 'extra' DNA, which in maize is still proliferating to this day. It should be stressed that not all this exogenous DNA is parasitic; indeed some of it seems to play an essential role in certain aspects of plant development. The 'extra' DNA may also play a role in enhancing variation, hence contributing to the evolution of new varieties and even new species (e.g. the various types of wheat). This may be especially true for plants such as maize, which is probably the most diverse of all crop species, largely due to the many active transposons that are able to move within and between chromosomes in its genome.[289] Movement of transposons within genomes often causes changes in the expression of genes in the vicinity of their new site of integration. For example it is the unpredictable movement of transposons that gives rise to the striking and unique colour patterns in the variegated flowers of plants such as morning glory (*Ipomoea* spp.) and ornamental petunias (*Petunia hybrida*).[290]

Gene transfer between plant and non-plant genomes

In the last few years it has become apparent that DNA is constantly travelling to and from the genomes of plants and animals in a process called horizontal gene transfer.[291] There are many examples of such spontaneous gene transfer between complex eukaryotes, including higher plants, fungi, and some animals. Genes can be transferred from one species to another by several mechanisms including: host–parasite exchange,[292] transfer via a plant virus,[293] transfer via pathogenic[294] or mycorrhizal fungi,[295] transfer from a biting insect,[296] and non-standard fertilization involving more than one pollen grain.[297] Recently discovered examples of horizontal gene transfer in plants include: movement of an isomerase gene from a member of the genus *Poa* (which includes meadow-grasses and bluegrasses) to the genome of an unrelated species, sheep's fescue, *Festuca ovina*;[298] movement of a transposon gene from rice to members of the *Setaria* genus;[299] and the transfer of a mitochondrial gene from an asterid (a group of flowering plants that includes the Solanaceae) to members of the gymnosperm genus, *Gnetum*.[300] As more plant genomes are sequenced and analysed in detail, it is becoming ever more apparent that interorganism gene transfer between unrelated species is a lot more common than was previously suspected.[301] There are even cases of plant genes being transferred to animals. For example the simple aquatic animal, *Hydra viridis*, contains a fully functional ascorbate peroxidase gene transferred to it from a former symbiotic partner, the alga, *Chlorella vulgaris*.[302]

Genes and other DNA elements can move between species, but they can also move around to different locations within the cells of a given organism. For example as well as the movement of DNA elements, such as transposons, within the main nuclear genome, we now know that entire genes or clusters of genes can be transferred between the various organellar genomes and the main nuclear genome. In both plants and animals, the vast bulk of genomic DNA resides in the nucleus, but mitochondria, which are responsible for respiration and ATP generation, also contain small residual genomes of about 200 to 600 kb. Plants have an additional, third, genome in their plastid organelles, which typically contains about 130 to 150 kb of circular DNA.[303] These vital organelles are present in every plant cell but are especially important in leaves, where they become pigmented to form the green chloroplasts that are the sites of photosynthesis. Plastids almost certainly originated from photosynthetic cyanobacteria that were engulfed by a much larger eukaryotic cell, resulting in the creation of the first plant cells more than one billion years ago.[304] Instead of being digested by the host cell, the bacterial guest was tolerated and eventually became indispensable by fixing atmospheric CO_2 to synthesize sugars and other organic molecules for the benefit of its new host.[305] The new organism formed by this symbiotic union was a green alga, and it is from such green algae that all of today's plants, and hence all of our crops, are derived.[306]

Recent molecular studies have shown that individual genes and larger DNA segments from the plastid genome are continually being transferred to the larger nuclear genome.[307] When this process began, many of the transferred plastid genes were successfully integrated into the nuclear genome, with the result that much of the original plastid genome (which is of bacterial origin) now resides in the nucleus. However, it seems that the process has now reached some sort of limit and further gene transfer from plastid to nucleus is no longer favoured. Gene transfer to the nucleus still occurs at a surprisingly rapid rate, but newly integrated plastid genes are broken up and eventually eliminated from the nuclear genome.[308] In other words, the balance of DNA between the plastid and nuclear genomes has now reached a state of dynamic equilibrium whereby further transfer appears to be maladaptive.[309] Although considerable gene flow still occurs within plant cells, the transferred DNA is gradually selected against and removed, albeit over an evolutionary timescale that is numbered in the hundreds of millennia. Such dynamic behaviour by DNA, and its propensity to be transferred between very different classes of organism is increasingly calling into question the already rather fragile concept of the genetically unique biological 'species', which we will now examine.

Biological species

The concept of a biological species is rather like that of the genome. Both concepts are human artefacts invented as convenient ways of classifying parts of the living world. However, like the genome concept discussed above, the species concept is now coming under critical review.[310] Indeed, as discussed in Box 5.1, there is no really satisfactory definition of what constitutes a species; and this is especially true when we study our crop plants with their now famously mobile and inconstant genomes.[311] Members of a given species have very similar genetic endowments (i.e. they have similar genomes), and their phenotypic similarity means that they normally (but not always, see Boxes 5.2 and 5.3) look similar and behave in a similar manner. In the case of most of our familiar animals, a species can seem like a clearly defined and relatively stable entity that persists over timescales measured in millennia and often over many millions of years. Hence, despite some similarities, lions are obviously a different species from tigers;[312] and humans are equally obviously a very different species to their nearest anthropoid relatives, the chimpanzees.[313] *H. sapiens* may be a relatively new species, but even we have probably been around for about one million

---

Box 5.1 Is there such a thing as a biological species?

The conventional definition of a biological species is something along the lines of: 'a group of organisms sharing a considerable measure of genetic and phenotypic similarity, coupled with the potential for interbreeding and producing fertile progeny'.

Unfortunately, this nice, tidy view of what constitutes a species is soon brought into doubt when we examine plants and animals in more detail. One problem that confronts many ecologists is how to decide when a population of closely related variants has split into different species. Consider, for example, a species of insect that exhibits so-called clinal variation. These insects are all from the same original species, but have gradually started to diverge from one another, as their respective populations spread geographically ever further apart. Insects in adjacent geographical areas will still be relatively similar to one another and will interbreed freely. But, insects in areas that are further away from their centre of origin may have diverged so much that, although they can still breed with their immediate neighbours, they cannot now interbreed with more remote populations. Because all of the insects, however remote, can still interbreed freely with their immediate neighbours the entire group can be said to constitute a single species. However, as soon as we look at non-neighbours, we see that they cannot necessarily interbreed. So, do have a single species here, or more than one? If the latter is deemed to be the case, how do we define what constitutes a different species and what are its boundaries?

There is no correct answer to these questions. In many cases, the definition of what constitutes a species will be based on the arbitrary, and sometimes contested, decision of an individual scientist (Pigliucci, 2003). Such considerations apply even more to plants than to animals. Hence a recent article in *Nature* began as follows: 'Many

botanists doubt the existence of plant species, viewing them as arbitrary constructs of the human mind, as opposed to discrete objective entities that represent reproductively independent lineages or "units of evolution" ' (Rieseberg *et al.*, 2006). In the context of an understanding of crop evolution, it is important that our view about what constitutes a species should be informed by the latest scientific, and especially genetic, knowledge.

The utility of the species concept lies in its convenience as a rather broad-brush method of distinguishing between different types of organism, both past and present. It has been much less useful in the study of many microbial organisms, and especially the prokaryotic Archaea and Eubacteria. In the light of our new knowledge of genetics, the utility of the species concept as a rigorous method of classifying more complex organisms is also becoming increasingly uncertain. For example the immense confusion that surrounds the classification of some of our major crops, and especially the wheat family (see Box 2.1), is symptomatic of a lack of utility in the species concept in these particular cases. Such uncertainties, and our ever-changing view as to what constitutes a species, should cause us to exercise caution in ascribing too much importance to this concept. The species idea has its merits in describing some aspects of the biological world, but it is neither a sacrosanct nor universally useful concept.

In Boxes 5.2 and 5.3, two contrasting examples are given to illustrate some of the problems with the species concept. In the first example a type of butterfly that we thought for centuries to be a single species now turns out to be made up of many different species. In the second example we will see how a group of very diverse-looking plants that are not obviously related at first sight, are actually genetically almost identical members of the same brassica species.

years. We also observe other present-day species, such as many ants, which have apparently remained virtually unchanged for over 50 million years.[314] However, as discussed in Boxes 5.1, 5.2, and 5.3 and illustrated in Figure 5.1, the reality of what constitutes a species, especially in the case of plants, is often more complex that is suggested by these seemingly straightforward examples.

Box 5.2 The skipper butterfly—one species or many?

The neotropical skipper butterfly, *Astraptes fulgerator*, is an attractive and colourful insect that was first described by Johann Walch in 1775. It is distributed over a vast, and very climatically diverse area of the Americas, from the southern USA to northern Argentina. Skipper butterflies in these ecologically distinct regions look very similar to one another, and detailed dissections of their genitalia have failed to reveal any significant differences. Sometimes, so-called cryptic species may resemble each other superficially, but their true identity is almost always expressed in the form of morphologically different genitalia. These distinct genital structures prevent breeding between the otherwise similar-looking members of different cryptic species. This phenomenon is especially common in insects, including many butterflies (Burns, 1994). It seemed clear, therefore, that the skipper butterfly was a single biological species.

However, entomologists recently became suspicious when they discovered that skipper butterflies in one small area of Northwestern Costa Rica were apparently able to feed off a huge range of different plants (Janzen, 2003; more of these interesting butterfly records are available online at: http://janzen.sas.upenn.edu). Such insects are normally extremely specialized in their food preferences, and many butterflies are only able to feed on a single plant species. This is because plants subject to persistent attack by herbivores tend to develop an ever-changing portfolio of chemical defences. These defences often include a cocktail of insecticides targeted at specific insect pests. In response, insect herbivores such as butterflies frequently develop the ability to break down such defences, e.g. by either detoxifying or sequestering the insecticidal chemicals. For example monarch butterflies, *Danaus plexippus*, have developed an ability to use their tissues for the safe storage of toxic cardiac glycosides, such as cardenolides, that are present in their favoured food plant, the milkweed *Asclepias syriaca*.

Not only does this strategy render the cardenolides toxins harmless to the monarch, but the toxins also make the butterflies unpalatable to potential predators who soon learn to avoid these distinctively coloured insects (Brower and Moffit, 1974). The evolution of such defence mechanisms normally takes a long time and is relatively costly in energy terms for the insect concerned. This means that over the course of many millennia a particular species of butterfly often becomes adapted to just one type of plant that it is able to feed off in safety. Meanwhile, other close relatives of the same plant that might happen to make a slightly different cocktail of insecticides, often remain toxic to the butterfly and hence cannot be part of its diet. It seemed very odd to the chemical ecologists in Cost Rica that a single species of insect, such as the skipper butterfly, could suddenly develop the dozens of detoxification and/or sequestration mechanisms needed to feed off the many species of plant that were apparently part of the diet of such butterflies in the region.

Eventually, the problem was solved in 2004 by using a novel DNA 'barcoding' technique to analyse the genomes of the butterflies. This revealed that what had hitherto been considered to be a single species of butterfly, as observed and studied for over two centuries, in fact consisted of at least ten apparently morphologically identical, but genetically quite distinct, species (Hebert *et al.* 2004). It therefore appears that although we humans may think that all skipper butterflies look virtually identical, the butterflies themselves can somehow distinguish their genetic differences and thereby do not interbreed with variants (i.e. different species) that are specialized to feed off different plants to themselves. In this example, what was thought for centuries to be one species of insect is now known to be at least ten, and probably more.

In 2007, there was an interesting twist to this story when anthropologist David Harrison reported that the Mayan-descended, Tzeltal-speaking indigenous people of the Mexican home range of the butterfly have distinctive names for the larvae of all ten cryptic species (Vince, 2007). The reason is that the larvae of each species affect a different wild or crop plant that is relevant to this human culture, hence justifying a separate name. So, while Europeans required the full gamut of modern genetic technologies to distinguish these species, the locals had known all along (but nobody had asked them).

Box 5.3 Brassicas—many forms in a single species

Our second example of the difficulty in assigning organisms to a human-invented species concept relates to a group of very different-looking vegetable crops that come from the brassica family. These plants include broccoli, the various forms of cabbage, cauliflower, Brussels sprouts, kale, and kohlrabi (Figure 5.1). The broccoli vegetable is dark-green with hundreds of tiny unopened flower buds, while cabbages consist of a large, tightly curled ball of pale green, dark green, or red leaves. Cauliflower has a white core that looks like many tiny bleached flowers, although it is actually a collection of rapidly dividing structures called an arrested meristem. The other brassicas are similarly diverse.

These different brassica vegetables are quite distinctive in their flavour, odour and even the time of year when they can be collected, and at first sight it seems inconceivable that they could belong to the same species. Yet, remarkably, they are all members of the same species, namely *Brassica oleracea*. It turns out that these very dissimilar-looking plants share an almost identical genome, save for a few tiny changes caused by a small number of mutations in key genes regulating plant development that have had extremely far-reaching, morphological consequences. Hence, a mutation in a single gene is enough to convert the wild brassica-like weed, *Arabidopsis thaliana*, into a plant that resembles a miniature cauliflower (Smyth, 1995).

The cauliflower phenotype is due to arrested development in the floral meristem that maintains it in a permanent vegetative state, instead of differentiating into normal floral structures. This produces the characteristic edible, white 'curd' of the cauliflower head. Meanwhile, in broccoli, the phenotype results from arrested development at a later stage of floral development, when the meristem has already differentiated into many small immature flower buds (Purugganan *et al.*, 2000). Brussels sprouts are compact bunches of unopened leaves, with a characteristic sharp taste that is not to everybody's liking. Both kale and kohlrabi are more conventional-looking leafy vegetables, and are not all that dissimilar in appearance to spinach.

The ability of *Brassica oleracea* to undergo so many radical developmental mutations, while remaining a single species, has been a great boon to the many hundreds of millions of people across the world who include these highly nutritious vegetables in their diet. The genetics and utility of the Brassicaceae family are discussed in more detail in Chapter 7. The extreme phenotypic diversity of these variants of a single brassica species contrasts with the seeming morphological identity of all forms of skipper butterfly (Box 5.2), even though the latter are made up of several genetically distinct species. One take-home message from these two examples is that external appearance can be very deceptive in enabling us to decide what is, or is not, a biological species.

Revising our concept of the 'species'

As we learn more about genetics and taxonomy, it is evident that the species concept is often imprecisely defined, as well as being rather elastic and continually subject to revision.[315] The plasticity and the essential artificiality of the species concept is shown by the repeated difficulty in classifying organisms into discrete species. We will see numerous examples of these challenges in defining many types of our major crop plants over the course of the next two chapters, and further examples occur in the scientific literature almost weekly, as ever more data from genomic analyses are published. As long ago as 1908, US naturalist, CF Bessey noted: 'Nature produces individuals and nothing more. Species have been invented in order that we may refer to great numbers of individuals collectively.'[316] For example, as we will see in more detail below, Asian rice is a single species and yet it contains several races that cannot interbreed, whereas most types of Asian rice can interbreed with *Oryza rufipogon*, which is a completely separate species. Also, the rice genome has expanded considerably in size and then contracted again, but we still classify all these very different historical forms of rice as a single species. It is evident, therefore, that a biological species is very much an *ad hoc* and provisional entity, that is itself constantly evolving, mutating, and being reclassified by biologists.[317]

This is a very far cry from the original idea of a species as an entity that was possibly divinely ordained, and either immutable or at least only subject to very gradual changes. The plasticity of

Figure 5.1 Diverse forms of a single crop species, *Brassica oleracea*. These very different looking forms of vegetable are all members of the same species, *Brassica oleracea*. Their dramatically divergent morphologies are due to a few minor mutations in key genes that regulate important developmental processes, such as flower and leaf development. (A) Curly cabbage, (B) cabbage, (C) broccoli, (D) ornamental red cabbage, (E) kohlrabi, (F) Brussels sprouts, (G) cauliflower, (H) curly kale. (B) and (C) courtesy of Dan Lineberger, Texas A and M University, USA.

biological species is very much a recent scientific viewpoint, only dating back to the nineteenth century evolutionary arguments of Lamarck and Darwin. Unfortunately, the word 'species' also tends to be used more generally in a more inflexible and somewhat mystical sense that, in Western philosophical tradition, dates back to Thomas Aquinas and even to Aristotle. In much of our contemporary discourse on genetics and biotechnology, additional, more recent notions such as 'species integrity' and 'intrinsic value' are frequently invoked. These apparently new ideas have distinct overtones of an outmoded and inappropriate (in a biological context) philosophical essentialism.[318] Rather than clinging to such unverifiable notions, it may be more productive to regard a biological species as a somewhat arbitrary and provisional, human-defined entity that can sometimes be of use in operational situations, such as classification, but has no intrinsic meaning of itself.

As we shall see later in Part IV, many scientists clung to the idea that species were fixed and unchanging, as late as the nineteenth century and beyond. In contrast, pragmatic early plant breeders such as Thomas Fairchild were able to create new hybrids that apparently broke all the rules of supposed 'species integrity'. Another example is US plant breeder, Luther Burbank, who was a pioneer of crop improvement in the late nineteenth century. Among other achievements, he was responsible for the Idaho potato, which is still a mainstay of the American market in French fries. He also built up the plum industry in California, where he developed eleven of today's most important varieties. Despite his strong Christian faith, Burbank had a more broadminded view than some of the scientists of his day, especially about the immutability of species. As well as improving existing crops, such as potatoes and plums, he experimented with wide crosses between plants of different species. For example he produced a potato/tomato hybrid, although he never developed these sterile plants any further. In a speech in San Francisco in 1901, Burbank stated that botanists had once: 'thought their classified species were more fixed and unchangeable than anything in heaven or earth that we can now imagine. We have learned that they are as plastic in our hands as clay in the hands of the

potter or colors on the artist's canvas, and can readily be molded into more beautiful forms and colors than any painter or sculptor can ever hope to bring forth.'[319]

The domestication syndrome

Having reviewed some of the rapidly accumulating evidence for the plasticity of plant genomes, and indeed entire species, we will move on to consider the relevance of this knowledge of plant genetics for an understanding of crop domestication. This was not a single event, but rather a series of processes whereby a few plants gradually adapted to new conditions imposed by cultivation and thereby became more suited to human exploitation, for example by yielding more and better-quality products.[320] We have seen that domestication is not necessarily a one-way process; under the right circumstances domesticated plants can revert to wild type. Also, the process can affect both partners, that is human domesticators as well as plant domesticants. As we will see in Chapter 9, humans have adapted in numerous ways, both genetically and culturally, to agriculture as well as effecting many profound genetic changes in their plant and animal domesticants. Moreover, even in our most ancient crops, the process of domestication still continues today; it is a dynamic, unceasing interaction between humans, domesticants, and the environment.

For example agricultural practices themselves are constantly evolving, as is the external environment (whether biotic or abiotic; anthropogenic or non-anthropogenic) experienced by a given crop. Hence an optimum wheat variety of a few decades ago may no longer be suitable under today's farming regimes, in today's climate, or with today's ever-evolving suite of pest and disease species. Nevertheless, we can still usefully talk about a few crucial initial changes that were necessary to begin the process of domestication of a wild plant. Once these changes were accomplished, the plant had become a crop, albeit a fairly rough and ready sort of crop, which may have needed many centuries of further selection before it was suitable to become a primary staple. In this section, we will be looking at these initial changes that are necessary to move a plant from being a wild species to one that is partially or totally dependent on humans for its growth and is also useful to them. It is apparent that not all plants are susceptible to easy domestication, so what is so special about our major crops?

Domestication-related genes

The key to successful domestication of a crop lies, of course, in its genetic endowment. To put it simply, if a plant happens to have the right genetics, it will be much easier to domesticate into a crop than most other plants. To a great extent, our most successful crops have been selected, not just because they are good sources of food or other products, but also because their genetic organization has lent itself to the selection of a limited number of traits that makes them easier to manage and cultivate than their wild ancestors. These traits are collectively known as the 'domestication syndrome', a concept developed by Jack Harlan[321] and others.[322] We will now look at these rather special domestication traits as a group, so that we can begin to see how crops and their wild ancestors might differ genetically from the thousands of other potentially useful plants that are less amenable to the domestication process.[323] Major domestication-related traits include non-shattering seeds, large seeds, high yield, synchronous flowering and seed set, loss of seed dormancy, and traits responsible for ease of harvesting and food preparation (Table 4.1).

It should be stressed here that crops are not the only plants containing genes that could potentially result in domestication-friendly phenotypes, such as seed retention and lack of dormancy. Indeed, it now appears that many, and perhaps all, major groups of higher plants have relatively conserved gene families that regulate such attributes as seed size, seed weight, short-day flowering, and seed retention. For example very similar domestication-related genetic loci regulating such traits have been found in a range of very divergent plant families, such as legumes (Fabaceae), cereals (Poaceae), and solanaceous vegetables (Solanaceae), including hundreds of non-crop species.[324] This begs the question: if the vast majority of the more than 100,000 species of higher plants contain very

similar domestication-related genes, why have so few crops been successfully domesticated? The emerging answer is that, in most crops that have been studied to date, it is not simply the presence of domestication-related genes that is the key to creating a successful domesticant; rather it is the chromosomal location and method of regulation of such genes.

Clustering and regulation of domestication-related genes

Three interrelated genetic factors have greatly facilitated the manipulation of domestication-related traits in the major crop plants: (1) some of the most important traits are regulated by just one or two genes; (2) many domestication-related genes are located in small clusters in the crop genome (see Table 5.1); and (3) even when a trait is regulated by many unlinked genes, it is commonly found that a very small number of 'master genes' can have a huge influence on expression of the trait. Some specific examples of these genetic factors as they apply to our major crops are as follows (see also Figure 5.2):

1. Regulation of key traits by one or two genes: Probably the most important trait for the early cultivators of grain crops was seed shattering. In the wild plant, much of the seed would be shed from the plant, and therefore lost, before it could be harvested. In our most successful crops, this crucial trait is regulated in a very simple manner by either one (rice, lentil) or two (wheat, barley, sorghum, oat, pearl millet) genes.[325]
2. Clustering of domestication-related genes: One of the best examples of this phenomenon can be found in maize. The tall, high-yielding, cultivated version of maize and the short, small-seeded, wild teosinte plant are mainly distinguished by differences in DNA sequences and expression patterns of five groups of genes. The five groups of genes regulate: (1) tendency of the ear to shatter; (2) percentage of male structures in the primary inflorescence; (3) internode length on the primary branch; (4) and (5) increased numbers of kernels per cob. In the maize genome, all five of these genetic loci are tightly clustered in a small region of chromosome

Table 5.1 Genomic regions showing QTL (quantitative trait locus) clustering for domestication traits

Crop	Reproduction	Mapping cross (domesticated × wild forms)	Location of QTL cluster	Attributes of corresponding traits
Maize [1]	Outcrossing, $2n = 4x = 20$	F2: *Zea mays* ssp. *mays* × *Z mays* ssp. *parviglumis*	Chr 1	Shattering (ear disarticulation), growth habit, branching pattern
			Chr 2S	(*tb1*), ear and spikelet architecture
			Chr 3L	Number of rows of cupules
			Chr 4S	Growth habit, ear architecture
			Chr 5	Glume hardness (*tga1*)
				Ear architecture
Common bean [2]	Self-pollinated, $2n = 2x = 22$	F2: *Phaseolus vulgaris* cultivated form × *P vulgaris* wild form	LG D1	Growth habit and phenology
			LG D2	Seed dispersal (pod dehiscence) and dormancy
			LG D7	Pod length and size
Rice [3]	Self-pollinated, $2n = 2x = 24$	F2: *Oryza sativa* × *O. rufipogon*	Chr 1	Growth habit (tillering and height), shattering, panicle architecture
			Chr 3	Shattering, panicle architecture, earliness
			Chr 6	Shattering, panicle architecture, earliness
			Chr 7	Panicle architecture
			Chr 8	Growth habit (height), earliness, shattering
Pearl millet [4]	Outcrossing, $2n = 2x = 14$	F2: *Pennisetum glaucum* ssp. *glaucum* × *P. glaucum* ssp. *monodii*	LG6	Shattering, spikelet architecture, spike weight, growth habit
			LG7	Spikelet architecture, spike size, growth habit and phenology
Sunflower [5]	Outcrossing, $2n = 2x = 34$	F3: *Helianthus annuus* var. *macrocarpus* × *H annuus* var. *annuus*	LG17	Shattering, apical dominance, achene weight, earliness
			LG09	Achene size and weight, growth habit, head size
Eggplant [6]	Self-compatible, $2n = 2x = 24$	F2: *Solanum melongenas* × *S. linnaenum*	LG06	Growth habit, achene size and weight, earliness, head size
				No obvious colocalization

This table, which is adapted from Poncet *et al.* (2004), shows the strong clustering of many major domestication-related traits in several important crops, as revealed by molecular genetic analysis. Such clustering would have greatly increased the likelihood of a comparatively rapid evolution of domesticated cultigens as the plants adapted to the new conditions imposed by early protofarmers.

Chr, chromosome; LG, linkage group.

[1] Doebley and Stec (1991); [2] Koinange *et al.* (1996); [3] Xiong *et al.* (1999); [4] Poncet *et al.* (2000); [5] Burke *et al.* (2002); [6] Doganlar *et al.* (2002).

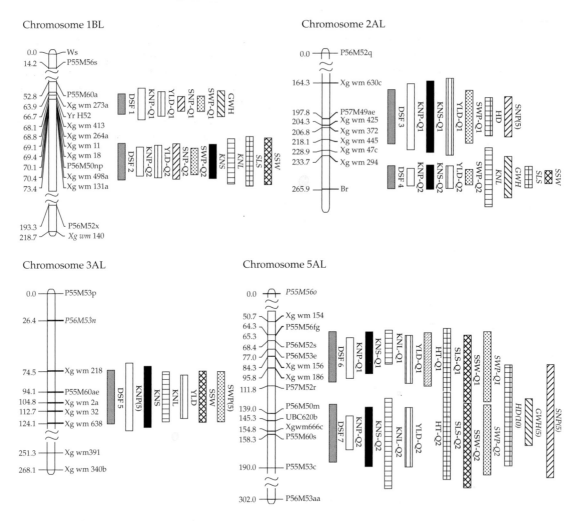

Figure 5.2 Clustering of genes associated with crop domestication traits. Tight clustering of domestication syndrome factors (DSFs) and their associated quantitative trait loci (QTLs) is shown for four of the chromosomes of wild emmer wheat, *T. dicoccoides*. DSFs and corresponding QTLs: ▪, DSF; ■, KNS; ▤, kernel number/spikelet (KNL); ▢, YLD; ▢, HT; ▦, spikelet number/spike (SLS); ▨, single spike weight (SSW); ▢, spike weight/plant (SWP); ▢, kernel weight/plant (KNP); ▦, HD; ▢, GWH; ▨, spike number/plant (SNP) (data from *Peng et al.*, 2002).

eight.[326] One of the key genes in this cluster is *tga* which, as we will see in the next chapter, was probably the one of the earliest traits to be modified during domestication as it rendered maize much more edible than its wild precursors.[327]

3. Dozens of separate, and not necessarily physically linked, domestication-related genes that control key traits such as seed dispersal, are now know to be regulated by just a few 'master genes'.[328] A good example of this form of gene control was discovered during a detailed study of the control of

domestication syndrome genes in the common bean.[329] The authors looked at 15 separate traits, ranging from pod number to seed yield. Each of these traits was regulated by up to several dozen individual genes. In every case, however, they found that between one and four genes controlled the vast majority of variation in the trait in question. As more crops are studied at the molecular genetic level, we are finding more and more examples of these sorts of 'master genes'. In genetic terms, such genes are often are associated with so-called quantitative trait loci (QTLs), and can be identified by molecular marker analysis. Important domestication-related QTLs have recently been found in dozens of crops from rice and sorghum to tomato and potato.[330]

One of the most telling features of the domestication process in the major ancient crops is its variability in terms of frequency, duration, and localization. For example some crops, such as potatoes,[331] barley,[332] emmer wheat,[333] einkorn wheat,[334] cassava/manioc,[335] maize,[336] and bottle gourd,[337] seem to have been domesticated only once. This contrasts with squash,[338] cotton,[339] millet,[340] and common beans,[341] where data suggest multiple domestication events. In the next two chapters, we will examine the nature of these domestications in more detail.

CHAPTER 6

The domestication of cereal crops

Ceres, most bounteous lady, thy rich leas
Of wheat, rye, barley, fetches, oats, and pease;

William Shakespeare (1564–1616) *The Tempest*

Introduction

Having introduced some of the intricacies of plant genetics in the previous two chapters, we will now move on to examine the genetic make up of our major crops. The aim will be to describe how it is that the particular genetic attributes of this small number of plants has made possible their domestication and exploitation via agriculture. As we will see, domesticated species possess several unique genetic attributes, the understanding of which is of key importance for efforts to effect their improvement via breeding. One of the most striking features of domesticated plants and animals is the relative genetic similarity of the present-day members of each of these species. In other words, domesticated plants and animals tend to be less genetically variable than most (but by no means all) wild species. In many cases, this genetic uniformity is due to so-called 'domestication bottlenecks' whereby all members of a domesticated species are often descended from a very few (and sometimes just one) selected individuals.[342] As we saw in Chapter 1, similar genetic bottlenecks can also occur in wild populations of both plants and animals, including our own species. For example almost all present-day non-African humans may be descended from relatively small groups of migrants that left Africa between 70,000 and 50,000 years ago. Even more dramatic is the recent evidence that almost all of the tens of millions of aboriginal Amerind people that originally populated the entire continents of North and South America may be descended from as few as 70 individuals.[343]

The kind of genetic bottleneck that results from domestication can be even more severe than these examples, with each of the tens of billions of members of a major crop species sometimes being descended from a single mutant plant. It is possible that maize is such a species; and all of our commercial bananas are genetically identical clones of a single triploid hybrid plant. One of the problems that this causes for the breeder is a lack of genetic diversity that, to varying degrees, affects almost all of our domesticants, whether plant or animal. Such genetic uniformity can render these populations more susceptible to new diseases and obviously reduces the essential raw material available to the breeder. In the case of animal domesticants, this was appreciated by ancient farmers, who would regularly leave tethered females out in the open. The farmers hoped that wild males, either from the same or from a closely related species, would mate with the domesticated females and hence augment the genetic diversity of the entire herd or flock.[344] In the case of crops, such cross fertilization with wild relatives would have been a regular occurrence in outbreeding species that were being cultivated near to their centre of origin. In the case of the temperate cereals, cross fertilization with wild relatives would have been commonplace in southwest and Central Asia, as long as the crops remained outbreeders. However, in several crops, such as rapeseed, the self-fertile genotypes that have now been selected by farmers are much less likely to interbreed, even when grown near wild relatives.

Even more problematic for the maintenance of genetic variation in crop species was the gradual spread of agriculture during and after the Neolithic period. By the Bronze and Iron Ages, the vast majority of crops were being grown in areas, such as Europe, northern Asia, or Africa, where no wild relatives existed. This geographical isolation precluded the introduction of novel variation from wild relatives, and led to a gradual increase in the genetic uniformity of the crop as a whole in such regions. It was the work of breeders and geneticists such as Nikolai Vavilov that simultaneously highlighted this problem and suggested a solution; namely to conserve and exploit wild relatives of crops for future breeding programmes. Before this can be done usefully, however, the modern breeder needs to know the extent and nature of the genetic uniformity of a crop species. Such information allows the breeder to identify traits where little or no useful variation exists. In turn, this might require a search for such genetic variation, either in wild relatives of the crop or in unrelated species from which the traits might be obtained via modern techniques such as wide crossing or transgenesis.[345]

Some of the most important agronomic traits for which many major crops are lacking genetic diversity relate to disease resistance and the ability to withstand abiotic stresses, such as drought or salinity. These are complex, quantitative traits that are regulated by many interacting genes. Therefore, the ultimate goal of the crop geneticist is not merely to describe and understand the workings of plant genomes *per se*, but also to use such knowledge to improve our crops as they face new challenges in our ever-changing world. Over the past decade, there has been tremendous progress in crop genetics, much of which has been due to the kinds of insights into plant genomes that we looked at in the last chapter.[346] In this chapter, we will focus on the cereals. A timeline of cereal evolution is presented in Figure 6.1. In the next chapter, we will go on to look at the other major groups of ancient crops, such as pulses and potatoes, before finishing with a brief examination of a much more recently domesticated group of crops with its own uniquely fascinating and convoluted evolutionary and genetic history, namely the brassicas.

Wheat

The wheat group of plants includes some of humanity's most important crop staples, but it can also be a geneticist's and taxonomist's nightmare (see Box 2.1).[347] The wheat group contains several dozen species that often hybridize with one other to create completely new polyploid species.[348] Some of these new hybrid species, including breadwheat and durum wheat, contain the entire genomes of two or three ancestral species. Even worse, a given hybrid species may have arisen independently on several different occasions, and in various different locations. For example we know that the hybridization to produce breadwheat has occurred several times, on each occasion yielding a slightly different sort of hybrid variety. Although, technically speaking, they might be members of the same species; the more recent hybrids may be genetically quite different from older hybrids, whose genomes have had several thousand generations to evolve new and distinctive patterns of gene expression.

We know this is the case because it is now possible to recreate new, so-called synthetic, breadwheat hybrids in the lab. A synthetic species of breadwheat can be recreated by hybridizing emmer wheat, *Triticum turgidum*, with the goat grass, *Aegilops tauschii*.[349] The newly created hybrid looks fairly similar to a rather primitive form of breadwheat in that it produces good flour, but it is hulled, and therefore not free-threshing like modern breadwheat. Such recently created, or synthetic, hybrid plants often behave differently to their older cousins that were formed by spontaneous hybridization in the field. This means that a crop such as breadwheat, which looks like a single species (at least to human eyes), may in fact contain a mixture of individuals that have very diverse origins and genetic constitutions. Sometimes such variants will not even interbreed, even though they are still classified as being members of the same species.

The three genomes of crop wheats

Probably the simplest way to understand the genetics of the wheats is to split them up into the three groups of diploid species, from which the polyploid wheats can then be derived (Figure 6.2). Because they are diploid, members of these three basic wheat species have two sets of chromosomes per genome, which are represented here by two letters. The first of the three groups is therefore

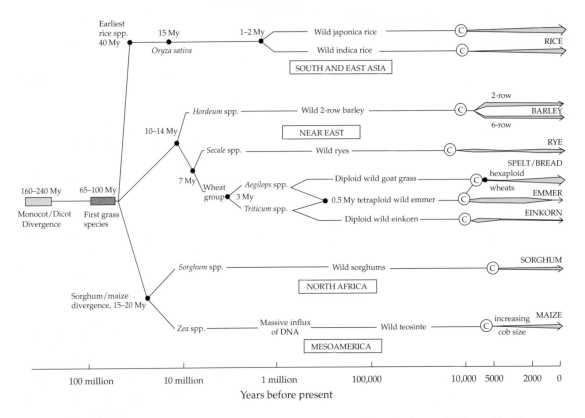

Figure 6.1 Evolution of the major cereal crops. Cereals are monocot plants, characterized by long thin leaves, which diverged from the broader leaved dicots about 200 million years ago. The first grasses date from almost 100 million years ago and the most ancient group of cereals are the rices, which appeared about 40 million years ago. Several groups of rice then developed in south and east Asia. In western Asia, the major group of cereals were wheat, barley, and rye, which diverged from a common ancestor about ten million years ago. The ancestors of maize and sorghum split about 15 million years ago and wild sorghum species became important edible plants in Africa, while teosinte served the same purpose in Mesoamerica until its mutation into the more cultivation-friendly form known as maize. The evolution of domesticated forms of these various cereals did not occur until the Palaeolithic/Neolithic transition, from about *13,000* to *5000* BP, and was determined by a new suite of human-created environmental conditions such as cultivation and harvesting. For reasons of space, the very diverse group of millet crops are not represented here but are described instead in the main text. Note the logarithmic scale of the time axis. My, million years before present; ©, beginning of cultivation and appearance of domesticated forms.

called AA, the second is BB, and the third is DD.[350] In the context of domesticated wheats, the key diploid species are einkorn wheat, or *Triticum monococcum*, which has an AA genome; a goat grass called *Aegilops speltoides*, which has a BB genome; and a second type of goat grass, called *Aegilops tauschii* (sometimes called *Aegilops squarrosa*), which has a DD genome. This is a bit like saying, for the sake of argument, that modern humans have an AA genome, gorillas have a BB genome and chimpanzees have a DD genome. Animals are much less likely to form fertile interspecific hybrids

than plants so there are no examples of new species made up of human–gorilla (AABB) or chimpanzee–human (DDAA) hybrids.[351]

It has long been known that interspecific animal hybrids can sometimes occur, including between large mammals, but such hybrids are invariably sterile. Humans have taken advantage of this to create several hybrid animals for various purposes, some useful and others more frivolous. Hence, that stalwart pack animal, the mule, is the sterile hybrid of a female horse, or mare, (*Equus caballus*) crossed with a male ass (*Equus asinus*). In zoos, hybrids of a

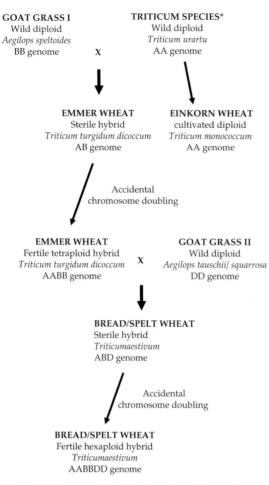

GOAT GRASS I
Wild diploid
Aegilops speltoides
BB genome
X
TRITICUM SPECIES*
Wild diploid
Triticum urartu
AA genome

EMMER WHEAT
Sterile hybrid
Triticum turgidum dicoccum
AB genome

EINKORN WHEAT
cultivated diploid
Triticum monococcum
AA genome

Accidental
chromosome doubling

EMMER WHEAT
Fertile tetraploid hybrid
Triticum turgidum dicoccum
AABB genome
X
GOAT GRASS II
Wild diploid
Aegilops tauschii/ squarrosa
DD genome

BREAD/SPELT WHEAT
Sterile hybrid
Triticumaestivum
ABD genome

Accidental
chromosome doubling

BREAD/SPELT WHEAT
Fertile hexaploid hybrid
Triticumaestivum
AABBDD genome

Figure 6.2 Recent evolution of the domesticated wheats. The various domesticated wheats were originally derived from wild diploid grasses of the *Triticum* and *Aegilops* genera. One of the earliest cultivated wheats was the diploid species, einkorn, but this soon gave way to the more versatile and productive tetraploid species, emmer, which was a mainstay of early agriculture in the Near East, Europe, and South Asia. A further round of hybridization and chromosome doubling after *10,000 BP* gave rise to the hexaploid bread and spelt wheats, which remain our major temperate cereal crops to this day. * *Triticum urartu* is considered as the most likely A-genome donor to the polyploid wheats and precursor to einkorn (*T. monococcum*), but related species such as *T. thaoudar*, *T. boeoticum*, and *T. aegilopoides* may have also been involved.

male lion and a female tiger (a liger) and, much less frequently, a male tiger and a female lion (a tigon) have been created for somewhat less obvious purposes than the mule. Unlike the situation in plants, these hybrid animals are unable to produce fertile

eggs or sperm and therefore do not yield any progeny. For this reason, polyploidy is not believed to be a particularly useful mechanism for the evolution of new species in animals. Unlike most animals, however, the vast majority of plants will quite readily hybridize to form fully fertile new polyploid species that contain several genomes.

The wheat group, including *Triticum* and *Aegilops* species, is especially good at forming inter- and intragenus hybrids that rapidly stabilize their divergent genomes so that they behave as diploids within a few generations.[352] So, now we can look at how the polyploid wheats arose. The first hybridization was of a diploid Triticum sp. (AA) and goat grass (BB) to produce emmer wheat, which is an allotetraploid with an AABB genome.[353] The transition from diploid to a stable tetraploid version of wheat may have occurred as early as 500,000 years ago.[354] Tetraploid emmer wheat was probably formed following a spontaneous hybridization event and went on to grow as a successful new species of wild cereal that spread throughout the Near East long before people began to collect or cultivate its seed. Wild emmer wheat was certainly growing profusely alongside the diploid einkorn wheat in the Jordan Valley and in Syria more than 23,000 years ago.[355] It is likely that wild emmer would have been made into a crude paste and eaten as a form of porridge before the development of baking techniques that made breadmaking possible.[356] A wild subspecies of emmer, called *Triticum turgidum dicoccoides*, still exists in western Asia.

Emmer was eventually domesticated into a cultivated subspecies known as *T. turgidum dicoccum*, which is still grown as a livestock feed in mountainous parts of Europe and the USA. Emmer wheat was grown extensively around the Mediterranean region and beyond, where it served as the main source of grain for making good quality bread and pastry. As we will see in Part III, emmer wheat and barley went on to become the twin crop staples of the early agrourban civilizations of Mesopotamia, Egypt, and the Indus Valley. Emmer was eventually superseded by modern breadwheat about 2000 years ago, mainly because the latter proved to be much easier to harvest and thresh. Another derived subspecies of wild emmer is durum wheat, or *T. turgidum durum*. Durum wheat is still widely grown today, especially around the Mediterranean region and in the USA. It is much easier to thresh than emmer, but its high gluten content makes durum paste sticky and unsuitable for bread making. However, this sticky paste is ideal for making a variety of foods including semolina, couscous (a short pasta popular in north Africa and the Near East), and the many versions of long pasta such as spaghetti, macaroni, and tagliatelle. The differences in the ear and grain structures of some domesticated wheats is shown in Figure 6.3.

The next stage in the evolution of polyploid wheats was the formation of modern hexaploid breadwheat, *Triticum aestivum*. Although *Triticum aestivum* is called 'breadwheat', it is of course possible to make various forms of bread from most of the main cereal crops. Hence we have rye bread from rye, corn bread from maize, and even a rather tasteless 'bread' that can be made from rice. The reason for calling *T. aestivum* 'breadwheat' is that it is the source of the most highly prized form of light, easily chewed, and readily digestible form of bread. From their earliest cultivation until comparatively recent times, the hexaploid breadwheats had a special social cachet because their narrow geographic range (compared to hardier cereals such as wheat, barley or emmer) meant that they was relatively expensive and therefore often only available to the wealthier classes. In contrast, the poorer classes were obliged to make do with heavier and darker mixed breads made from whatever cheaper cereals were available.[357] The new hexaploid breadwheat species, *T. aestivum*, was probably formed in the region of Transcaucasia, as a result of several spontaneous hybridizations between a cultivated form of the tetraploid emmer wheat (AABB) and one of the wild goat grasses with a DD genome.

The diploid donor of the DD genome that created breadwheat is most probably, *Aegilops tauschii*.[358] This goat grass was probably the male parent, while a cultivated variety of emmer wheat, *Triticum turgidum* subsp. *dicoccum*, was the female parent. The evidence for this is that the plastid genome (which is derived from the female parent) of all polyploid wheats appears to originate from *Aegilops speltoides*. Since *A. speltoides* was the BB genome donor to emmer wheat, it follows that emmer must be the female parent of the original hybrid that led to breadwheat, *Triticum aestivum*.

(A) (B) (C)

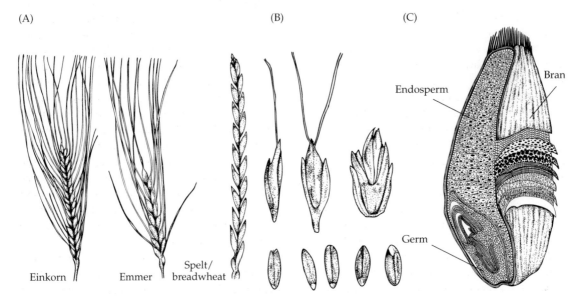

Figure 6.3 Structures of the major domesticated wheats. The structure of ears (A) and grains (B) of the three major historic cultivated wheats; from left to right in each case are einkorn, emmer, and spelt/bread wheats. (C) Section through a grain of breadwheat, showing the large starchy endosperm and the germ (kernel), which contains proteins and oils that are important for breadmaking quality.

Therefore, breadwheat has the genome, AABBDD; in other words it contains the complete, albeit somewhat rearranged, genomes of three original parental diploid species. In animal terms, this is analogous to the creation of a fully fertile hybrid species containing the full human/gorilla/chimpanzee genomes—which is quite a thought and perhaps not the sort of thing to bump into on a dark evening! The analogy illustrates the distinctly bizarre genetics of one of our major crops, and such genetic oddities are now turning out to be rather common in agriculture as a whole.

In modern farming, there are numerous cultivated subspecies of hexaploid breadwheat, which is the main form of commercially grown wheat. Although one of its parental species was a cultivated form of emmer wheat, the hybridizations to create breadwheat were almost certainly spontaneous, rather than man-made. Also, despite its seeming improbability, successful hybridization between a domesticated tetraploid species and a wild diploid occurred on numerous occasions after about *11,000 BP*.[359] Whenever one of these hybridizations occurred, a slightly different form of hexaploid wheat was formed. In one case the hybrid was a different type of breadwheat called spelt, *Triticum aestivum* subsp. *spelta*, which is still grown in parts of the Mediterranean Basin. While it seems very unlikely that early cereal farmers contributed to the actual creation of breadwheat, once the new plants appeared in the field, they were soon recognized as improvements over emmer wheat and widely adopted for cultivation.

In addition to breadwheat, several other types of hexaploid wheat were created by spontaneous hybridization on different occasions in various parts of Central and western Asia. An example of the latter is *Triticum vavilovii* in Armenia.[360] Modern breadwheat was already being cultivated in Anatolia by *9000 BP* and may have been present at Abu Hureyra in northern Syria as early as *11,000 BP*.[361] Hexaploid breadwheat soon spread across central Asia. However, as discussed above, it took much longer for breadwheat to displace emmer as the major wheat crop in the more conservative Mediterranean/European regions. Indeed, emmer was still the favoured wheat crop grown by the Romans, less than 2000 years ago, and breadwheat did not became a major crop in Britain until the fourth century CE.[362] Today, breadwheat is cultivated in temperate climates throughout the world and has acquired considerable cultural significance among European and Near Eastern societies. From the Near East we have the Christian New Testament saying 'man doth not live by bread alone', and the prayer 'give us this day our daily bread'. In both examples, the word 'bread' stands for food in general. The continued cultural significance of bread is apparent in the modern, albeit slightly dated, Anglo-American slang usages where bread and dough each mean money—that ultimate medium of contemporary material value. As we will see later in Chapter 8, the word for 'rice' is literally synonymous with 'food' in several Asian languages.

Wheat adapts to cultivation

We now know that all of our current varieties of wheat probably originate from wild diploid wheat species similar to einkorn. As noted in Chapter 2, wild einkorn wheat was being harvested by the Natufians at least *23,000 BP*. The origin of domestication of einkorn wheat is unclear. A wild group of einkorn that may be the ancestral variety has been traced to southeast Anatolia and it has been suggested that all of the tetraploid wheats may have originated in the vicinity of the Karacadağ Mountains, not far from the present border with Syria.[363] However, it is also possible that there were other, now-extinct, ancestral populations with a wider range that were cultivated further south, in the Levant.[364] Cultivated einkorn continued to be a popular cultivated crop from *12,000* to *6000 BP*, before giving way to the cultivated form of emmer wheat by the mid-Bronze Age. From this time onwards, breads made from diploid einkorn were generally considered inferior to those made from the tetraploid, emmer, or hexaploid, spelt, wheats.[365]

Einkorn cultivation continued to be popular in isolated regions from the Bronze Age until the early twentieth century, especially in areas where cultivation of the polyploid wheats was more difficult. Today, einkorn production is limited to small, isolated regions in India and around the

Mediterranean.[366] But although they have been out of favour as a major cultivated species of wheat for over 6000 years, the ancient einkorn wheats may yet stage a comeback. Intolerance of wheat gluten, often known as celiac disease, is a troublesome condition that rules out many wheat-based foods in the diet of those affected.[367] Einkorn flour can often be consumed by celiac sufferers without the troublesome side-effects of breadwheat products. Meanwhile, studies in Canada and Europe have also emphasized the nutritional qualities of this ancient diploid wheat. As a result, einkorn and other now-rare wheats, including emmer and spelt, are currently being assessed as possible alternative crops by breeders and nutritionists in Europe and the USA.[368]

Hexaploid breadwheat has one of the largest genomes of any crop, which at 16,000 Mb of DNA is more than five-fold larger than the human genome. While the size and complexity of the breadwheat genome has greatly complicated its molecular analysis, more progress has been made with the smaller, tetraploid genome of emmer wheat. Recent studies of the genetics of domestication-related traits in emmer wheat show that there are many genes involved in the process, but that they are highly clustered in a few areas of the genome.[369] For example the *Br* trait (see below) is closely linked with a cluster of eight additional major loci that regulate various domestication syndrome characters. Another finding from the same study is that most domestication traits are present in the A genome and are therefore derived from the einkorn-like ancestor, rather than the wild goat grass ancestor (which donated the B genome). Particular traits of interest were seed shattering, seed size, flowering time, and overall grain yield. As with rice (see below), the seed-shattering trait in wheat is regulated by a single gene (*Br*) and can therefore be readily selected against by farmers.[370] Wild-type wheats would have a functional *Br* gene and would easily lose their seed. Farmers would have tended to unconsciously select mutant plants that kept their seeds due to the presence of a non-functional form of the *Br* gene. Moreover, since the *Br* gene is closely linked to eight other DNA regions that regulate additional domestication-related traits, selection for *Br* mutants would be more likely

to enable farmers to 'accidentally' select for favourable variants of these other traits as well.

So, perhaps all the early farmers had to do was select non-shattering seeded varieties (which they would do automatically as most seeds from shattering varieties would be lost before harvest) and they would also automatically have selected eight other useful traits 'for free'. As we will see in subsequent sections, it is precisely this kind of genetic linkage between domestication-related traits that appears to have been one of the key factors that favoured the cultivation and successful domestication of most of our ancient crops. The timescale of wheat domestication is still uncertain. According to some studies, wild wheats may have been cultivated for as much as several thousand years before farmers were able to select domesticated forms.[371] This contrasts with other findings from wheat and other cereals, which suggest that domestication traits could have evolved rapidly without conscious selection. Clearly, further studies are required to resolve this important question in the case of wheat.

One of the best-studied domestication-related traits in wheat is the free-threshing phenotype. Wild wheats and the earliest domesticants produce a thick-hulled grain in which the starch-rich seed is enclosed within a tough coating, or glume. Considerable force, from pounding or milling, is required to release the seed from the glume. Soon after initial cultivation of emmer and einkorn wheat, new variants (such as durum) appeared in some regions, which had more fragile glumes and were therefore much easier to thresh to extract the grain. These 'free-threshing' wheats were rapidly adopted by farmers. The origin of the free-threshing trait has yet to be full resolved, but we know that several genes are involved and that it possibly evolved more than once.[372] With their higher yields and greater ease and efficiency of harvesting, free-threshing wheats soon became the dominant form of cereal crop wherever the climate allowed their cultivation. By *8000 BP*, emmer had spread westwards throughout the Mediterranean Basin and, by *6000 BP*, free-threshing wheats had reached the northern foothills of the Alps.[373] Today, the many and varied forms of wheat continue their evolution and manipulation by breeders, while foods made from these crops remain for many people the 'staff of life'.

Barley

Barley is a much simpler crop than wheat, both genetically and in terms of its domestication history.[374] Unlike the hexaploid breadwheat, barley is a diploid species with 14 chromosomes. However, the barley genome contains much more DNA (about 5000 Mb) than other diploid cereals such as maize (2400 Mb) or rice (430 Mb). Barley is now regarded as being so close to its wild ancestor that they are classified as variants of the same species. The cultivated form is *Hordeum vulgare* subsp. *vulgare* and the wild form is *H. vulgare* subsp. *spontaneum*.[375] Unlike wheat, barley was probably domesticated only once, in the Jordan Valley of the Near East, and all subsequent forms of cultivated barley may be descended from this one event.[376] Because barley is mostly self-pollinating, it is relatively easy to fix new genetic variants into discrete breeding lines and there are hundreds of modern varieties and thousands of land races of the crop known today.[377] Domesticated forms of barley tend to have shorter stems, larger grains, and more robust structures to hold the grains on the ear of the plant. The latter 'brittle rachis' (*Bt*) trait means that cultivated barley does not shed its seed as readily as the wild form.

The rachis holds the grains onto the stalk of cereal plant and in wild plants it normally becomes brittle as the ears mature. This allows the grains to readily break off from the plant, to fall into the soil or be otherwise dispersed. This trait made it difficult for people to harvest grain from wild cereals such as barley. Even if early farmers or cereal gatherers found intact ears of barley still on their stalks, once they attempted to harvest the grain, the brittle rachis trait could cause the ears to shatter and the grains to be lost. Repeated harvesting by cutting plants at the base of the stalk (this would have been done with flint-bladed sickles) would select for barley variants with tougher, non-brittle rachises, as has been demonstrated in field experiments with wild barley.[378] The brittle rachis trait in barley is controlled by the *Bt* locus, consisting of two tightly linked genes on chromosome 3.[379] The likelihood of

a prefarming plant gatherer/manager unknowingly selecting this desirable character, simply by collecting wild barley over an area of about 200 ha, is surprisingly high.[380] Assuming a fairly conservative mutation rate of one per million plants, the non-brittle rachis form would become the dominant cultivated form of wild barley within as little as 20 years of human management.[381]

Wild barley holds its grains in two parallel rows on the ears, while many (but not all) domesticated forms have six rows (Figure 6.4). In general six-row barley has a higher overall grain and protein yield, although the two-row forms have larger individual grains and are more resistant to lodging.[382] Two-row barleys are still grown today for the brewing

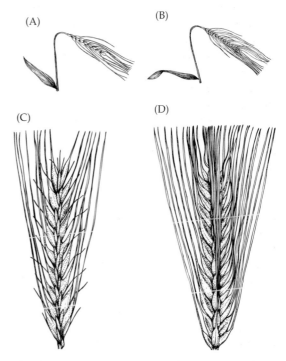

Figure 6.4 Wild and domesticated forms of barley. (A) Wild barley is a smaller plant with fewer ears and smaller grains that are readily shed from the mother plant. (B) Domesticated barley has more numerous, larger grains that tend to be retained on the plant until harvest. (C) Early forms of barley had two rows of grains but, a few millennia after its first domestication, a mutant with six rows of grains (D) was found by Mesopotamian farmers. This higher-yielding, six-row form of barley was particularly amenable to cultivation under intensive irrigation systems and became the staple foodstuff of Mesopotamian and Egyptian civilizations for many thousands of years.

industry, where they produce the best single-grain malts. The difference between two-row and six-row barley is controlled by just two genes, *Vrs1* and *Int-c*, of which *Vrs1* is the most important. The identity of the *Vrs1* gene, which lies on chromosome 2, is still unknown but, as with so many other domestication-related genes, it seems to have many other effects (i.e. the gene is pleiotropic), that are useful for the cultivation of barley in addition to just regulating row number.[383] Another striking mutation, in another gene called *Mlo*, occurred after domestication, resulting in a new type of barley that was resistant to one of the major pathogens that can afflict the plant, the fungus powdery mildew or *Blumeria graminis* spp. *hordei*.[384] The *Mlo* mutation results in a defect in gene expression, which, for reasons as yet unknown, enables the barley plant to become resistant to all known forms of this normally virulent fungus.[385] Powdery mildew is still a widespread disease of other crops, and particularly favours the relatively damp and cool conditions in which some forms of barley also thrive. We now know that this spontaneous mutation only occurred once and that most modern barley varieties are therefore descended from a single mutant plant. It is quite possible that, were it not for this fortuitous spontaneous mutation, and its equally lucky recognition by an especially observant early cereal farmer, we would not be growing barley as a crop today.

Barley was the principal cereal crop throughout the Near East in prehistoric times and was a major dietary staple of the early Mesopotamian and Egyptian civilizations. Wild barley was being collected and ground to produce flour in the Levant (at Ohalo near the Sea of Galilee) by semisedentary cultures as long ago as *23,000 BP*. At this stage, barley was the principal cereal being used by these people, although wild emmer wheat was also found at the same site. Later finds in the same region show that wild barley continued to be collected and used for food production, in places such as Ohalo (*19,000 BP*),[386] Abu Hureyra (*11,000 BP*),[387] and Mureybit (*10,000 BP*).[388] The discovery of non-brittle, larger-grained barley at various Levantine and Mesopotamian sites dating from *11,000–10,000 BP* is evidence of the spread of domesticated forms of the crop, although the two forms

probably overlapped as the newer varieties were slowly disseminated throughout the region.[389] As we will see in Chapter 10, intensively farmed barley monocultures were the principal sustenance of ancient southern Mesopotamian civilizations for several thousand years.

For many millennia, wild barley was harvested from mixed cereal stands with wild wheats and other grasses. As domesticated varieties of barley were adopted in the millennia after *11,000 BP*, the crop was still commonly grown alongside the domesticated wheats, einkorn and emmer. In some regions, however, barley gradually decreased in importance as a staple crop, as the new forms of wheat started to provide better yields and superior grain, especially for breadmaking. Nevertheless, barley regularly made a comeback in preference to wheat during several episodes of climatic- and anthropogenic-related salinity and aridity that afflicted much of Mesopotamian agriculture (especially in the south) after *6500 BP* (Chapter 10). While emmer was the preferred crop where climate and soil conditions permitted, there were many episodes of drought and/or soil salinity when barley was the only worthwhile, staple crop to cultivate. Indeed, barley can still thrive today in conditions that are too cold even for that other hardy stalwart, rye. By the Classical era, barley was generally regarded as fit only for the poor and animals. Eventually the crop was largely relegated to marginal areas, including the extremities of northwest Europe, or was grown as animal feed—especially for horses. Barley is still grown in many temperate climates today, but is mainly used for livestock feed or beer making, rather than as an edible grain crop for human consumption.

Rye and oats

Rye and oats are relatively minor temperate cereal crops in comparison with wheat and barley. Until recently, it was thought that rye and oats were merely weeds of the major cereal crops that did not evolve domesticated forms until many millennia after the start of wheat and barley cultivation.[390] During the 1990s, this notion was challenged in the case of rye by the surprising discovery from the Syrian village of Abu Hureyra of seemingly domesticated, large-seeded forms of rye, dating from as early as *12,000 BP* (Chapter 3).

Rye

Cultivated rye, *Secale cereale*, is a Near Eastern plant that probably arose from its wild relative, *Secale montanum*.[391] The other two members of the genus *Secale*, *S. iranicum* and *S. sylvestre*, are relatively distant relatives of the crop species and probably not involved in its domestication. All four *Secale* species are diploids with 14 chromosomes. Rye is related to the diploid wheats as shown by the close similarity of their respective genomes. Although rye and diploid wheats have the same number of chromosomes, rye has a much larger genome size due to the massive amplification of repetitive DNA regions. Like other cereals, rye is anemophilous, or wind pollinated. In contrast, many dicotyledonous crops, such as the brassicas, are entomophilous, or insect pollinated. Unusually for a grain crop, rye plants are incapable of self-pollination and therefore an important factor in determining the ultimate grain yield is the efficiency of wind pollination during flowering. As with barley and wheat, the brittle rachis trait in rye, which is controlled by a single gene, is a key attribute for a domestication-friendly plant.[392] The first putative rye domesticants from Abu Hureyra had larger seeds than typical existing wild varieties and the non-shattering trait also quickly became fixed in cultivated populations of the plant (Figure 3.2).[393] Wild rye still grows today in dense stands on Mount Ararat and on the Karacadağ slopes of present-day Turkey.[394]

Much less is known about rye genetics and the mechanism of domestication compared with the other major cereal crops.[395] This is largely due to the fact that, while rye may have been one of the earliest domesticants, it never became established as an important human dietary staple and therefore has not received as much attention in the scientific literature. Although it was one of the earliest crop domesticants, rye was never as productive as wheat or barley; the quality of its flour was markedly inferior for breadmaking; and the dark, heavy bread eventually became associated with lower social status wherever alternative types of

breadmaking cereals were available. Rye was quickly supplanted by wheat and barley as the preferred cereal crops in the Near East during the Neolithic period. As we saw in the case of early cereal cultivation at Abu Hureyra (Chapter 3), rye soon became superseded by barley and emmer wheat as the favoured crops due to their combination of better yields and superior flour quality. Rye grains contain about 13% protein, plus some gluten, although not as much as wheat, so it produces heavier, less satisfying bread.

Despite these drawbacks, rye has always managed to maintain a foothold as part of temperate-zone farming. Several millennia after it was domesticated and then largely abandoned, the crop staged something of a resurgence, as agriculture spread to the cooler climates of northern and eastern Europe. Here, the cold hardiness and drought tolerance of rye, which outperforms many other cereals in this regard, made it a useful and resilient crop. During the Hallstatt period of *3200–2500 BP*, rye became established in such regions, where it was better adapted to the relatively poor, light soils and the harsher winters. Rye bread soon became a popular staple, surviving today in the numerous dark breads of Central Europe. Some of the enduring prejudice against rye bread is summed up in the name 'pumpernickel'. This name comes from the German *'pumpern'*, meaning to break wind and *'nickel'*, which refers to the devil or 'Old Nick' in English. So, to put it crudely, pumpernickel means 'devil's fart'; a name that is doubtless connected with one of the digestive consequences of eating unrefined, high-fibre food. This may also explain at least one of the attractions of wheat-based white breads over the darker rye breads. More recently, there has been a rebirth of interest in rye as an alternative to wheat, due to its high fibre and other nutritional benefits (e.g. for diabetics), and rye is increasingly being promoted as a health food.[396]

Oats

Oats are members of a large group of related species that, as with the wheats (see above), has yet to be classified in a universally agreed manner. Various authors have divided the genus, *Avena*, into anything from seven to 30 different species,[397] according to morphological[398] or interfertility[399] criteria. Wild *Avena* species can be diploid, tetraploid, or hexaploid, containing respectively 14, 28, or 42 chromosomes, and are nearly all located around the Mediterranean Basin and Near East. Cultivated oats are all members of the hexaploid species, *Avena sativa*, with the exception of some localized cultivation of the tetraploid *A. abyssinica* in parts of Ethiopia.[400] The three diploid genomes present in *A. sativa* are termed A, C, and D, with the D genome possibly derived via autopolyploidization of the A genome, rather than hybridization with a different species.[401] The original diploid donor of the A genome was probably *A. canariensis*. The tetraploid species that provided the combined A+C genomes was probably either *A. murphyi* or *A. insularis*.[402] Although cultivated hexaploid oats have been separated into as many as three species by some authors, with additional species said to be the wild progenitors of cultivated oats, more recent evidence suggests that all of these forms of oat can be regarded as comprising a single species complex, namely *A. sativa*.[403]

Oats were domesticated much later than the other temperate cereal crops of Near Eastern origin. It is likely that the grains of wild oats were collected alongside wheat, barley, and rye by Palaeolithic and Neolithic hunter–gatherers, as all of these cereals grew together in mixed stands across wide areas of the Near East. However, the other cereals then became favoured over oats as potential crops, because they had larger and heavier seeds and were less prone to dormancy than wild oats. During the later Neolithic period, oats were mainly present as weedy admixtures in the cultivated cereals, while barley and wheat were the major crops that gradually spread from the Near East across Europe towards the Atlantic coast between *9000 BP* and *3000 BP*. As this process continued, the cooler and moister conditions of northwest Europe sometimes favoured oats over the other cereals. Oats gradually made the transition from minor weed to valued domesticated crop, and they were being cultivated as a single crop in Germany by about *4000 BP*.

As with wheat and barley, the major domestication-related traits in oats involved a breakdown of the original method of seed dispersal. Wild and weedy oats readily shed seed immediately after maturation, with the seeds tending to insert into the ground via specialized drill-like structures. In contrast, cultivated oats tend to retain their grain-bearing organs on the plant after maturation, where they are more readily available for harvesting.[404] It is likely that these mutations became more favoured by European farmers as oats began to out-perform the more established cereal crops in cooler and damper regions or climatic periods. In the warmer, drier Mediterranean climate, the ancient Greeks and Romans still considered oats to be weeds and used the grains to prepare medications rather than for food. The vigour of weedy varieties of oats was noted by Pliny, who, like Theophrastus several centuries beforehand, regarded them with a prejudicial contempt as a diseased variety of wheat, fit only for animals and barbarians.[405] Oats grew well along the Atlantic littoral and, following their introduction into Britain by the Romans, they soon became a staple cereal in the damp and misty climates along the Celtic Fringe of Europe, where they are still consumed with enthusiasm today, for example as a porridge.[406] Oats were spread across the temperate regions of the world by European colonists after the sixteenth century CE and had reached Australia and the Americas well before they were eventually adopted for cultivation as a stand-alone crop in their original homeland of the Near East.

Millets

Millet is a catchall term that applies to any one of a diverse group of small-seeded cereal plants of the subfamily Panicoideae, which is part of the grass family, or Poaceae (Figure 6.5). Although the various millets are not closely related genetically, they are similar in their agronomic characteristics and uses. In order of current worldwide production, the major millet crops are pearl millet (*Pennisetum glaucum*), foxtail millet (*Setaria italica*), broomcorn or proso millet (*Panicum miliaceum*), and finger millet (*Eleusine coracana*).[407] There are at least six additional crops that are classified as millets, but these are not grown on an extensive scale.[408] All the millets are warm-weather crops that are sensitive to late frosts, but many of them are also efficient users

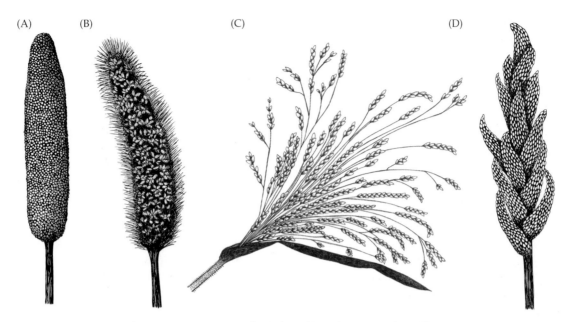

Figure 6.5 The millet group of crops. Millets are not a single family of cereal, but a diverse range of very different panicoid species that have been independently domesticated in Europe, Asia, and Africa. There are four major millet crop staples: (A) pearl millet, (B) foxtail millet, (C) broomcorn or proso millet, (D) finger millet.

of water and are therefore commonly grown in more arid regions where other cereals such as wheat will not thrive. Compared to the other major cereal crops, the genetics and evolutionary origins of the millets have been relatively little studied, but it is believed that pearl and finger millets arose in Africa while foxtail and broomcorn millets are of Chinese origin.

Pearl millet is a diploid plant with 14 chromosomes; it belongs to a highly heterogeneous group that includes both wild and cultivated forms. At one time these were divided into 15 separate species, but more recently have been designated as the single species, *Pennisetum glaucum*, in view of their mutual sexual compatibility.[409] Cultivated varieties probably originate from north Africa and/or tropical west Africa where they may have been grown since about *8000 BP*.[410] Key domestication traits in the newly cultivated varieties of pearl millet were the retention of the grain-bearing structures on the plant, that is a non-shattering character, and the partial exposure of the grains rather than their enclosure in a hard coating.

Cultivated forms of finger millet are allotetraploids, with 36 chromosomes, that arose from hybridization between two wild diploid species, *Eleusine indica* and *Eleusine floccifolia*.[411] Tetraploid finger millet is largely confined to eastern and southern Africa and the cultivated varieties probably arose in north Africa and/or Uganda around or before *4000 BP*. By *3000 BP*, finger millet had arrived in India where it differentiated into numerous, distinct local varieties, many of which are still grown today.[412]

Foxtail millet is a diploid with 18 chromosomes that probably arose from a closely related wild species, *Setaria viridis*, which is distributed across the whole of Eurasia. Indeed, the wild and cultivated plants can probably be regarded as two forms of the same species. Foxtail millet has generally been regarded as a crop of Chinese origin, due to the resemblance between cultivated and local wild populations from eastern Asia. The crop was always known to be of early origin but there is recent evidence of cultivation in northern China as long ago as *10,500 BP* or even earlier, which would place this millet alongside other ancient cereal crops such as wheat, rye, and barley.[413] Foxtail millet was also cultivated in Neolithic Europe and it now appears that the European cultivars most resemble local wild millets, and hence may have been domesticated independently of the Chinese varieties. If this finding is confirmed, foxtail millet would be one of the very few indigenous European domesticants, and perhaps the only cereal crop to enjoy this status.[414]

Broomcorn millet is a tetraploid plant with 36 chromosomes. The cultivated form is of unknown origin, although closely related wild populations of the same species are native to central China. Domesticated varieties of broomcorn millet probably date back to at least *10,000 BP* in northern China, and were first grown in southern Europe about *3000 BP*.[415]

Millets produce small starchy grains that can be processed to make a nutritious flour, which is almost as rich in protein as wheat. The flour is especially rich in B vitamins, such as niacin, B6, and folic acid, as well in as the minerals calcium, iron, magnesium, potassium, and zinc. Millets were sometimes grown alongside wheat, for example in parts of northern China, and the two cereal flours can be combined to make a form of leavened bread, although millets alone can only be used to make flat breads. Millet crops are still grown as subsistence staples in some parts of Africa and Asia, but elsewhere tend to be used as feed or forage crops, rather than for human consumption.

Rice

The Asian rice plant has the smallest genome of any of the major crops. At about 430 Mb, the rice genome is one-fortieth the size of the wheat genome. Despite this small genome size, Asian rice has 24 chromosomes, which suggests that it may be an ancient polyploid species that now behaves as a diploid.[416] And although its genome is also seven times smaller than the human genome, rice plants probably have more than twice as many genes as people.[417] Thanks to its small size, the rice genome was one of the early model plants for molecular geneticists, and in December 2004 it became the first crop plant genome to be fully sequenced.[418] These data are already shedding much light on genomic architecture, some of which may be applicable to the wider field of crop genetics. However, despite knowing the sequences of the

45,000 to 56,000 genes of rice, only half of these genes have been assigned even tentative roles, and we still only know definitely the function of a paltry 100 rice genes.[419] The full analysis of such vast amounts of genomic data and its extrapolation to the behaviour of complex traits in crops will probably take many more decades.

The taxonomy of rice is complicated because, while all cultivated Asian rice is from the same species, *Oryza sativa*, this species had already differentiated into three separate and rather distinctive races long before its domestication. The more northerly race is japonica, which is a short-grained form that is well adapted to warm-temperate climates. The two more southerly races, indica and javanica, are longer-grained and are better suited to damp, tropical climates. Despite being members of the same species, indica and japonica rice cannot interbreed with each other and forced hybrids are sterile.[420] Perversely, all races of *Oryza sativa* are able to interbreed with a separate species, *Oryza rufipogon*, a perennial wild plant, which is now known to be its wild progenitor.[421] *Oryza sativa* and its close wild relative *Oryza nivara* may have begun to diverge from their common ancestor as long as 15 million years ago. Some forms of *Oryza sativa* were then able to migrate from the Asian mainland, which at that time was joined to Australia and New Guinea. Once this land bridge was inundated, the now-isolated Australian types of wild rice followed a different evolutionary path to the mainland varieties and none of them were ever domesticated.

Meanwhile, the south Asian and Chinese forms, which were the ancestors of what are commonly referred to today as indica and japonica (or sinica) rice, began to diverge from each other between one and two million years ago.[422] Subsequent genetic divergence of the two races of rice in different parts of the Asian mainland was facilitated by the increasingly impassable geographical barrier of the Himalayas. Prior to this period, the proto-Himalayas would have been a far less formidable barrier than they are today. During the summer, the range would have regularly been traversed by animals, some of which may have carried the small rice seeds in their fur. It seems that the two races of rice were domesticated independently of one

(A)

(B)   (C)

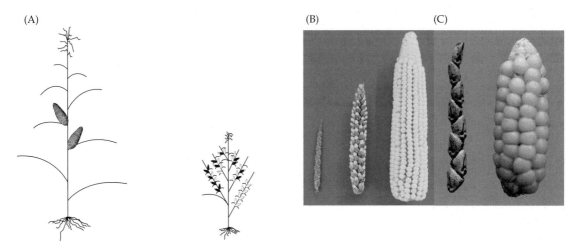

Figure 6.6 Evolution of teosinte into maize. (A) Differences between maize (left) and teosinte (right); note that maize is less branched and has far larger cobs than its wild ancestor. (B) Teosinte ear (left), modern maize ear (right), and their F₁ hybrid (centre). (C) Teosinte ear (left) and 'reconstructed' small, primitive maize ear (right). This small-eared form of maize was bred by George Beadle by crossing teosinte with Argentine popcorn maize and selecting the smallest progeny to reconstruct a primitive, small-eared maize that would resemble the earliest maize samples recovered from sites in the Tehuacán Valley in Mexico. It took over 3000 years for this small-ear form of maize to develop into the familiar large-cob form that became the staple food crop of Mesoamerican civilizations, from the Maya to the Aztecs. (B) and (C) courtesy of John Doebley, University of Wisconsin, USA.

the short, highly-branched forms characteristic of undomesticated maize and teosinte.[438] In maize, this gene has mutated to an inactive form, resulting in a suppression of the normally prolific branching of the main stem, so that the plant now has just a single main stem and much larger cobs. The change wrought by the mutation of this single gene is visually quite dramatic in terms of the overall plant architecture, but more importantly it also results in a plant that produces an increased food yield due to the larger size of its seed cobs.[439]

The second of these agronomically important genetic alterations in maize is due to a mutation in the gene, called *tga*, which controls the formation of a hard seed case around the seed kernels. The *tga* gene stands for *teosinte glume architecture* and it played a vital role in the domestication of maize.[440] In teosinte, a particularly hard external casing makes the grains very difficult to digest, so most of them pass through the stomach and out in the faeces. From the point of view of the wild teosinte plant, this is a good strategy to promote grain dispersal via animal vectors that might be duped into eating the seeds. But it is very bad news indeed for hungry humans who wish to derive nutritional

benefit from the plant. Before the mutation of this gene, the original teosinte seed would have been virtually inedible unless it was vigorously and repeatedly milled. If this version of teosinte was ever gathered or cultivated by ancient Mesoamericans, it was probably used for its sweet stalks, rather than the virtually indigestible seeds.[441] In maize plants, the mutated version of the *tga* gene no longer functions to produce a seed coat, so that the kernels are now bare on the cob and hence much easier for humans to digest.

The change to a naked, exposed kernel may, at first sight, seem disadvantageous for the maize plant because its seeds will now be completely digested by people (and other animals) who eat them, rather than being disseminated via their droppings. In fact, however, this mutation has resulted in a much improved dispersal mechanism for the seeds. Rather than relying upon the vagaries of the digestive systems of the odd passing animal, maize could now exploit the far more effective propagation skills of its human partners in this particular domestication dyad. The original human guardians of the newly mutated maize plants would have carefully saved some of the best seed for replanting in the most favoured locations, and then protected their bountiful new food source as it grew to maturity. As we will see in Chapter 8, this interaction was reinforced by the evolution of a religiocultural tradition that ascribed a specially protected, mystical status to the maize plant in its new domestic context (wheat, barley, rice, etc. were also treated thus in other societies). We can therefore more usefully regard the process of maize–human interactions as a mutually beneficial coevolutionary relationship between two biological species, rather than a series of anthropocentric domestication events (Box 2.3).

Some other important domestication-related genes in maize include *ramosa-1* and *ramosa-2*, both of which affect inflorescence architecture;[442] *su1*, which affects the texture of the commonest form of maize bread, the tortilla;[443] and *pbf*, which regulates the accumulation of seed storage proteins.[444] It is likely that the initial domestication-related selections, that resulted in the evolution of maize from a teosinte-like ancestor and the subsequent cultivation of the new crop, involved a small number of these key genes. However, selection of these genes would have also led to selection of many other genes that were physically linked with the original domestication-related genes. Furthermore, during the subsequent millennia of increasingly widespread cultivation, many other domestication-related genes of lesser importance would also have been selected. It is now estimated that at least one thousand of the 40,000-odd genes in the maize genome have been affected in some way by the domestication process.[445] The maize varieties cultivated by early Mesoamerican farmers were probably hybridized, either deliberately or spontaneously in the field, with wild teosinte. This would explain the presence of as much as 77% of the genetic diversity of teosinte in the genome of cultivated maize.[446] As with the other cereals surveyed in this chapter, cultivation-suitable maize varieties of wild teosinte developed into a true crop via a small number of chance mutations. The initial domestication of maize occurred relatively quickly, thanks to an unusual clustering of domestication-trait genes and the careful selection of such mutants by human societies that were searching for better food production strategies.

Sorghum

Sorghum is an important tropical and subtropical cereal crop. Members of the *Sorghum* genus all grow as cane-like grasses ranging in height from 50 cm to 6 metres. Numerous members of the genus are distributed throughout Africa and Asia of which three are of particular interest.[447] The cultivated crop is *Sorghum bicolor*, an allotetraploid with 20 chromosomes (Figure 6.7).[448] A closely related species is *Sorghum halepense*, better known as Johnson grass, which is currently one of the most aggressive and persistent weeds in the world. Ironically, it now appears that Johnson grass has recently evolved from the crop species, *Sorghum bicolor*, following a further round of allopolyploidization to produce an octoploid plant with 40 chromosomes. This is interesting in showing how a species can adapt in different ways to human disturbance of ecosystems. In the first instance, *Sorghum bicolor*, developed the normal set of domestication traits that facilitated its selection and propagation by farmers to be a major crop across Africa and Asia. In the second case, the same species hybridized with a related wild species to produce a new form of

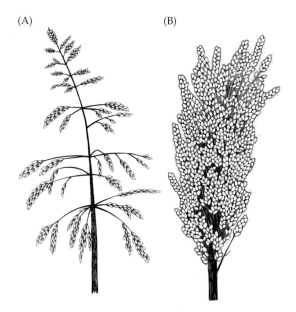

(A)　　　　　　　　　　　(B)

Figure 6.7 Sorghum: an important African cereal crop. Wild (A) and cultivated sorghum (B), *Sorghum bicolor*. The notoriously invasive weedy species Johnson grass, or *Sorghum halepense*, is a hybrid of domesticated sorghum and a wild relative.

hyperactive weed. The third species of interest in this genus is *S. propinquum*, another allotetraploid with 20 chromosomes that is found throughout Asia and is the likely second parent of Johnson grass.[449]

The latent weediness of domesticated forms of sorghum is seen whenever any of its five modern races are grown under cultivation. Within a short time, a weedy race is invariably found in the vicinity of the crop, with which it freely hybridizes.[450] This is unusual for a domesticated crop because most of them make very poor weeds indeed, rapidly becoming extinct outside cultivated ecosystems. Cultivated forms of *S. bicolor* are probably derived from wild forms of the same species that were selected for non-shattering seed heads, large seeds, ease of threshing, and synchronous maturation. Molecular genetic analysis shows the by-now expected pattern seen in other cereals whereby many key domestication traits are most regulated by a small number of genes that are often physically linked to each other.[451] The site(s) and date(s) of sorghum domestication have yet to be conclusively demonstrated. Some of the oldest archaeological remains are from India, but there are strong cases for other, and perhaps earlier, domestications in several parts of Africa, including the Sahara, Ethiopia, and Central Africa.[452] The distance between these putative centres makes it likely that sorghum was domesticated independently on several occasions some time earlier than *3000 BP*.

According to a recent review, one of the primary centres of origin may have been a 1000-km belt between latitudes 10° and 15° north, running through modern Ethiopia, Sudan, and Chad.[453] It is hypothesized that the major bicolor race of sorghum was domesticated in this region as early as *5000 BP*, followed by its dissemination to the Indus Valley within a few centuries. Some authors favour an earlier date for sorghum domestication, possibly as early as *8000 BP*,[454] but others suggest dates around *3000 BP*.[455] The Indus Valley may have served as a secondary centre of origin, from which the durra race of sorghum was spread to the Near East by about *4000 BP*. Other smaller and later centres of origin may include Guinea and Zambia. The latest consensus is that sorghum is probably a crop of African origin, and that its introduction into regions such as China and India occurred well after *4000 BP*, although this conclusion may be subject to change as the pace of archaeobotanical research and discovery continues to accelerate in both Africa and Asia (see Chapters 11 and 12).

CHAPTER 7

The domestication of non-cereal crops

Al the povere peple tho pescoddes fetten;
Benes and baken apples thei broghte in hir lappes,
Chibolles and chervelles and ripe chiries manye,
And profrede Piers this present to plese with Hunger.

William Langland, *c.* 1380, *Piers Ploughman,*
Passus 6

Introduction

Although cereals are by far the most important crops in most parts of the world, farmers rarely choose to grow cereals alone. In the long term, a broader portfolio of crop species is desirable both for nutritional reasons and to hedge one's bets against the caprices of climate, disease, and pests. From the earliest days of plant cultivation in the Near East, the Indus and Nile Valleys, China, and Mesoamerica, the important cereal staples were normally supplemented by various types of pulse crops, which are invaluable dietary sources of essential amino acids that are deficient in most cereal crops. Other cultures grew root crops, such as potatoes and yams, for many centuries before adopting cereals as additional staples. The reasons for the selection of these and other ancient crops were broadly similar to those that governed the selection of the cereal species that we surveyed in the previous chapter. In particular, it will become apparent that although the non-cereal crops constitute an extremely heterogeneous group of plants, they share many of the same genetic attributes that facilitated cultivation of the major cereal domesticants.

Pulses

Pulses are annual legumes cultivated for their seeds, and accompany cereals as major crop staples in most regions of grain agriculture. They include many different types of beans, lentils, and peas. Other legumes, such as peanut or soybean that are used for oil extraction and cover crops such as alfalfa or clover, are not normally regarded as pulses. Evidence from the Near East suggests that certain pulses were adopted as crops at about the same time as the earliest cereal domesticants, barley, wheat, and rye. Prior to this, wild pulses, including pea, lentil, chickpea (*Cicer arietinum*), bitter-vetch (*Vicia ervilia*), and grass pea (*Lathyrus sativus*) had formed part of the assemblage of informally managed plant resources that were exploited by hunter gatherers in this region. In the Americas, several types of beans were also eventually domesticated, but not until several millennia after the initial domestications of maize, squash, and potatoes.[456] Pulse crops are useful to farmers because of their ability to restore soil fertility by fixing atmospheric nitrogen to complex nitrates. This characteristic, coupled with their high protein levels, means that pulses are an ideal complement, both nutritionally and agronomically, to the higher yielding but relatively protein-deficient and nitrate-requiring starchy cereals.

Lentils

The lentil genus, *Lens*, includes seven diploid species, each of which has 14 chromosomes. The commonly cultivated form is *Lens culinaris*. The wild progenitor of *L. culinaris* is *L. orientalis* and the two species still readily interbreed. *L. culinaris* sometimes hybridizes with more distantly related members of the genus, but such crosses frequently involve embryo abortion, albino seedlings, and chromosomal rearrangements resulting in hybrid sterility, even if the seedlings reach maturity.[457] The wild progenitor, *L. orientalis*, is morphologically similar to the crop species, apart from being much smaller. The rich diversity of chromosomal types in the wild species compared to the cultivated lentil suggests that this crop was only domesticated once.[458] A major physiological difference between wild and cultivated lentil species is that wild plants bear pods that burst open to release their seeds immediately after maturation. Domesticated lentils retain their seeds for some time after maturation and this trait of non-dehiscence is due to a single mutation that would have been soon observed and exploited by early farmers. Other changes that occurred after domestication include larger seed size, and the development of a more robust stem that is able to grow unsupported in open fields; this contrasts with the trailing, vine-like trait of wild lentils. These characters are under relatively simple genetic control and their readily observable phenotypes would have facilitated selection by Neolithic farmers.

Wild lentils are found in the earliest preagricultural grain assemblages in the Near East, and can probably be considered as one of the 'founder crops', along with barley, emmer, and einkorn wheats. For example lentil seeds dating from *11,500 BP* were found, together with wild cereals, in prefarming sites ranging from Mureybit on the Euphrates in northern Syria to Netiv Hagdud in the Jordan Valley.[459] The domestication of lentils involved two stages, loss of seed dormancy and development of non-shattering seed pods, each governed by a single mutation.[460] Loss of dormancy, probably occurred between *11,000* and *9000 BP* in the core habitat of the wild progenitor, *L. orientalis*, namely the region now occupied by southeastern Turkey and northern Syria.[461] These non-dormant varieties rapidly spread south to the Jordan valley and it was here that the second stage of domestication, non-shattering pods, had already occurred by *8800 BP*, as attested by the huge hoard of fully domesticated seeds of *L. culinaris* at Yiftah'el, near Nazareth.[462] By *8000 BP*, lentils were present throughout the Near East, from Anatolia to the Levant and from Mesopotamia to Central Iran, and carbonized lentil grains are invariably found together with cultivated wheat and barley.[463] Lentils then appear to have travelled as part of a cereal-dominated suite of crops that spread to southeast Europe and predynastic Egypt by *6000 to 5000 BP*, and eastwards to Afghanistan and the Indian subcontinent by *4000 to 3000 BP*.[464] The grains of cultivated lentils are especially rich in protein, which at 25% of the seed weight makes it the most protein-rich crop after soybean. Although their grain yields are only about one-third of most cereal crops, lentils can usefully complement the starchy cereals to provide a balanced diet.[465] In particular, lentils would have substituted for animal protein in early farming cultures as opportunities for hunting became more limited.[466]

Peas

Peas are members of a small genus with just two members, the common pea, *Pisum sativum*, and a wild species, *Pisum fulvum*. Both species are self-pollinating diploids with 14 chromosomes, originating in the Near East and Mediterranean Basin. The domesticated pea is particularly celebrated by geneticists due to its use by the Austrian monk, Gregor Mendel, for his pioneering series of experiments that established the principles of heredity in the mid-nineteenth century.[467] Wild forms of *P. sativum* still occur in the Near East and eastern Mediterranean, where they have either bushy or vine-like growth habits. As with lentils, the major domestication trait in pea is seed retention within the mature pod, a trait that was the consequence of a single mutation around the time of early cultivation. Two additional key traits were a gradual increase in seed size from 3–4 mm to 6–8 mm and the reduction of the thick texture and rough surface of the seed coat. The doubling of seed size occurred over several millennia because, unlike many other domestication traits, seed size in peas is a complex character regulated by many genes. Presumably, farmers preferentially selected any slightly larger seeds for propagation, and the predominantly self-pollinating nature of peas assisted the fixation of the new variant in subsequent populations. Their smoother and thinner seed coat improved the edibility of peas as well as enabling the seeds to germinate immediately without a period of dormancy, hence ensuring a good crop stand in which the plants matured at the same time.

Wild peas with the closest genetic similarity to domesticated varieties are found from eastern Anatolia to the southern Levant, and this region can be regarded as the most likely centre of origin for peas.[468] Pea seeds have about 22% protein, plus a useful amount of starch, and the crop is well adapted to a range of climates, from the warm Mediterranean to the cooler maritime regions of north-west Europe. This ecological versatility favoured the widespread adoption of peas as a pro- teinaceous staple by European farmers, often in preference to lentils, which have a more restricted climatic range. Pea seeds dated to about *9500 BP* were found at Çayönü in the Taurus foothills, but these were rough-textured forms that had probably been gathered from wild stands.[469] The earliest remains of peas in a definitive farming context are found at a slightly later period than lentils, at about *8500 BP* at Çayönü and then at Tell Aswad in the Damascus Basin,[470] and Jericho in the Jordan Valley.[471] However, rough-textured forms were still being gathered in places such as Hacilar in south-west Anatolia as late as *7400 BP*,[472] indicating that pea cultivation may have spread more slowly, at least in some regions, than that of the major cereals. Smooth-coated, domesticated peas were present in Greece by *7500 BP*[473] and had reached the *Linearbandkeramik* cultures of the lower Rhine Valley by *6400* to *6000 BP*.[474] As with lentils, peas took longer to spread to eastern Asia, but had reached Afghanistan and the Indian subcontinent by *4000* to *3000 BP*.[475]

Beans

Several types of bean crop have been domesticated in various regions of the world. The broad bean, *Vicia faba*, originated from the Near East, while the common bean, *Phaseolus vulgaris*, is indigenous to the Americas. Broad beans are diploid plants with 12 chromosomes, and are unusual among crop species in not being self-pollinated.[476] One of the problems in trying to produce genetically fixed varieties from such crops is that enforced self- pollination often leads to the phenomenon of inbreeding depression, and a consequent loss of yield. Wild varieties of *Vicia faba*, such as *minor*, have readily shattering pods and smaller seeds

than modern domesticants. It is likely that seed retention was an early trait favoured by cultivation, but seed size remained relatively small until Roman times. The broad bean is regarded as a close relative of a group of large-seeded wild vetches, also in the genus *Vicia*, that are distributed across the Near East and Mediterranean. The exact ances- tor of the crop species is as yet unknown: one of the problems being that the wild vetches have a very different genomic organization to the crop species, with 14 chromosomes rather than the 12 in *Vicia faba*. Geneticists continue to hunt for the as-yet elusive 12-chromosomed wild ancestor, although this plant may now be very rare or even extinct. Broad beans are relative newcomers compared to peas and lentils, and were not cultivated as crops until about *4500 BP*. Broad bean cultivation was centred on the Mediterranean, from Iberia to the Aegean,[477] rather than the Fertile Crescent of the Near East like the other Old World pulses (see above).

The common bean, *Phaseolus vulgaris*, is a mor- phologically diverse species that includes pinto beans, kidney beans, black beans, haricot (white) beans, and numerous green beans. The large genus, *Phaseolus*, contains some 50 wild species, and also contains four other domesticated species, namely lima bean, *Phaseolus lunatus*; runner bean, *P. coc- cineus*; tepary bean, *P. acutifolius*; and year bean, *P. polyanthus*. All are diploid species, mostly with 22 chromosomes. The common bean has one of the smallest genome sizes in the legume family, at 625 Mb. After pea, the common bean has been one of the most important plants for plant geneticists and it was used by Mendel to confirm the genetic data that he obtained from his experiments with peas. Although most wild *Phaseolus* species now occur in Mesoamerica, genetic evidence suggests that the wild progenitor of *P. vulgaris* came from the eastern flanks of the northern Andes, in present- day Ecuador and northern Peru.[478] Common beans appear to have been domesticated several times, with one major centre in Mexico and another in the Andes, plus some additional minor centres.[479]

As with the other pulses, the major domestica- tion traits of common beans are pod dehiscence, seed dormancy, and growth habit. Other desirable traits selected by farmers are seed size and colour, and synchronous flowering (due to photoperiod sensitivity). Molecular genetic analysis shows that each of these apparently complex domestication traits in common beans is regulated by no more than one to four genes.[480] Wild beans have been found dating from *10,900 to 8500 BP* in the lower levels of the Guilá Naquitz cave site in southern Mexico (Oaxaca), where the remains suggest that the plants had been gathered for their tender shoots and pods.[481] Genetic data imply that wild *P. vulgaris* populations near Guadalajara, in the west-central Mexican state of Jalisco, are the progenitors of domesticated cultivars in Mexico.[482] However, there is a considerable spatial and temporal gap before the first directly dated common beans appear in Mesoamerica. The earliest findings date from only *2300 BP* and come from Coxcatlán in the Tehuacán Valley, more than 700 km southeast of Guadalajara.[483] It is likely that there was an earlier, unrelated domestication of another type of *P. vulgaris* in the Peruvian Andes at about *4400 BP*.[484] However, compared to most Old World crops, very little is known about the origin or spread of *Phaseolus* crops in South America, although they were amongst the primary staples in the region by the time Europeans arrived. As we will see in the next chapter, *Phaseolus* beans went on to form part of the trinity of crops (with maize and squash), known as the *milpa* system, that were grown together by the ancient Mesoamericans to provide an ideal agronomic and nutritional mixture for sustainable agriculture.

Potatoes and other Solanaceae

Potatoes

Potatoes are one of those crops that, like wheat, have an infuriatingly complex genetic endowment. Although their basic diploid chromosome number is 24, there are many closely related variants of the potato, with 36, 48, and even 60 chromosomes. There is a group of about 20 wild solanaceous species, known as the *Solanum brevicaule* complex, which morphologically resembles cultivated potatoes. Many of these lesser-known members of the potato family are still grown by local farmers in montane regions of South America. The most

commonly cultivated form of potato, and the one that has now been adopted worldwide as a staple crop, is *Solanum tuberosum* subsp. *tuberosum*, which is an autopolyploid with 48 chromosomes. At least four different diploid *Solanum* species have been implicated in the formation of *S. tuberosum* and there seem to have been repeated hybridization and chromosome multiplication events during the evolution of this crop.[485] Despite its large chromosome number and complex origins, the genome of *S. tuberosum* has the relatively modest size of about 840 Mb. The complexity of present-day potato genetics has been ascribed, in part, to multiple instances of domestication in different regions of the Andes. However, recent genetic studies have challenged this notion, and have led instead to a proposal that cultivated potatoes may have had a single origin, and hence were domesticated only once.[486]

Potatoes are quite unlike the other major crops that we have surveyed so far, in that they are grown for their starch-rich roots and only rarely propagated from seed. The edible part of the potato, selected by early Andean farmers, is a modified starchy root, called a tuber. This means that aspiring potato farmers would have been interested in very different genetic traits compared to grain farmers. For example traits such as seed shattering, synchronous flowering, or grain size would have been irrelevant. Instead, the most important traits would have centred on the potato tuber itself. Because the tubers of wild potatoes normally contain bitter-tasting and potentially toxic alkaloids, the primary trait of interest to farmers would have been low alkaloid content. One wonders how many hundreds, or possibly thousands, of people were poisoned by toxic wild potatoes before a chance mutation resulted in a low-alkaloid variety that would have been edible.[487] Because wild tubers tend to be very small compared to domesticated varieties, another important trait would have been tuber size. This latter trait would have been a lot easier, not to mention rather less dangerous, for the early farmers to select.

Potatoes rarely set fertile seed and normally propagate vegetatively rather than undergoing sexual reproduction. This means that all of the vegetatively propagated progeny from a particular plant will be clones that are genetically identical to each other and to the original parent plant. The first stage in the domestication of potatoes was probably the selection and propagation of a clonal line that had a low alkaloid content.[488] The cultivation of genetically identical clones makes for a uniform and predictable crop, but also carries the risk of susceptibility to new pests and diseases. This risk is minimized in the Andean centre of domestication, where potatoes have grown wild for millions of years, have been cultivated for over 13 millennia, and have therefore been able to build up resistance to local pests and diseases.[489] But when potatoes were taken overseas, they encountered new pathogens against which they had little or no resistance. This means that if a single potato is attacked by new disease or pest, to which it does not have any resistance, then the entire crop, and perhaps an entire country, is at risk.

Potatoes were brought into Europe by the Spanish in 1537, in the form of the Andean clonal variety, *S. tuberosum* subsp. *andigena*. However, the plant was not well received and for several centuries potatoes were regarded by many ignorant and conservative folk as being ruinous of the soil and unfit for human consumption.[490] In fact, potatoes are exceptionally high yielding in most European soils, achieving as much as 50 tonnes/ha, and are also one of the most nutrient-rich vegetable crops. Quite apart from their very high amounts of complex starchy carbohydrates, potatoes are rich in vitamins B$_6$ and C, as well as folate, niacin, protein, iodine, and many other minerals.[491] Gradually, the Andean potato became more accepted by farmers and by the nineteenth century it was an important crop in northern Europe. Unfortunately, this variety proved to be unsuitable for cultivation in Europe because of its sensitivity to fungal pathogens and consequent catastrophic crop losses.

It was this Andean potato variety that suffered a series of infestations by the oomycete mould, *Phytophthora infestans*, throughout Europe during the mid-nineteenth century.[492] Because it had become the sole staple crop of most of the relatively impoverished rural population of many parts of Ireland, the failure of the potato harvest for several years in succession caused a ruinous famine.[493] As a result of this disaster, the Andean variety was largely abandoned in favour of a more resistant variety from Chile. Most of the cultivated potatoes grown today in Europe and North America are based on a single clone of *S. tuberosum* subsp. *tuberosum* that was introduced from Chile soon after the potato blight. This means that potatoes are still, genetically speaking, a dangerously uniform crop. Fortunately, however, the Chilean lines seem to be relatively resistant to fungal pathogens—at least so far.

Other solanaceous crops

In addition to potatoes, other important solanaceous crops include: tomatoes, *Lycopersicon esculentum*; eggplant or aubergine, *Solanum melongena*; and peppers (e.g. chilli, bell, and jalapeño), *Capsicum annuum*. These are all fruit, rather than grain or tuber, crops and therefore have different domestication-related traits. Instead of traits such as pod shattering, would-be farmers would have focused on fruit size and taste, as well as harvestability traits such as the absence of thorns on the main plant or the fruit case. Most solanaceous species have the same basic diploid chromosome number of 24. Genetic studies with eggplant suggest that many of the solanaceous crops have similar genomic distributions of key traits that regulate the dramatic phenotypic differences in fruit weight, prickliness, shape, and colour that distinguish cultivated plants from their wild relatives.[494] In these studies, 62 domestication traits were localized to only six genetic loci with major regulatory effects. In the case of tomatoes, although there are as many as 30 traits that regulate fruit size and shape, a single locus, called *fw2.2*, accounts for as much as one-third of variation in fruit weight.[495] Almost half of these major loci in the eggplant genome had counterparts in either the potato, pepper, and/or tomato genomes. The data are consistent with a similar mode of genetically-driven domestication (with humans as the selection agents) in these otherwise unrelated crops that were independently brought into cultivation on different continents, by different human cultures, and at very different times in the past.

The wild form of a tomato fruit is a small (1–2 g), round, seed-dense berry that is ideal for reproduction and dispersal in the wild. The combination of favourable genetics and human intervention enabled the conversion of these tiny berries into the vast range of large-sized (50–1000 g), fleshy, seed-poor domesticated fruits that can be round, oblate, pear-shaped, or torpedo-shaped. Naturally, the new cultivated forms are very poorly adapted for seed dispersal in the wild and now rely instead on humans for propagation. Wild forms of eggplant, which originated in east Asia, are unpleasantly prickly, and have small, bitter-tasting fruits.[496] Domestication resulted in much larger, palatable fruits with fewer, softer prickles.[497] Eggplant was domesticated at an unknown period in the region encompassed by modern China, India, and Thailand and did not reach western Asia, Europe, and Africa until post-Roman times.[498] There are four domesticated species of pepper within the *Capsicum* genus, but *Capsicum annuum* is by far the most important crop.[499] All four cultivated *Capsicum* species originate from South America but the location of their domestication has yet to be determined conclusively. Until recently, the consensus was that wild *C. annuum* spread to Mesoamerica, which was an important centre of diversity for the species, as well as its most probable first site of domestication, possibly as early as *9000 BP*.[500] However, recent data suggest very early cultivation of three domesticated Capsicum species in the Peruvian Andes, where they were already being grown alongside maize and potatoes by *4000 BP* during the preceramic period at sites such as Waynuna.[501]

Evidence of the location and chronology of domestication of these fruit crops is more difficult to acquire than for grain crops because, unlike grains, the relatively soft fleshy fruits are hardly ever preserved for posterity. Therefore we can only infer their domestication route via indirect methods such as human cultural records. For example textiles, language, and written texts have each shed light on the possible domestication of tomatoes. Tomatoes originated in the Andean region of South America, where all of their wild relatives are still located, but cultural data point to a Mesoamerican centre of domestication. Pre-Columbian cultures in the Andes region often decorated textiles with depictions of their most important crops,

but tomatoes are absent.[502] On the other hand, ancient Mesoamerican peoples knew of tomatoes, which they called '*tomati*' or '*xitomatl*'.[503] And, whereas Peruvian texts do not mention the crop, Mesoamerican writings include tomatoes in recipes for dishes that include what we now know as salsa. Finally, recent genetic studies indicate that the accessions of *L. esculentum* var. *cerasiforme*, that were imported to Europe in the sixteenth century, had already reached an advanced stage of cultivation in Mexico. No genetically similar variety to *cerasiforme* has been found in South America, which strengthens the case for tomato domestication being restricted to Mesoamerica. However, we still do not know when, how many times, and in how many places the crop was domesticated within this relatively large botanically and culturally diverse region.

Brassicas

The brassica crops are of particular interest to the geneticist and plant breeder alike. Members of this genus probably originated in the Mediterranean–Near East region, where many of their wild relatives still flourish today. There are dozens of varieties of brassicas that are grown as vegetable crops all over the world. Brassica vegetables are especially popular in the Far East, for example in China and Korea, where the leafy kales are greatly prized. In Western countries, mutated forms of a single species (*Brassica oleracea*) have given rise to such commonplace vegetables as broccoli, cabbage, and Brussels sprouts (Figure 5.1). However, in terms of economic value, the most important cultivated species today is *Brassica napus*, or oilseed rape (known as canola in North America and Australia), which is the second most important, global oilseed crop. Geneticists are particularly interested in the *Brassica* genus because it is closely related to the model species used for much of the research into modern plant genetics, namely *Arabidopsis thaliana*. In 2001, amid great international fanfare, the completion of the sequencing of the *Arabidopsis* genome was announced.[504] Even before this news, we had already started to realize that the *Arabidopsis* research could tell us a lot about brassica genetics and hence inform our efforts to improve these crops.[505]

The multiple genomes of the brassicas

The genetic history of the brassicas turns out to be just as convoluted as that of the wheat and potato families, which we have just considered. The original breakthrough that provided the most important insight into brassica genetics occurred over 70 years ago. Back in 1935, a Japanese geneticist called Naga-hara U (*sic*) proposed that the major brassica crops are all derived from three diploid species, namely *Brassica rapa*, *B. oleracea*, and *B. nigra*. These brassicas respectively contain 10, 9, and 8 chromosomes.[506] U then proposed that these diploid species had spontaneously hybridized with each other in three different combinations to create three additional allotetraploid species, namely *B. napus* (19 chromosomes), *B. juncea* (18 chromosomes), and *B. carinata* (17 chromosomes). This genetic model of the brassicas, known as U's triangle, is depicted in Figure 7.1. U's triangle has allowed brassica breeders to recreate new versions of the allotetraploid crops, by constructing new hybrids from their diploid parental species via artificial genetic crosses. In the case of a resynthesized oilseed rape, for example, this would mean the creation of a new hybrid by combining *B. rapa* with *B. oleracea*. This approach has allowed breeders to bring in useful genes for traits, such as disease resistance, from the wild populations of the diploid brassica species and to transfer them, via the newly created or 'resynthesized' hybrids, into the genomes of the crop species.[507]

During the 1990s, a combination of research on *Arabidopsis* and brassica genetics revealed an unexpected additional complexity in this genetic saga. It seems that each of the genomes of what we had considered as the three basic 'diploid' brassica species might in fact contain three partially, rearranged copies of a much older genome that was extremely similar to that of *Arabidopsis*.[508] These three ancient genomes have become reshuffled over the past 10 to 20 million years, but are still evident within the so-called 'diploid' brassicas of today. In other words, the 'diploid' brassica are, in reality, derived from ancient hexaploid plants. This in turn means that the 'tetraploid' brassicas, such as oilseed rape, are actually dodecaploids with no fewer than 12 residual genomes lurking within their DNA. This was bizarre enough, but the genetics of the brassicas then became even more convoluted with the discovery that *Arabidopsis thaliana* itself is almost certainly an ancient tetraploid.[509] So now the 'diploid' brassicas have become dodecaploids and the 'tetraploid' brassicas have become 24-ploids![510]

The precise nature of the events that created this remarkable genetic architecture in the brassicas species is not yet clear, but things might have unfolded something along the following lines (see also Figure 7.2). At some time, over 40 million years ago, there was a small plant of the mustard family (Brassicaceae) with a tiny genome of about 60 Mb. About 38 million years ago, this little diploid, cress-like plant either spontaneously doubled all of its chromosomes or hybridized with a close relative to create a new tetraploid cress plant with a double-sized genome of 120 Mb.[511] The modern species of thale cress, *Arabidopsis thaliana*, is the direct descendant of this plant and still has a similarly sized genome of just over 100 Mb, although it now

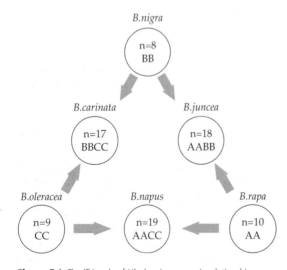

Figure 7.1 The 'Triangle of U', showing genomic relationships between *Brassica* species. The 'Triangle of U', named after Japanese geneticist, Naga-hara U, is an important insight into the origin and genetics of crops in the *Brassica* genus. The major *Brassica* crops are derived from genetically complex polyploid species. For example the commercially important vegetable oil crop, oilseed rape (*B. napus*), which has an AACC genome, is a hybrid of cabbage (*B. oleracea*, CC genome) and turnip (*B. rapa*, AA genome). Likewise, Ethiopian mustard (*B. carinata*, BBCC genome) and Indian mustard (*B. juncea*, AABB genome) are also hybrids of pseudodiploid *Brassicas*. *n* = haploid chromosome number.

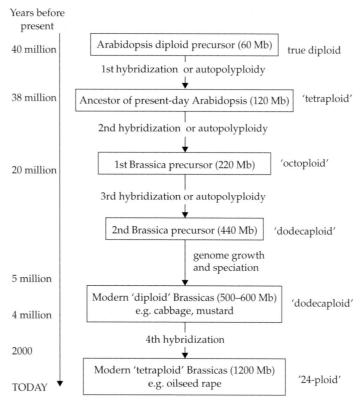

Years before present

40 million — Arabidopsis diploid precursor (60 Mb) — true diploid

1st hybridization or autopolyploidy

38 million — Ancestor of present-day Arabidopsis (120 Mb) — 'tetraploid'

2nd hybridization or autopolyploidy

20 million — 1st Brassica precursor (220 Mb) — 'octoploid'

3rd hybridization or autopolyploidy

— 2nd Brassica precursor (440 Mb) — 'dodecaploid'

genome growth and speciation

5 million
4 million — Modern 'diploid' Brassicas (500–600 Mb) e.g. cabbage, mustard — 'dodecaploid'

4th hybridization

2000
TODAY — Modern 'tetraploid' Brassicas (1200 Mb) e.g. oilseed rape — '24-ploid'

Figure 7.2 Evolution of the *Brassica* genomes. Present-day brassica crops are products of a complex evolutionary process that may have involved up to four successive rounds of hybridization to produce ever more complex polyploid species. According to modern genomic studies, the most important *Brassica* crop, oilseed rape, may contain the partially rearranged remains of as many as 24 ancestral genomes. This means that, external appearances notwithstanding, the genome of oilseed rape might contain ten-fold more genes than the human genome.

behaves like a conventional diploid species. About 20 million years ago, some of the tetraploid *Arabidopsis*-like cress plants formed new polyploid hybrids that contained first two 'tetraploid' genomes and then three 'tetraploid' genomes of 250 to 400 Mb. By 5 million years ago, one of the new hybrids with three 'tetraploid' genomes became a successful new species that was the progenitor of today's 'diploid' brassicas. Its genome, which had now grown to about 500 to 600 Mb, rearranged itself from being a relatively unstable hexaploid into a more stable pseudodiploid configuration. As we saw above in the case of the polyploid wheats, this sort of functional diploidization can start immediately after the initial hybridization event that produced the new polyploid. Brassicas appear to have an analogous mechanism that often, although not invariably, ensured a rapid diploidization of the genome of a new allopolyploid hybrid.[512]

About four million years ago, the ancestral brassica species diverged into several closely related species, including our familiar 'diploid' brassica crops, such as *B. rapa*, *B. oleracea*, and *B. nigra*,[513] all of which still have genomes of 500 to 600 Mb.[514] The final chapter in this genetic saga happened much more recently, and probably occurred well after people had begun to cultivate the 'diploid' brassicas as crops. Some of the 'diploid' brassicas formed yet another series of interspecific hybrids between themselves to create the 'tetraploid' brassica crops, including oilseed rape, which may have arisen as recently as 2000 years ago. The genome of oilseed rape is about 1200 Mb and still contains the fully intact and essentially unrearranged genomes of its two recent parental species. This remarkable plant has therefore undergone at least four rounds of hybridization and polyploidization over the past 20 million years and its genome has grown 20-fold. Unlike most of the cereal crops that we have already looked at, however, brassica genomes do not contain massive amounts of repetitive DNA. Hence the brassica genomes are relatively large, not due to the presence of 'extra' non-coding DNA of exogenous origin, but rather because they contain a lot more genes. It is largely due to their complex multiple-polyploid origins that the brassicas are such a diverse and useful group of crops.[515]

A uniquely versatile group of crops

One benefit of having so many additional genes is that the brassicas are capable of a huge amount of metabolic flexibility. They use this ability to synthesize a vast range of secondary compounds, which mainly serve to deter or poison the many would-be pathogens and herbivores that would otherwise damage or even destroy the plants. As well as deterrents and toxins of many hues, brassicas also produce volatile chemicals that attract beneficial insects that help to rid them of their pests. For example, the cabbage white caterpillar, *Pieris brassicae*, is a serious pest of brassicas that has become adapted to the normal deterrents and toxins produced by the leaves, so the plants are rendered defenceless against attack. Some brassicas have responded to the threat posed by this noxious caterpillar pest by releasing volatile chemicals that attract tiny parasitic chalcid or braconid female wasps. Once they reach the brassica plant, the wasps quickly locate any caterpillars that might be present, paralyse them, and proceed to lay dozens of eggs inside their bodies. This leads to the eventual, and not very pleasant, death of the caterpillars as they are consumed from within, while still very much alive, by dozens of minute but voracious wasp larvae.[516]

Probably the most characteristic secondary compounds made by brassicas are the glucosinolates. Glucosinolates are toxic to many pest species of brassica crops, and a few forms can also cause goitre and other problems in animals and humans.[517] However, the edible brassicas, and especially the vegetable varieties, contain other, more desirable types of glucosinolates that cause the slightly sharp taste that is common to all brassicas from broccoli to Brussels sprouts. Glucosinolates are also the main flavour ingredient of mustard, *Sinapis alba*, which is a close relative of the *Brassica* genus. The characteristic sharp flavours of brassica vegetables may not be to everybody's taste, but the compounds causing them are responsible for some of the most important, positive nutritional qualities of these oft-maligned vegetables.[518] According to medieval folklore, some brassicas, especially broccoli, are efficacious in warding off various diseases including various forms of cancer. Recent laboratory studies have confirmed the anticarcinogenic activity of broccoli glucosinolates in cultured human cells. There is now a great deal of interest in breeding new varieties of broccoli and other brassicas that contain higher levels of these potentially health-enhancing compounds.[519]

The first cultivated brassica crops were probably varieties of *B. rapa*, or turnip, that were grown for their seed oil about *4000 BP*.[520] *B. rapa* was domesticated repeatedly from wild populations that occurred across Eurasia, from the Mediterranean to India. It is likely that turnips first came to the attention of early cereal farmers as commonly encountered weeds in fields of wheat and barley. The leafy vegetable kales of *B. oleracea* were probably the second brassica species to be cultivated and there are written accounts from ancient Greek sources, such as Theophrastus, that record them from at least *2500 BP*.[521] Kales were also a popular crop in ancient China and elsewhere in the Far East. Oilseed rape, *B. napus*, may not have been brought into widespread cultivation until well after the end of the Roman Empire. We know that all the brassica crops were widely grown in western Europe during the medieval period, either as edible vegetables or as forage for animals. In some cases, the crops were also grown for their oil-rich seed. Oilseed rape in particular was used as a source of oil, although this was mostly used as a fuel for lighting rather than for its present uses for margarine and cooking or salad oils.

Although oilseed rape is now a major crop, its relatively recent provenance as a domesticant is evident from its many persistent wild, or 'weedy', traits. For example it still tends to lose its seeds before harvest due to premature pod shattering.[522] Despite several much effort by breeders, this pod-shattering trait has proved remarkable difficult to control.[523] It seems that, unlike in the majority of older crops, the pod shattering/seed retention trait in the brassicas behaves as a more complex, multigenic character that will require several mutations to alter to a more human-friendly configuration. Another unusual feature of some brassica crops is that they can readily escape from farmed areas to survive as free-living plants. For example, in the UK, feral rape now grows on riverbanks, roadside verges, and field margins, often located many kilometres from its cultivation site. The persistence of

such wild traits means that much work remains to be done by geneticists and breeders before they will be able to domesticate oilseed rape to the same extent as more established crops such as wheat, rice, or maize. As we saw in Chapter 4, radish is a similarly 'weedy' crop that can interbreed with wild relatives to produce invasive new hybrid species. Radish is of Near Eastern origin and was probably domesticated in the eastern Mediterranean some time after *8000 BP*, with evidence of cultivation by *4780 BP* in Egypt, reaching China by *2500 BP*.[524] Despite this long history as a crop, however, radish is still only semidomesticated in comparison with the likes of wheat or peas.

PART III

People and plants in prehistoric times: ten millennia of climatic and social change

Civilization exists by geological consent, subject to change without notice

Will Durant (1885–1981) attributed

CHAPTER 8

People and the emergence of crops

Our fathers planted gardens long ago
Whose fruits we reap with joy today;
Their labour constitutes a debt we owe
Which to our heirs we must repay;
For all crops sown in any land
Are destined for a future man.

Nizami Ganjavi (1141–1204) Azeri Persian Poet[525]

Introduction

In Parts I and II, we examined the human, environmental, and genetic contexts of agricultural development during the late Palaeolithic and early Neolithic Eras. In the following five chapters of Part III, we will focus on the consequences of agriculture for the people involved, whether as individuals or societies. We will begin with an overview of the various crop domestication processes in the areas of origin of the major crops. This will be followed in Chapter 9 by a review of the biological consequences of farming for people, which were often far from benign. Chapters 10 to 12 will then relate the fascinating story of how different forms of agriculture led to the evolution of some quite distinctive societies in various parts of the world. As we will see, the eventual fates of these civilizations depended on complex interactions between social, environmental, and biological factors—one of the latter being the nature of the major crop(s) being cultivated by each society. In this first chapter of Part III, we will survey how human societies interacted with their protocrops as the latter were first brought into informal cultivation and then more fully domesticated into true crops in various parts of the world.

During these processes, there was no sudden, global agricultural revolution. Rather, there were numerous, gradual, localized processes whereby certain wild plants were increasingly managed by one or more human cultures. For each crop, domestication occurred independently, often on different continents. In some cases, a crop might be domesticated at several different times in widely separated localities. Hence, common beans were domesticated at least twice, two millennia apart, first in Mexico and then in Peru. Rice was probably domesticated many times in several regions of Asia. As far as we can tell, some of the domestication processes may have been linked to varying degrees with sudden climate change, such as the Younger Dryas. However, in all cases of crop cultivation, the process was also triggered by a series of other factors, possibly including sedentism, cultural developments within societies, population increases, and technological advances.[526] The relative importance of these factors varied from place to place and from crop to crop, but one of the most important factors in, and indeed the prerequisite for, successful domestication was the genetic constitution of the plants themselves.

Emergence of cereal crops in the Near East

During and immediately after the Younger Dryas, there was a very gradual transition to domesticated cereals at Natufian sites in the greater Levantine region. Wild-type seeds of cereals, and other smaller-grained, starchy grasses, continued to be used, albeit in gradually diminishing quantities, for a further 2500 years before there was a more or less complete dependence on the new cultivated forms of the large-grained cereals (Figure 8.1B).[527] This may have been due to the spread of cultivated family plots of cereals from several small, localized centres of origin to a much wider range of sites in the southwest Asia. There would also have been a gradual diffusion in the use of better-performing seeds as people compared their cereal plots with those of their neighbours. They would have noticed differences in factors such as plant vigour, grain size, ease of harvesting, processing performance and, possibly, the taste of the resulting foodstuffs. The important insight that like (normally) gives rise to like applies as much to plants as it does to animals and people, would have led early farmers to preferentially propagate seeds from better performing plots. The custodians of the best plots might even have bartered their superior grain for goods, services, or future favours, hence becoming the first seed merchants.

Such activities would have greatly accelerated the dissemination of crop variants that had favourable characteristics, as defined by the newly emerging groups of empirical farmer–breeders. Whereas wild cereals normally rely on the vagaries of wind pollination or seed dispersal to colonize new areas, the newly favoured domesticated varieties had a much faster and more efficient dispersal mechanism, that is human beings. Dispersal was no longer limited by normal ecological mechanisms and the new seeds were eventually transported across mountain ranges, continents, and oceans, until their habitat frequently encompassed the entire globe. One of the factors that led to the dispersal of cereals, such as wheat, across the world is that most humans find it easy to digest and assimilate nutrients from wheat products. In contrast, liquid milk still cannot be tolerated by most adults around the world. Hence, the use of liquid milk as a dietary staple is pretty much restricted to those populations (such as northern Europeans and some African pastoralists) that carry lactose-tolerance mutations (see Chapter 9).

Even after the cultivation of cereals led to the favouring of new agronomically adapted varieties, the original wild-type cereal populations persisted in many regions and some of them still survive to this day.[528] In many cases, wild cereals will grow in the immediately vicinity of a related crop. If the crop is an outbreeder, this can result in cross fertilization, which would have been a double-edged sword for early farmers. On the one hand, the wild cereals may have been useful repositories of valuable traits such as disease resistance. But, interbreeding of the crop with neighbouring wild

species could also dilute the effect of agronomically useful traits, such as large grain size or synchronous development and seed set. This may have caused early farmers to weed out wild cereals from the vicinity of their crops, creating a genetic barrier between them. Gradually, the differences between the cultivated cereals and the wild forms became more and more marked until they reached a point where we can truly speak about 'domesticated' plant varieties that sometimes developed into separate species from their wild progenitors. It should be repeated, however, that throughout this early period of agricultural evolution, it is not necessary to invoke any conscious attempt at breeding desirable traits on the part of the first farmers. The very practice of clearing, sowing, weeding, harvesting, and storing grain would have provided the conditions that favoured the evolution of what we now know as 'domesticated' varieties of each type of cereal and even the evolution of new species (Figure 8.1).

The lack of a conscious human effort to breed improved cereal varieties does not mean that early farmers failed to recognize and exploit superior varieties emerging from their fields. Clearly, they were observant and experienced cultivators who would have been quick to capitalize on any opportunity to augment their staple food supply. Gradually, this new knowledge about cultivation, and the improved seeds themselves, would have been disseminated from multiple sites within the greater Levantine region.[529] It is likely that different cereals were cultivated at different sites, depending on the local soil conditions, climate, and pre-existing knowledge of the human population. For example, as discussed in Chapter 3, the Abu Hureyra people were somewhat unusual in their initial reliance on rye. The more southern Natufians and other groups focussed more on wheat and barley, but different types of wheat tended to be used in different areas. Hence, emmer was the main crop at Aswad, while einkorn was preferred at Mureybit.[530] The dissemination of cereal agriculture is therefore not like a simple recipe. Rather, it involves a complex set of options involving numerous crops and varieties that could potentially be grown in a wide variety of climatic zones from the Arctic Circle to the Sahara. This was the unique power of the temperate

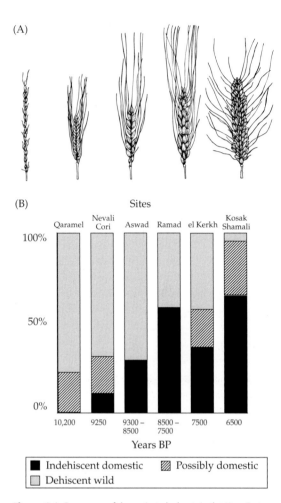

Figure 8.1 Emergence of domesticated wheats in the Near East. Domesticated forms of wheat gradually emerged over several millennia as human activities such as gathering and cultivation created new niches for mutations such as non-shattering (indehiscent) ears and larger grains. (A) Selection and breeding of wheat from the earliest wild einkorn varieties (left) to the most modern high-yielding cultivars (right). Note the progressive increase in grain number and size. Other key traits that are less obvious include stronger retention of the grains on the head, rapid germination, and improved flour quality. (B) Archaeological data from several human-occupied sites in the Levant show the very gradual supplanting of wild-type dehiscent wheat with domestic indehiscent cultigens over a period of almost 4000 years (data from Tanno and Wilcox, 2006).

cereals, enabling their cultivation to spread far beyond their Levantine and Near Eastern centres of origin.[531]

Over the next three or four millennia, the cereal/pulse farming package spread throughout central and western Eurasia, as shown in Figure 8.2.[532] Cultivation of the Near Eastern cereals also spread to west Africa and the Nile Valley,[533] and by *5000 BP* wheat and barley had reached China.[534] It is likely that the major mechanism for the spread of the temperate cereal crops across Eurasia was via the transfer of seeds and of farming expertise from one group to another, possibly in the context of reciprocal trade. It has also been proposed that agriculture may have been spread by the physical replacement of non-farming cultures by farming cultures. While there may have been several instances of such forcible spreading (possibly including millet farming in northern China—see Chapter 11), this is not now regarded as the principal mode of agricultural dissemination across Eurasia. As we will discuss later in Chapter 12, recent genetic evidence suggests that, while there was some migration of people from the Levant into Europe in the early to mid-Neolithic, this made only a minor contribution to the current European gene pool, which is mostly derived from mid-Palaeolithic migrants who had arrived tens of millennia previously.[535]

It is likely, therefore, that agriculture spread across Eurasia just as much, or more, due to transfer of knowledge and the trading of seeds, than via migration and/or conquest. Such a process would be analogous to the way many other technological innovations, from gunpowder to plastics, have been transferred around the world over the past few millennia without significant population movement. It is also evident that agricultural diffusion did not necessarily occur separately for each type of crop. For example, in the case of Eurasia, a package of crops including emmer and einkorn wheat, barley, peas, lentils, and flax, was disseminated as a group.[536] Again, this is reminiscent of the global spread of other 'bundles' of related technologies, such as ancient metallurgy, or more recent examples such as electronics and information technology.

Rice and millet come to eastern Asia

Rice

Rice is by far the most important crop in the world today in its contribution to human nutrition. It is

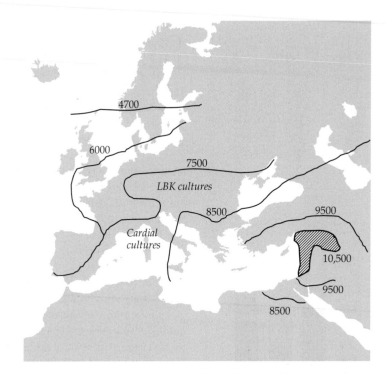

Figure 8.2 Spread of agriculture into Europe from the Near East. The map shows the approximate trajectory of the spread of agriculture during the Neolithic Era. The lines represent the approximate boundaries of widespread agriculture at various dates BP. At this time, farming was frequently restricted to more favourable zones such as river valleys, and was often intermingled with pastoral and/or hunter–gatherer communities with whom farmers traded and exchanged marriage partners. The northwesterly progression of agriculture was by no means either smooth or continuous. In particular, there was a mid-Holocene hiatus before the selection of new cold-hardy cereal mutants enabled cereal farming to spread to cooler Atlantic littoral regions by *6000 BP*, as discussed in more detail in Chapter 12. Map based on data from Zohary and Hopf (2000) and Diamond (1997). LBK, *Linearbandkeramik*.

the staple food of more than two billion people and a regular dietary component of billions more. In contrast to temperate cereals, rice is a warm-climate, water-requiring plant, grown primarily in tropical and subtropical countries. By far the most commonly cultivated species is *Oryza sativa*, also known as Asian rice, while a much less common species is *O. glaberrima* or African rice. The origins of Asian rice are much less clear than the various Near Eastern cereals that we have just considered. The two main reasons for this are the extremely large area of potential rice cultivation, and a dearth of archaeological evidence. Whereas the early precursors of the temperate cereals, such as wheat and barley, were restricted to a relatively small area of southwest Asia, wild rice was present across a vast area extending throughout the whole of eastern and southern Asia. Hitherto, this region has been

relatively neglected by archaeologists, who have tended to focus much more on Near Eastern sites. Even now, with more attention being paid to east and south Asian archaeology and palaeobotany, the warm, wet climate typical of rice-growing areas means that ancient samples are much less likely to be well preserved than in the drier Levantine sites.

These factors mean that it is not yet possible to define where and when rice was first cultivated in Asia. Most existing evidence points to the independent cultivation of rice in several widely separate locations across the continent.[537] For example researchers have made out fairly robust cases respectively for India, central China,[538] and South-East Asia[539] as cradles of rice cultivation. A multisite origin for Asian rice is also supported by the fact that there are two quite distinct races of cultivated rice, namely *indica* and *japonica*.[540] There has been much discussion about whether these races had a single common ancestor that was the original form in which rice was domesticated.[541] However, recent DNA-based evidence seems to suggest that *indica* and *japonica* rice varieties diverged as early as one to two million years ago.[542] If confirmed, this would mean that *indica* and *japonica* were already separate varieties when they were domesticated around ten millennia ago. Therefore, the two races must have been domesticated independently and hence modern Asian rice has at least two, and possibly many more, centres of origin.[543] Most authorities estimate that rice was first grown as a crop at or before *10,000 BP*,[544] but its spread from its centre(s) of origin was slower than that of the temperate cereals and it was not widely adopted as a major food source for several millennia after its initial domestication. Rice did not become a widespread dietary staple until about *7000 BP* and did not reach the Asian littoral regions of Korea and Japan until much later, at about *3000 BP*.

It is likely that, as with the temperate cereals in the Near East, Asiatic hunter–gatherers began collecting wild rice to supplement their diet long before they started to cultivate it as a crop.[545] The stimulus for systematic crop cultivation is not known, but the process is not necessarily linked as closely to sudden climate change as was cultivation of the temperate cereals. For example we cannot establish a close correlation with the Younger Dryas, because this climatic episode ended at *11,500 BP*, which is at least a millennium or two before our earliest firm evidence of rice cultivation. Furthermore, the severity of the Younger Dryas was much less pronounced in the rice-growing regions of Asia than it was in the Near East and North America.[546] Once people started systematic cultivation of wild rice, it is likely that there was a fairly rapid change in its genetics, with more cultivation-suitable variants being favoured over weedy-like variants. Evidence for this comes from a Japanese study showing that once wild rice is cultivated, domesticated-like varieties arise spontaneously within a few years.[547] For example, in some experiments, it was found that characters controlling seed shedding and seed dormancy became more closely associated simply by cultivating wild rice, and without any sort of deliberate selection by the scientists.[548]

This study implies that early rice farmers did not necessarily need to know what they were doing as regards selection of domestication traits in their first wild crops. The mere fact of cultivating wild rice, and allowing it to interbreed with other wild varieties, would have led to the emergence of something resembling a domesticated, higher yielding, and more easily cultivated form of rice, possibly within as little as a few decades. This process would have been greatly assisted by another recent discovery about rice genetics, namely that many important domestication-related genes in rice are clustered together on a few chromosomes in the rice genome.[549] For the early farmers, this clustering would have meant that, if just one of these useful characters were selected, whether deliberately or not, several other valuable traits would also be likely to be selected at the same time. One final point should be emphasized about the domestication of rice. Although Asian rice became the major staple throughout much of southern and eastern Asia, its dissemination took many millennia. In the meantime, other crops such as barley and beans were also being domesticated in the some of the same parts of Asia. Gradually, rice began to out-perform the other grain crops and eventually it partially or completely replaced most of them as the preferred dietary staple across much of the region.

The immense and abiding importance of rice to the peoples of Asia is reflected today in their languages. A few examples can be used to illustrate this point: in traditional speech, Japanese people do not use terms such as breakfast, lunch, and dinner; instead they use the word for *rice* to mean *meal*. Hence, breakfast is 'morning rice' (*asa gohan*); lunch is 'afternoon rice' (*hiru gohan*); and dinner is 'evening rice' (*ban gohan*). Even the names of some of the best-known Japanese companies are ultimately derived from the word for rice; for example, 'Toyota' means *bountiful rice field* and 'Honda' means *main rice field*.[550] In Bangladesh, China, and Thailand, instead of the greeting: 'How are you?' people ask: 'Have you eaten your rice today?' Several Asian languages use the same word for rice and food; or for rice and agriculture.[551] Use of a crop name in such intimate discourse is not found in anywhere else. It reflects the unique historic dependence of these Asian cultures on this single, and very special, crop—a dependence that largely persists to the present day.

The cultural importance of rice throughout Asia is also demonstrated by its ancient mythicoreligious status. Hence, rice is an integral part of many deistic creation myths in regions as far apart as Burma and Bali. In Chinese myth, rice was donated to people by an animal instead of a god. The myth involves the common narrative of a massive flood that destroyed much of the vegetation that had sustained the population. Hunting was difficult as game was scarce due to lack of plant life and people were on the verge of starvation. The myth relates how they were saved by a dog that wandered into their village with long yellow seeds hanging off its tail. Once planted these seeds developed into the productive rice crop that has sustained the Chinese to the present day. Even in modern China, it is still said that 'the precious things are not pearls and jade but the five grains', of which rice is foremost. Meanwhile, across the East China Sea, in Japan, the Shinto religion has always regarded the Emperor as a living embodiment of the rice god, *Ninigi-no-mikoto*.[552]

Millets

Millets are warm-season cereals that are relatively tolerant of the dryer climates found during several periods of the Holocene. The discovery in northern China of domesticated varieties of broomcorn and foxtail millet from *10,500 BP*, or earlier, suggests that millet cultivation might have predated that of rice in parts of Asia.[553] The main Neolithic agricultural zones of Northern China are the north China Plain, which extends from present-day Nanjing in the south to Beijing in the north, and the Loess Plateau region, immediately to the west (see Figure 11.2). The Loess Plateau region covers a vast area of more than 640,000 square kilometres, from the arid borderlands of Mongolia to the relatively fertile Yellow River Valley. Millet cultivation seems to have started in the northern part of the Loess Plateau region during a particularly humid period at about *11,000 BP*.[554] This followed an as-yet unspecified transitional period between hunter–gathering and farming during the Palaeolithic/Neolithic overlap,

locally termed the Tengger period, which corresponds to the Younger Dryas phenomenon.[555]

Recent findings (see Chapter 11) point to a rapid introduction of millet farming into the Loess Plateau region, possibly by migrants from the northwest, into relatively unpopulated areas. In contrast to Near Eastern cereal agriculture, which largely developed in already well-populated regions, the relatively sudden introduction of farming into north China by external migrants has been cited to explain the much more rapid adoption of intensive millet cultivation compared to the much slower indigenous intensification of barley and wheat cultivation in the Near East. Frustratingly, there are relatively few finds to connect this fascinating period to the later and much better-characterized millet farming cultures that were so prominent in the Loess Plateau region by the main Chinese Neolithic Period of the mid-Holocene at about *8000* to *5000 BP*.

Maize arrives in Mesoamerica

Maize is currently the third most important cereal crop in the world (after rice and wheat). The word 'maize' is based on one of the many Mesoamerican names for the crop, as also reflected in its botanical name, *Zea mays*. Most Mesoamerican maize-growing cultures had their own name for the crop. Whereas maize itself comes from the Taino word *mahis*, the Maya called the crop *ixim*, the Zatopec *rxoa*, and the Nahuatl (Aztec) *centli*.[556] When the first English colonists arrived in North America in the sixteenth century, they noticed a local crop that slightly resembled their own European corn (wheat) and christened it 'Indian corn'. Maize was also called Indian corn when it was first imported into Europe, for example as a belated form of food aid during the Irish potato famine of the mid-nineteenth century. In North America itself, maize soon became known simply as 'corn', although in Britain the word corn has retained its original meaning as a grain crop, and is mainly applied to wheat. Today, maize is a staple food across much of the Americas and in many parts of Africa, where it is often made into unleavened bread or baked to make cakes. Maize is also the main ingredient in the most popular group of breakfast cereals in the west, as well as providing the basic ingredient for numerous popular snack products, ranging from popcorn to corn chips.

In addition to its use in a host of human foods, maize is a versatile and highly nutritious animal feedstock. Although maize originated as a subtropical crop, there are now cool/temperate varieties that can be grown commercially as far north as latitude 55°, in the Scottish border regions of the UK. In some ways, maize is a more versatile food crop than wheat or rice because, as well as its high quality starch and protein, the grain contains a substantial amount (from 5–16% of total grain weight) of edible oil. This polyunsaturate-rich oil can be extracted from the maize grain and either used on its own in salads and for cooking, or in the manufacture of processed foods such as pastries and cakes. Compared to the pulses, maize is relatively lacking in protein (about 9–10% of grain weight) and is deficient in the essential amino acids, tryptophan and lysine. Maize starch is converted into a sweetening agent, called corn syrup, used in many processed foods, especially confectionary products. Finally, maize starch can be fermented to produce many industrial products, ranging from textiles and biopolymers to a form of alcohol, termed 'gasohol', used in the USA and South America as a vehicle fuel.[557]

Cultivated maize, *Zea mays* spp. *mays*, is derived from a group of very different-looking wild cereals, known collectively as teosinte (see Figure 6.6). A typical teosinte plant has many branches and relatively few grains on its small seed-cobs, while cultivated maize is a much taller plant with few branches and just a few, large, grain-rich cobs. This means that the cultivated maize yields ten-fold more grain per unit area than the wild teosintes. There are several less-closely related forms of teosinte that include other species within the genus *Zea*,[558] but none of these plants were ancestors to maize. The ancestor of domesticated maize was one of the many teosinte varieties now classified as subspecies within the same overall species as cultivated maize, namely *Zea mays* spp. *parviglumis*. Recent evidence from a US/European collaboration has come to the surprising conclusion that most of the initial changes needed to convert a wild teosinte to domestic-type maize can be achieved by the modification of only three major genes.[559]

This genetic alteration could have been effected quite readily if early maize cultivators could recognize the physical differences in the mutant plants. By selecting seed from these mutated plants for subsequent sowing, the mutation would become fixed in the crop population over subsequent generations. Mutations in any of these three key genes would have led to easily observable and obviously desirable traits in the crop, including reduced branching, softer kernels, and tighter adhesion to the cob. This means that the Mesoamerican preagriculturalists would have been able to select agronomically superior triple-mutants of maize in a relatively short time. It has been estimated that this selection process may have taken as little as a decade and it seems likely that it occurred prior to the systematic cultivation of maize as a crop.[560] Therefore, preagricultural maize gatherers may have unknowingly acted as plant breeders and crop domesticators, long before they became actual tillers of the soil.

Genetic data also suggest that all our contemporary maize varieties derive from a single domestication event.[561] It remains possible that maize could have been domesticated more than once in different localities in Mesoamerica. But if multiple domestications of maize did occur, it seems that the progeny of only one of them has survived to the present day. The early teosinte-like plants that were cultivated ten millennia ago (before the key mutations occurred) would have yielded far less grain than the domesticated form of maize. Would this lower yielding crop have still been worthwhile for the Mesoamerican people to plant, nurture, and harvest? The answer appears to be a resounding 'yes'. Field experiments show that a family of five could grow enough grain (0.7 tonnes) on a 1.5-ha plot to provide themselves with a quarter of all their annual caloric needs.[562] Harvesting would not have been a problem as a family group could collect over a tonne of teosinte seed in just 3 weeks.[563] Note that this estimate is based on a modern attempt to grow teosinte by agronomically inexperienced scientists. It is quite likely that the ancient Mesoamericans, with their vast empirical knowledge of teosinte cultivation, would have achieved significantly higher yields than this, and therefore needed even less than 1.5 ha of maize per family.

During this initial period of cultivation, people would not have needed their new maize crop to provide all their food needs. Such communities were still in transition from full hunter–gathering to full sedentary–agrarian status. Therefore the 25% of their calories supplied by teosinte would have been supplemented by hunting and foraging, as well as by other experimental crops, such as squashes and gourds (see later in this section), which were also cultivated in the region during this period. Anthropologists have estimated that a typical community at this time in ancient Mexico would have numbered between 25 and 40 people, requiring a relatively modest 5 to 15 ha of teosinte crops for their sustenance.[564] Therefore, both botanists and anthropologists agree that even the earliest protocultivars of teosinte would have yielded enough grain, on a relatively small and easily worked area, to provide a worthwhile food source for the kinds of communities present in ancient Mesoamerica.

In contrast to the growing amount of genetic data, archaeological evidence for maize domestication remains scanty. A major challenge in identifying the origins and development of maize cultivation is the poor preservation of ancient samples in the relatively moist, warm Mesoamerican climate. Another limitation is that studies of Mesoamerican prehistory are of relatively recent provenance and are much fewer in number than those of the Near East. However, there has been an enormous burgeoning of interest in maize domestication over the past decade. The application of new analytical methods has given us powerful insights into this fascinating episode in human/plant interactions.[565] We now know that the cradle of maize cultivation was in southwestern Mexico, possibly along the drainage area of the Rio de las Balsas, in the present day states of Michoacán, Oaxaca, and Guerrero.[566] The earliest known samples of fossilized maize, from the Oaxacan highlands, have been dated to about *7000 BP*,[567] although recent DNA analyses suggest that the domesticated version of maize may have diverged from its wild teosinte ancestor as long ago as *9200 BP*.[568] Therefore there is an interesting agreement between the genetic and archaeological evidence that places the initial domestication of maize in the cooler highlands of

Oaxaca from which it would have spread along the river system to the lowlands at a later date.[569]

The evolution of domesticated forms of maize was facilitated by the fact that, although many genes can contribute to domestication-related traits, the most important differences between maize and teosinte appear to be controlled by just a few key genes.[570] It is likely that the early pioneers of maize cultivation began the process of domestication in a similar manner to early wheat farmers. Hunter–gatherers would have periodically collected the grain from wild stands of teosinte and processed it to flour, breads, and cakes. We have no record of these early stages of preagricultural maize/teosinte use, but it seems probable that Mesoamericans went through a similar extended process of familiarization with their plants to that undergone by the early wheat/barley/rye users in the Near East. Central Mexico appears to have had a more prolonged cold, dry phase than the Near East, possibly extending to *10,500 BP* or beyond in some areas.[571] But this is still over a millennium before the earliest domestication date of *9200 BP* implied by genetic evidence. Therefore, there are not enough data of sufficient precision to allow us to link maize domestication with this or any other episode of climatic change, in the same way as we can with Younger Dryas-related cultivation of cereals in the Near East.

Although the earliest fossil maize remains date from about *7000 BP*, there is evidence from pollen and starch analyses that people as far south as Panama were already cultivating maize crops at around *7800–7000 BP*.[572] It is possible that societal demands, rather than climatic or population pressures, may have been important contributors to the development of maize and other crops in Mesoamerica.[573] Following the initial domestication of maize in this small area of southern Mexico, the crop was spread throughout Central America over the next few millennia. Intensive cultivation of maize was the staple form of agriculture and provided the subsistence base of the later Olmec, Mayan, Toltec, Aztec, and related civilizations. To a great extent, it was the advances in maize cultivation by farmers that allowed the development of these and other complex human societies in Mesoamerica, who never forgot their debt to maize.

Accordingly, the domestication and many uses of the maize crop are central features of Aztec, Toltec, and Mayan mythologies.

The most common mythical story about the origin of maize concerns a fox that follows an ant and discovers a hoard of golden seeds. The fox eats the maize but, as he digests it, flatulence betrays his marvellous discovery to other animals in the vicinity. Many small animals can now eat the maize, but the plant still does not become available to humans until it is later released to them by divine intervention. One can interpret this story as reflecting a time when the ancestor of maize (i.e. teosinte) was only edible by animals until a sudden series of events, such as a set of mutations, 'released' the newly mutated maize as a domesticated plant that was now edible by people.[574] *Quetzalcoatl*, the Toltec and Aztec god of wisdom and knowledge, was regarded by these cultures as the discoverer of maize, and is also credited with devising the method of making maize meal into tortillas.[575] The Aztecs had separate male and female gods of young maize plants, called respectively, *Xochipilli* and *Chicomecoatl*. In Mayan mythology, there is a maize god called *Yum Kaax*, while in Yucatán the maize god is combined with the god of flora, *Yumil Kaxob* (Figure 8.3).[576] These diverse cultures had numerous other maize gods, alas far too many to describe here.[577]

From its Mexican centre of origin, maize gradually spread through the rest of the Americas along two different routes. The first pathway ran from Mexico to Guatemala, the Caribbean Islands, and thence to the South American lowlands and Andes foothills.[578] The second dispersal route was via northern Mexico into the southwestern United States and on to the eastern USA and southern Canada.[579] Maize was quickly taken up in those regions of South America where the climate was suitable for its cultivation. But it took a much longer time for it to become a staple crop in North America, possibly due to a combination of climatic factors and the availability of more attractive food sources.[580] The first archaeological records of maize in the southwestern USA date from between *4000* and *3200* BP (see Chapter 12).[581] It was not until *1200* BP that maize was more widely adopted by semisedentary Amerindian cultures as far away as southern Canada and New England.[582] However, maize was not always a successful crop in North America. As we shall see in Chapter 9, its adoption

Figure 8.3 Gods, maize, and chocolate in Mesoamerica. (A) *Yum Kaax*, god of maize and wildlife and protector of farmers, was a principal deity in ancient Mayan culture. Here, the stylized god figure holds a bowl containing several large maize cobs. (B) *Xocalatl*, a form of drinking chocolate was a greatly prized, Mesoamerican beverage normally only affordable by the wealthy. The drawing shows a thirteenth century CE bridal couple sharing a goblet (normally gold) full of frothing *xocalatl*, which was drunk to mark the sealing of a marriage union.

had drastic health consequences for at least one group of people at the Dickson Mounds in Illinois. Amerindians in the southern and eastern Great Basin region of North America temporarily adopted maize farming about 1000 years ago, but then reverted to hunter–gatherering.[583]

Other cultures, other crops

We have focussed up to now on cereal crops, because they are by far the major staples across much of the world. Cereals are also relatively straightforward crops to grow under intensive agricultural conditions. But there are many other important crops, of which two groups merit special mention here; namely the starch-rich, non-grain crops such as squash and potatoes, and the protein-rich pulses such as the beans. We can list the earliest crops in terms of their plant groups as follows: the grains or cereals, such as wheat, barley, maize, and rice are from the *Gramineae*; the seed pulses, such as chickpea, lentil, garden pea, and beans are from the *Leguminoseae*; the squashes and gourds are from the *Cucurbitaceae*; several types of fleshy tubers, including potatoes, and fruit crops such as tomatoes and peppers are from the *Solanaceae*; yams come from the *Dioscoreaceae*; and sweet potatoes from the *Convolvulaceae*.[584] In this section, we will briefly survey some of the many non-cereal crops cultivated by ancient cultures across the world. There is insufficient space to discuss all the crops domesticated by our ancestors over the past dozen or so millennia, but several recent reference texts are recommended for the interested reader.[585]

Squash

There are several domesticated species of the squash family, or *Cucurbitaceae*. Such crops include the pumpkin, bottle gourd, melon, loofah, cucumber, plus many species of squash itself.[586] The pumpkin squash from Mesoamerica may be one of our oldest cultivated crops. It almost certainly predates maize and is possibly a close contemporary of Near Eastern wheat. Excavations at the Mexican cave sites of Guilá Naquitz in Oaxaca and at Ocampo in Tamaulipas have revealed evidence of *Cucurbita pepo* domestication as early as *10,800 BP*,[587]

while samples at the Coxcatlán cave site in the Tehuacán Valley have been dated to *7920 BP*.[588] However, these Mesoamerican events may have been preceded a much earlier domestication in Ecuador, where a related cultivated squash, *C. ecuadorensis*, has been dated to *12,000* to *10,000 BP*.[589] Genetic evidence suggests that several species of the genus *Cucurbita* were domesticated on many separate occasions in locations ranging from Andean South America, through Mesoamerica, to eastern regions of North America.[590]

These findings indicate that Mesoamerican and Andean cultures may have cultivated non-cereals as their first edible crop. However, early Mesoamericans did not used squash for its flesh in the way that we do today; rather the plant was used for its seeds, somewhat more akin to the use of cereals. Nowadays the carbohydrate-rich, but relatively tasteless, flesh of the squash fruit is often baked to accompany a meat dish or spiced to add more flavour. However, the earliest domesticated varieties of squash had relatively little flesh, and even that was bitter and unpalatable due to the toxic terpenoid, cucurbitacin, which is unique to the Cucurbitaceae.[591] Prolonged boiling was needed to remove the bitter toxins and make the flesh edible, and there was so little of it in the early fruits that it was not worth using. Hence, ancestral squash crops were grown principally for their nutritious, edible seeds and the paltry, unappetizing flesh may have been only used *in extremis*, for example as a starvation food.[592] Squash seeds are rich in oil, and hence high in calories and lipophilic vitamins, as well as having a good protein and starch content. They can be eaten raw or roasted and would have provided a nutritious, abundant, and easily managed food resource for their cultivators. It is therefore not surprising to find that one of the initial signs of squash domestication was a substantial increase in seed size.[593] It was only the later chance appearance of more fleshy and palatable mutations that altered the use of squash and enabled it to be grown alongside maize (and, later, common beans) as part of the *milpa* system of ancient American agriculture. While most crops of the squash family are edible, the bottle gourd is an extremely useful non-edible species and may be one of the earliest domesticated plants (Box 8.1).

Box 8.1 Bottle gourds and dogs—the first non-food domesticants and early migrants to the New World

The bottle gourd, *Lagenaria siceraria*, is a relative of squash that, unlike most other ancient domesticates, was not grown as a staple food crop. Instead, its large, tough, durable, hard-shelled fruits were highly prized as containers for various liquids and solids, particularly foodstuffs, as well as being used to make musical instruments and fishing floats.

The origin of this crop has for long been an enigma. The bottle gourd originated in Africa as a thin-shelled wild species that was initially of little interest to humans. More useful thick-shelled versions of the plant, probably selected and grown by people, were present in East Asia by about *9000 BP*. However, domesticated bottle gourds have also found associated with cultures throughout the Americas from as early as *8000 BP*. During this period, we have no evidence of intercourse between Old and New World cultures, which had been long separated by the inundation of Beringia several millennia beforehand. So how did bottle gourd get from Asia to the Americas? Was it transported there as a wild plant (perhaps by birds carrying seeds), and independently domesticated by people in the New World? Or was there a single very early origin of domestication in the Old World followed by deliberate transportation by human migrants to the Americas?

A combination of archaeological and genetic evidence has recently suggested an answer to this long-standing conundrum (Erickson *et al.*, 2005). These data show that domesticated forms of bottle gourd were already present in the Americas as far south as Mexico by *10,000 BP*, and that the crop was definitely of Asian, not African, origin. This means that it must have been domesticated in Asia well before *10,000 BP*, and

probably by *13,000–12,000 BP*, in order to make the lengthy journey with its human cultivators from Siberia, across Beringia, and down the length of North America to Mexico. It also raises the fascinating possibility that some of the Asiatic migrants, who made the trek across the Beringia land bridge before it was flooded between *11,000 and 10,500 BP*, may have brought at least one domesticated crop with them.

This runs counter to our prevailing notions of these proto-Amerindian migrants as exclusively hunter–gatherers. It also puts bottle gourd alongside the Near Eastern cereals as one of mankind's earliest domesticants. The site of its putative Asian domestication is unknown, but is likely to be towards the east and north of the continent and well away from the other early domestication centres in China or the Fertile Crescent. Meanwhile, a separate population of African bottle gourd was independently domesticated in the Nile Valley, but this only occurred many millennia later, at about *4000 BP* (Schweinfurth, 1884; Crawford, 1992).

In addition to bottle gourd, some of the Palaeoindians may have brought domesticated dogs with them on their migrations from Asia. It has been suggested that dogs had been domesticated as early as *12,000 BP* in Eurasia and that at least five different breeds accompanied the various waves of human migrants (Wayne *et al.*, 2006). Therefore the dog and the bottle gourd may be examples of two principally non-food species that were domesticated and transported around the world by highly mobile late-Palaeolithic cultures, long before they adopted other, edible domesticants in an agricultural context (Erickson *et al.*, 2005).

Potatoes

Another important non-cereal crop domesticated more than ten millennia ago is the potato. Like squash, potatoes were first grown in the Americas, but they are not warm-weather crops. Their centre of domestication probably lies in the temperate zone of the Andean highlands, especially in the high plateau region shared by present-day Bolivia and Peru.[594] While there are very few edible grain plants in this region, many plants produce starch-rich, fleshy tubers. Unfortunately, these wild tubers are also normally extremely bitter, often producing

toxic quantities of alkaloids in order to deter herbivores and various microbial pathogens. The potato belongs to the family, *Solanaceae*, which also includes tomatoes. Solanaceous plants are notorious for the presence of highly poisonous alkaloids, for example the deadly nightshade, *Atropa bella-donna*. Even our present-day, highly domesticated potatoes and tomatoes may contain enough alkaloids to make a person very ill, if you are unlucky enough to eat the wrong part of the plant.[595]

For example potato tubers left in the light eventually turn green and sprout shoots. These green tissues can contain unhealthy amounts of alkaloids, such as solanine and chaconine. The same is true for very young, green tomatoes, which contain the alkaloid, tomatine. These solanaceous alkaloids act as cholinesterase inhibitors in the nervous system, causing muscle weakness, drowsiness, paralysis, and even death. Luckily, most alkaloids taste very bitter and are therefore readily detected before swallowing, but this is not always the case, so green solanaceous tissues are best avoided.[596] Occasionally, even relatively small doses of solanaceous alkaloids that are well below the threshold of taste can be dangerous. For example it was recently reported that some glycoalkaloids from supposedly safe, freshly purchased potatoes (i.e. non-green) might be implicated in inflammatory bowel disease. Given our current preoccupation with food safety, it is quite likely that, were potatoes or tomatoes to be introduced as new foods today, they would be banned by most national regulatory agencies due to the risk of accidental alkaloid poisoning. This is not just a theoretical risk. During the 1980s, a potato variety called Lenape was withdrawn from sale because of its potential toxicity due to high levels of alkaloids such as solanine.[597]

Luckily for us, ancient Andean peoples had fewer scruples than we do about experimenting with potential new foods. The exigencies of their food supply in the high mountains would have meant that they had good reason to repeatedly sample any potential edible plants, even the bitter and potentially toxic tubers with which they would have already become familiar. It is likely that, at some point, there arose a mutated form of potato with a reduced tuber alkaloid content. For a wild plant this would normally be a distinctly disadvantageous mutation, because the newly sweet tubers would become palatable to animals that would normally avoid their bitter taste. The mutant tubers, therefore, would have been rapidly eaten out of existence by opportunistic herbivores. However, if some of these sweeter tubers were recognized by humans as a useful food source before the animals got to them, they would have been protected and cultivated. Potatoes can be readily propagated asexually from tubers without the need to collect or plant seeds, which is useful to the would-be farmer because potato plants do not readily set seed.

Moreover, by propagating tubers, rather than seeds, the early potato farmers would have been carrying out a very different, and potentially more powerful, form of plant reproduction than the seed-propagation by early cereal farmers.

The progeny of the original low-alkaloid tubers would have been genetically identical clones of the parental mutant plant because they had been vegetatively propagated. This means that a genetically uniform clonal variety, derived from a single mutated low-alkaloid potato plant, could be propagated very efficiently to produce a highly nutritious and relatively safe staple food crop. There is evidence that people were consuming potatoes in the Andes as long ago as *13,000 BP*.[598] It took a long time for potatoes to be adopted outside their centre of origin and they did not reach Mexico until *3000 to 2000 BP*.[599] This lag in the uptake of potatoes may have been due to the presence of already well-established and successful crops, such as maize and squash, in Mesoamerica, coupled with a warmer climate and more varied topography that was generally unfavourable to potato cultivation.[600] Following their introduction into Europe and Asia over the past two centuries, potatoes have gone on to be one of the most successful global crops.[601]

Pulses

The pulses include beans, lentils, and peas, and are members of the legume family, or *Leguminoseae*. Many legume seeds are especially rich in protein and therefore can complement the carbohydrate-rich cereals to provide a well-balanced diet. Although the protein of legume seeds tends to be deficient in several essential amino acids,[602] the latter can be obtained from other plant sources, such as nuts. This means that a diet of cereals and legumes, supplemented by nuts and fruits, could dispense altogether with the need to hunt game or consume any expensive (in terms of time and effort to secure) animal protein. Human groups across the world were familiar with pulses long before they began to cultivate them. During the Palaeolithic Era, wild lentils and peas were harvested alongside wild cereals in the Near East and many other nutritious legumes would have supplemented human diets from Asia to the Americas.[603] Other important

Old World legumes include chickpea (from central and western Asia), cowpea (mostly from Africa), faba beans (from the Near East), and soybean (from northern China). The most important type of bean from the New World is the *Phaseolus* genus, which includes dozens of species, of five of which are cultivated.[604]

There is good evidence that the common bean has been cultivated for several millennia in the same region of southwestern Mexico as maize and squash. These three crops are regularly referred to as the maize-beans-squash trinity.[605] Beans were domesticated much later than maize or squash and archaeological and genetic studies suggest that common beans may have had at least two independent centres of origin.[606] In Mexico, the earliest evidence for bean cultivation is about *2300 BP*.[607] Although the location of these finds is very close to the centres of origin of both maize and squash, the domestication dates are eight millennia apart. It also seems that there was an earlier, completely unrelated, domestication of another type of *P. vulgaris* in the Peruvian Andes at about *4400 BP*.[608] Once the common bean, or one of the other New World legumes, had been domesticated, it is likely that the new legume crop would have been integrated into a combined cropping system with maize and squash, both of which were spreading across the Americas by the time that the legumes were first cultivated. This three-fold cropping system, known as *milpa*, is still practised today by traditional societies throughout Latin America. The term '*milpa*' means 'maize field' but refers to something more complex. A *milpa* is a field, often recently cleared, in which a farmer plants several crops, such as maize, squash, and beans, at once. Sometimes other crops such as jicama (*Pachyrhizus erosus*), tomatoes, melon (*Cucumis* spp.), chillies, sweet potato (*Ipomoea batatas*), amaranth (*Amaranthus* spp.), and mucuna (*Mucuna* spp) are also included.

Milpa crops are both nutritionally and environmentally complementary. Hence, maize is rich in carbohydrates and oils, but is deficient in the essential amino acids lysine and tryptophan, which are required to make proteins, and the vitamin, niacin. Beans are protein-rich with an abundance of lysine and tryptophan, but lack the essential amino acids

cysteine and methionine, which are provided by the maize. As a result, beans and maize make a nutritionally complete meal. Squash is rich in carbohydrates and many vitamins.[609] A combination of the three crops in the diet therefore gives a good diversity of essential nutrients, as well as mere calories. The nutritional qualities of a *milpa* diet may be one of the factors behind the lack of animal farming in most of Mesoamerica. Cultures such as the Aztec raised a few animal domesticants, such as turkeys and dogs, they obtained fish and seafood from lakes, and they even ate larger insects such as crickets and maguey worms. But these were only used as supplements or for special occasions. The versatile *milpa* cropping system acted as a hedge against diseases or other problems that might afflict one of the crops during a particular season, but were very unlikely to affect all three crops at once. Most crop diseases are caused by viruses, bacteria, or fungi that are specific for a single type of crop. Hence, a cereal disease such as wheat leaf rust does not affect legume crops, and so on. Thanks in part to the *milpa* system; there are regions of Mesoamerica and South America where intensive agriculture has been now practiced continuously for at least four millennia.[610]

Soybeans

Although most new domesticants eventually moved far beyond their regions of origin, some crops remained highly localized until modern times. This applies to many members of the bean family, including one the most widely grown present day crops, namely soybeans, *Glycine max* (Figure 8.4). Cultivation of soybeans began in the eastern half of northern China about *3000* to *4000 BP*,[611] but is almost certainly much older.[612] Soybean is derived from a wild relative, *Glycine soya*, which is still found throughout northeastern Asia. This form of wild soy plant grew as a recumbent vine with small black or brown seeds. As with many beans, wild soy seeds require considerable preparation in order to be rendered at all digestible by humans. And as with most of the other beans, soybeans are notorious for their flatulence-inducing properties. Flatulence is mainly caused by oligosaccharides in the beans that are broken down by intestinal

(A) (B)

Figure 8.4 Soybean: a uniquely versatile legume. The versatile soybean was first cultivated in eastern China about 4000 years ago, but only reached the West during the last century. (A) Soybean plant (courtesy Oklahoma Farm Bureau, USA). (B) Loose soya beans.

microbes, with the release of foul-smelling gases.[613] This unpleasant property of beans has been tolerated, if not exactly welcomed, by bean-eating cultures for many millennia. Recently, however, breeders have started using new genetic techniques to reduce levels of flatulence-inducing components in these otherwise highly desirable crops.[614]

For would-be bean eaters in the ancient world, flatulence would have been the least of their worries. Of much greater import was the deadly cocktail of toxins present in many legumes, which deter herbivores, pests, and pathogens from eating or attacking the seeds. Despite millennia of breeding, toxins are still present in many of our major pulse crops today. For example soybeans contain more than 15 toxins that must be heat-treated before they become edible.[615] Chickpeas contain neurotoxic lathyrogens, while other legumes contain poisonous lectins and cyanogenic glycosides. In the UK alone, between 1976 and 1989, red kidney beans

were implicated in 50 cases of poisoning.[616] Such beans must be carefully prepared by prolonged soaking to leech out the toxins, strained to remove the extracted toxins, and cooked to soften remaining seed tissue and to inactivate any residual toxins.[617] These procedures were doubtless arrived at by trial and error by many cultures as they sought to harness the nutritional value of the wild beans. It probably then took a millennium or more of empirical experimentation and selection before an erect soybean plant, producing much larger seeds than the original wild forms, eventually emerged. The rewards for this prolonged effort were considerable. The domesticated version of soybean is one of the best sources of protein of any crop, and far better than the mainly starch-rich cereals.

Soon, Asian farmers were using soybean to make a host of food products: the beans can be fermented into a paste (miso) or a sauce (soy sauce), or used to prepare curds (tofu), dissolved flour (soy milk), and vegetable oil. Soybeans are, indeed, one of the most versatile of the major crops. As well as being protein-rich, they contain considerable amounts of starch and oil, thereby supplying the three macronutrients (protein, carbohydrate, and fat) required in our diet. Soy protein extracts, often termed 'textured vegetable protein' are nowadays commonly used in meat substitutes, in products such as vegetarian burgers and sausages. It is estimated that as much as 60% of all processed food products in a typical Western supermarket contain components derived from soybeans.[618] Some of the delight felt by Chinese soybean farmers can be gleaned from the names that they still gave to their favourite varieties, including: *Great Treasure, Brings Happiness, Yellow Jewel*, and *Heaven's Bird*. Contrast these sublimely evocative names with the more prosaic varietal names of contemporary Western crops, such as: Creso (wheat); Maris Piper and Russet (potato); or Westar and Tower (rapeseed).

A further advantage of soybeans, which must have soon become apparent to early farmers, is the ability of the crop to grow in soils too depleted of nitrates to support other types of crop, such as cereals. The reason is that soybeans are legumes and can therefore fix their own nitrogen, rather than relying on nitrogen already present in the soil. The growth of most plants is limited by the availability of nitrogen. Plants require nitrogen for the synthesis of proteins and nucleic acids in order to support their growth and development. Even if nitrogenous compounds are present in the soil, they are mostly unavailable to plants. They may be sequestered in compounds that plants cannot use or be trapped in decaying vegetation or animal manure that requires microbial breakdown to render it available. Nitrogen-fixing plants, such as the legumes, contain bacteria, most commonly of the *Rhizobium*

genus, that are able to fix gaseous nitrogen into soluble nitrates. The bacteria live as symbionts in specialized swellings of the plant roots, called nodules. Inside the nodules, the bacteria receive nutrients from the host plant while the plant in turn uses the dissolved nitrates to support its own growth.

This means that, not only can legumes be grown in nitrogen-depleted soil, their cultivation actually enriches the soil for the next crop. For this reason, legumes are now commonly used as so-called 'break crops' that are grown every 3 to 5 years to re-enrich the soil after cereal cultivation. This practice is called crop rotation. Continuous cultivation of cereal crops results in the steady depletion of soil-borne nitrogen and eventually this is reflected in diminishing crop yields. It would have been possible for early farmers to supplement the depleted soil-borne nitrogen by adding fertilizers, such as animal or human dung, but this was not always practical. Therefore, domestication of a self-fertilizing crop, which is also rich in scarce proteins, would have been a considerable boon for the early Chinese soybean farmers. By *1500 BP*, soybeans were being cultivated in much of eastern Asia, but the crop did not move beyond this region until well into the twentieth century CE. This was despite the fact that Europeans and Americans had been aware of this crop for centuries beforehand.[619] Ironically, following the collection of hundreds of soybean seed samples in the late nineteenth century by prospectors from the US government, the crop has now been adopted with a vengeance throughout the Americas. For the past 50 years, this legume has been the second most important grain crop in the USA (after maize), and over the past decade Brazil and Argentina have also emerged as major centres of soybean cultivation.

CHAPTER 9

Agriculture: a mixed blessing

Do not arouse a disdainful mind when you prepare
a broth of wild grasses;
Do not arouse joyful mind when you prepare a fine
cream soup.

Dogen Zengi, 1200–1253 CE

Introduction

We have seen that the increasing reliance of a few
semisedentary cultures on cultivated cereals and
pulses resulted in unconscious selection for certain
genetic changes in these plants that led, in turn, to
the evolution of what we now know as 'domesticated' varieties, with profoundly altered morphologies and reproductive mechanisms. But what were
the impacts of this switch to crop cultivation on the
people involved? The obvious impacts were at societal level. One of the earliest effects of having to
manage and protect their new crops on a more
intensive basis would have been to reinforce ten-
dencies towards sedentism. As crop yields rose
during the early Neolithic, human settlements
increased significantly in size and sophistication.
Small settlements slowly grew into larger villages,
then into towns, and eventually cities. There was
also a parallel development of a range of new technologies and cultural artefacts. These developments underpinned the evolution of increasingly
organized and complex, technologically advanced
societies that gradually spread across the world,
supplanting most hunter–gatherer cultures.

But some of the most profound effects of agriculture on people have been largely overlooked until
recently. The switch to a narrower, largely cereal-
based diet and more crowded, sedentary living
conditions has also had dramatic effects on our
bodies. Just like our crops, we have responded to
these external challenges both in terms of our
behaviour and via genetic changes that have made

us better adapted to our new domesticated lifestyle.
Over the past ten millennia, our external appearance has changed as we became smaller and less
robust, especially in the facial area. In addition,
numerous, more significant but less physically
obvious genetic changes have spread through
human populations, gradually locking us into an
ever-closer association with, and dependence on,
our domesticated plants and animals. In this chapter, we will focus on the surprisingly profound
impact of agriculture, and its associated lifestyles,
on many aspects of internal human biology, and
especially on our genetic endowment (see Box 9.1).
We will see that we too have been genetically modified and to some extent 'domesticated' such that
most late-Holocene humans differ in many respects
from our Palaeolithic ancestors.

Early agriculture and human nutrition

It is important to appreciate that, although they
evolved gradually, the lifestyle and nutritional
status of early Neolithic, agriculturally based
societies was a fairly radical departure from conditions experienced by previous human societies. The
received wisdom that agriculture was a 'great leap
forward, the advance that catapulted us out of the
hand-to-mouth, day-to-day existence of hunter–
gatherers . . . and into the complex, cultured, literate existence of modern human beings' is still
prevalent in much of the popular literature of
today.[620] However, while many agriculturally
related developments, such as civilization, sophisticated technologies, etc., can perhaps be interpreted
as 'progressive' and 'a good thing', it appears that
the overall impact of farming on human well-being
was often far from benign. The progressivist view
of agriculture can be criticized on several levels,
including its correlation with the rise of elites,

Box 9.1 *Homo sapiens* continues to evolve—at an ever increasing rate

There is a common perception that human development is now overwhelmingly influenced by cultural factors, and that our species is no longer directly subject to change via biological evolution. It is certainly true that, thanks to our many technological innovations, we can live almost anywhere on earth, from tropical rainforests to the Arctic tundra, without undergoing the lengthy and complex biological adaptations required by other organisms. However, *H. sapiens* is still very much a biological species; and as such it remains subject to a wide range of selective pressures whereby certain variants will be reproductively favoured over others. This point is often overlooked when it is stated that humans no longer 'need' to evolve. There is no direction or predetermined endpoint in evolutionary processes and selection does not stop just because we now have access to central heating or nuclear weaponry. What has changed is the type of selective pressures that act on contemporary *Homo sapiens*, which are very different from those that affected our Palaeolithic ancestors.

These new selective pressures often arise from the very societies that we have created, and include many factors that have arisen as we seek to adapt to the consequences of embracing agriculture and its associated lifestyles. Over the past three decades, and especially in the last 5 years, research into human genetics has significantly altered our perspective on the evolutionary impacts of the new, and largely human-created, environment in which most of us now live. The emerging picture is of a hominid species that is still fully subject to a wide range of evolutionary pressures that have sometimes resulted in startlingly rapid genetic changes over the past few thousand years. Recent genetic changes have altered our appearance, making modern humans shorter, more gracile, and much more prone to dental problems than our ancestors. The changes in our craniofacial regions, which have greatly accelerated since the adoption of farming ten millennia ago, are largely an adaptation to softer-textured food Some people also carry a mutation that causes their skin colour to revert to the unpigmented state found in hirsute apes, instead of the darker colour developed by early humans as they lost much of their body hair.

Some of the more recent genetic changes have previously gone unremarked because they do not affect our external appearance, and were only uncovered by sophisticated DNA analysis. For example two mutations in the lactase gene among some Northern European pastoralists between five and ten millennia ago resulted in the paedomorphic retention of lactose tolerance into adulthood, i.e. such people could now drink milk throughout their lives. These mutations soon spread through adjacent populations and are now carried by the vast majority of Northern Europeans, but are much rarer in other groups. A second set of mutations probably occurred about five millennia ago in West Africa as new farming practices favoured the spread of malarial mosquitoes. Mutations in at least three genes predisposed people to chronic anaemia, but also protected them from the even worse scourge of malaria.

Many farming cultures went on to live in highly crowded conditions that favoured the spread of animal-borne diseases, or zoonoses. Such societies often adapted to these chronic infections by developing a partial, genetically based, tolerance, e.g. a smallpox infection might wipe out a proportion of the population but there were normally enough survivors to carry on. Other societies not exposed to these diseases did not develop this type of partial immunity and were all but wiped out when eventually confronted by human carriers of these infections, as occurred in the post-Columbian Americas. Some of the most recent studies of Asian, European, and African populations reveals that, far from ceasing to evolve, the genomes of humans are currently evolving much more rapidly than before, as new selective pressures constantly arise from our unprecedented changes in habitat, food, population density, and pathogen exposure (Voight *et al.*, 2006).

---

exploitation, and profound inequalities in human societies. These arguments have been much explored elsewhere, but another less well-aired aspect of agriculture is its long-term biological implications for modern-day humans, including anybody reading this book.

The thesis that agriculture was not necessarily a 'great leap forward' for individual human health was first raised seriously in the 1970s by scholars such as Mark Cohen and George Armelagos.[621] Many subsequent studies have shown the often-adverse impacts of a cereal-based diet on the health of some early farmer societies.[622] It is now clear that moves to a more monotonous diet had far-reaching nutritional and genetic consequences for humankind. In short, although they were now getting food from a more efficient source than previously, it seems that folk in some of the new farming communities were often not nearly as well nourished as neighbouring hunter–gatherers.[623] The more restricted diet of many farming cultures led to a series of vitamin deficiencies that severely impacted on the their well-being and life span. In particular, the diet of the average early farmer/urban dweller contained a lot more starchy carbohydrate and less animal protein than that of a typical hunter–gatherer. If they depended excessively on a few staple crops, the early farmers might not have met their vitamin requirements as easily as foragers who had access to a much broader range of foodstuffs.[624]

The main evidence for a decline in health among some of the early agricultural communities comes from analysis of skeletal remains in gravesites of known provenance.[625] An example is the study of eastern Mediterranean populations by Lawrence Angel.[626] This is one of the most comprehensive, early analyses of the physical effects on people of the transition to farming in the Palaeolithic and Neolithic periods, extending from about *30,000 BP* to the present day.[627] Compared to immediately preceding and contemporaneous hunter–gatherer specimens from the region, agricultural populations showed many pathological changes. Perhaps the most striking of these was a drastic reduction in stature (height), plus a host of skeletal and vitamin-related disorders. For example dental pathologies, such as hypoplasia of tooth enamel, increased in frequency after the Natufian period. The incidence of dental caries also increased steadily from the Natufian through the Neolithic Era and beyond. In the next section, we will look at the data on changes in human stature as an indicator of overall nutritional status, but we will also see that there are many other useful indicators associated with the skeletal remains. Smith *et al.* have used these data to support the theory that there was a marked decline in health that is correlated with the aftermath of agriculture and animal husbandry.[628]

While this view of decreasing individual fitness (but not necessarily societal fitness, as discussed below) in early agrarian societies is widely accepted, it is not without its critics.[629] For example, in his impressive recent synthesis on early

agriculture, *First Farmers: The Origins of Agricultural Societies*, Peter Bellwood has made the following comments: 'The early centuries of agricultural development were probably fairly healthy, in the sense that the major epidemic diseases of history, many known to have derived from domesticated animals, had probably not yet developed. Neither, perhaps, had crop diseases.'[630] This statement is almost certainly true, as we will discuss in relation to the Abu Hureyra culture in Chapter 10. For example the earliest farming cultures did not yet live in close proximity to domesticated livestock. Hence, the many zoonoses, or animal-derived infections, that were to plague humankind and reduce the average life expectancy had yet to be transmitted to people. Examples of zoonoses from domestic livestock include anthrax, salmonellosis, toxoplasmosis, brucellosis, and trichinosis. Other diseases, including diphtheria, influenza, measles, smallpox, and tuberculosis, probably also originated as zoonoses. Two very recent zoonoses from domestic livestock (in both cases, poultry) are SARS and avian influenza.[631]

Many other communicable diseases became much more prevalent once people adopted high-density, urban lifestyles, at least five millennia after the initial development of agriculture. Although there were many small and large villages in the wider Near East from about *9000 BP*, the first recorded cities only date from the late Uruk period at about *5800 BP*. In Mesopotamia, these new cities soon absorbed most of the population, which became progressively ever more urbanized until, by *4000 BP*, an astonishing 90% of Sumerians were city dwellers.[632] For example by *4700 BP*, the city of Uruk had a population of at least 50,000; most of whom lived in exceptionally crowded conditions.[633] Population densities of 100 to 200 persons per hectare can be inferred from the number and nature of the buildings at Uruk, which compares with densities of only 80 to 100 persons per hectare in the most heavily built-up areas of modern European cities.[634] Such crowded conditions were often coupled with a lack of sanitation which favoured the spread of infectious diseases, such as typhus, plague, smallpox, chickenpox, and measles.[635] These pathogens were much rarer in the more dispersed human populations prior to urbanization.

a genetically based tolerance, if not an outright immunity, to the disease.

Ironically, the presence of endemic diseases in a human society, while obviously debilitating to the many individuals who were directly affected, also conferred a selective advantage to the group as a whole in relation to other societies.[659] Hence, although smallpox and influenza were serious, episodic diseases for most Eurasian societies, these populations had built up a level of tolerance sufficient to ensure that most people survived such infections. Other Old World diseases that were not present in the Americas until contact with Europeans include bubonic plague, measles, mumps, chickenpox, cholera, diphtheria, typhus, malaria, leprosy, and yellow fever.[660] The more highly dispersed Native American populations did not develop either tolerance or any form of acquired immunity to any of these diseases. The result was that tens of millions of indigenous people perished within decades of their first contact with European visitors in the sixteenth century. This created a niche into which the Europeans moved to establish new populations that survive to this day. It is possible that similar mechanisms acted on a smaller scale in other parts of the world, hence conferring a powerful selective advantage on disease-tolerant migrants moving with their 'semi-domesticated' pathogens into the territory of more susceptible native populations.

The sickle-cell trait and other antimalarial mutations

The sickle-cell trait is a genetic adaptation to an unexpected consequence of agriculture, that is malaria, and is caused by a mutation in the human genome that occurred a few thousand years ago in western Africa. This hypothesis was first proposed by Frank Livingstone, who showed that slash-and-burn agriculture exposed human populations to *Anopheles gambiae*, a mosquito that is a major vector for the parasite *Plasmodium falciparum* that causes malaria.[661] The agricultural practices of expanding populations of yam cultivators, especially in the Middle Niger region, created numerous stagnant pools of water, greatly increasing the number of potential breeding sites for mosquitoes. This form

of agriculture first arose in Africa at about *3000 BP* and was followed by a much elevated incidence of malaria in the region. However, within a few hundred years, local human populations were already showing the first genetic adaptations to the disease.[662] The most effective adaptation was the sickle-cell anaemia trait, a mutation in the haemoglobin gene that rapidly spread through populations in malarial regions and is still commonly found today in west Africa.

People carrying two copies of this mutation (one from each parent) produce an aberrant form of haemoglobin that causes their red blood cells to be sickle-shaped, rather than the more rounded biconcave shape of normal erythrocytes. Such people suffer from so-called sickle-cell anaemia due to less efficient blood oxygen transport and consequently have a reduced life expectancy. In normal circumstances such a harmful mutation would tend to be eliminated from the population within a short time. However, people who inherit a single copy of the mutation produce relatively few sickle-shaped red blood cells and are normally asymptomatic despite having some aberrant haemoglobin. If such people are infected with malaria, the plasmodial parasite is unable to reproduce in the presence of the aberrant haemoglobin. Therefore these people are more likely to recover from malaria and hand on the mutation to their descendents. A second series of antimalarial mutations results in a deficiency in the activity of the enzyme, glucose-6-phosphate-dehydrogenase, or G6PD. This deficiency is caused by mutations in either or both of two different alleles, called Med and A-, of the G6PD gene. The effects of the mutations are similar to the sickle cell trait in that they lead to anaemia in affected individuals but also confer much increased protection against the malaria parasite. In the absence of malaria, these mutations are clearly bad for the individuals concerned, especially the carriers of two copies of the mutated gene who have greatly reduced fitness. However, when malaria is common, it is better to be slightly less healthy and resistant to the parasite (i.e. to carry a single mutation) than to carry a normal haemoglobin gene and hence be more vulnerable to the parasite.

The DNA profiling of human populations in malarial regions of Africa has shown that these two mutations arose between 1600 and 11,700 years ago.[663] Such deleterious mutations could only have been maintained under conditions of severe and widespread malarial infection. Further genetic analysis has shown that mosquitoes have also changed over the past 6000 to 10,000 years to become more both virulent and more adapted to human hosts.[664] It has been suggested that a period of climatic warming at about *8000 BP* may have favoured this new strain of mosquito over other variants in the population.[665] However, it seems that the newly virulent genotype of mosquito did not become a widespread scourge of human populations until its dissemination was enhanced via agricultural practices that were introduced in west Africa some 5000 years later. Malaria is still so prevalent in west Africa that, despite the dire effects of the sickle cell and *G6PD* mutations on affected individuals, these two traits remain sufficiently advantageous at the population level that they continue to be maintained as adaptive genetic characters for human populations in this particular region.

Vitamin D, pale skin, and lactose tolerance

Another set of genetic changes in humans relates to our requirement for vitamin D, which is needed for the formation and maintenance of healthy bones, plus other functions relating to calcium homeostasis. Cereal-based diets are deficient in vitamin D, but humans can avoid this problem by synthesizing the vitamin endogenously, providing they are exposed to sufficient sunlight. Vitamin D synthesis occurs in the skin and requires ultraviolet radiation from the sun. The skin of early hominids was uniformly dark, due to high levels of the pigment melanin in their epidermal cells. Interestingly, it seems likely that the more primitive form of human skin pigmentation is a light colour. The skin of most apes and early hominids was (and still is in the case of chimpanzees and gorillas) virtually unpigmented, but was covered instead with a profusion of dark hairs. When African hominids lost much of their bodily hair (probably as a way of facilitating thermoregulation as they moved from their original woodland locations to more open savannah habitats), it became adaptive to develop a darkly

pigmented skin as a protection against UV radiation.[666] The pigment melanin shields the epidermis from potentially harmful ultraviolet radiation, but at the cost of reducing its capacity for endogenous vitamin D biosynthesis. This was not an issue in Africa, even for the original dark-skinned human populations, because the high levels of year-round sunlight enabled them to synthesize sufficient endogenous vitamin D.

However, dark skin became a potential problem for those people who migrated out of Africa into less sunny, northerly latitudes, especially if they failed to maintain a vitamin D-rich diet. In Eurasian farming-based cultures, the dearth of dietary vitamin D would have led to an explosion in the rates of debilitating deficiency diseases such as rickets. This set up a powerful selective pressure for a reversion to the much lighter skin pigmentation of our early anthropoid ape ancestors.[667] Lighter-skinned people are able to synthesize their own vitamin D thanks to their greater ability to absorb UV light, which is used to convert provitamin D_3 into the active form of the vitamin. This has led to two very important genetic changes in one particular group of humans in northern Europe. Firstly, a much lighter skin colour was selected for. Although this was potentially maladaptive in terms of skin damage by intense sunlight, their paler skin enabled these people to synthesize sufficient vitamin D, even in cloudier temperate regions.[668] When they move back to sunnier regions, these light-skinned people often experience significant problems, for example increased incidence of skin cancer, as their phenotype then becomes maladaptive.[669]

The second genetic adaptation to vitamin D deficiency was a mutation that allowed the pale-skinned, northern Europeans to digest milk sugars. Full-fat milk is a rich source of vitamin D, calcium, and phosphate, but most adults across the world today cannot drink milk because they are lactose-intolerant, that is they cannot efficiently digest the milk sugar, lactose. Most humans still lose this ability in infancy, soon after weaning, and become lactose intolerant as adults. There is a very good reason for this. The energy cost of breastfeeding human infants is high, especially for mobile females in hunter–gatherer groups. Therefore, nursing mothers must wean their growing off-spring off breast milk before beginning a new pregnancy. The onset of lactose intolerance in older infants is a useful mechanism that facilitates their move to solid foods, and frees up their mother to invest in further offspring. Hence we find very high levels of lactose intolerance amongst adults in the vast majority of the world's population. For example, 85% of Australian aborigines, 93% of Chinese, 98% of Thais, and 100% of Amerindians are lactose-intolerant.[670] Lactose intolerance does not preclude the eating of sold milk products, such as cheese or butter, because the latter contain little or no lactose. However, it does mean that a potentially calorie-rich foodstuff, namely lactose-rich liquid milk from domesticated mammals, is unavailable to such people.

In contrast to most of the world population, the incidence of lactose intolerance in Swedes is only 2%, because these people are descended from a population that developed a specific mutation that enabled them to maintain the infantile trait of lactose tolerance. The lactose tolerance mutation is therefore another recent paedomorphic, domestication-related trait that has some similarities with what we saw previously with the Siberian fox study in Chapter 3. Lactose tolerance mutations probably arose in northern European pastoralists as recently as 5000 to 10,000 years ago. In genetic terms, acquisition of adult lactose tolerance seems to be due to one or two simple mutations in the promoter region of the lactase gene, as described in 2002 by a group from Finland.[671] These mutations result in the persistence of lactase activity into adulthood and allow such people to drink liquid milk from mammals. Not only did the milk provide useful calories from its sugars and fats, it was also a rich source of protein and vitamin D.

The population-wide benefits conferred by the lactose tolerance trait far outweighed disadvantages caused by older infants retaining the ability to digest their mother's milk (and therefore being able to compete for maternal milk with younger unweaned infants). Besides, these older infants and children could now be fed milk from cows or goats instead of relying on maternal milk. This extremely adaptive mutation soon spread through populations in northern Europe, although its incidence in other human groups around the world has remained very low. Analyses by molecular geneticists indicate that the recent selection for this trait in north European, dairy-dependent cultures is one of the most powerful such events that they have measured in the human genome.[672] Separate lactose-tolerance mutations have arisen independently in a few other highly pastoralist groups, such as the Bedouin and Hausa in Africa, but otherwise this trait is rare in most human populations.[673]

As a cultural aside, it is interesting that adult lactose intolerance is still regularly described in Western scientific and medical literature as if it were a disease or even a genetic 'defect' whereas, on the contrary, it is the original and arguably the 'normal' and most adaptive condition for most of humanity. In contrast, most indigenous northern and western Europeans (including the author) are the progeny of a few 'aberrant' mutated individuals. Their mutant status enables this atypical group of humans to constitutively express an otherwise infantile trait throughout their lives and allows them to drink milk produced by several other mammalian species.[674] Lactose tolerance is also an interesting example of reciprocal genetic adaptation as part of a domestication dyad. In this case, the dyad is humans and domesticated mammals. Mammals such as cattle and goats were selected by human pastoralists for genetic traits such as docility and high milk and meat yield. However, the new availability of highly nutritious, vitamin-rich milk to human adults would in turn have set up strong selection pressures favouring a mutation that enabled human adults to tolerate liquid milk—something that only infants had been capable of hitherto. Eventually, therefore, the humans also became genetically adapted to their animals, and in a sense were therefore also 'domesticated'.

This reinforces the view of domestication as a special case of the wider phenomenon of coevolutionary development of mutually interacting species (as discussed in Chapters 1 and 3). Hence, we find that sometimes the domesticator becomes genetically modified to suit the domesticant, as well as *vice versa*. In the present example, one group of northern European pastoralists acquired a paedomorphic, lactose-tolerant mutation, greatly reinforcing their tendency to keep and protect immense herds of their favoured milk-producing herbivores, such as cattle and goats. Their cattle were of course assiduously guarded from predation by, or competition from, other wild carnivores and herbivores, most of which have now become extinct. Therefore, it may be equally valid to say that such humans have been 'domesticated' because they have been genetically transformed in a way that favours the prospects of these particular herbivorous mammals. The close association of northern Europeans with cattle has resulted genetic modification of the vast majority of this particular population into lactose-tolerant mutants who are atypical of 'normal' humans. This reciprocal domestic association of cattle with humans is undoubtedly beneficial to the humans, but it has also distinctly favoured the survival prospects of hundreds of millions of cattle around the world.

A final twist to this story comes from the discovery that many European cats have a similar (but independently acquired) mutation to their owners, which enables adult cats to digest milk. In contrast, most non-European cat breeds, and almost all other mammals, are lactose intolerant as adults. The lactose-tolerance mutation probably arose relatively recently in European cats and would have conferred a significant advantage in enabling them to use a plentiful new source of food, namely milk supplied by (or stolen from) their commensal human partners. In contrast, non-European cats rarely have access to milk (which their owners or commensal human partners cannot drink) and therefore there is no adaptive advantage in developing adult lactose tolerance. In this case, we have a novel form of domestication triad, with three mammalian species interacting both behaviourally and genetically in ways that reinforce their mutual relationships in an evolutionarily adaptive manner. If cattle and other sources of milk (e.g. goats) were to suddenly die out, lactose-tolerant humans (and cats) might be at a selective disadvantage compared to lactose-intolerant populations, unless alternative forms of infant food could be found. In this sense, most northern European humans have become, to some extent, genetically 'hard-wired' in favour of living in association with their plant and animal domesticants.

Dental changes and the recent 'maxillary shrinkage'

As we saw above, the transition to agriculturally derived diets was frequently accompanied by dental pathologies, related either to dietary deficiencies or to adverse food texture. The entire skeleton of *Homo sapiens* has undergone considerable gracilization during its million-odd years of evolution. For example most of our bones are considerably more slender than those of our close Neanderthal cousins. However, over the past ten millennia a much more rapid change in one part of our skeleton has occurred. This is the so-called 'maxillary shrinkage', involving a generalized reduction of our facial structures. It is a very recent phenomenon that is correlated with the adoption of agriculture and the consequent shifts in diet.[675] This correlation has been observed in farming cultures across the world, from the Near East and Europe to the Americas and the Far East.[676] Experiments with mammals fed from birth on diets of soft foodstuffs have revealed that there is an overall reduction in the craniofacial apparatus in all species tested.[677] In one case, minipigs given a diet of soft food were found to consistently develop smaller faces and jaws compared with conspecifics fed with hard foods.[678]

These responses to dietary change have short-term and long-term components. In the short term, there is a decrease in the rate of deposition in the masticatory bones during early life, resulting in the irreversible reduction of final bone mass in the adult animal. In the longer term, over numerous generations, there is a tendency towards a genetically determined reduction in facial bone size, even if a diet of hard foods is resumed. The latter effect is what we see in the early agriculturalists. This genetic change in our facial structure is not directly due to the consumption of plant products *per se*, but it is related to the adoption of a cereal-based diet. As we saw previously, cereal grains are very difficult for non-specialist herbivores to grind without rapid and irreversible tooth wear. Humans have relatively small teeth and, despite the plant-based diet of our ancestors, there has also been a significant reduction in human tooth size over the past ten millennia.[679] However, our mouths have shrunk even more rapidly than our teeth, leading to chronic and ever-growing problems associated with dental crowding in modern human populations. So why are our teeth and other facial bones getting smaller if we are eating hard-textured plant-based foods? The answer is, of course, that we learned to use grinding stones to process hard cereal grains into soft flour way back in the Palaeolithic Era, well over 20,000 years ago and long before the beginnings of formal agriculture. Therefore, the human diet came to resemble that of the minipigs fed on soft foods and we responded by the still ongoing process of maxillary shrinkage.

How did the sickly Neolithic farmers prevail?

The seemingly disastrous health impact of cereal farming on many mid-Neolithic cultures begs several questions. Why did they persist in these seemingly deleterious and maladaptive practices? How did relatively stunted, frail, and sickly agriculturalists so decisively out-compete their taller, sturdier, and generally healthier hunter–gatherer neighbours? The answers may lie in a cost–benefit analysis of the impacts, not just of farming itself, which was obviously maladaptive in terms of individual health, but of all the other associated factors that were part of the emerging agricultural lifestyle. These additional factors include the ability to support greater numbers of people in a smaller area; opportunities to construct sturdier, more permanent habitations; and the impetus to technological development and greater social complexity. Members of the roaming bands of hunter–gatherers may have been, on average, rather better nourished and healthier than their agrarian neighbours. But, in the longer term, the relatively healthy nomads would have been unable to compete as a group with the scrawnier, but far more numerous, better equipped (with both tools and weapons), better housed, and better organized farming communities. Hence the main selective advantages of agriculture lie at the level of human societies, rather than at the level of the individual.

Neolithic farmer societies did not just compete with local hunter–gatherers. As farming communities expanded, they also met neighbouring agriculturalists, some of whom may have developed different, and possibly superior, forms of crop husbandry. For example the cultivation of rye, which as we have already seen started as the main cereal at Abu Hureyra during the Younger Dryas, was soon supplemented by, and then completely supplanted by, the use of wheat and barley. Neighbouring communities, such as the Natufians, had not grown rye but used emmer and einkorn wheat and barley instead. As it became apparent from contact with their neighbours that wheat and barley were often better crops in this region, the Abu Hureyra people soon adopted these agricultural innovations. A further innovation that greatly improved the relatively impoverished diet of early farmers was the domestication of certain animals to provide food. This did not happen until well into the Neolithic period, several millennia after crop-based agricultural communities, such as Abu Hureyra, had already become established.

Livestock domestication

Soon after the first crops were cultivated, people began to experiment with the domestication of some of the local livestock, as an alternative to the costly and frequently unsuccessful practice of hunting. However, the first animals to be domesticated were not used for food, but served other purposes. We have already seen that dogs had been domesticated in what may have been a relatively rapid process in the Near East by *12,000 BP*, and quite possibly several millennia earlier. Cats probably coevolved with people into their present semidomesticated association in a somewhat analogous manner to dogs. The value of cats in protecting agricultural produce is encapsulated in the following quote from the ancient Egyptian scribe, Ahmose, dated about *3650 BP*: 'In each of 7 houses there are 7 cats; each cat kills 7 mice; each mouse would have eaten 7 grains; and each grain would have produced 7 hekat of wheat. How much wheat is saved by the cats?' The answer is 7^5 or 16,807 hekat. Since a hekat of grain weighs about 3.75 kg, the total amount of grain saved amounts to over 63 tonnes (or about 1.3 tonnes per cat). Therefore, each of the industrious Egyptian felines described by Ahmose saved more than enough grain to make almost 3000 loaves, which is enough to feed several dozen people for a year. This may explain the almost ubiquitous presence of cats as often uninvited, but nevertheless generally welcome, guests in human habitations for much of human history.

While cats, dogs and horses seem to have been adopted on many occasions by cultures in different regions, the main species of edible domestic livestock seem to have tamed on only a few occasions.[680] According to genetic evidence, cattle, sheep, pigs, donkeys, and water buffalo may each have only two origins of domestication,[681] while goats may have three such origins.[682] Goats were probably the first herbivores to be domesticated, possibly as early as *9000* to *10,000 BP*.[683] The progenitor of the domestic goat, *Capra hircus*, was a Eurasian species, the bezoar, *Capra aegagrus*, and the earliest domestication probably occurred in the Near East. As with other instances of both plant and animal domestication that we have considered here, there are good reasons to believe that livestock domestication was initially an unconscious process on the part of the people involved.[684]

Archaeological and genetic evidence suggest that people started keeping small herds of domestic livestock to complement other food sources, such as crops.[685] The risk of inbreeding depression seems to have motivated people to periodically refresh the gene pool of their domesticated animals, either via trade with other livestock keepers, or by (accidentally or deliberately) allowing the domesticants to outbreed with wild relatives.[686] This has enabled humans to maintain relatively homogeneous herds of domesticants, but with sufficient genetic diversity to adapt to new diseases and environmental changes as they were introduced into new habitats. Cattle and sheep have now been taken across the world from their Near Eastern centres of origin, to areas as diverse as the tropics and the arctic fringes of Eurasia.[687] The two most important advantages of animal domestication are elimination of the time-consuming and potentially dangerous need to hunt game, and ability to utilize more efficiently a resource that is normally unavailable for human nutrition, namely cellulose.

Cellulose, which is the major structural component of plant cell walls, is the most abundant organic molecule on earth and is made up entirely of the simple sugar, glucose. Unfortunately the cellulose molecule is also very resistant to breakdown and only a few bacteria and fungi possess this ability. Grass-eating ruminants, including the most common domesticants such as cattle, sheep, and goats, can digest cellulose thanks to symbiotic bacteria and fungi in their rumens. In essence, these animals act as bioreactors that convert useless (to humans) cellulose into extremely valuable products such as meat and milk. The acquisition of this new resource was so advantageous that it enabled some human societies, either partially or completely, to abandon arable farming and to readopt seminomadic lifestyles based on pastoralism. However, very few nomadic societies live solely on the produce of their animals. More commonly, they cultivate gardens within their home range or trade for crop products with nearby agrarian communities. Over the past two centuries, most remaining nomadic hunter–gatherer or pastoral societies have been compelled, often by force, to adopt more sedentary lifestyles, as the agrourbanists completed their domination of the earth.[688] In the next three chapters, we will examine the development of the major early agrourban cultures in greater detail.

CHAPTER 10

Evolution of agrourban cultures:
I The Near East

That on the banks of the Tigris and Euphrates evil
  weeds grow,
That the hoe does not tackle the fertile fields,
That the seed is not planted in the ground,
That the city and its surrounding lands are razed
  to ruins

Ancient Sumerian poem, c. 4000 BP, *Lamentation
over the Destruction of Ur* [689]

Introduction

The cultivation of a small number of plant species
with genetically-linked domestication-related traits
occurred independently in Eurasia, Africa, and the
Americas at several periods from about *12,000 BP* to
about *4000 BP*. We have seen in the previous chapter
that dependence on a relatively restricted range of
food plants sometimes had adverse consequences
for individual human health. However, such
drawbacks were counterbalanced by society-level
benefits that enabled farming cultures to compete
more effectively against animal and human rivals.
In the following three chapters, we will resume our
examination of the interlinked processes of agricultural and societal development in the ancient
world. Some more general links between agriculture and the development of complex urban
societies are explored in Box 10.1. We begin, in the
present chapter, with a detailed look the Near East
before moving on in the next two chapters to
survey other important centres of agrosocial development, including east and south Asia, Africa,
Europe, and the Americas.[690] While stressing the
important societal and biological dimensions that
framed these events, the discussion will also be
informed by recent discoveries on the impact of

regional and global climatic processes that have
profoundly affected human development during
the Holocene Era.[691]

Some of the earliest instances of successful and
enduring domestication of food crops occurred in
south-west Asia during the Palaeolithic/Neolithic
transition at about *13,000–11,000 BP* (see Table 10.1
for a list of events in the Near East until *5000 BP*,
and Table 10.2 for the major crops). Despite many
local reversals, agriculture gradually became a
more enduring phenomenon as it spread across the
area traditionally known as the 'Fertile Crescent'.
This region stretches from the Mediterranean
littoral in the Levant, north-eastwards across
Anatolia, then south-eastwards across the Tigris/
Euphrates basins and the Zagros Mountains
towards the Persian Gulf (see Figure 10.1 for the
principal locations mentioned in the text and
Figure 10.2 for maps depicting the spread of agriculture across the Near East). The consolidation
and spread of farming across this vast region
occurred synergistically with the evolution of new
cultures and technologies.[692] These mutually
dependent processes facilitated the more efficient
exploitation of cereal-dominated crops via more
intensive forms of agriculture, which in turn
generated increasing amounts of storable food
surpluses. By *7000 BP*, these surpluses were large
and reliable enough (at least over a period of
decades) to fuel the growth of ever-larger villages
and small towns, such as Tell Hammam[693] and
Umm Dabaghiyah.[694]

By *5500 BP*, as agricultural intensification and
more effective management of both crop and
human resources continued, it was possible to
sustain development of the first cities, such as

Box 10.1　Are technology, cities, and empires inevitable consequences of agriculture?

We have seen that one of the most pervasive features of agricultural societies across the world is the tendency towards increasing urbanization, social inequality of wealth and power, and the eventual rise of authoritarian elites that are often dominated by a single ruler. It is widely assumed that the change to a sedentary, agricultural lifestyle led, more or less automatically, to the generation of food surpluses that freed part of the population to develop more complex technologies. These in turn made agriculture yet more efficient and able to support ever-larger towns and cites with their burgeoning complement of bureaucrats, priests, warriors, and kings. But is this necessarily an inevitable process? I will argue here that there is no inevitability about increasing societal complexity arising as a consequence of an agrarian lifestyle. Instead, I suggest that the evolution of such complexity is an adaptive response to a particular set of socioenvironmental conditions, and that it can become maladaptive if conditions change.

First, there is no absolute requirement for agriculture in the development of sedentism. As discussed in Chapter 2, several early Neolithic societies became semi- or completely sedentary long before they adopted agriculture. The Jomon culture in Japan, which dates from c. 16,000 BP, soon developed a sedentary lifestyle with advanced pottery, but did not make a transition to full-scale rice farming until about 2500 BP, when the technology was probably imported from the Mumun culture in neighbouring Korea (Habu, 2004). In south-east Turkey, the early Neolithic settlement of Göbekli Tepe, dating from 11,000 BP, was created by hunter–gatherers long before agriculture came to the region (Schmidt, 2001).

There are also more recent instances that provide useful counter-examples to the supposed linkage between agriculture and complex sedentary societies. Until about a century ago, the Amerindians of the Pacific Northwest exploited a bountiful environment with an abundance of fish, game, and plant resources (Ames, 1999; Ames and Maschner, 2000; Deur, 1999, 2002). Despite not developing agriculture, they lived in permanent towns of over 1000 people, including craft specialists such as wood

carvers and canoe makers, as well as social castes such as slaves and a hereditary ruling elite. These well-organized, urban dwellers practiced warfare and long-distance trade, but despite all of these trappings of 'civilization', they never practiced farming.

So, people can develop urban cultures in the absence of farming, but does the presence of farming guarantee urban development? The answer is: obviously not. As illustrated by many examples in the main text, numerous farming cultures lived for many millennia as dispersed groups of hamlets and small villages with no tendency to coalesce into larger units. It seems that there were at least two prerequisites to the emergence of larger and more complex social groups, namely access to very high-yielding crops, and an extended period of social stability.

The mild, moist climatic period of the mid-Holocene (7300–5000 BP), combined with high-yielding emmer and barley crops, facilitated a long period of social stability and large, reliable crop surpluses that underpinned the development of the first cities and kingdoms in Mesopotamia, the Indus Valley, and Egypt. In contrast, low yields for several millennia after its domestication precluded production of sufficient surplus maize to support cities in Mesoamerica. However, soon after maize cobs had reached a sufficient size to provide reliably large surpluses of storable food (at about 2200 BP), there was a flowering of urban cultures throughout the region.

However, the development of such cities could also be halted, sometimes permanently, by the often-linked processes of social instability, crop failure, and climate change. As we explore further in Box 11.1, one can regard social complexity as an adaptation to socioenvironmental stability, with the corollary that socioenvironmental instability will tend to favour the evolution of simpler social units. As explored in Chapter 17, the possibility of increasing climate change in the future, with the attendant likelihood of greater social instability, may favour a drastic simplification of our current social order, and a possible reversion to non-agrarian, dispersed lifestyles.

Tell Hamoukar,[695] Uruk, and Eridu.[696] Further intensification of the agricultural economy and increased urbanization culminated in the establishment of a string of city-states, such as Ur and Nippur, which by 5000 BP dominated large

hinterlands of dependent villages and farmland. This process culminated in the establishment of the first transregional empire, that of Akkad, by 4300 BP.[697] During the past few decades, increasingly detailed physical evidence has been

17,000–15,000	Early Kebaran	Throughout Near East	**Heinrich Event** Very cold and severe aridity	Isolated wood; steppe; and desert	Small game, wild fruits, large-seeded grasses, e.g. barley and emmer	Geometric microlithic tools	Nomadic and seminomadic, dispersed
15,000	Kebaran	Kebara	Rapid warming, sudden increase of 40% in atmospheric CO_2 levels	Prolific woodland Grassland	Prolific fauna, less use of plants	Geometric microlithic tools	Nomadic and seminomadic, dispersed
14,500	Kebaran	Throughout Near East	Warm and moist	Prolific woodland Grassland	Prolific fauna, less use of plants	Geometric microlithic tools	Nomadic and seminomadic, dispersed
13,500	Early Natufian	Abu Hureyra I	Warm and moist	Prolific woodland Grassland	Prolific fauna, use of wild plants including fruits, rye, barley and pulses	Geometric microlithic tools	Nomadic and seminomadic, dispersed villages, pit-houses
13,000	Early Natufian	Abu Hureyra I	**Younger Dryas** Cold and dry	Much reduced woodland, steppic grassland	Much reduced fauna and flora, more use of wild rye, einkorn and barley	Loss of game and and many wild plants, earliest tillage	Nomadic and seminomadic, dispersed villages, pit-houses
12,500	Natufian	Mureybet IA Abu Hureyra I	**Younger Dryas** Cold and dry	Patchy relict woodland, reduced steppic grassland	Much reduced fauna and flora, more use of wild einkorn and barley **RYE**[†]	Adaptation of existing tools for rainfed farming	Multifamily small villages of 100–200 people
12,000	Natufian Khiamian	Mureybet IB-II Abu Hureyra I Shanidar	**Younger Dryas** Cold and dry	Patchy relict woodland, reduced steppic grassland	Wild einkorn, RYE Stock manipulation Wild sheep herding	Grinding querns, Sickles adapted for harvesting	Multifamily small villages of 100–200 people
11,400	PPNA	Jericho Netiv Hagdud	**Younger Dryas** Cold and dry	Patchy relict woodland, Steppic grassland	2-ROWED BARLEY Wild plants, gazelle, fish, waterfowl	Grinding tools, flint-bladed sickles	Small–medium villages of 100–400 people
11,000	PPNA	Sheikh Hassan Jerf al-Ahmar Mureybet III	Younger Dryas ends Warming and wetter	Partial return of woodland Rich grassland	EMMER, EINKORN RYE, 2-ROWED BARLEY, livestock	Improved farming tools, some pastoralism	Small–medium villages of 100–400 people
10,500	Early PPNB	Dja'de Harif Mureybet IVA	Warm and moist	Mixed dispersed wood/grassland	EMMER, EINKORN, SGCP	More intensive tillage, manure, brick-making	Formalized multiroom brick houses, small towns
10,000	Middle PPNB	Halula, Aswad Mureybet IVB Abu Hureyra IIA	Warm and moist	Mixed dispersed wood/grassland	EMMER, PULSES, EINKORN, RYE, 2-ROWED BARLEY, SGCP	First use of clay tokens as precursor to writing	Spread of trade and village/town-based agrourban culture
9600	Middle PPNB	Netiv Hagdud	Warm and moist Euxine Lake inundated to form Black Sea	Mixed dispersed wood/grassland	EMMER, PULSES, EINKORN, RYE, 2-ROWED BARLEY, SGCP	Agropastoral with seasonal game hunting	Increase in societal complexity, specialized crafts
9500 9000	Late PPNB	Bouqras 11–8 Abu Hureyra IIB	Start of Holocene climatic optimum Warm and moist	Mixed dispersed wood/grassland	EMMER, PULSES, EINKORN, RYE, 2-ROWED BARLEY, SGCP	Agropastoral with seasonal game hunting	Modest towns 2500–3000 people, central planning

Table 10.1 (Continued)

Years BP	Human species/culture	Major sites	Climate	Dominant flora	Human food resources‡	Technologies	Lifestyles
8800	Final PPNB	Bouqras 7–1	Warm and moist	Mixed dispersed wood/grassland	(EPEB) EMMER, PULSES, EINKORN AND 2-ROWED BARLEY	First pottery	Modest towns 5000–6000 people, central planning
8200	Early PN	Qal'at Jarmo Abu Hureyra IIC					
	Pre-Halaf	Nebi Mend Umm Dabaghiyah	**8200 Drought** Cool and dry	Much reduced woodland, Steppic grassland	Reduced EPEB crops Increased pastoralism SGCP livestock	Long-distance trade, clay seals, elaborate pottery	Many towns/villages abandoned, population decline
8000	Hassuna and Samarra						
8000	Halaf	Arjoune Choga Mami Umm Qseir	Warm and moist	Mixed dispersed wood/grassland	New varieties of EPEB 6-ROWED BARLEY SGCP livestock	Intensification, water management, early irrigation	Rapid recovery, migration to South, population growth
7500	Ubaid						
7000	Early Chalcolithic	Tel Halaf, Arpachiyeh, Amuq E, Kurdu	Warm and moist	Mixed dispersed wood/grassland	Higher yield EPEB crops SGCP livestock	Improved irrigation, canals, copper working	Settlement of South with irrigation farming
6500	Chalcolithic	Ras Shamra III, Hammam IVC, Ziyadeh 2–12	Warm and moist	Mixed dispersed wood/grassland	High yield EPEB crops SGCP livestock	Potters wheel, ploughs, mass-production	Centralization of urban–industrial production
6000	Early Uruk	Eridu, Uruk, Ur, Hammam IVD, Tell Afis	Warm and moist	Mixed dispersed wood/grassland	High yield EPEB crops more use of BARLEY SGCP livestock	Large temples (Eridu) manage via bureaucracy	Population increase, first cities, elites control food supply
5500	Late Chalcolithic Ubaid Middle Uruk	Tell Hamoukar, Tel Leilan, Brak	Warm and moist	Mixed dispersed wood/grassland	High yield EPEB crops more use of BARLEY SGCP livestock	Animal traction, monumental building, writing	Hyperurbanization in Sumer, partial colonization of North
5200	Late Uruk	Qraya, Habuba Kabira, Tel Leilan	**5200 Drought**	Much reduced woodland, steppic grassland	Reduced EPEB crops Increased pastoralism SGCP livestock	Simplified in North, intensified in South	Collapse of agrourbanism in North, continuity in South

This chronology is focussed on agrosocial development in the Levantine and Mesopotamian regions of the Near East. Before the Younger Dryas, optimal food production strategies varied greatly as the climate changed, and with it the availability of such resources as megafauna, small game, fish, and edible plants. People probably had both the intelligence and technology for farming from Middle Palaeolithic times, but did not have access to the right plants, and/or were better served by hunter gathering. Once they started to cultivate food plants, it sometimes became adaptive to set up larger communities, which in turn established a set of social dynamics that often favoured increasing urbanism, stratification, and technosocial complexity. But this was not a unidirectional process and its trajectory was constantly modulated by social and environmental factors.

PPNA, Prepottery Neolithic A; PPNB, Prepottery Neolithic B; SGCP, domesticated sheep, goat, cattle, and/or pigs.

‡ Wild plants shown as lower case, domesticated crops are in capitals.

† First evidence of any domesticated form of a crop.

Table 10.2 Presence of wild and domesticated versions of the major cereals, pulses and tree species from archaeological sites throughout the Near East

Archaeological Site§	Date, years BP	Einkorn w†	Emmer w†	Barley w†	Einkorn d†	Emmer d†	Naked Wheat d†	Barley 2r d†	Barley 6r d†	Aegilops w†	Lentil ?†	Pea ?†	Bitter vetch ?†	Oak w†	Almond w†	Pistacia w†	Flax ?†	Reference
Ohalo II	19,000	—	O‡	O	—	—	—	—	—	—	O	—	—	A‡	O	O	—	Kislev et al. 1992
Franchthi	12,400–9000	—	—	O	—	—	—	—	—	—	O	O	O	—	O	O	—	Hansen 1991
Hayonim	12,300–11,900	—	—	O	—	—	—	—	—	—	O	—	O	—	—	—	—	Hopf and Bar Yosef 1987
Wadi Hammeh	12,200–11,900	—	—	O	—	—	—	—	—	—	O	—	—	W‡	—	O	—	Willcox 1991b; Colledge 1994
Abu Hureyra 1	11,000–10,000	O	—	O	—	—	—	—	—	—	O	—	O	W	W	O	—	Hillman et al. 1989
Hallan Çemi	10,600–9900	—	—	O	—	—	—	—	—	—	O	O	O	O	O	O	—	Rosenberg et al. 1995
Mureybit I–III	10,200–9500	O	—	O	—	—	—	—	—	—	O	—	—	W	O	O	—	van Zeist and Bakker-Heeres 1984
Qermez Dere	10,100–9700	—	—	O	—	—	—	—	—	—	O	—	O	—	—	O	—	Nesbitt 1995
Netiv Hagdud	10,000–9400	—	—	O	—	—	—	—	—	—	O	O	O	W	O	—	—	Bar-Yosef et al. 1991
Jerf el Ahmar	9800–9700	O	—	O	—	—	—	—	—	O	O	O	O	W	O	O	—	Willcox 1996
M'lefaat	9800–9600	o‡	—	O	—	—	—	—	—	O	O	O	O	—	O	O	—	Nesbitt 1995
Tell Aswad Ia	9700–9600	—	o	O	—	o	—	o	—	—	O	O	O	—	O	O	—	van Zeist and Bakker-Heeres 1984
D'jade	9600–9000	O	o	O	—	o	—	—	—	O	O	O	O	W	O	O	—	Willcox 1996
Çayönü	9500–9200	o	o	O	—	—	—	—	—	—	O	—	O	W	O	O	—	van Zeist and de Roller 1994
Jericho PPNA	9500–9000	—	—	o	o	o	—	—	—	—	O	—	—	—	O	O	—	Hopf 1983
Mureybit IV	9400–8500	O	—	O	—	-	—	o	—	—	O	—	—	W	O	O	—	van Zeist and Bakker-Heeres 1984
Cafer Höyük XIII–X	9400–9000	O	O	—	o	o	—	—	—	—	O	O	O	W	O	O	—	Willcox 1991c; de Moulins 1993
Tell Aswad Ib	9300–8800	—	—	O	—	o	—	?	—	—	O	O	O	—	—	O	—	van Zeist and Bakker-Heeres 1984
Çayönü	9200–8500	O	O	O	o	o	—	?	—	—	O	O	O	W	O	O	O	van Zeist and de Roller 1994
Nevali Cori	9200	—	—	o	o	—	—	—	—	O	O	O	O	O	O	O	O	Pasternak 1995
Ain Ghazal	9000–8500	—	—	O	O	O	—	O	—	—	O	O	O	W	—	O	O	Rollefson et al. 1985
Jericho PPNB	9000–8500	—	—	O	O	O	—	O	—	—	O	O	—	—	—	—	O	Hopf 1983
Cafer Höyük IX–VI	9000–8400	O	—	—	O	O	—	—	—	—	O	—	O	W	W	O	O	de Moulins 1993
Nahal Hemar	9000–8200	—	—	—	—	O	—	O	—	—	O	—	—	A	O	O	—	Kislev 1988
Beidha	8900–8700	—	—	—	O	O	—	o	—	O	—	—	O	W	O	O	—	Helbaek 1996

Table 10.2 (Continued)

Archaeological Site§	Date, years BP	Einkorn w†	Emmer w†	Barley w†	Einkorn d†	Emmer d†	Naked Wheat d†	Barley 2r d†	Barley 6r d†	Aegilops w†	Lentil ?†	Pea ?†	Bitter vetch ?†	Oak w†	Almond w†	Pistacia w†	Flax ?†	Reference
Ganj Dareh	8900–8200	—	o	o	—	—	—	o	—	—	o	—	—	—	o	o	—	van Zeist et al. 1986
Ali Kosh BM	8800–8000	o	—	o	o	o	—	o	o	—	—	—	—	A	—	o	—	Helbaek 1969
Jilat 7	8800–8400	o	—	o	o	o	—	—	o	—	—	—	o	—	o	o	—	Colledge 1994
Asikli	8800–8400	o	—	—	?	o	o	o	o	—	o	o	o	—	o	o	—	van Zeist and de Roller 1995
Abu Hureyra PPNB	8800–8000	o	—	o	o	o	o	o	—	o	o	—	—	W	W	o	o	de Moulins 1993
Tell Aswad II	8700–8400	o	—	o	o	o	o	o	—	—	o	o	—	—	—	o	o	van Zeist and Bakker-Heeres 1984
Ghoraifé I	8700–8100	—	—	o	—	o	o	o	—	—	o	o	—	—	—	o	o	van Zeist and Bakker-Heeres 1984
Abdul Hosein	8700–7500	—	—	—	—	o	—	o	—	—	o	—	—	W	W	o	—	Hubbard 1990; Willcox 1990
Halula	8700	o	o	o	o	o	o	o	—	o	o	o	o	W	—	o	o	Willcox 1996
Magzalia	8600–7800	—	—	o	—	o	o	o	—	o	o	—	—	—	—	—	o	Willcox, unpublished
Gritille	8500–7700	—	—	—	—	o	—	o	—	—	o	—	o	—	—	—	—	Voigt 1984
Can Hassan III	8500–7600	o	—	o	o	o	o	o	—	—	o	—	o	W	o	o	—	French et al. 1972
Jarmo	8500	o	—	o	o	o	—	o	—	—	o	—	—	W	o	o	—	Braidwood and Braidwood 1983

There is considerable chronological overlap between sites, particularly for the later periods. Note that both wild and domesticated versions of each of the major crops are found at most sites for much of this period, indicating that domestication was a gradual process, even at a single site, and that wild versions of the crops continued to be important sources of food for several millennia after the appearance of the first domesticated forms. Domesticated versions gradually appeared over a wide area during the last half of the *tenth millennium BP*. In addition to the cereals, lentils are commonly found at most sites, although tree nuts are also well represented, indicating the continuing importance of the gathering of wild foodstuffs for many millennia after the beginning of full-scale farming (data are from Table 1 in Wilcox, 1998, as adapted from original data of Nesbitt and Samuel, 1996).

† w = wild; d = domestic; ? = wild and/or domestic.

‡ O = present; o = identification based on small number of poorly preserved finds; W = identification based on wood; A = acorn

§ see Figure 10.3 for location of archaeological sites (based on http://www.ipgri.cgiar.org/Publications/HTMLPublications/47/ch06.htm).

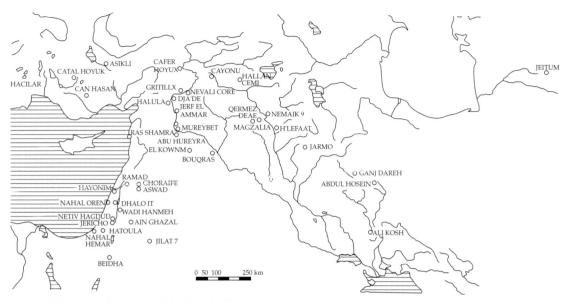

Figure 10.3 Location of archaeological sites listed in Table 10.2.

extents in different regions (Figure 10.5).[703] In many respects, the intensity of this hitherto largely unremarked climate shift was second only to that of the Younger Dryas.[704]

One of the possible causes of the *8200 BP* event, which was felt across the northern hemisphere, was a sudden collapse of the dome of the Laurentide ice sheet covering Hudson Bay in North America.[705] This released vast quantities of fresh water into the North Atlantic, perturbing the thermohaline circulation and temporarily displacing the Gulf Stream that brings mild, damp conditions to western Eurasia. A recent study by a group of British palaeoceanographers has provided convincing evidence that Laurentide meltwater was implicated in a cooling/drying phase that had effects throughout western Eurasia, beginning about *8260 BP* and lasting for several centuries.[706] As rainfall levels declined across the northern hemisphere, pistachio and oak forests receded in the Near East and Mediterranean and cereal yields fell in rainfed farming regions throughout south-west Asia.[707] Most of the Levantine prepottery Neolithic farming villages were abandoned, while in northern Mesopotamia the Umm Dabaghiyah culture was dispersed to a few refuge areas.[708] New cultures

arose that were more dependent on sheep pastoralism, which may have been an adaptation to lower crop productivity.[709] In the Tigris Valley, the appearance and subsequent expansion of the Hassuna and Samarran cultures coincided with the end of this cool, dry climatic interlude.

The post-*8,200 BP* era marks the beginning of a rich period characterized by a great deal of agricultural and technological innovation and the expansion of complex societies throughout the Near East. It was during this period of societal 'rebound' that many significant agrotechnological achievements occurred. Pottery was developed; irrigation and more complex forms of water management allowed both the expansion and intensification of arable farming; new varieties of wheat and six-rowed barley were selected; larger villages and towns sprung up across the region; and by the end of this favourable climatic interlude, metal working had started as the Chalcolithic period saw the first use of copper tools. During the Ubaid Period (*7500–6000 BP*) towns grew into small cities such as Eridu, which had a population of about 4000 people. This era also witnessed the emergence of hierarchies, state-organized religions, and more rigid forms of social stratification, and, by *5400 BP*,

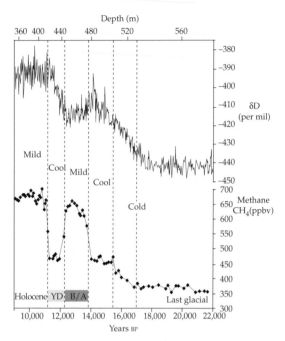

Figure 10.4 Plant growth during the Pleistocene to Holocene transition. The last major Ice Age (Last Glacial), which ended about *15,000 BP*, was marked by cold, arid conditions and low atmospheric CO_2 concentrations, all of which negatively impacted plant growth. It was followed by the relatively mild and wet Bølling/Allerød (B/A) interstadial period, until a sudden return to cold and dry conditions during the Younger Dryas (YD), from *12,800–11,600 BP*. Since the end of the Younger Dryas, the climate of the Holocene has been relatively stable and its generally warm, moist climate and higher atmospheric CO_2 concentrations have greatly favoured plant growth. The early Holocene, from *11,600–8,200 BP* was an especially benign period that witnessed the domestication of most of the major crops and the rise of increasingly complex agrourban cultures in various parts of Asia and Africa. Data are from the Antarctic ice dome where the solid lines show δD values as a proxy for local temperature, and diamonds show methane levels, which are related to vegetation abundance; ppbv, parts per billion by volume; δD per mil, parts per thousand difference from the isotope ratio of the reference standard. (Redrawn from Monnin *et al.*, 2001.)

written scripts had appeared. However, in the midst of this epochal period of progress, at about *5200 BP*, the development of all Near Eastern cultures was significantly affected by a second momentous post-Younger Dryas climatic event.

The climatic event of *c. 5200 BP*

The *5200 BP* cool/arid event lasted between one and two centuries (depending on the region) and

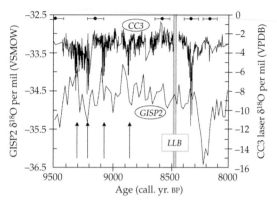

Figure 10.5 The *8200 BP* climatic event. In much of the northern hemisphere, the Holocene Climatic Optimum was abruptly punctuated at about *8200 BP* by a locally severe cold/dry period. Geophysical measurements from Greenland and Ireland suggest a sudden cooling of 7° ± 3°C (δ^{18}O lines) at roughly *8300–8200 BP*. This occurred within a few centuries of the break-up of the Laurentide Lakes ice dome and the two events may be related. The graph shows climatic data between *9500* and *8000 BP* where CC3 is the record from an Irish site and GISP2 is from a Greenland ice core. The vertical shaded line at *8470 BP* labelled LLB denotes the approximate timing of the Laurentide Lakes event. U-series ages and their 1σ error bars are shown in the upper part of the diagram. Arrows denote timing of δ^{18}O fluctuations in the new high-resolution record, interpreted as meltwater release events (δ^{18}O per mil, parts per thousand difference from the isotope ratio of the reference standard). (Data from McDermott *et al.*, 2001).

affected many areas of the world, but especially south-west Asia.[710] This event was probably linked to the slightly earlier climatic changes that terminated the African Humid Period, resulting in desertification of the Sahara region (see Figure 10.6 and Chapter 12).[711] Thanks to the existence of written records, we know that from about *5200 BP*, crop yields in many areas of southern Mesopotamia were declining, possibly linked in part with periodic episodes of salinity. The latter may have been due to over-irrigation as people in the densely populated Tigris and Euphrates Valleys struggled to cope with the drier climate. One response mentioned in numerous written records is a switch from wheat to barley as the principal crop staple, especially in Sumer. Evidently Mesopotamian farmers knew that barley was more tolerant of saline soils. The crisis in wheat cultivation continued long after the climate returned to a more favourable state, and by *4400 BP* matters were especially severe in the southern

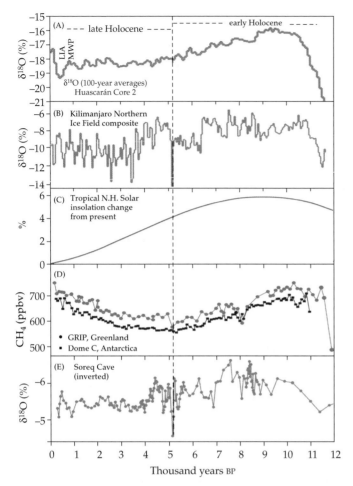

Figure 10.6 The *5200 BP* climatic event. At about *5200 BP*, there was a second major Holocene climatic perturbation that affected much of the world. The event coincided with major cultural discontinuities in Near Eastern civilizations, especially in the rainfed farming regions of Northern Mesopotamia where many towns were abruptly abandoned. The graphs show data from around the world, but the *5200 BP* event (vertical dotted line) is particularly prominent in Africa (Kilimanjaro) and the Near East (Soreq). Graphs show $\delta^{18}O$ ice-core records from (A) Huascarán (Peru); and (B) Kilimanjaro (Africa); (C) Northern hemisphere (N.H.) insolation change from present values; (D) methane records from Greenland and Antarctica; (E) is a $\delta^{18}O$ record from the Soreq Cave (Israel). LIA, Little Ice Age; MWP, Medieval Warm Period; GRIP, Greenland Ice Core Project; ppbv, parts per billion by volume (data from Thompson *et al.* 2006).

Sumerian heartland that included the rapidly developing urban centres of Eridu, Uruk, and Ur.

The prolongation of salinity problems, as originally precipitated by the *5200 BP* event, was probably at least partially due to the increasingly unsustainable forms of agricultural intensification required to support emerging city-states in the region. It has been suggested that the *5200 BP* event had a key impact on the urbanization and expansion processes then underway in the late Uruk cultural period.[712] For example, as we will discuss later, this climatic change may have affected societal processes including: the final Uruk Period of urbanization in southern Mesopotamia;[713] the expansion of late-Ubaid/early-Uruk settlements in northern Mesopotamia;[714] and the establishment and collapse of Uruk 'colonies' in the far north of

the region.[715] Radiocarbon dating reveals a series of demographic, economic, and political crises during the late Uruk Period across the Habur (Khabur) and Assyrian Plains of northern Mesopotamia.[716] These crises either extinguished or severely checked the growth of many settlements in the region for as long as 400 years.[717] The drought probably also affected the Anatolian and Iranian Plateau regions.[718] Despite its adverse effects on crop production, and the retardation of technological innovation for several centuries, in the longer term, the *5200 BP* event may have accelerated development of more intensive agriculture, and more complex, highly stratified urban societies, in southern Mesopotamia.

Following the return of more reliable rainfall, recovery in some regions was swift and new cities sprang up throughout the Sumerian heartland of southern Mesopotamia. The invention of writing a few centuries before enabled the keeping of systematic records, especially relating to collection and distribution of the vast, new urban grain stores.[719] In the cities, the first self-declared kings arose during the Early Dynastic Period of *4900* to *4334 BP*. Thanks to plentiful cereal yields from the intensively worked farmland, the elites of the city-states were able to mobilize and feed large pools of labourers, who in turn worked on complex and elaborate irrigation projects to extend further the area of cropland and ensure its continued productivity. The cities of Ur, Kish, and Uruk vied with each other for control of farmland and its population. Eventually, in about *4334 BP*, King Sargon of Kish emerged as pre-eminent ruler of a vast region extending far beyond Sumer, from the Persian Gulf to the Taurus Mountains of Anatolia. The new dynasty was called Akkadian, after the newly built capital city of Akkad (or Agade) that Sargon established near the Euphrates about 40 km north-west of Kish. The Akkadians were particularly interested in seizing control of the rainfed farming areas of northern Mesopotamia, ushering in the first experiment in large-scale imperial agriculture. However, this first empire in human history did not endure for long. Despite the vigorous activities of his grandson, Naram-Sin, who briefly extended the boundaries of the Akkadian realm, Sargon's empire and dynasty both collapsed abruptly around *4193 BP*. This disaster coincided with a third major climatic event that would have widespread and enduring effects well beyond the Near East.

The climatic event of *c. 4200 BP*

This third major Holocene climatic event lasted for about 400 years and was probably a global phenomenon (Figure 10.7). While it significantly affected Mesopotamian cultures, its effects in locations as far flung as the Indus Valley and northwest China were even more profound. For example, as we will see in the next chapter, the *4200 BP* event was probably responsible for the permanent extinction of the highly developed network of cities, towns, and productive farmland that made up the Harappan civilization of the Indus Valley.

Climatic records show similar aridification events throughout the Americas, from the Great Salt Lake in Utah to Belize in Mesoamerica and Peru to the south.[720] More precise dating techniques, such as dendrochronology and ice-core analysis, have enabled investigators to track near-simultaneous climate-related phenomena starting about *4290 BP*, lasting for 250 to 300 years, and characterized by high aridity.[721] The mechanism(s) of these global events remain to be determined conclusively, but there is emerging evidence of a link with periodic oscillations in the earth's orbit and solar activity. Such phenomena can trigger changes in oceanic currents and the North Atlantic thermohaline circulation.[722]

In the Near East, the *4200 BP* event was marked by a sudden increase in aridity. This was mirrored by falling lake levels, a sharp decline of large woodland species such as oak, and their replacement by smaller plants such as the burnet shrub, *Sanguisorba minor*, which is normally a denizen of particularly arid zones.[723] Average rainfall in locations such as the Levant declined by 20 to 30% causing a sharp drop in the level of the Dead Sea.[724] The increasing aridification was also seen in greatly elevated amounts of dust, possibly caused by localized loss of arable topsoils or even more serious desertification across the region. We will examine the social effects of the *4200 BP* event in more detail later in this chapter, but they can be briefly summarized as follows.[725] Within about a decade of the sudden drop in rainfall, farmlands in the Habur Plains of north-east Syria, such as the rich city of Leilan and its dependent villages, were abandoned.[726] In the next few decades, huge regions of northern Mesopotamia became depopulated, from the Upper Euphrates Valley in the west to the Assyrian Plains to the east. Farming ceased over much of the region as people either emigrated or adopted pastoral lifestyles. Large numbers of dispossessed farmers and city dwellers attempted to move to those remaining areas where agriculture was still possible.

The population of southern Mesopotamia suddenly doubled during the immediate post-Akkadian Period, almost certainly due to immigration from the arid north.[727] These people were coming from former rainfed farming areas

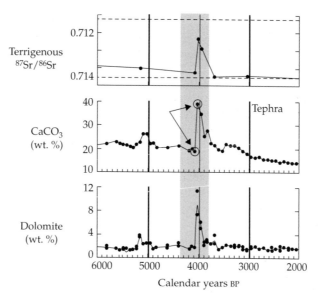

Figure 10.7 The *4200 BP* climatic event. The third and most serious, major climatic episode of the Holocene occurred soon after *4200 BP* and had profound societal effects in the Americas, Eurasia, Africa, and China. The subsequent long-term drought is associated with regression or collapse of advanced societies in Mesopotamia, the Nile and Indus Valleys, and Northern China. The three panels show proxy records of Mesopotamian aridity reconstructed from wind-borne sediments of strontium, calcium carbonate, and dolomite in a deep-sea core from the Gulf of Oman. The shaded grey area highlights a drought period of about 300 years, centred on *c. 4100 BP* that coincided with the Akkadian collapse. Data adapted from deMenocal, 2001; available from NOAA (http://www.ncdc.noaa.gov).

where cereal agriculture was no longer possible. They were seeking out the alluvial soils of the south where the extensive irrigation works still made it feasible to grow crops despite the much-reduced rainfall. Even pastoralists tried to migrate southwards as the remaining ground cover in northern and western Mesopotamia became insufficient to feed their animals. The successor culture to the short-lived Akkadian hegemony occupied what is known as the Ur III Period, which lasted about a century. Despite their best efforts, especially in bureaucratic organization, the Ur III dynasty eventually fell victim to the long-term effects of the aridification that had started almost two centuries before. By *4004 BP*, the city of Ur lay in ruins together with much of the intensive agrourban system so laboriously constructed over the preceding two millennia of Sumerian civilization. In the remainder of this chapter, we will look at these events in more detail in the context of the latest findings from palaeoresearch using both physical and biological approaches.

Establishment and spread of farming: *11,000* to *8200 BP*

Beginnings—from Abu Hureyra to Çatalhöyük

At the end of Chapter 3, we left the early Abu Hureyra community as its people adapted to the impoverishment of their food supply during the Younger Dryas by developing the first recorded example of agriculture. By *11,000 BP*, their diet had been reduced to those few grain-bearing plants that were still able to grow under the prevailing cool, arid conditions. They collected everdwindling amounts of wild grasses such as clubrush (*Scirpus maritimus/ tuberosus*) and Euphrates knotgrass (*Polygonum corrigioloides*) throughout this period. But their real salvation came in the form of a larger-seeded version of rye that only survived because the people learned how to actively cultivate it, rather than simply allowing it to grow wild and then collecting it. Even after all the other food grains had died out, this new form of cultivated rye was still available. Lentils may have also been cultivated during this time. This small village managed to eke out an existence by cultivating mainly rye for several centuries more, before it was largely (but not totally) deserted from *11,000* to *10,500 BP*. The causes of this depopulation of Abu Hureyra are unknown, but may be related to the harsh climate, impoverished diet, and/or to emigration and a resumption of a more mobile lifestyle.

The small residual population at Abu Hureyra continued to cultivate rye and lentils. They also acquired (probably from neighbouring Levantine groups) other cultivated cereals, including emmer, einkorn, and breadwheat, as well as barley. By now, all of the original wild cereal species that had once been such an important food resource in the area had completely died out. These wild cereals never returned and therefore the people at Abu Hureyra were obliged to continue the cultivation of their newly domesticated, large-grained cereals, even after the sudden climatic improvement at the end of the Younger Dryas. Because they were now domesticated, the remaining cereals at Abu Hureyra were no longer adapted to grow in the wild. Any attempt to sow these seeds without careful cultivation, including weeding and protection from pests and herbivores, would have resulted in loss of the crop. So, unless they were prepared to eschew their settled life and resume foraging, the villagers were now committed to remaining as farmers. Many people left the village, but luckily for the few who stayed, the return of warmer, wetter conditions suddenly transformed the yields of their crops, and agriculture once again became a highly adaptive practice for food production.

Rye was grown as a major crop until it was overtaken by domesticated forms of einkorn and emmer wheats some time after *10,500 BP*. Legumes, such as lentils, chickpeas and field beans were cultivated at Abu Hureyra from this period and would have greatly improved the nutritional quality of the diet. By *9600 BP*, the collection of wild plants had virtually ceased at Abu Hureyra and the population was almost completely dependent on domesticated species for plant-derived foodstuffs. But this had been no overnight revolution. The transition from entirely wild, collected plants to entirely cultivated plants took over 2500 years at Abu Hureyra, with many setbacks along the way. One of the main advantages of the new crop-based economy was its potential (given the right climatic conditions) for far greater *per capita* food output than hunter-gathering. This in turn provided the impetus for the suite of technological and cultural changes that eventually enabled the development of increasingly complex and stratified societies and, by *5000* to *4000 BP*, of large-scale urbanized cultures across much of Mesopotamia.[728]

Some of the first steps towards urbanization were made at Abu Hureyra soon after *10,500 BP*. Its increasing agricultural output enabled the population to expand and devote more time to constructing what amounted to a modest town of several thousand inhabitants. Instead of the former small-scale, somewhat disorganized village of simple reed huts, a new brick-built town emerged. The new town shows some evidence of centralized planning, with rectilinear mud-brick houses aligned approximately north-west by south-east along narrow lanes interspersed by open courts. This layout was maintained for long periods, with out-of-use buildings being replaced in the same position by similar structures to preserve the overall organization. By *9500 BP*, the town was already about 8 ha in area and supported 2500 to 3000 people (Figure 10.8). This was still the prepottery stage in the region and farming was supplemented by seasonal hunting of the abundant herds of gazelle, which had returned to the Upper Euphrates as the climate improved. During the early Neolithic, Abu Hureyra farmers were comparatively healthy compared to their rye-dependent ancestors of the Younger Dryas. They were also distinctly better off than their more urbanized and dietarily impoverished descendents in the state-farming epoch that was yet to come.

The people of Abu Hureyra were relatively short, with a mean adult height of about 155 cm for females (5 ft 1 in) and 162 cm for males (5 ft 4 in); they were also rather gracile with only a modest degree of sexual dimorphism. Their size reduction

Figure 10.8 Artist's impression of Abu Hureyra *c. 9500 BP*. This imaginative reconstruction depicts the settlement as a medium-sized town of some 2500 to 3000 people, more than two millennia after the emergence of the first cultivated food plants. Much of the surrounding land in the Upper Euphrates Valley has been cleared for rainfed cultivation of emmer, barley, and legumes. At this time, pottery had not been introduced in the region, but the emerging societal complexity is evident from the well laid out urban and farming landscape. Adapted from Moore *et al.* (2000).

was probably associated with a greater dietary reliance on cereals and legumes, instead of animal flesh. Decreased sexual dimorphism compared to hunter–gatherers may be due to an increased female workload, more equal access to foodstuffs (which were largely prepared by women), and reduced nutrition for both sexes. These trends, and especially the more equal access to food resources, favoured the evolution of more similarly sized adults. Skeletal samples show comparable evidence

of general wear and tear between males and females, and of more serious pathologies, as seen in similar, traditional farming cultures of recent times. One difference is the excessive tooth wear, leading in many cases to complete loss of teeth by midadulthood. The high incidence of chipped and severely worn teeth, even in younger adults, is indicative of a very hard, gritty diet. Factors responsible for excessive tooth wear include a diet of larger, coarser, domesticated cereal grains and methods of cereal preparation.[729] The latter involved repeated grinding of grains with heavy, coarse stone tools that left numerous fine, hard stone chippings in the flour. Repeated chewing of such abrasive food resulted in extensive tooth wear that could leave a person in chronic pain, and eventually toothless before the age of 30.

Soon after *9000 BP*, Abu Hureyra entered its most prosperous stage as it grew to 5000 to 6000 people and occupying 16 ha. Several innovations were made at this time. Barley became a more prominent crop, along with einkorn and breadwheat, supplemented by several domesticated legumes. On the animal side, gazelle hunting was almost totally replaced by the keeping of (presumably domesticated) sheep and goats. Towards the latter part of this high point in the evolution of Abu Hureyra, a further innovation appeared that was to transform the quality of life of its inhabitants. This was the making of pottery ware, which for the first time allowed food to be soaked and cooked in water before eating. Cooking of cereal grains to make porridge is an alternative to grinding them to flour. The effect was to markedly reduce the amount of tooth wear in the population as a whole.[730] Another indirect consequence was an increase in female fertility. Prior to this time, mothers had been obliged by the lack of soft food to breastfeed children until the age of 4 or more. Tooth-wear patterns suggest that 4-year olds were then weaned onto a diet of prechewed food before assuming an adult diet. Thanks to pottery, mothers could wean children at a much earlier age by giving them a diet of very soft, cooked cereal mush. Having weaned an infant, a woman would then be able to bear and suckle another child, and female fertility soon rose sharply at Abu Hureyra.[731]

The final phase of occupation at Abu Hureyra occurred around *8200 BP* and witnessed a decline in the size of the settlement to about 3000 people. The inhabited area also contracted but the community was still productive and adopted innovations such as keeping of cattle and pigs to supplement existing livestock resources. One reason for the decline in population may have been a reduction in the carrying capacity of the basic agricultural system, namely cereal farming. The timing of this setback in the development of what had been a thriving, vigorous, and innovative community is tantalizingly close to the recently discovered climatic event of *c. 8200 BP* in which the majority of prepottery farming villages in the Levant, and many in northern Mesopotamia, were abandoned (see above). For 400 years, rainfall decreased significantly, which would have first reduced cereal yields and perhaps pushed the population of Abu Hureyra into livestock grazing as a partial alternative. This may have sufficed for a few centuries but eventually the rains may have been insufficient either for crops or pasture. Whatever the sequence of events, this unique settlement that had endured for over 3000 years; that had lived through the worst of the Younger Dryas; that had perhaps been the first community to domesticate a crop; and that was one of the earliest adopters of pottery; was then abandoned for good. Abu Hureyra was never resettled. Soon after the archaeological excavation of the site in the 1970s, the immense Taqba Dam was constructed across the Euphrates, and today the entire settlement, with its remaining secrets, is submerged at the bottom of the new Lake Assad.

Elites, cities, and irrigation: *8000* to *5200 BP*

The cultural development of Near Eastern societies from *8000* to *5000 BP* is often depicted in terms of regional pottery styles.[732] These descriptions have their limitations, with such cultures often overlapping both in time and space, but with this caveat in mind, we will follow a similar terminology here. To summarize, the Hassunians were the first distinct pottery culture, occupying part of modern Kurdistan. They overlapped with the Samarrans, who migrated southwards and probably invented irrigation agriculture when they entered the more arid parts of central and southern Mesopotamia. During the millennium from *7500 BP*, the Hassunian culture in the north was gradually supplanted by that of the Halafians. Meanwhile the Samarrans and related cultures, using increasingly intensive forms of irrigation agriculture, colonized the south as far as the Persian Gulf, where they gave way to the more urbanized Ubaid culture, from *7300* to *5500 BP*, which then spread to northern Mesopotamia. Finally, the Uruk culture emerged in

Sumer about *6000 BP*, and with it the first true city-states, such as Eridu, Uruk, and Ur.[733] The peoples that made up these cultural groups were linked by linguistic and cultural affinities. The two main language groups were Sumerian in the far south, and Semitic languages such as Akkadian, which became the *lingua franca* in much of Mesopotamia. One interesting aspect of Mesopotamian society is the extent of intermingling and cultural heterogeneity that occurred after urbanization. Most preurban communities appear to have been fairly homogeneous in their ethnicity and language before *c.* 5500 *BP*, but after that time there was extensive cultural and racial mixing as mixed populations moved to the growing urban centres.[734] This phenomenon is probably linked with the expansion of southern Mesopotamian influence during the Ubaid and Uruk Periods, and with the *5200 BP* climatic event that marked their end.

The Hassunians

The Hassunian culture lasted from about *8000* to *7250 BP* and was the first coherent, farming-based society established in northern Mesopotamia. The Hassunians originated in the Assyrian highlands and the eponymous settlement of Tell Hassuna is just south of the modern city of Mosul. Their distinctive pottery, with its reddish linear designs, can be found throughout the region. The population tended to live in small villages of 1 to 3 ha and few, if any, had more than 500 inhabitants. The economy was based on rainfed barley and emmer wheat cultivation, supplemented by sheep, cattle, and goats. Apart from pottery, an important technological innovation used by Hassunians was a system of permanent recording of data. This was done using wax or clay seals stamped with symbols to indicate the nature, provenance, and perhaps ownership of the contents of a container.[735] The seal could be used to cover a jar or other vessel to prevent tampering. Even doorways could be sealed and labelled to mark entire rooms or chambers, such as grain stores or burial sites.

This innovation, together with the development of pottery containers, opened the way for the traceable communal (rather than individual) storage of all kinds of items, including individually owned

caches of grain and tools, in more secure, purpose-built central storehouses. It also facilitated trade over longer distances, as goods could be sealed and recorded on dispatch, in transit, and on arrival. The Hassunians were neither the earliest nor the largest cereals-based culture in the wider Near East. Even the largest Hassunian villages of *7500 BP* were much smaller than the Levantine prepottery town of Jericho had been over 1000 years previously. And, a few hundred kilometres to the north, the enigmatic Anatolian city of Çatalhöyük (which had already been abandoned for 400 years) was over 1500 years older, and at its height had been 20-times larger than any Hassunian settlement (Figure 10.9).[736] Both Hassunians and Samarrans overlapped with the Halafians, who were based further west in northern Syria. Eventually the Hassunian pottery style was superseded by the Halafian, which then continued for a further millennium. There is no evidence of any discontinuity in the sorts of agriculture practiced by the Hassunians and Halafians. Both cultures practiced rainfed farming using emmer and two-row barley, plus a few pulses such as lentils, and fibre crops such as flax.

Halafian culture

The Halafian culture used new pottery styles and burial practices that may have been introduced into northern Mesopotamia by migrants from outside the region. The site of Tell Halaf lies in northern Syria, close to the modern border with Turkey. This rainfed farming settlement flourished from about *7050* to *6300 BP* and similar pottery has been discovered throughout the wider region from Anatolia to northern Mesopotamia. This region received sufficiently reliable rainfall for rainfed, or dryland, farming. Although not as productive in terms of crop yield per hectare as irrigation farming, rainfed cultivation is a lot easier to manage and has lower labour costs. From *7000* to *5200 BP*, numerous, small city-states were established across the Halafian region, with the beginnings of greater social complexity and stratification than the Hassunians. For example in a storehouse at Arpachiyeh (near Nineveh) there were samples of jewellery, sculpture, and high-quality pottery, which suggested

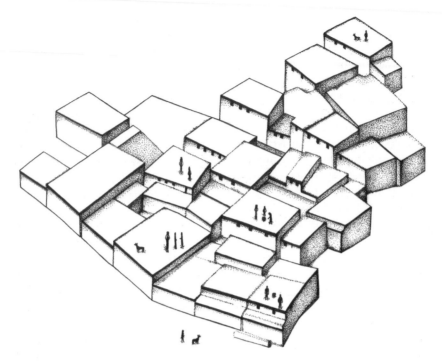

Figure 10.9 Artist's impression of Çatalhöyük, *9400–8000 BP*. This enigmatic Central Anatolian town housed as many as 3000–8000 people for about 1400 years, at a time when there was very little large-scale human settlement outside the core agricultural heartland of the Levant and Northern Mesopotamia. The town was probably an important trading centre, and its dependence on cultivated cereals is attested to by the prevalence of substantial storage bins in many buildings. However, unlike other early urban communities, much of the land around Çatalhöyük was unsuitable for either farming or wild plant collection, so food may have been imported from more distant grain-growing regions. Adapted from Hodder (2006).

that local chiefs were able to appropriate much of the surplus wealth of the community, possibly by controlling the communal storehouses that held the central grain supply.[737]

The Samarrans

From about *8000* to *6800 BP*, the Samarrans gradually occupied a region of central Mesopotamia that extended beyond the limit of reliable rainfed farming. The middle and lower reaches of the Tigris and Euphrates Basins are very low lying (rising to only 30 m above sea level as much as 500 km inland) and flood prone. This endowed the region with a profusion of rich alluvial soils that would be ideal for crop cultivation, were it not for the erratic rainfall patterns and the fact that these valleys lie below the 200 mm isohyet regarded as the practical limit for rainfed farming (see Box 2.2). Hence these

otherwise arid river valleys were pregnant with agricultural potential, if only the Samarran farmers could discover a mechanism to bring water to the latently fertile land. The Samarrans would have already been familiar with the need to keep plants watered on a small scale, for example in herb gardens. What was needed in the more arid environment of the south was a larger-scale method of irrigation that could canalize entire rivers and bring water to hundreds of hectares of farmland. This need precipitated social changes that emerged gradually over the succeeding centuries as irrigation agriculture continued to develop in its ambition and technological complexity.[738]

The earliest Samarran farmers adapted in two ways to their more arid conditions—first by growing new types of drought-tolerant grains, and then by developing the first medium-to-large scale irrigation systems ever seen. By *8000* to *7500 BP*, farmers at Choga Mami and Ali Kosh in the Zagros foothills were building canals as much as 5 km in length.[739] This is indicative of a much closer form of social organization than was necessary in earlier, rainfed farming cultures, but we have no evidence of how this was manifest, for example whether there was an early development of resource-controlling elites. As its potential for food production became apparent, canal building for crop irrigation gradually spread throughout the region.[740] To the north-west of the Choga Mami site, in the relatively arid Mandali and Deh Loran Plains at the base of the Zagros Mountains, almost all of the settlements from Samarran times are close to ancient canals.

The Samarrans then moved into the lowlands of southern Mesopotamia, laying the foundations of the later Ubaid culture that went on to produce the Sumerian civilization of Uruk, Ur, and Akkad. Canal building started on a relatively small scale in central/south Mesopotamia, gradually becoming more complex and ambitious as sufficient surplus labour, and the organizational ability to control and manage it, slowly evolved during the Ubaid and Uruk Periods.[741] Most of the canals of this period were more like large ditches, and were not built as permanent features. The practice of irrigation may have favoured mutations in the crops that enabled them to grow faster and yield more under the new conditions. In this regard, it is interesting that the six-row form of barley made its appearance at roughly the same time as crop irrigation was practiced on a wide scale.[742] While there is no evidence that the farmers using irrigation consciously selected six-row barley, plants with six-row mutations would have been much more likely to survive under such conditions. This may be an early example of inadvertent genetic engineering by farmers seeking to control and improve their agricultural environment.

Ubaid culture

The Ubaid culture arose about *7900 BP* and completed the earlier Samarran farming settlement of southern Mesopotamia.[743] This culture originated from a society firmly based on irrigation agriculture and used a new pottery style with concentric and wave-like decorations. The influence of Ubaid material culture, and especially pottery, was felt

right across Mesopotamia and the Levant, although this did not involve direct political intervention beyond southern Mesopotamia, until the Uruk expansion into the north after *5500 BP*. The combination of the increasing use of more elaborate irrigation networks, and more efficient harnessing of the spring flooding of the Euphrates, enabled Ubaid populations to greatly improve crop yields, and thereby to increase their size and geographical range. Two important technological innovations made by the Ubaid people were the invention of the plough and the potter's wheel, possibly as early as *6500 BP*. The plough facilitated crop production by making it possible to cultivate much larger areas per labourer, although the actual yield per hectare was generally lower. Therefore ploughing was more suitable for larger scale, state-controlled systems of irrigation farming that were evolving in Sumer at this time, where field size was not as much a limitation as was the supply of manpower.

The advent of the potter's wheel marked another stage in the reduction of individual autonomy in Ubaid society. The transition from the late-Ubaid to early-Uruk Periods is characterized by the gradual replacement of domestically produced, painted pottery using a slow wheel. Instead, people no longer made their own pots but rather used more uniform types of unpainted pottery that were mass produced by specialists in the emerging urban centres. This entailed a sacrifice of the aesthetic aspects of pottery design and decoration in favour of pragmatic, utilitarian demands. One reason for this cultural shift, which may have affected other aspects of life as well, was the increasingly pressing need for all available labour to be mobilized to maintain the vital irrigation networks and to tend and harvest the crops, as well as working on the vast civil engineering projects such as construction of granaries, palaces, temples, fortifications, etc. The need for regional-scale organization and strategic decision making led to the evolution of ever-larger urban units. These cities were centres for the large-scale storage and processing of the precious grain crops that were transported from the field to be stored in heavily protected citadel-like granaries. The same Ubaid cities carried out the centralized, state-controlled, mass-production of pottery, tools, and weapons.

While they started in the Ubaid Period, these cultural and societal trends were greatly reinforced and accelerated during the later Uruk Periods, eventually being exported to the new urban centres developing right across the Fertile Crescent. It is likely that the emergence of chiefs and kings during this period was at least partially a response to the need for efficient administration and executive management of the complex social units that were required to exploit the immense agricultural potential of southern Mesopotamia.[744] For many millennia, basic social units within large villages and town had already progressed beyond mere clan and kinship bonds, but most communities prior to *6500 BP* remained relatively egalitarian, with communal ventures being on the basis of mutual self interest rather than coercion. However, the new Ubaid, and later Uruk, communities were both larger and more regimented places than anything that had existed before. There was no sudden appearance of the trappings of elite power and royalty in these communities. Rather, there was a gradual emergence, during this relatively stable social and climatic period, of increasing inequality and more stratified social systems. Some people, such as canal-digging peasant-labourers, had a manifestly worse deal than others, such as grain-store administrators or skilled artisans. Nevertheless, it was in the wider interests of the society as a whole that everybody more or less accepted their unequal lot in life; the alternative being the ever-present danger of serious social disruption that could endanger the welfare of the entire community.

This led to the evolution of ideologies, such as specific forms of religion, which both explained and justified the new *status quo*. Such ideologies can therefore be considered as adaptive responses that enhanced the competitive potential of the community as a whole, despite often reducing the individual fitness and quality of life for many of its members. One by-product of the new social organizations of the Ubaid Period was the concentration of power in the hands of a single person, the king. The Sumerian word for king, *lugal*, means 'great man' and was originally given to any local ruler. Initially at least, the king was a sort of chief bureaucrat who ensured crops were planted, tended, harvested, stored, and that grain was distributed to

all (although not necessarily equally). This early description of kingship seems to have endured throughout the Ubaid and early Uruk Periods, as summarized by the rubric: '(he) who administered in the god's name the large farm that (is) the city-state'.[745] This neatly encapsulates the central position of agriculture at the core of much of Mesopotamian society. Although the mighty cities and associated artefacts are nowadays thought of as crowning achievements, to the ancient Mesopotamians themselves the city was, at least initially, simply a convenient mechanism to organize and direct the 'state farm'. It was the latter that was the real focus of these city-states, and their kings were simply the head managers of the farm.

The early Uruk Period

The early Uruk Period extended from about *6300* to *5100 BP* and is named after the city of Uruk (also called Warka), which lies on the Lower Euphrates immediately to the north-west of its two great rival cities, Eridu and Ur.[746] It is thought that the land of Iraq derives its name from Uruk/Warka, called Erech in the Hebrew Bible.[747] This was a period of increasing urbanization in southern Mesopotamia, marked by what would later be seen as a momentous social achievement, namely the invention of writing (see Box 10.2 and 10.3). Writing did not emerge overnight, but evolved over several millennia. Several Near Eastern cultures had kept records of stored or traded crops since as early as *10,000 BP*, but this involved the use of simple systems of marked labels or tokens. The earliest tokens appeared about *10,000 BP* at the same time as crop domestication at several sites, including Mureybet and Tell Aswad.[748] Later on, large numbers of shaped clay tokens dating from *8500 BP* were found at the prepottery Neolithic village site of Ain Ghazal in the Jordan Valley.[749] Gradually, the markings became more elaborate and developed into a mixed abstract–pictographic system able to represent not only concrete objects like stored barley but also more abstruse concepts such as divinity, law, and kingship.[750]

By *5400 BP*, the Uruk culture had invented a durable system of clay tablets, which were initially used almost exclusively to record transactions

Table 10.3 Late Chalcolithic–Iron Age chronology in Mesopotamia, *5500* to *2500* BP*

Date, BP	Period/culture	Archaeological sites	Climate	Major cropping system	New technologies	Social systems
5500	Late Chalcolithic Middle Uruk	Tell Hamoukar, Tell Leilan, Brak	Warm and moist	High yield EPEB crops, more use of BARLEY, SGCP livestock	Temples and places, logographic‡ texts, clay tablets	Hyperurbanization in Sumer, colonization of north
5200	Late Uruk	Qraya, Habuba Kabira, Tell Leilan	**5200 Drought** End of Holocene climatic optimum	Reduced EPEB crops Increased pastoralism SGCP livestock	Simplified in north, intensified in south	Collapse of agrourbanism in north, continuity in south
5000	Iranian Plateau	Khuzistan sites Susa Elam	Warm and moist	High yield EPEB crops, SGCP livestock	Metal working, mineral resources, trade entrepôts	Uruk-influenced increase in social complexity and urbanization
5000	Early dynastic Sumer and Akkad	Sumer and Akkad	Warm and moist	High yield EPEB crops SGCP livestock, Huge irrigation projects	More complex texts, increased bureaucratic efficiency	Spread of walled cities ruled by kings, 80–90% urbanization, corvée system
4700	Early dynastic Sumer and Akkad	Shuruppak III Abu Salabihk Ur	Warm and moist	High yield EPEB crops SGCP livestock, Huge irrigation projects	Urban-based centralized industrial production	Population of Uruk reaches 50,000, temples dominate agroeconomy
4500	Early dynastic Sumer and Akkad Early Bronze	Shuruppak III Eresh, Kish	Warm and moist	High yield EPEB crops SGCP livestock, Huge irrigation projects	Qanat irrigation system	Warring city-states vie for resources, libraries at Shurrupak and Eresh,
4350	Early Bronze	Uruk, Kish, Lagash	Warm and moist	High yield EPEB crops, SGCP livestock, Huge irrigation projects	Urban-based centralized industrial production	Lugalzagesi, king of Uruk, briefly unites Sumer and Akkad
4340–4180	Akkadian dynasty	Tell Leilan, Nineveh, Mozan, Brak	Warm and moist, then increasing drought	High yield EPEB crops but focus on tribute-barley in north and south	Imposition of Akkadian technologies on North Mesopotamia	Sargon of Akkad conquers Sumer
4180–4112	Interregnum	Beidar, Tel Leilan, Mozan, Brak, Ebla	**4200 Drought** Cool and dry	Partial collapse of rainfed farming, Irrigation survives	Regression in north, Continuity and innovation in south	End of Akkadian empire, Collapse of northern cities Mass-migration to south
4112	Ur III	Ur, Uruk, Eridu, Nippur	Cool and dry	High yield EPEB crops, but much more focus on barley	Huge investment in massive defence and irrigation projects	Foundation of Ur III dynasty by Ur–Nammu, increase in bureaucratic control
4004	Ur III	Ur, Uruk, Eridu, Nippur, Lagash, Larsa	Cool and dry	Decreasing yield of all crops and trees, virtual barley monoculture	Collapse of some city-states but little loss of technological momentum	Loss of food supply, societal disruption, invasion of nomads, collapse of Ur III
4000	Middle Bronze North Mesopotamia	Tell Leilan Tell Brak Mozan	Cool and dry	Pastoralism Isolated rainfed farming	Technological simplification, loss of trade links	Cultural simplification, Cities and towns abandoned, extensive depopulation
3900	South Mesopotamia	Larsa, Isin, Lagash, Babylon	Warmer and wetter	High yield EPEB crops, SGCP livestock	Rapid agrourban recovery based on smaller, more sustainable systems	Warfare between small city-states (Larsa, Eshnunna, Isin), emergence of Babylon

Table 10.3 (continued)

Date, BP	Period/culture	Archaeological sites	Climate	Major cropping system	New technologies	Social systems
3800	North Mesopotamia	Mari, Alalkh, Ebla, Tell Leilan	Warm and moist	Rainfed cultivation of high yield EPEB crops, SGCP livestock,	Slower recovery as rainfed farming resumes and trade links re-established	Start of recovery in north, new city-states, Shamshi-Adad founds first local empire
3850	South Mesopotamia	Babylon, Nippur,	Warm and moist	High yield EPEB crops, sesame, dates, vines, SGCP livestock	First texts on practical agronomy and lawmaking, new new irrigation schemes	Hammurabi founds Early Babylonian Empire, revival of bureaucracy and learning
3600	Late Bronze North Mesopotamia	Mari, Alalkh, Ebla, Tell Leilan	Warm and moist	Rainfed cultivation of high yield EPEB crops, SGCP livestock,	Widespread trade, complex bronze tools	Northern recovery accelerates with many competing city-states, some ruling to the Levant
3400	North Mesopotamia	Assur, Nineveh	Warm and moist	High yield EPEB crops sesame, dates, vines, SGCP livestock	Improved transport and trade	Rise of early Assyrian Empire under Ashur-uballit I, reunification of Mesopotamia
3200	Iron Age North and South Mesopotamia	Babylon, Nineveh	Warm and moist	High yield EPEB crops, sesame, dates, vines, SGCP livestock	Iron-tipped ploughs and seeders improve crop yields	Tiglath-pileser I in Assyria and Nebuchadnezzar I in Babylon rule mature urban empires
2934–2539	Iron Age North and South Mesopotamia	Sippar, Nineveh	Warm and moist	High yield EPEB crops, sesame, dates, vines, SGCP livestock	Botanical gardens, libraries, herbals, rise of private owners of farmland	Neo Assyrian and Neo Babylonian Empires with interests in agrobotany
2539–2000	North and South Mesopotamia		Warm and moist	High yield EPEB crops, sesame, dates, vines, SGCP livestock	Greek scientific period, many herbals, first accounts of systematic botany	Persian and Hellenistic periods, much dissemination of cultural and technical innovations
2000	North and South Mesopotamia		Warm and moist	High yield EPEB crops, sesame, dates, vines, SGCP livestock	Much rewriting of Greek works, little new knowledge created	Roman imperium

The major highlights of this period are the pervasive trends towards societal complexity (despite much warfare and upheaval), first in the southern Mesopotamian regions of intensive irrigation farming and then in less productive rainfed regions of the north. These trends were interrupted twice, most seriously in the north, by long-term droughts c. 5200 and 4200 BP. In both cases, recovery was aided by the eventual resumption of trade and cultural links with the less-affected south. Interestingly, it appears that Sumer may have over-urbanized immediately before the crisis of 4200 BP, and subsequent development occurred at a more sustainable level with smaller cities and a larger rural population.

* Sources include Nissen (1990); Pollock (1999); Akkermans and Schwartz (2003); Yoffee (2005); plus many primary sources as cited in the Notes.

† Cuneiform is an example of a logographic written text. Logographs, or ideograms, represent ideas directly rather than words and are non-alphabetic. Invented by the Sumerians, cuneiform was used for 3000 years by cultures such as the Akkadians, Babylonians, Hittites, and Assyrians. After 2700 BP, it was gradually replaced in Mesopotamia by the more versatile Aramaic alphabetic script.

EPEB, emmer, pulses, einkorn and barley as staple crop package; SGCP, domesticated sheep, goat, cattle, and/or pigs

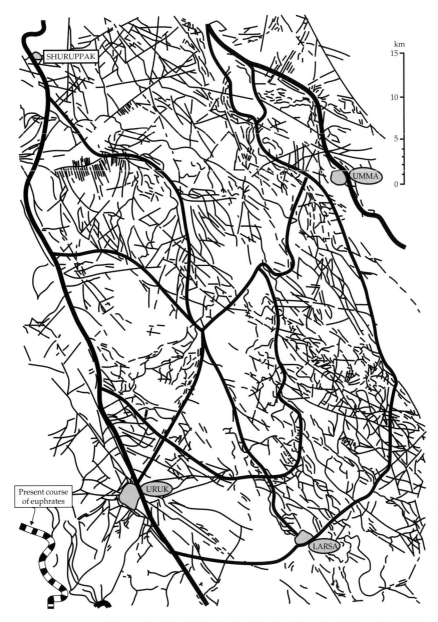

Figure 10.10 Irrigation systems around the city of Uruk, *c. 4400 BP*. Ancient Sumerian cities such as Uruk were invariably located on sites adjacent to major waterways such as the Euphrates and Tigris rivers or principal tributaries thereof. The immediate region around a city such as Uruk was capable of immense productivity providing it was adequately irrigated. Over the past few millennia, however, the rivers courses have changed, and the ruins of once wealthy cities such as Uruk and Larsa are now stranded incongruously in an arid, inhospitable desert landscape. Luckily, many traces can still be discerned of the intricate agrourban irrigation networks that once brought forth such a bounteous harvest to feed these formidable cities, and fuelled the extension of their influence over many hundreds of kilometres. This drawing gives an impression of the intricate network of major waterways (thick lines) and canals (thin lines) that were excavated and maintained at such immense cost by the inhabitants of the city-states of Sumer. Only a small proportion of the waterways have been identified to date and most of the gaps (e.g. to the east and north of Uruk) are due to the presence of seasonal swamps or dunes that have prevented their identification. In some areas, such as south of Shuruppak, a regular series of parallel canals can be readily discerned while in other areas the canals follow the convoluted meanders of the alluvial terrain. The drawing is based on a series of maps and textual data from Adams and Nissen (1972), Nissen (1988), and Roaf (1990).

Figure 10.11 The agricultural landscape of southern Mesopotamia. An imaginative reconstruction of the landscape of Sumer and Akkad at the height of the early flowering of complex agrourban cultures during the fifth millennium *BP*. The landscape was dominated by an all-pervasive network of waterways, canals, and irrigation ditches maintained at vast expense by the labour of much of the male population (bottom centre). Others tended the crops until harvest time when much of the population was mobilized to collect (left) and transport the grain to barges (top left) for transportation to the massive granaries of the local city, whose towering ziggurat would be visible on the far horizon of the flat plains (top right). Virtually the whole population lived in such cities, with labourers venturing out daily to the fields for their agricultural work.

organized construction of a vast canal to supply waters from the Tigris to irrigate of an area to the east of the city.[769] Due to poor design, water from the canal seeped into fields, flooding them and raising the water table across a huge area.[770] The rising groundwater leached salts from the soil, as noted in state records kept by surveyors from the central temple at Lagash.[771] The dire results of this and similar environmental disasters are described succinctly in the *Atrahasis Epic* of ancient Sumer:

> The black fields become white
> That broad plain was choked with salt[772]

As we will see below, this kind of soil salinization was probably one of several factors that exacerbated the effects of the later and much more serious drought that started about *4200 BP*.

Recovery in the north

The post-*5200 BP* collapse of village society in much of the rainfed-farming belt of northern Mesopotamia was far from total. As cereal crop yields declined, there were smaller surpluses to support the large, complex urban structures that had come to overlay the basic primary production of land-tied farmers. But while the larger towns in the north were soon abandoned, many smaller settlements simply dwindled in size to fit the reduced crop output. A small population of farmers remained throughout this period, although the majority probably either migrated south or adopted pastoral lifestyles.[773] In many areas, it took almost 500 years for a full recovery to occur and for predrought levels of trade and communication with the south to be re-established. Slowly, conditions in the north improved but, in contrast with the barley near-monoculture of the south, the more variable, rainfed farming in the north necessitated the cultivation of a mixture of cereals, lest any single crop might fail in a given year.[774] For example, in Leilan, farmers normally grew three cereals: emmer wheat, durum wheat, and two-row barley.[775] Although two-row barley is less productive than the six-row form, it is more suited to the unpredictability of rainfed farming.[776] Lentils were grown as the favoured pulse, alongside smaller amounts of chickpeas and field peas. As settlements expanded and demand for food grew, farming moved from the fertile wadi valley bottoms to the drier soils higher up. This entailed a greater focus on barley, which was tolerant of the more marginal, drier soils.[777]

By *4500 BP*, settlements such as Leilan in the Habur Plains were showing complex features indicative of renewed agricultural surpluses and the return of social elitism.[778] These include communal sewage and drainage systems, standardized state-controlled styles of architecture, construction of centralized grain stores and acropolis cultic platforms, and various forms of iconography (including written texts) that both depict and celebrate state power.[779] During the next few centuries, settlements across the region, such as Brak, Ebla, Leilan, Mozan, and Nineveh, grew and prospered, although these small towns did not approach the size and degree of organization of the huge Sumerian cities to the south.[780] Even under the more favourable climatic conditions of the fifth millennium *BP*, the rainfed farming of northern Mesopotamia and the rest of the Fertile Crescent was never able to generate the level of crop surpluses produced by the irrigation agriculture that was exploited so effectively in Sumer during the late Uruk Period. Nevertheless, a significant degree of urbanization was achieved in the north at this time. In some areas, such as along the Habur River, even relatively modest-sized communities built impressively large, centralized grain storage and processing centres that have been interpreted as evidence of a tight control of agriculture by the state.[781]

Ironically, the renewed efficiency of northern agriculture, coupled with the concentration of its controlling elites in a few medium-sized urban centres, made the region an attractive and vulnerable target for the burgeoning power and emerging imperial ambitions of Sumerian kings who ruled the city-states of the south. The north was initially spared from invasion by the chronic infighting among the powerful city-states of Sumer, and the reluctance of any of its rulers to expose an unguarded flank to his enemies by taking his army up north. However, in *4334 BP*, King Sargon of Kish succeeded in overthrowing or pacifying the other Sumerian kings, and for the first time Sumer was united under a single leader. To stress the new beginning that was inaugurated by this development, Sargon established a new capital city for his empire at Agade (or Akkad), from which his dynasty received its name.[782]

Rise of the Akkadian Empire

The Akkadian Empire was immediately preceded by several short-lived attempts by local kings to establish hegemony over the warring city-states of Sumer/Akkad. Although Akkad was located to the north of Sumer, and the people spoke a Semitic language unrelated to Sumerian, the two regions were parts of a single cultural group of highly urbanized and centralized city-states that depended on intensive, state-managed irrigation agriculture. Individual cities across Sumer and Akkad vied with each other for control of resources, including land. The eventual unification of the cities of Sumer and Akkad gave King Sargon access to huge potential wealth, mainly in the form of crops, tools (including weapons), and manpower, which he used both to build up his new capital at Agade and to expand his empire to the north. As discussed above, by this time the north was becoming significantly richer, but its urban centres were much smaller, and its society was less centralized and less able to field large armies than the south. It was therefore ripe for plucking, and Sargon duly obliged. The Akkadians left a swathe of destruction as they sacked and pillaged their way through northern cites such as Ebla, Brak, Tuttul, and Mari, as well as many smaller towns throughout the region.

The advent of the Akkadian *imperium* involved imposition of centralized state control system on northern societies. There is evidence of agricultural intensification during this period, presumably to generate increased crop surpluses for shipment to feed the growing urban population of the south.[783] A greater focus on barley cultivation may have been partially due to changes in soil conditions, and also to the greater familiarity of the Akkadians with barley as a portable means of wealth and the principal ration of their labouring classes. There seems to have been a deliberate switch from a broad production system based on a fairly equal mixture of emmer, durum, and barley, to a more precarious, but higher yielding system dominated by barley.[784] As discussed in Box 10.4, the Akkadians also brought their barley-rationing system to the north, along with many other cultural features.[785] At well-documented sites such as Leilan, huge numbers of mass-produced Akkadian-style *sila* ration bowls have been found.[786] But not all northern produce was appropriated as tribute for the south and it is likely that there was a mixed agricultural economy during the 140 years of

Akkadian rule. The state sector dominated, with its coercive focus on barley monocultures grown as tribute by subject farmers. But evidence from Brak and Leilan suggests that there was a much smaller parallel, household-based cultivation of a more diverse range of crops, such as pulses, presumable grown 'on the side' by farming households for their personal consumption.[787]

In addition to being a better-yielding and more familiar crop for the Akkadian rulers, barley was a more efficient form of tribute than emmer wheat because it was less labour intensive to process and transport. However, this reliance on a single crop became increasingly dangerous as the climate in the north became steadily more arid. The Akkadian empire reached its maximum extent under the rule of Sargon's grandson, Naram-Sin, but remained vulnerable to pressure from nomadic pastoralists and hunter–gatherers who threatened the rich and fertile lowland realm from all quarters.[788] Of course, this external threat had been faced by agricultural city-states of Mesopotamia for millennia. Indeed, the need to protect their grain stores from outsiders was one of the original motivations for Mesopotamian agrarian cultures to establish towns and cities in the first place. The ability to store crop surpluses in secure silos and citadels in well-fortified cities gave such cultures a strong selective advantage over those that remained more dispersed and vulnerable in isolated villages. However, this urbanization strategy also involved farmers ceding much or all of the control of their crops to the elites that controlled the city-states.

Providing there was sufficient food available and living conditions were not overly harsh for the urban and rural labour forces of these city-states, it was in everybody's interest to collaborate, both to maintain food and industrial production, and to defend the state against external invaders. The domination of a local elite by a more remote group of outsiders, that is the Akkadians, need not have prejudiced these arrangements providing the incomers did not cause too much disruption to existing social and economic structures. It seems likely that the Akkadians managed this balancing act for much of the 140-odd-year period that their empire endured. For example one of the most successful Akkadian kings, Naram-Sin, married one of

Box 10.4 Dada and his 40,800 litres of barley

With the right combination of plants, management, and climate, agriculture gradually gave people the potential to generate food surpluses that could support ever-larger populations. The farming settlement of Abu Hureyra started as a tiny hamlet and over the course of three millennia grew to a medium-sized town. But for thousands of years, agriculture remained an essentially regional activity, whereby the relatively bulky grain surpluses produced by farmers were only used within a restricted hinterland of a few dozen kilometres. All of this changed in *4334 BP* with the establishment of Akkadian hegemony over Mesopotamia by King Sargon.

Sargon and his descendents conquered a large swathe of the Near East, including much of the rainfed farming regions of northern Mesopotamia. Here, they encountered a more extensive and diverse form of agriculture compared with the barley-dominated, irrigated plains of Sumer. In the north, the Akkadians established the first true imperial administration with the imposition of their material culture (pottery, architecture, etc.), social organization, and an elaborate system of agricultural tribute. For the first time in recorded history, much of the farming produce of a conquered region was shipped over many hundreds of kilometres to supply the needs of a remote, dominant power.

This process is epitomized in a small fragment (7.5 × 4 cm) of clay tablet now held at the British Museum (King, 1896). The tablet records the measuring, by an Akkadian official called Dada, of a large consignment of barley and wheat that was to be shipped from the northern city of Nagar (Tell Brak) to the Sumerian city of Sippar. The cargo included 30 tonnes of barley and over 3 tonnes of emmer wheat. This grain was the enforced produce of northern farmers who were coerced into producing barley as their primary crop, to be supplied as tribute to Akkadian officials.

Barley and emmer grains can be stored for long periods but are very heavy to transport. The Akkadians built huge grain stores in conquered cities such as Nagar, where the tribute was held until ready for loading onto long mule trains for passage to a convenient river port for further shipment down the Euphrates to Sippar. The ability of the Akkadians to organize such complex logistical operations was due to a mastery of bureaucracy made possible by the invention of writing. The Akkadian, and most subsequent empires, were dependent for their very existence on similar systems of tribute and bureaucracy.

The key part of this cuneiform tablet reads:

> *40,800 litres of barley,*
> *4,200 litres of emmer*
> *for Sippar from Nagar,*
> *measured by Dada*

The remainder of the tablet records several other large crop shipments, including 50,000 litres of barley (37 tonnes) from a place called Shar-Sin, which was measured by another official called Warad-Shamash (Sommerfeld *et al.*, 2004).

his daughters to the ruler of the city of Mozan (near Leilan). These cities were in the region known as Subir, which was at the northern limits of the empire and at the outer edge of rainfed farming belt. Presumably this and similar acts were designed to seal the co-operation of local elites in the new imperial project. Cities such as Mozan and Leilan were incorporated into an emerging imperialized rainfed agricultural economy that serviced both the local elites and the imperial centre of Agade, which lay some 400 km to the south. The transport of such goods to imperial centres was recalled in *The Curse of Akkad*, a poem written less than a century after the collapse of the empire:

She* then filled Agade's stores for emmer wheat
  with gold,
She filled its stores for white emmer wheat with silver;
She delivered copper, tin, and blocks of lapis lazuli to
  its granaries
and sealed its silos from outside . . .
Ships brought the goods of Sumer upstream to
  Akkad, . . .
Elam and Subir carried goods to her with pack-asses,
All the provincial governors, temple administrators,
and land registrars at the edge of the plains
regularly supplied the monthly and New Year offerings
  there.[789]

* the goddess Inanna

The fall of Akkad and Ur

The Akkadian empire lasted for well over a century and might have endured longer were it not for the onset of a chronic drought at *4200 BP*, which was especially severe in northern Mesopotamia.[790] This event drastically reduced the quantity of grain shipments from rainfed farming areas such as Subir, upon which the empire had come to depend so much (see Box 10.4). The resulting lack of food and loss of prestige would have fatally weakened the power of the Akkadian state. Hence, the son of Naram-Sin, Shar-kali-sharri, who ruled the empire for a further 25 years, faced increasingly serious dissent from southern cities such as Kish and Uruk, as well as mounting external pressures from nomadic groups in the north, east, and west. A few years after the death of Shar-kali-sharri, in *c. 4198 BP*, the empire started to collapse as the

north suffered an agricultural and societal disaster and the cities of the south went their own way as independent states. When the Akkadian empire eventually fell in *4193 BP*, it was probably due to a combination of internal weakness exacerbated by crop failures in the north, and the pressure of external enemies who themselves may have been driven to attack the empire because of a lack of food resources in their own drought-stricken lands.

The abandonment of agricultural areas began gradually in the more marginal lands around the outer periphery of Mesopotamia. These regions received less rainfall and less seasonal runoff from the rivers and canals and were therefore more susceptible to any sustained decline in precipitation. It took one or two decades for the effects of the drought to be felt in their full force and for the stocks of food and water in such regions to be exhausted. But eventually the city of Leilan was completely abandoned along with most other population centres in the Habur Plains of northeast Syria.[791] Throughout the north, many hitherto thriving towns and cities were deserted within a decade or so of the demise of the Akkadian empire. In an echo of the prolonged drought of *5200 BP*, there is no evidence of large-scale warfare or wanton destruction of cities after the *4200 BP* event. Rather, the cities lost most of their population, or were abandoned simply because they could no longer be sustained by the reduced crop yields of the now-parched soils in their arable hinterlands.

As with the Mayan and Harappan urban collapses that will be examined in the next two chapters, the north Mesopotamian collapse did not involve the total depopulation of the region.[792] Rather, the population was adjusted downwards by mortality and emigration until it reached the carrying capacity of the new climatic conditions. In some cases, much-reduced populations clung onto parts of the otherwise deserted cities, but mostly people reverted to simpler, small-scale, village-based rural economic and social units. There is evidence of cultural continuity in these smaller communities. An interesting phenomenon here is the persistence of some elements of the urban-based hierarchies that existed before the disaster of *4200 BP*. Hence, small localized entities calling themselves 'kingdoms' sprung up amidst the ruins of cites such as Brak in

Subir.[793] It is also important to bear in mind that, although these environmental events may appear relatively sudden and drastic on a geological scale, they would have appeared much more gradual and patchy in nature to the people involved. Hence, although the rains were much reduced, they did not fail totally and a few favoured areas may have escaped the worst of the drought. The initial effects of a decrease in food production would have been economic, such as a dramatic rise in food prices. As stated in *The Curse of Akkad*:

At that time, one shekel's worth of oil was only
  one-half quart,
One shekel's worth of grain was only one-half quart. . . .
These sold at such prices in the markets of all the cities!

Eventually, grain tributes were withheld from regional centres as local populations struggled to feed themselves. This led to an exodus, first from the larger cities, and later from smaller cities and towns as the productive capacity of the local agricultural hinterland continued to decline. Eventually many of the agricultural units were unable to sustain even the local farming population, and the peasantry joined the exodus southwards towards the still relatively prosperous cites and farmlands of Sumer. It was largely due to this exodus from the north that the population of southern Mesopotamia doubled over a relatively short period soon after the Akkadian collapse.[794] By this time the rich alluvial soils of the Lower Tigris and Euphrates Valleys were watered by an elaborate series of permanent irrigation works (Figures 10.10 and 10.11). The use of canalized water to irrigate large areas meant that the Sumerians were once more insulated from the initial effects of the drought that was devastating much of the north. This complex agricultural system required detailed organization and detailed control of labour and logistics by the state. Perhaps the most impressive of these bureaucratic structures was the immediate successor to the Akkadian hegemony in Sumer, namely the city of Ur.

Following the Akkadian collapse, there was a period of about 80 years during which Sumerian cities struggled with the population influx from the north and their own need to sustain food production. The city of Ur developed a particularly effective system of control over every facet of its productive apparatus, especially its agriculture. This allowed the state to increase its agricultural output, and hence its power. Gradually the success of Ur enabled the city to extend its sphere of influence to the whole of Sumer and beyond, from the Persian Gulf to Nineveh in the north, and from the Zagros foothills in the east to the Syrian Desert in the west. In *4112 BP*, the local ruler, Ur-Nammu became the first king of the Third Dynasty of Ur (often abbreviated as Ur III). Ur-Nammu ruled over a large and diverse territory governed by means of a meticulously organized and all-pervasive bureaucracy.[795] Largely thanks to the efforts of his officials, this state endured and even flourished, with Ur-Nammu reigning for 18 years and his son, Shulgi, for 47. The city of Ur expanded and the newly enlarged labour force was used to build a vast new ziggurat, which acted both to propitiate the gods and further to reinforce the power of the king and his officials.[796]

These impressive material achievements were all built on the foundation of state-regulated cereal production. Despite the continuing northern drought, the effects of which gradually spread to the south, crop production was further intensified in the Ur III Period. Inevitably, the carrying capacity and sustainability of the system were approached, and in places exceeded, with the result that cereal yields began to fall as the effects of salinization and drought tightened their grip across the region. Across the south, river levels fell and irrigation became progressively more difficult. Farmers attempted techniques such as fallowing and leaching to reduce the effects of salinization, and these may well have been effective in enabling some crops to be grown, but at much reduced yields.[797] Food shortages started to grip the swollen urban population as crop yields faltered. In the city-state of Lagash, fields that had produced as much as 2537 litres of barley grain per hectare in the halcyon days of *4400 BP* were, by the Ur III Period, only yielding 1460 litres per hectare.[798] In neighbouring Larsa, a paltry 897 litres per hectare was all that was possible.

During the Ur III Period, the rulers of south tried to adapt to the increasing water shortage by shortening their canals, reducing food rations, and asserting ever more intrusive state regulation of the populace. Hard-pressed bureaucrats tried to maintain control by redoubling their bookkeeping efforts, recording even the tiniest transactions of goods and services.[799] Records show that the food crisis at Ur III was mirrored in an almost fanatical measuring and itemizing of grain distribution in ever-smaller amounts, down to single cupfuls of barley.[800] Official surveyors were dispatched to measure the dimensions of each barley field and estimate the future grain yield. After harvest, a separate group of accountants recorded the actual yield.[801] Wheat cultivation almost ceased during this period and even barley yields declined by two-thirds. By this time, it is recorded that all the fruit trees had disappeared from the once-rich land of Sumer, and the intricate but inherently fragile irrigation system, which made possible the increasingly precarious barley monoculture, was virtually all that was left to sustain the population. Some idea of the agricultural hinterland required to sustain an ancient city such as Ur can be grasped by considering the case of medieval London, which would been broadly comparable in its food demands and technological sophistication to fifth millennium BP Ur. Like Ur at its height, medieval London had a population of about 80,000 people, most of whom subsisted on a grain-based diet. The city drew on an area of 10,000 square kilometres for its grain supplies, and this area expanded considerably in years of poor harvest.[802] In good times such cities would act as an economic stimulus to greater productivity in their hinterlands, but in bad times the power of a city would drain resources from, and possibly compromise the future productivity of its vital agricultural base.

Beyond the delicate oasis of civilization that was Sumer of the Ur III Period, matters were even worse in the drought-hit surrounding regions. In northern Mesopotamia, the Assyrian King List records at this time: 'seventeen kings living in tents'.[803] Those referred to as 'kings' were normally urban-based leaders who would not be expected to live in tents, so these peripatetic rulers may have been refugees from abandoned cities, or pastoral chiefs who had moved into the region as its agricultural economy collapsed. By 4030 BP, pressures from migrants and nomadic raiders necessitated construction by King Shu-Sin of Ur of a massive 180-km-long wall across central Mesopotamia. This wall was built to the north of Sippar and extended between the Tigris and Euphrates, from Badigihursaga to Simudar (see Figure 10.1).[804] It became known as 'Repeller of the Amorites' after what the Ur III record keeper regarded as the most dangerous of the many nomadic groups. The typical Amorite was despised and stereotyped by Sumerians as one 'who knows no grain, who does not bend his knees (to cultivate the land), who eats raw meat, who has no house during his lifetime, who is not buried after death'.[805] In fact, by no means all Amorites were uncivilized nomads, and many had settled in cities across southern Mesopotamia. Indeed, following the collapse of Ur, Amorites became some of the most successful kings and administrators in the region.

While the Amorites may not have been the real target of the eponymous wall, this impressive feat of engineering, which took 24 years and the toil of untold thousands of labourers to build, was certainly meant to stem nomadic incursions into the vulnerable Sumerian heartland. The last Ur III king, and son of Shu-Sin, was Ibbi-Sin, who maintained an increasingly insecure diplomatic balance for a decade before his state finally succumbed. A key vulnerability was always the food supply, and a series of letters shows how Ibbi-Sin was held to ransom by a subordinate governor, who withheld grain shipments destined for Ur.[806] Soon famine broke out in Ur and its empire collapsed. Even worse, in 4004 BP, the city of Ur itself, the pride of Sumer and a byword for enduring power and material achievement across the region, was captured and thoroughly despoiled by the Elamites from the east: 'Ur—its weak and its strong perished through hunger, O Nanna, Ur has been destroyed, its people have been dispersed'.[807] This epochal event marked the end of Sumerian domination in Mesopotamia and ushered in a lengthy period of chaos, opening the way for groups such as the Semitic-speaking nomads from the north and west to invade, settle, and dominate much of the region.[808] The climate started to improve again after 3900 BP, but by then the centre of Mesopotamian agricultural production and political power had decisively shifted northwards from Ur and Uruk to the new imperial powers of Babylon and Assyria. Meanwhile the passing of Ur and Akkad was lamented across Mesopotamia for centuries to come:

For the first time since cities were built and founded,
The great agricultural fields produced no grain,
The ponds produced no fish,
The irrigated orchards produced neither syrup nor wine,
The gathered clouds did not rain, the macgurum did
  not grow.[809]

Renewed recovery

The dramatic events at the end of the fifth millennium BP mark the end of the Early Bronze Age in the Near East. The collapse of the Akkadian and Ur III states, the abandonment of agriculture, and depopulation of much of northern Mesopotamia, were serious disasters that could have, but did not, extinguish civilization in the region. As we will see in the next chapter, the Harappan civilization suffered a comparable series of calamities at about the same time, but never recovered and soon disappeared from human memory. In contrast, the Mesopotamians not only recovered relatively quickly from their disasters, they went on to forge two new imperial states, Babylon and Assyria, that dominated the region for another 1200 years. It appears that the worst effects of the drought soon passed in southern Mesopotamia and that irrigation agriculture was able to resume, albeit on a reduced and more sustainable scale. The bloated cites of Ur and Uruk were not revived, and the other Sumerian cites were of a more manageable size and more in keeping with the sustainable outputs of their agricultural hinterlands. An example of post-Ur III recovery comes from the small successor city-state of Larsa whose king, Gungunum, captured the remains of the city of Ur in about 3930 BP. Gungunum and his successors deliberately destroyed the canals of the neighbouring city of Isin, resulting in its ruin. Perhaps in atonement for such an unusually wanton act of vandalism, a later king of Larsa, Rim-Sin, set about building a new canal as he proclaimed in the following inscription:

Enlil (the main god) the Great Mountain, gave me the great mission of digging a canal to bring abundant waters to Sumer and Akkad, to make their fields grow the dappled barley . . . the numerous people whose shepherd-ship Enlil has given me . . . I made them work with my great power. I called the canal Tuqmat-Erra, and thus restored the eternal waters of the Tigris and Euphrates.[810]

Recovery in the north took longer but already, by 3800 BP, the Amorite-named ruler, Shamshi-Adad had founded the first northern Mesopotamian empire at about the same time as his more famous and slightly younger contemporary, Hammurabi, founded the Babylonian empire in the south. Before this, however, the north had endured several centuries of impoverishment and famine. Many Early Bronze Age cities as far afield as Anatolia and the Levant lay in ruins that were intermittently occupied by small bands of squatters who built very basic shelters amongst the rubble of their ancestors' once fine homes.[811] In a few places, including Ebla, which had been laid waste by Sargon of Akkad, the city had recovered sufficiently by 3800 BP to rebuild its defensive walls and a new administrative centre.[812] Gradually, over the next few centuries, agriculture recovered as rainfall became more reliable, and soon a network of new or rebuilt cities sprang up across the region.

These cities, such as Mari, Alalkh, and Leilan controlled small states and went on to spawn a series of elaborate bureaucracies. Their meticulous scribes have left us a wealth of written records on tens of thousands of clay tablets, many of which have yet to be translated. Among the larger northern city-states were Mani, Yamkhad, and Qatna, but the region never went on to generate southern-style unstable megacities such as Ur or Uruk.[813] Rather, there was a hierarchy of small villages, towns, and modest cities that mostly reoccupied the same regions that had been farmed and urbanized before the collapse of 4200 BP. These resurgent northern societies were more stratified than before, and readily adopted southern innovations such as mass production of pottery and other wares. In Chapter 13, we will resume our survey of Near Eastern historical and agricultural development. Meanwhile, in the final two chapters of Part III, we will examine the trajectory of early agricultural and societal development in some of the other important centres of crop domestication elsewhere in Asia, as well as in Africa, the Americas, and Europe.

CHAPTER 11

Evolution of agrourban cultures: II South and east Asia

Cutting stalks at noontime,
Perspiration drips to the earth.
Know you that your bowl of rice,
Each grain from hardship comes

Cheng Chan-Pao, Chinese philosopher

Introduction

For much of the twentieth century, the Near East was regarded as the ultimate cradle of agriculture and urbanization, almost to the exclusion of other centres of crop origin and diversity. This picture has started to change radically over recent decades as much more attention has been paid to the development of other agricultural societies across the world. The emerging view is of a series of processes, whereby predominantly cereal-based forms of agriculture evolved on most continents in the millennia after the Younger Dryas Interval. In this chapter, we will examine complex, agriculturally based societies in two other large regions of Asia. We will begin with the newly discovered Indus Valley civilization that may have domesticated barley independently of the Near Eastern cultures discussed in the previous chapter.[814] In the second part, we will compare the development of the northern millet-growing cultures with the southern rice farmers of ancient China. One especially striking feature about all three of these otherwise unrelated civilizations is their sudden visitation by a series of environmental disasters shortly after *4200 BP*. As we saw with the almost simultaneous Akkadian and Ur III collapses in the Near East, there is growing evidence that a climatic shift to cooler and more arid conditions was a significant, common factor in these social catastrophes, which

affected regions many thousands of kilometres apart. One notable difference between these ancient Asian cultures and Mesopotamia is an apparent absence of the coercive state management of agriculture as described in the previous chapter.

The Indus Valley

The Indus Valley region extends for more than 1000 km, in a roughly north-easterly direction from the river delta on the coast of the Arabian Sea to the foothills of the Himalayas in Kashmir. This fertile river valley, which is over 300 km wide in places, is bounded to the south-east by the Thar Desert of Rajasthan and to the north-west by the arid mountains of Balochistan (formerly Baluchistan) and Afghanistan. The region is on the margins of a late-summer monsoon belt that has fluctuated constantly over the past 20 millennia, sometimes bringing plentiful, if seasonal, rainfall, and sometimes leading to more arid conditions that did not favour domesticated cereal species. Although the climate of the Indus Valley today is rather dry, with an average annual rainfall of 130 mm, conditions in the early- to mid-Holocene were considerably wetter and rather more conducive to the growth of dense stands of wild grain-bearing grasses.[815]

Beginnings

Farming did not arrive in the Indus Valley proper until about *8000 BP*, when nomadic pastoralists established small villages that served as bases to grow barley and wheat at places such as Mohenjo-Daro, some 400 km north of the Arabian Sea (see Table 11.1 for chronology of major events and

Table 11.1 Chronology of the Indus Valley farming cultures, 9000 to 3000 BP

Date BP	Location	Climate	Food plants	Agricultural technologies	Other technologies	Social systems
10,000	Balochistan, Iran/Afghan border	Warm and moist	Wild fruits and cereals	None	Stone tools	Nomadic and semisedentary hunter gatherers, increase in use of cereals
9000	Mehrgarh, Kachhi Plain	Warm and moist	Emmer/einkorn, barley, wild dates, cherries, *Ziziphus*	Extensive rainfed farming	Prepottery	Small-scale villages of mud bricks
8000	Mehrgarh, Kachhi Plain	Warm and moist	Six-row naked barley, emmer/einkorn and free-threshing breadwheat	Extensive rainfed farming	Prepottery	Small-scale villages of mud bricks
7000	Mehrgarh, Kachhi Plain	Warm and moist	Six-row barley and emmer, breadwheat, pulses, grapes	Extensive rainfed farming, large storage granaries, cotton	Distinctive pottery Sheep, cattle and goats Trade: Iran and Arabia	Larger towns with increasingly elaborate architecture
6000	Balochistan, Indus Valley	Warm and moist	Six-row barley and emmer, breadwheat, pulses, grapes	Spread of rainfed farming from Mehrgarh, ploughs	Wheel pottery	Rapid development of large towns and small cities
5500	Harappa, Mohenjo-Daro, and Ganweriwala Indus Valley	Warm and moist	Increase in breadwheat, emmer, barley, pulses	Local irrigation schemes, intensive farming	Pottery, defensive works, dikes	Emergence of elites, huge central grain stores
5000	Harappa, Mohenjo-Daro, Ganweriwala, etc.	Warm and moist	Sorghum, pearl millet introduced from Africa	Highly intensive irrigation farming	Long-distance trading with Mesopotamia, Early writing	Harappan Ravi period of regional urban cultures, large well-planned cities
4500	Harappa, Mohenjo-Daro, Ganweriwala, Banavi, Kalibangan *etc*	Warm and wet	Cereals-based but increasingly diverse, cotton, millets, dates	Highly intensive, locally based, irrigation farming	Novel water technologies, writing, binary measuring	Integration of regions, many large planned cities; absence of fortifications or palaces
4000	Dozens of cities across the Indus Valley	**Drought** Cool and dry	Rapid disappearance of crops	Collapse of irrigation systems as Saraswati and Indus change course	Collapse of civic water systems, cities fall into disrepair	Rapid depopulation of cities, mass hunger and unrest, more village-based economies
3800	Mohenjo-Daro	**Drought** Cool and dry	Absence of crops	Irreversible damage to irrigation networks	Terminal decline of infrastructure	City reduced from 250 to less than 0.5 ha, chronic hunger, no leadership
3800–3300	Dozens of cities across the Indus Valley	Warm and moist	Small-scale pastoralism some plant gathering, emigration East and South	Sporadic, localized, small-scale farming and pastoralism	Pottery and simple bronze tools	All cities abandoned for good, population crash, village-based or dispersed nomadism
3300	Ganga-Yumana Plain Gandhara	Warm and moist	Mainly pastoral remnants of Harappans	Sporadic, localized, small-scale farming and pastoralism	Pottery and simple bronze tools	Ochre Coloured Pottery (OCP) Gandhara grave (Swät) cultures retain some Harappan features

The trajectory of agrosocial development in the Indus Valley is intriguingly similar to that of the Near East in many respects, but also has some profound differences, not least of which is its complete and final extinction after 4000 BP. From 6000–4000 BP, the suite of wheat/barley/pulse crops, irrigation systems, and large number of sophisticated city-states was very similar to the contemporary situation in southern Mesopotamia. The major differences between the civilizations were social, most notably the apparent absence of monumental buildings and other elite structures in Indus Valley cities, which nevertheless were highly organized, complex societies capable of managing intensive and intricate agrourban enterprises. Sadly, their relatively egalitarian social systems did not prevent the catastrophic collapse of these cities following a sudden environmental crisis after 4000 BP.

Figure 11.1 for a map of the region). Immediately prior to this, several small-scale farming cultures had already been established in the Kachhi Plain to the north-west of the Indus Valley. Excavations at Mehrgarh, near the modern city of Quetta, have revealed evidence of wheat and barley farming[816] and cattle domestication[817] in this region dating back as far as *9000 BP*. Both wheat and barley occur here as indigenous wild plants and may have been exploited by hunter–gatherers for a long time prior to their eventual cultivation, perhaps even as early as the Younger Dryas or before. Already by *8000 BP*, farmers at Mehrgarh were growing a form of six-rowed barley as a major crop.[818] The early

date suggests that this type of barley may have been domesticated *in situ* rather than being imported from the Near East, which is consistent with recent genetic data.[819] While its crops may or may not have been indigenous, the Mehrgarh culture certainly was. These people started by constructing mud-brick buildings at *9000 BP* and went on to develop larger, urban centres and a distinctive style of pottery by *7500 BP*, which is about a millennium after the pre-Halaf pottery cultures of northern Mesopotamia. Ceramic and Chalcolithic cultures then developed and the town of Mehrgarh grew, becoming more elaborate in its architecture and organization.

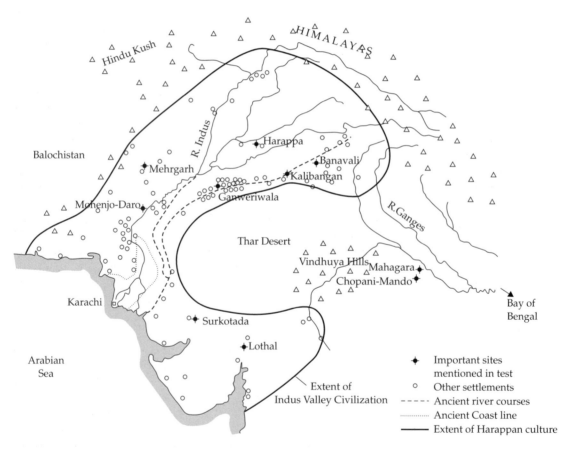

Figure 11.1 Locations of North Indian agrourban cultures. Agrourban cultures originally spread from Mehrgarh in the northwest to the Indus Valley proper. Note the two main clusters of cities along the lower reaches of the River Indus, one to the south of Mohenjo-Daro, and the other along the upper reaches of the River Saraswati from Ganweriwala to Banavali. Following the aridification episode after *c. 4000 BP*, rainfall throughout the region declined and the River Saraswati suddenly dried up, leading to the abandonment of agriculture and the demise of the entire Indus Valley civilization. Meanwhile, far to the southeast on the upper reaches of the Ganges, communities such as Mahagara and Chopani-Mando were some of the earliest centres of rice cultivation in southern Asia.

By *6000 BP*, cereal-based agriculture had spread beyond Mehrgarh to other parts of Balochistan and the Indus Valley, eventually reaching eastwards across the subcontinent to the Indian Ocean. It was in the Indus Valley and its immediate environs that some of the greatest early Asian civilizations were established, in cities such as Harappa, Mohenjo-Daro, and Ganweriwala.[820] The fertile, alluvial soil of the plains and benign climate of the period encouraged development of more elaborate and intensive forms agriculture than had been possible at Mehrgarh. Irrigation and the use of the plough by pre-Harappan cultures, such as the Amri and Rehman Dheri, enabled cultivation of larger areas and the generation of sufficient surplus food to sustain ever-larger urban populations. By *5000 BP*, there is evidence of long-distance trading for artefacts such as lapis lazuli and turquoise, from as far afield as Iran. Between *5500* and *4500 BP*, many hundreds of village-based farming communities were established across the Indus Valley. These pre-Harappan cultures had extensive trade networks and had broadened their agricultural base to include new crops such as peas, sesame, dates, and cotton. The major animal domesticant was the water buffalo, which is still the cornerstone of intensive farming in the region today. Soon after *5000 BP*, a new plant domesticant had appeared in the Indus Valley. This was a race of sorghum, which was probably introduced from east Africa, although it underwent considerable additional development after its arrival in southern Asia.[821] Pearl millet was also imported from Africa and, while its cultivation overlapped with that of wheat and barley, it was mostly grown to the south of the Indus Valley, for example at locations such as Surkotada and Lothal.

Rise of the Harappan cities after *5500 BP*

As the pre-Harappan period progressed, the domination of agriculture and society by small villages gradually gave way to larger towns and even cities.[822] As in the Near East, by *5000 BP*, the Indus Valley supported flourishing urbanized cultures based on intensive agricultural management. The earliest, truly urban civilization in the Indus Valley, which was extant during the Harappan Ravi Phase,

endured for over five centuries, from *5300 to 4800 BP*, and possibly longer. By *4600 BP*, there were dozens of cities in the Indus Valley region, of which the larger conurbations, such as Harappa and Mohenjo-Daro, extended over hundreds of hectares and supported populations of 30,000 to 40,000 people, making them comparable in size and organization with the largest cities in Mesopotamia. The greater Indus Valley region extended for about one million square kilometres, making it larger in area than ancient Egypt and Mesopotamia combined.[823] Unlike the largely unplanned and chaotically set-out cites that later grew up in medieval Europe, the Indus cities were carefully planned in advance. They had orderly street networks and standardized house designs. The various social and occupational groups were housed in distinctive dwellings in different quarters of the city. Assiduous attention to water control was a particular feature of Harappan cities, with sumps, drains, wells, baths, and toilets supplementing the impressive public drinking water and sewage systems. Large public buildings were built for ceremonies and/or entertainment, plus a huge central complex that housed the all-important cereal granaries.

As in the Near East, physical control of the large quantities of agricultural produce stored in the central granaries, suitably augmented by religiomythical symbolism, would have been a potent manifestation of the power of the local elites.[824] Notable by their absence in Harappan cities, however, are the more overt signs of substantial palatial or military architectural features, such as ziggurats or fortresses. During this period, there is the first evidence of writing in the so-called Indus script. The emergence of writing in south Asia occurred almost simultaneously with the transition in the Near East from pictographic/logographic to more complex written texts (Box 10.2 and 10.3). Harappans also developed and widely disseminated an innovative binary counting system, used to calculate weights and measures, and to standardize architectural features, such as brick size for urban construction projects.

It has often been assumed that Indus Valley civilizations were broadly similar to their contemporaries in the Near East, both with regard to their agricultural management and their social structures. However, unlike the city-states and empires then emerging in the Near East, there is no evidence that the Indus Valley civilization practiced the sort of highly organized and centralized coercive state management of its agricultural system seen in Mesopotamia.[825] Instead of the vast state-organized canal networks seen in the Near East, the Harappans and their neighbours seem to have adopted a more bottom-up system of agronomic management based on the long-term building-up and elaboration of small-scale irrigation and field-development schemes, as exemplified later by terrace agriculture.[826] Thanks to well-established trade links between the Indus Valley and Mesopotamia, the two civilizations were aware of one another, with the Sumerians calling the Indus Valley 'Meluhha' and the people 'Meluhhaites'.[827] During the later Harappan period there was a considerable diversification in crop use, in contrast with the trend to barley monocultures in contemporary imperializing societies of the Near East.[828]

While the barley/wheat-based Indus Valley civilization was technologically the most advanced ancient society in the Indian Subcontinent, a smaller but important rice-based culture arose in the Vindhyan Hills to the south-west (see Figure 11.1). This relatively verdant region is separated from the Indus Valley by the vast expanse of the Thar Desert, which today is over 500 km wide in places, although it was much less extensive in the Harappan period. Several wild cereals, including rice, grew in the Vindhyan Hills and rice cultivation, at sites such as Chopani-Mando and Mahagara, may have been underway as early as 7000 BP. The relative isolation of this area and the early development of rice farming imply that it was developed indigenously. Probably because high-productivity intensive farming was not possible in the Vindhyan Hills, these settlements never developed into sophisticated urban centres as seen in the Indus Valley or Mesopotamia. However, they were responsible for the dissemination of rice farming over much of southern Asia. Chopani-Mando and Mahagara are located on the upper reaches of the Ganges drainage system and it is likely that migrants from this area spread rice farming down the Ganges Valley into the fertile plains of Bengal, and beyond into south-east Asia. This was also the route of the main Aryan incursions from Central Asia into India, which started after 3500 BP, and soon penetrated to the Ganges delta. This Vedic culture adopted farming by 3000 BP, but writing and urbanization did not re-emerge until after 2500 BP, almost 1500 years after the demise of the Indus Valley civilization, which we will now examine.

The collapse of *c. 4000 BP*

The first signs of a decline in Indus Valley civilization came about 3900 BP, when the cities became progressively depopulated and the remaining residents showed increasing signs of malnutrition.[829] By 3800 BP, most of the great cities had been completely abandoned and agriculture had ceased in the region. The reasons for this collapse, and its almost unprecedented completeness—the cities were never reoccupied and urban culture disappeared from the subcontinent for more than a millennium—have always been mysterious (Box 11.1). As with all such events, there were probably several causative factors, but recent physical measurements show strong links with a wider series of environmental events that include perturbation of the summer monsoon cycle, soil aridification, and rerouting of major rivers. These phenomena immediately preceded, not only the collapse of the Indus Valley cities, but also similar disasters in China and the Near East. Other factors, possibly including social conditions, epidemics, warfare, etc. were undoubtedly involved in destabilizing all of these societies, but the loss of their primary crop base due to a lack of water would have been an especially devastating blow, from which some of them never recovered.

During the most intensive period of urban and cultural development, which witnessed the invention of writing and complex societies across the Indus Valley region, the climate was exceptionally moist and biotically productive. Indeed, there was so much rainfall in the period 4500 to 4000 BP that flooding was a periodic problem and some lower-lying settlements were inundated on several occasions. The hills along the Indus Valley were richly forested and teemed with such a profusion of wildlife that people were able to supplement their crop-based diet with seasonal hunting forays.

Date (BP)	Culture	Climate	Region	Crops	Farming	Technology	Social organization
8000	Dadiwan	Warm and moist	Northern China	Broomcorn and foxtail millet, pulses	Intensive rainfed farming	Specialized pottery	Well-organized sedentary cultures, pit houses with large crop stores
7000	Hemudu	Warm and wet	Yangtze Basin	Rice as major staple	Irrigation farming	Cooking in pots	Small villages, craft specialization
7000	Dadiwan	Warm and moist	Northern China	Broomcorn and foxtail millet, wheat, pulses	Intensive rainfed farming	Protowriting	Emergence of elites, huge central grain stores
7000–5000	Yangshao	Warm and moist	Henan, Shaanxi, and Shanxi	Millets, wheat, rice, pulses	Intensive rainfed and irrigation farming	Long distance trade, e.g. jade	Village and town-based hierarchical chiefdoms, some pastoralism
4500	Longshan	Warm and wet	Lower Yellow River Valley	Millets, wheat, rice, pulses	Intensive rainfed and irrigation farming	Skilled pottery	Village and town-based hierarchical cultures, religious centres
4400	Qijia	Warm and wet	Gansu	Millets, wheat, rice, pulses	Intensive rainfed and irrigation farming, Horses domesticated	Bronze tools and artefacts	Larger towns
4000	Many cultures including Qijia and Yangshao	**Drought** Cool and dry	Throughout Northwest China	Loss of both wild and cultivated food plants	Cessation of rainfed farming	Increasing reliance on cattle and sheep	Sudden depopulation, cultural simplification, nomadic pastoralism
4000	Many cultures	Warm and wet	Southern China	Rice and millet staples, pulses, other crops	More intensive irrigation farming	Increasingly sophisticated bronze tools	Unaffected by drought, cultural continuity and increasing complexity
3000–2000	Many cultures	Warm and moist	Most of northern China	Pastoralism with some plant gathering	None	Pottery and simple bronze tools	Widespread nomadic pastoralism in much of region

Agrosocial development in China is similar to the Near East in the relatively rapid and widespread domestication of local food plants within a few millennia of the Younger Dryas, but its subsequent trajectory was markedly different. Despite the early invention of pottery and localized intensification of crop production, there was no corresponding emergence of urbanized civilizations before the Shang Dynasty after *3500 BP*, almost three millennia later than the earliest city-states of Mesopotamia. One of the reasons for this delay might have been the lower yields of early millet and rice crop varieties, compared with wheat and barley, which could not generate sufficient surplus production to sustain large urban centres. Climatic amelioration and improved higher-yielding crop varieties after the *4000 BP* drought could have opened the way to increased food production and the subsequent emergence of complex urban cultures.

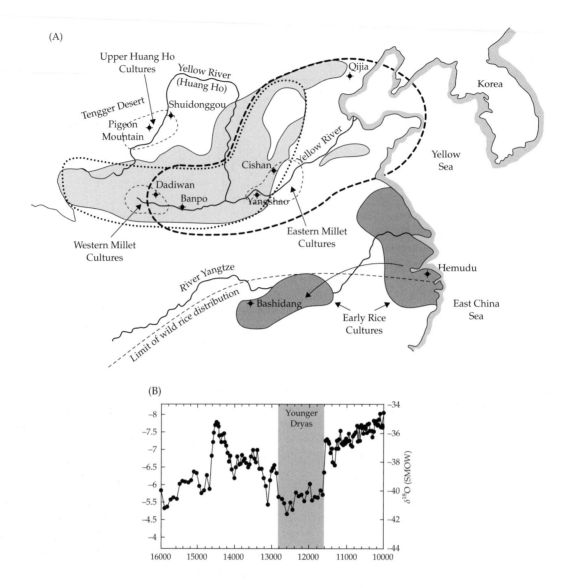

Figure 11.2 China: cradle of millet and rice farming cultures. (A) Following its initial domestication after 10,800 *BP*, large-scale millet farming started on or near the Loess Plateau (grey shading) *c. 8000 BP*, and spread eastwards along the axis of the Yellow River (Huang Ho) to the North China Plain. One of the earliest sites of millet domestication was the Pigeon Mountain area on the edge of the Tengger Desert. Following a cold/arid period, the centre of farming switched to the Western and Eastern Millet Cultures based respectively at Dadiwan and Yangshao, where highly sophisticated pottery styles and early forms of writing were developed. The limits of the later Dadiwan and Yangshao cultural groups are shown respectively by a bold dotted line and a bold dashed line. The lower yield of millet farming compared to barley or wheat precluded agricultural intensification on a sufficient scale to support the kinds of large urban centres that had arisen in the contemporary Near East and Indus Valley cultures. Further south, large-scale rice farming started between *7000* and *5000 BP* in two major centres, first around Hemudu in the lower reaches of the Yangtze and later around Bashidang in the middle Yangtze Valley. Both centres were close to the limit of naturally occurring wild rice (dashed line). Following its domestication, rice cultivation gradually spread northwards where the exceptionally high yields under irrigation eventually underpinned the development of highly urbanized Chinese civilizations, beginning with the Shang Dynasty in the *fourth millennium BP*. (B) Climate in late Pleistocene China as reconstructed from the Hulu Cave near Nanjing (data expressed as oxygen isotope ratios based on Vienna Standard Mean Ocean Water, or SMOW). Recent studies indicate that the climate in China may have been affected profoundly by the Younger Dryas event. Note the mild period from *14,500* to *13,000 BP*, during which rice and millet may have been subject to some domestication-related changes before cultivation may have been temporarily abandoned during the Younger Dryas (data from Wang *et al.*, 2001), available from NOAA (http://www.ncdc.noaa.gov).

Beginnings of agriculture

The earliest experiments in plant cultivation in China may have taken place at about the same time as post-Younger Dryas protofarming started at Near Eastern sites such as Abu Hureyra, or maybe even earlier.[846] As in the Near East, it is suggested that, in north-west China, wild cereals such as broomcorn (proso) millet may have become increasingly important food resources both before and during the Younger Dryas.[847] The magnitude of the Younger Dryas was less marked in China and plant assemblies were very different, but it is likely that there was an analogous process of diminished returns from game hunting, coupled with an increased availability of domestication-ready cereals, which facilitated a shift away from hunter gathering towards a more sedentary lifestyle of plant cultivation. This would have been promoted by the increased geographical range and biological productivity of wild millets immediately after the Younger Dryas. By this time, local human populations would have had many millennia of familiarity with, and increasing dependence on, millets as a source of edible grain. By analogy with the Near East, one could expect that technologies involved in grinding the grain to flour, and perhaps further processing, were developed during the predomestication, prefarming phase of millet exploitation.[848]

It has been proposed that one of the earliest centres of Chinese farming was in the relatively dry region of the Upper Yellow (Huang He) River. During the late Pleistocene, this region was significantly moister than today and early experiments in crop cultivation may have already been in progress by the Pleistocene/Holocene transition. During the early Holocene, at about *10,800* to *10,000 BP*, there was a post-Younger Dryas aridification event in northern China during which the Upper Yellow River region may have become too dry for reliable crop cultivation.[849] Agriculturalists in this deteriorating climate may have responded by migrating southwards and occupying the region around Dadiwan, where a longer lasting and more intensive form of millet farming developed. More arid conditions and poorer soils favoured broomcorn millet and foxtail millet, both of which are relatively drought-tolerant (Figure 11.3). During this

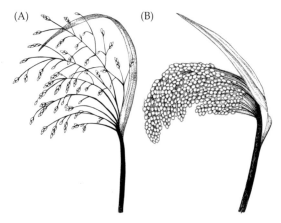

(A)   (B)

Figure 11.3 Broomcorn (proso) millet: one of the founder crops in China. Comparison of (A) wild and (B) domesticated forms of proso millet, which was the major food staple of many farming cultures in Northern China during the early Neolithic Era. The domesticated form has shorter but more profuse ears and would have yielded much more grain than its wild ancestor. However, even relatively intensive millet farming does not produce anything like the yields of other major cereal staples, such as barley, wheat, or rice.

period, extending to *9000 BP*, archaeological evidence of the less intensive phase of millet farming is tantalizingly sparse.[850] The spread and subsequent intensification of millet farming may have been precipitated by a climatic perturbation some time after *9000 BP*, which adversely affected millet yields. This forced farmers to migrate southwards from the original centre of broomcorn millet cultivation into the more agriculturally productive fluvial deposits of the Chinese Loess Plateau.[851] It is possible that the climatic event in question is related to the *c. 8200 BP* episode discussed in the previous chapter.[852]

Dadiwan, Yangshao, Longshan, and Qijia cultures: *8000 to 4000 BP*

During the middle part of the Chinese Neolithic Period (*8000–5000 BP*), intensive cereal cultivation developed in two areas in the north-central region, respectively termed the Western and Eastern Millet Cultures.[853] The western group occupied parts of what is now Gansu Province.[854] This culture and the associated archaeological period are often called the Dadiwan after the main excavation site in the area. The major crops of these people were the two cereals, broomcorn millet and foxtail millet, both of which had by this time been domesticated into large-seeded, higher-yielding varieties. These two crops were grown separately and the relative lack of weeds, whether from wild relatives or non-related opportunistic species, is suggestive of intensive cultivation with a high degree of varietal selection and crop management by the early farmers of the region. Beginning at *8000 BP*, there is evidence of a transition to an organized, sedentary, crop-based culture with pithouses and some pottery. By *7000 BP*, a well-organized millet-based society had developed at Dadiwan, complete with storage facilities, harvesting tools, and fully domesticated seeds. It has also been claimed that pottery inscriptions from as early as *7000 BP* may represent a precursor of written Chinese, although these finds are not as well documented as the later and better known Banpo inscriptions discussed below.[855]

Several hundred kilometres to the east, along the Lower Yellow River Valley and beyond, the Eastern Millet Cultures probably developed similar forms of cereal agriculture, which were then exported further afield. The earliest domesticated millet in this region, dating from *8000* to *7700 BP*, was found slightly to the north of Yangshao at the rich site of Cishan.[856] The later Yangshao culture flourished from about *7000* to *5000 BP*. These people mainly cultivated various forms of millet, although some villages also grew rice and wheat, possibly imported from outside the region.[857] By *6000 BP*, most Yangshao areas were using an intensive form of foxtail millet cultivation, complete with storage pits and finely prepared tools for digging and harvesting the crop.[858] It was in the Yangshao village of Banpo that some of the earliest inscriptions were found on pottery, dating from *6000 BP*, which then developed into the recognizable Chinese written characters of today.

In some areas the Yangshao were succeeded by the Longshan culture (*5000–4000 BP*), which was especially noted for its highly skilled pottery technology. The Longshan culture was hierarchical, with differentiated classes with regard to wealth and prestige, but these people did not build the kinds of large cities that were characteristic of their contemporaries in the Near Eastern or Indus Valley

civilizations. A slightly later development was the Qijia, an early Bronze Age culture in modern Gansu Province that was most active from *4400* to *3900 BP*.[859] The Qijia used domesticated horses and made some of the earliest known bronze and copper mirrors. However, like many other cultures across Asia, the Qijia and their immediate neighbours in north-west China suffered a serious reverse and possibly collapsed as a civilization soon after *4000 BP* as we will now discuss.

The collapse of *c. 4000 BP*

Chinese societies from the Yellow River to the Yangtze Valley underwent dramatic changes at about *4000 BP* that may be linked to a climatic shift from relatively moist to much more arid conditions. As with other regions of Eurasia, this part of China had undergone several previous cycles of aridity, followed by the return of more agronomically friendly humid/wet conditions. For example archaeological data from the Tengger Desert in north-west China are consistent with aridification episodes at about *8200 BP* and *5200 BP*, but these appear to have been more localized and were certainly less consequential for societal development than the *4000 BP* event.[860] The momentous events around *4000 BP* in China are probably linked with the near-contemporary social disasters in the Near East and Indus Valley as described previously. Each of these widely dispersed events, in otherwise unlinked cultures, appears to have a common factor, namely a climatic shift that affected crop production. The phenomenon in China has been best studied in the drier western region of the Chinese Loess Plateau which, as we have just seen, was the original cradle of drought-adapted intensive cereal farming in China, which expanded so successfully and underpinned so much cultural development during the Dadiwan period and beyond.

There is archaeological and palaeoclimatic evidence of a relatively wet period before *4500 BP*, followed by a much drier interlude with a drastic reduction in all types of indigenous vegetation.[861] This ecological disaster during the centuries around *4000 BP* was closely followed by an abandonment of the more drought-affected western areas of the Qijia region. As the drought progressed there seems to have been a virtual collapse of the Qijia and Yangshao cultures as reflected in a cessation of rainfed agriculture, followed by extensive depopulation and cultural simplification. Eventually, a new, seminomadic, pastoralist, subsistence culture emerged that lasted for more than 1600 years.[862] The extent of the north Chinese collapse at about *4000 BP* was much more far-reaching in its length and severity than the demise of the Akkadian Empire and the fall of Ur in the Near East, and is comparable with the end of the Indus Valley civilization. It is possible that millet farming was more vulnerable to the sort of climatic change in the region and/or the social structures of the Qijia and neighbouring agricultural societies were less resilient than those of the other, more westerly, Asian societies.

Another factor in promoting an early recovery in the Near East may have been a more rapid return of rainfall. Historical records show that by *3900 BP*, the seasonal rains had returned to Mesopotamia, crop production and populations soon rose, and two new Amorite dynasties emerged in the states of Babylon and Assyria, that were to dominate the Near East for almost 1500 years. In some of the main millet growing regions of north-west China, agriculture would not return until as late as *2000 BP*, that is at the same time as the height of the Roman imperium in western Eurasia. However, there was a more rapid recovery in the middle valley area of the Yellow River, where tradition tells of the establishment of the Shang dynasty at about *3600 BP*, although there were no written records of the dynasty until *3500 BP* when a pictographic script was (re?)invented.[863] The Shang dynasty eventually dominated much of northern China and was the earliest of a series of such dynasties that ruled large parts of north-east Asia for the next three and a half millennia.[864]

The success of the early Chinese millet farmers is still reflected today in the DNA of many east Asian populations. Surprisingly the type of DNA that has told us so much about the origins of modern Chinese populations is not human, but bacterial. Many people, as much as one-third of the world population, carry latent (dormant) forms of the tuberculosis pathogen, *Mycobacterium tuberculosis*.[865] Different human populations carry

different forms, or haplotypes, of the bacterium, so the genomic structure of these microbes can be used to trace population movements as far back as 100,000 years ago. Such studies have shown that the ancestors of northern Chinese millet farmers probably arrived in the area between *30,000* and *20,000 BP*. Following the development of agriculture after *10,000 BP*, this group gradually spread from their core area in the Upper-Middle Yellow River Basin, and their bacterial haplotypes are now found in populations throughout eastern Asia.[866] This genetic evidence of population spread is mirrored by linguistic data that show the radiation of the proto-Sino-Tibetan languages during the same period.[867]

Rice farming in southern China

Rice farming in southern China was initially based on endemic wild varieties that grew about as far north as the 30th parallel. In 2002, a Chinese/Japanese group reported the discovery in eastern China of fossilized phytoliths of domesticated rice apparently dating back to *13,900 BP* or earlier.[868] However, phytolith data are controversial in some quarters due to potential contamination problems.[869] Should this report eventually be confirmed, however, it would be of great interest because the putative domesticated rice from *13,900 BP* was not present in the later strata that coincide with the most severe phase of the Younger Dryas Interval, although similar types of rice then reappeared at about *10,000 BP* in the early Holocene. This may imply that some crop domestication occurred in the late Pleistocene, but was then discontinued (albeit temporarily) as the climate worsened with the onset of the Younger Dryas.[870] This timescale is somewhat different from the early appearance of domesticated rye at Abu Hureyra in the Near East.[871] In the latter case, the rye grains were dated to *11,800 BP*, which is close to the abrupt end of the Younger Dryas in the region.[872]

As in the Near East, there are indications that climatic changes associated with the transition from the Pleistocene to the Holocene in the middle Yangtze region may have shifted the density and distribution of human food resources, both plant and animal. In turn, this may have led to increased use of, and ultimate dependence on, rice, resulting to its relatively swift domestication in one or two key areas, followed by a much slower spread across the region as a whole. It has been suggested that the trajectory of rice farming and associated technology in China may differ in one fundamental respect from that in the Near East, namely in the timing of the invention of pottery. Pottery was probably used in the region from at least *10,000 BP*, and possibly several millennia earlier.[873] Associated with pottery at Yuchan, were remains of wild rice grains, but it could not be demonstrated that these had been cooked.[874] It is likely that domesticated rice was cultivated in the middle Yangtze Valley by *9000* to *8000 BP*, as shown in finds from the Pentoushan culture at Bashidang.[875]

By *7000 BP*, there is proof from a site at Hemudu near the east coast that rice had been domesticated and was being cooked in pots.[876] This raises the interesting point that, in China, a greater interest in rice as a food resource, and the impetus for its domestication, may have been provided by the availability after *10,000 BP* of pottery technology for its processing into a more palatable foodstuff via cooking. In contrast, as we saw in the previous chapter, cereal domestication in the Near East occurred several millennia before the invention and spread of pottery. The early cereal farmers in these regions were forced to grind and bake their grains, with disastrous results for their dentition, until finally by about *8500 BP* they were able to cook their wheat and barley into a soft porridge. Although rice was probably first cultivated in Central China at or before *10,000 BP*, it was several millennia before it became a staple crop capable generating surpluses able to sustain large non-agrarian (i.e. urban) populations. One of the earliest sites of proven rice cultivation is at Jiahu in Henan Province. This region is interesting because it marks the overlap of millet and rice farming, both of which were occurring in the area by *8500 BP*.[877]

Jiahu is also the site of one of the earliest examples of what may be another form of protowriting, dating from *8500 BP*, showing similarities with later Chinese characters.[878]

By *8500 BP*, farmers were cultivating several domesticated varieties of rice in what was an almost ideal habitat in the Yangtze Basin wetlands and upper reaches of the Huaihe River.[879] But rice did not become a true dietary staple until about *7000 BP*, and probably did not reach Korea and Japan until about *3000 BP*. As with wheat and barley in the Near East, domesticated rice took several millennia to diffuse from its centre(s) of origin and become a dietary staple across a wider region.[880] Rice is a much more productive crop than millet, but wild varieties cannot grow in northern China and even the domesticated crop required too much water for the kind of rainfed farming that could be practiced in most of the region. Despite the long period of early rice cultivation across southern China, there is little evidence of culturally and technologically sophisticated societies like those already developed by *7000 BP* by the Millet Cultures of the north. The aridification event of *4000 BP* affected the development of the more southern rice-growing cultures of China much less than their millet-growing neighbours. This was probably because the seasonal rainfall did not stop altogether but was displaced to the south, leaving millet farmers 'high and dry' but largely sparing warm-season rice growers. By *3000 BP*, Chinese rice farmers had discovered the merits of more intensive, paddy-based cultivation. Soon after *2300 BP*, the introduction of early-maturing Champa varieties from Vietnam enabled two rice crops to be grown per season while propagation by transplanting seedlings, rather than sowing seed, improved efficiency and yields still further.[881] These developments set the stage for the emergence of one of the most enduring and advanced of our civilisations, which today constitutes one-fifth of humanity.

CHAPTER 12

Evolution of agrourban cultures: III Africa, Europe, and the Americas

Today as in the time of Pliny and Columella, the hyacinth flourishes in Wales, the periwinkle in Illyria, the daisy on the ruins of Numantia; while around them cities have changed their masters and their names, collided and smashed, disappeared into nothingness, their peaceful generations have crossed down the ages as fresh and smiling as on the days of battle.

Edgar Quinet, 1825, *Philosophy of Human History*

Introduction

One of the greatest surprises of recent research into agricultural development in different societies has been the extent and antiquity of several hitherto unsuspected crop domestications in Africa. As with the rediscovery of the Harappan cities of the Indus Valley, we are now becoming aware of a series of innovative and complex agrarian cultures that once inhabited what is now known as the Sahara Desert. Although these cultures disappeared after the rains failed about *5500 BP*, they may have helped to spread sorghum and millet farming to many other parts of Africa as they trekked southwards and eastwards away from their doomed homeland. Indeed, it is likely that Saharan farmers influenced the later and much better known Nile Valley cultures of Sudan and Egypt almost as much as the latter were influenced by the ancient civilizations of the Near East.

In this chapter, we will also trace the spread of cereal/pulse farming from the Near East to Europe. The introduction of agriculture into Central Europe may have been accelerated by the sudden inundation of the Euxine Lake to form the Black Sea in about *7600 BP*, and the resulting migration of displaced farmers along fertile river valleys such as the Danube. European agriculture remained relatively small scale for several millennia and, although villages, complex technologies, and social hierarchies gradually developed, no urban cultures arose in the pre-Classical period that compare to those of Mesopotamia or the Indus Valley. In the Americas, several different forms of agriculture developed independently in centres such as the Andes and Mesoamerica. But it took many more millennia before crops such as maize became sufficiently productive to be grown in the kind of intensive cultivation systems required to sustain the impressive cities and empires that eventually evolved across the region.

Africa

With the exception of the Nile Valley, research on crop domestication and the development of complex societies in Africa was largely neglected until very recently. The continent is nowadays largely dominated by two types of ecosystem, a relatively arid north, and a tropical/subtropical centre/south that is mainly forest or savannah. Neither ecosystem is particularly conducive to the evolution of intensive agriculture. There has also been an impression that there was a dearth of indigenous plants possessing the right combination of nutritional qualities and genetic predisposition to domestication. As we will now see, there is an element of truth in these impressions, but recent research has shown that they fall far short of telling the whole story of African agriculture. In fact, there are several important crops, including sorghum and millet, that were independently domesticated in Africa. But, possibly the most surprising thing to emerge from archaeological and palaeoclimatic studies is the key role played by the Sahara region in the emergence and subsequent dissemination of agriculture in much of Africa.

The Sahara

Today, it seems curious that the Sahara, the largest desert on earth and one of the driest, hottest, and least hospitable places for life, was probably the earliest centre of agriculture in Africa.[882] However, after the Younger Dryas, the area now known as the Sahara Desert was a very different place indeed. Across the northern Africa, from the Atlantic coast to the Red Sea, the post-*11,000 BP* climate was moist enough to sustain a series of huge lakes and a profusely vegetated landscape (Figure 12.1). The area

of true desert receded to less than 5% of its current size, with much of the North African interior covered in an extensive belt of grassland that included numerous wild cereal species. In places, the grasslands extended deep into the central massifs, while at slightly higher altitudes, Mediterranean-type woodlands flourished. These relatively benign conditions persisted on and off for the next 5000 years in what is termed the African Humid Period.[883] This period was not uniformly humid and there were major aridification events at about *10,700 BP* and *7600 BP*. However, recovery from these events was relatively fast, as measured by the progressive increase in the maximum depth of Lake Chad after the end of each event. Hence, the lake depth was 30 metres at about *11,700 BP*, 40 metres at *9200 BP*, and almost 70 metres at *6000 BP*.[884]

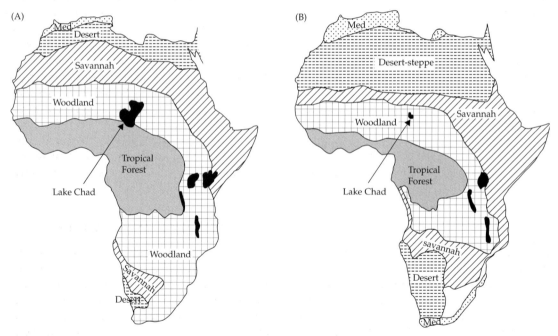

Figure 12.1 Vegetation patterns in mid-Holocene and modern Africa. (A) Generalized vegetation pattern at about *6000 BP*. Note the much reduced extent of the Sahara Desert compared to today, and the domination of most of North Africa by savannah and woodland ecosystems, as well as the greatly enlarged system of lakes, especially Lake Chad in the centre of the region. These habitats supported numerous human cultures during the early- to mid-Holocene, including sedentary cereal and tuber farmers and agropastoralists who occupied much of the fertile swathe of savannah and grassland across this immense continent. (B) The vegetation pattern of modern Africa reflects the consequences of a severe aridification event soon after *5450 BP*, which led to the desertification of most of the Sahara region and a continent-wide retreat of more complex vegetation communities such as woodland and tropical forest. The Saharan farming cultures abandoned their increasingly unproductive land, some moving east where they played a part in establishing the enduring agrourban civilization of the Nile Valley, while others moved south to bring agropastoralism to the rest of Africa. Med, Mediterranean zone. Both maps have been simplified for greater clarity but more detailed information is available from: Environmental Sciences Division, Oak Ridge National Laboratory, USA, http://www.esd.ornl.gov/projects/ qen/nercAFRICA.html.

Almost immediately after the end of the Younger Dryas, people from the south colonized the vast area of the newly verdant Sahara, settling especially around the belt of large lakes that extended across the continent at about the line of the 20th parallel (see Table 12.1 for chronology of major events). These lush savannah and lacustrine habitats supported a diverse and prolific megafaunal assemblage, including giraffe, hippopotamus, antelope, and elephant, in an area that nowadays has almost no measurable precipitation. Over the next 5000 years, the periodic bouts of aridity were neither severe nor lengthy enough to affect long-term settlement by the human groups that colonized the region so rapidly after the Younger Dryas. In the lakeland areas of the interior, migrating hunter–gatherer groups soon developed into large, mostly sedentary cultures with an economy based initially on exploitation of the rich lacustrine resources, including fish, waterfowl, and a lush flora. By *11,000 BP*, lacustrine societies had spread across 5000 km of Saharan North Africa and were the most numerous group of people in this part of the continent.

Some of the Saharan hunter–gatherers and lacustrine residents collected fruits and seeds, including cereal grains, from the prolific vegetation of the region. The most common wild cereals gathered as early as *11,000 BP* included sorghum, pearl millet, and fonio (*Digitaria exilis*). Edible tubers such as wild yams, *Dioscorea* spp., were also gathered.[885] Grinding stones were in use as early as *12,000 BP* and grain stores from about *9300 BP*, but it is not clear whether these were used for wild cereal grains or cultivated crops.[886] For example, as we saw in Chapter 2, Levantine hunter–gatherer cultures were grinding wild cereal grains and baking the flour in ovens as early as *23,000 BP*, but did not go on to develop agriculture for another ten millennia. According to Christopher Ehret, cattle rearing, crop cultivation, and pottery manufacture may have started in the fertile Sahara of the early Holocene well before the conventionally accepted dates, and possibly as early as *11,000 BP*.[887] Some of these claims remain controversial, but emerging evidence supports the thesis of a relatively early, independent emergence of agriculture and related technologies in this region, without necessarily

placing it quite as far back as Ehret has suggested.[888] In the more open areas of the Sahara, there is evidence of the beginnings of cereal cultivation and cattle herding at *9000 BP*.[889] By this time, central Saharan farming cultures were probably already producing the first African pottery, which was later exported westwards to the Niger and eastwards to the Nile.

It is possible that sorghum, finger millet, and pearl millet (sometimes also called bulrush millet), were domesticated independently in the North African region, and that both crops were already being cultivated there by *8000 BP* (Figure 12.2).[890] Between *8500* and *7500 BP*, as several of the lakes temporarily retreated, the lacustrine societies that had hitherto dominated much of the region steadily declined in numbers. Despite the return of wetter conditions after *7500 BP*, lacustrine societies did not recover in a landscape now dominated by agropastoralists. Barley was introduced from the Near East into the more easterly parts of the Saharan farming belt at about this time, but indigenous millet and sorghum continued to hold sway as the major crops further west. The relatively humid conditions continued until about *6000 BP* when Lake Chad reached its maximum postglacial extent.[891] This lake alone, and it was just one of several in the region, may have had an area as large as one million square kilometres, which is about 740-times greater than its much-reduced extent today.[892]

The Great Drought of the mid-Holocene

By the mid-Holocene, the Sahara had been supporting thriving and productive agricultural and pastoral societies for well over 4000 years. But events then took a dramatic and unexpected change for the worse for the Saharan agriculturalists. Within as little as a few decades, at about *5450 BP*, there was a sudden and catastrophic climatic shift across the whole of North Africa, whereby the summer monsoon moved several hundred kilometres southwards, leaving much of the land bereft of its life-giving seasonal inundation. Palaeoclimatic and modelling studies confirm that this shift was a rapid event, probably precipitated by a gradual increase in summer insolation due to fluctuations in the Earth's orbit.[893] Summer

Table 12.1 Chronology of African farming cultures, *14,000* to *3800* BP

Date BP	Location	Climate	Food plants	Agricultural technologies	Other technologies	Social systems
14,000	Upper Nile Valley Ethiopia	Seasonal rains	Wild tree fruits and and small-grained grasses	None	Specialized grindstones for small grasses, stone tools	Indigenous, highly dispersed hunter gatherers
12,000	Sahara region	Hyperarid	Virtually devoid of vegetation	None	No human population	No human population
11,500	Sahara region	Increasing warming and return of rainfall Increasing vegetation	Tree fruits, wild cereals including sorghum, pearl millet, fonio	None	Grindstones Stone tools	Hunter gatherers migrate from south to colonize newly fertile land of lakes, woods, and savannah
11,000	Sahara region	Warm and moist, many productive lakes, profusely vegetated	Tree fruits, wild cereals including sorghum, pearl millet, fonio	None	Grindstones Stone tools	Widespread semisedentary lacustrine societies Hunter gathering and fishing
11,000–10,000	Sahara region	Warm and moist, many productive lakes profusely vegetated	Tree fruits, wild cereals, cultivated millets (and sorghum?)	None	Pottery? Stone tools	Widespread semisedentary societies, cattle rearing Hunter gathering and fishing
10,500	Eastern Sahara	Seasonal rainfall,	Fruits, tubers, wild grains	None	Stone tools	Non-farming sedentary cultures
9300	Sahara region	Warm and moist, many productive lakes, profusely vegetated	Spread of millet and sorghum cultivation	None	Pottery? Stone tools	Midsized villages with grain stores away from lakes Hunter gathering and fishing
9000	Sahara region	Warm and moist, many productive lakes, profusely vegetated	Widespread cereal cultivation, wild fruits	Localized rainfed farming	Simple pottery Stone tools	Midsized villages with grain stores, cattle rearing Hunter gathering and fishing
9000	Eastern Sahara	Seasonal rainfall	Fruits, tubers, wild grains	None	Undecorated pottery	Non-farming, hunter–gathering, semisedentary cultures,
8400	Eastern Sahara	Seasonal rainfall	Fruits, tubers, wild grains	None	Undecorated pottery	Non-farming sedentary cultures, increasing pastoralism (sheep, cattle, goats)
8500–7500	Sahara region	**Partial drought** Retreat of lakes	Widespread cereal cultivation, fewer wild plants	Spread of rainfed farming	Pottery Stone tools	Decline of lacustrine societies
8000	Ethiopia	Seasonal rains	Domestication of local cereals: tef, noog, ensete	Dispersed rainfed farming	Grindstones Stone tools	Midsized villages with grain stores and broad-based cereal economy
7500	Sahara region	Warm and moist, enlarged lakes, profusely vegetated	Widespread cereal cultivation, wild fruits	Widespread rainfed farming	Pottery Stone tools	Midsized villages with grain stores, cattle rearing
7300	Eastern Sahara	**Progressive drought**	Decline in all vegetation	Abandonment of farming	Pottery Stone tools	Emigration of pastoralists to South and protofarmers to Nile Valley
7000	Nile Valley	Seasonal rains and river floods	Barley, emmer wheat	Small-scale irrigation	Early pottery Chalcolithic tools	Sedentary farming cultures spread from Eastern Sahara

Date	Region	Climate	Crops	Farming	Technology	Society
6500	Nile Valley	Seasonal rains and river floods	Barley, emmer wheat	Larger-scale irrigation	Naqada pottery, Cosmetics	Spread of sedentary farming cultures, increasing complexity and diversity
6000	Nile Valley	Seasonal rains and river floods	Barley, emmer wheat	Intensive irrigation farming	New pottery styles throughout Egypt, trade with Levant	Upper Nile densely populated with sedentary farming cultures, large towns
6000	Sahara region	Warm and moist, profusely vegetated Lake Chad = maximum	Widespread cereal cultivation, wild fruits	Widespread rainfed farming	Pottery, Stone tools	Midsized villages with grain stores, cattle rearing
6000	Sub-Saharan region	Hot and wet	Cultivation of yams and oil palm? Many wild plants	Small-scale farming and wild plant gathering	Pottery, Stone tools	Small-scale, dispersed, semisedentary village societies
5500	Nile Valley	Seasonal rains and river floods	Barley, emmer wheat	Intensive irrigation farming	Diverse pottery styles, beer manufacture	Large planned cities, Hierakonpolis, centralized brewing and pottery industries
5450	Sahara region	**Prolonged drought** Failure of mansoon	Beginning of crop failure, reduction in wild plants	Progressive abandonment of farming	Pottery, Chalcolithic tools	Collapse of sedentary farming and lacustrine societies
5400	Sahara region	**Prolonged drought** Failure of monsoon	Total crop failure, huge reduction in wild plants	End of farming	Pottery, Chalcolithic tools	Extensive depopulation, emigration of pastoralists and farmers to South
5000	Upper Nile Valley Ethiopia	Seasonal rains	Domestication of sorghum	Localized rainfed farming	Pottery, Chalcolithic tools	Small-scale, dispersed, semisedentary village societies, strong Egyptian influence
5000	Nile Valley	Seasonal rains and river floods	Barley, emmer wheat	Intensive irrigation farming	Hieroglyphic writing	Merging of chiefdoms into two kingdoms, 'divine' kingship, mummification of dead
4500	Nile Valley	Seasonal rains and river floods	Barley, emmer wheat	Centrally planned intensive irrigation farming	Pyramids of Giza, Bronze tools	Unification of Upper and Lower Egypt, increased centralization and bureaucratic control
4200	Nile Valley	**Drought**	Barley, emmer wheat	Failure of irrigation system	Bronze tools	Collapse of Old Kingdom, famine and some cultural simplification but recovery after rains return
4000	East and West Africa	Tropical: hot and wet	Domestication of pearl and foxtail millets	Localized rainfed farming and wild plants	Stone tools	Small-scale, dispersed, village-based and nomadic agropastoralist societies
3800	Nile Valley	Seasonal rains and river floods	Barley, emmer wheat	Rebuilding of larger grain storage and irrigation systems	Bronze tools	Rapid recovery of fully urbanized culture, New Kingdom, increasing bureaucratic control

One of the most important aspects of our new understanding of agrosocial development in Africa is the recognition of the important role of early Saharan societies. Following a series of locally catastrophic droughts, the easternmost Saharan cultures gave rise to the great Nile Valley civilization, while other groups trekked south and west into tropical Africa, taking their agropastoralist cultures with them.

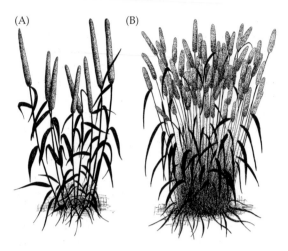

(A)   (B)

Figure 12.2 Pearl millet: one of the founder crops in Africa. Wild (A) and domesticated (B) forms of pearl millet. The domesticated form is much bushier and gives a far higher grain yield than its wild progenitor. This cereal crop, together with root crops, such as yams, was a key staple of many early farming cultures in Africa. As with proso millet in China and the early forms of maize in Mesoamerica, African pearl millet was only a modest yielding crop and was incapable of generating the kinds of surpluses necessary to support large agrourban cultures as found in the Near East and the Nile and Indus Valleys.

insolation increased gradually for several millennia without affecting monsoon patterns. But eventually a 'tipping point' was reached, triggering a rapid climatic shift that may have occurred in less than a century.[894] The Saharan monsoons moved steadily southwards and this once fertile region turned into the hyperarid desert of today.

By *5400 BP*, the organized, sedentary agricultural and lacustrine cultures had totally collapsed, to be replaced by much smaller numbers of highly mobile pastoralists.[895] Although this event was obviously highly detrimental to the development of arable farming in the affected region, there may have been at least two side effects that benefited populations further afield. Firstly, it possibly led to the domestication of the only large animal on the African continent, namely the donkey, *Equus asinus*.[896] And, secondly, at least some of the Saharan agriculturalists migrated to other parts of Africa, bringing with them their domesticated crops and agronomic know-how. Unlike the later, more widespread aridification episodes around *4000 BP*, the end of the African Humid Period was localized to

the greater Sahara region, with little evidence of effects in adjacent areas such as the Nile Valley and the sub-Saharan tropics of West Africa.

At the eastern edge of the Sahara region bordering on the Nile Valley, agriculture seems to have been a minor activity, even during the millennia of warm/damp conditions of the mid-Holocene.[897] This region remained hyperarid after the end of the Younger Dryas and appreciable rainfall did not return until about *10,500 BP*, which was rapidly followed by its recolonization by plants and animals, including some human migrants from the north and south. In contrast to many cultures of the mid-Sahara and the Near East, many these eastern groups adopted sedentary lifestyles without becoming farmers. By *9000 BP*, they were producing undecorated pottery and, by *8400 BP*, domesticated sheep, cattle, and goats had appeared. Although largely sedentary pastoralists, these people supplemented their diet by hunting, fishing, and plant-gathering in a landscape dominated by well-watered savannah and woodland that was ideal for livestock and game alike. It seems that there was no impetus to develop cereal farming under such favourable conditions that not only supplied ample faunal resources but also produced an abundance of wild grains, fruits, and tubers. Following the progressive aridification of the entire region from *7300 BP*, there was a large-scale population exodus in two directions. A more southerly group became specialized nomadic cattle-herding pastoralists, establishing was to become a major way of life throughout sub-Saharan Africa that has persisted to the present day.[898] Meanwhile, several more northerly groups settled in the Nile Valley, which had been only sparsely populated before this time. This coincides with the development of sedentary cultures along the Lower Nile and, by *7000 BP*, wheat and barley introduced from the Near East were being cultivated along the Valley. During this 3200-year Neolithic occupation of the eastern Sahara by human groups, the key to their success was adaptability in choosing differing strategies of resource exploitation as environmental conditions fluctuated in the region. This was coupled with a high degree of mobility and a readiness to migrate to other regions where new lifestyles might be necessary.

The Nile Valley

In early Egypt, the major crop staple was barley, which had been introduced from the Levant by *7000 BP*. Cultivation of other cereals, legumes, and fruit crops soon followed, although barley kept its status as the principal staple of the Egyptian civilizations for many millennia to come. Egypt was also influenced by other agrarian cultures to the south, including the grain-growing regions of the mid-Sahara (prior to *5400 BP*), Sudan, and Ethiopia. Egyptian agriculture was one of the earliest to become highly became intensified, as their barley and wheat crops responded well to artificial irrigation. This led to the coevolution of a highly organized, urbanized society and an equally well-organized system of high-yielding, state-organized cereal farming that had some parallels with the emerging agrourban complexes in southern Mesopotamia. In Egypt and Nubia, intensification took the form of irrigation systems to exploit the annual Nile floods and extend the area of high-output cultivation. By *6500 BP*, the Upper Nile Valley had become densely populated and the first use of ceramics was recorded at Naqada, near the later city of Thebes. Early Naqada-style pottery spread across Upper Egypt, but a newer style appeared at about *6000 BP* and was adopted in both Upper and Lower Egypt. This latter style shows early evidence of contact with Levantine cultures to the north.[899]

The earliest, complex urban centres in the Nile Valley grew up during the Naqadan cultural period, culminating in the great city of Hierakonpolis (also called Nekhen). By *5500 BP*, Hierakonpolis was well established, with the earliest, large temple complex in Egypt. The city also contained a segregated industrial zone that included a brewery capable of processing barley to produce more than 1400 litres of beer per day for a population that probably peaked at about 5 to 10,000 people at around *5400 BP*.[900] The Nile Valley appears to have escaped the effects of the aridification events of *8200* and *5200 BP*, which had such drastic impacts on agriculture in northern Mesopotamia (see Chapter 10).[901] This was partially due to the fact that, as in southern Mesopotamia, Nile Valley agriculture relied on irrigation rather

than direct rainfall. As long as sufficient rain fell in the upstream catchment areas that fed the rivers in these regions, irrigated crops would flourish. Centralized social structures continued to evolve in the Nile Valley enabling a synergy between increased crop output, higher populations, and the harnessing of labour for both agricultural and non-agricultural tasks. The First Pharaonic Dynasty emerged soon after *5000 BP* and by *4500 BP*, the Pharaoh Khufu (Cheops) had built the still awe-inspiring Great Pyramid at Giza. During this period of the Old Kingdom in Egypt, the productive barley-based farming system continued, and was supplemented by crops such as wheat, pulses, and numerous fruit trees planted along the rivers and canals.

The first major break in the development of Egyptian civilization occurred soon after *4200 BP* with the collapse of the Old Kingdom and two centuries of violent economic and social upheaval known as the Intermediate Period. A few centuries later, matters were described thus in the *Admonitions of Ipuwer*:

The fruitful water of Nile is flooding,
The fields are not cultivated,
Robbers and tramps wander about and
Foreign people invade the country from everywhere.
Diseases rage and women are barren.
All social order has ceased,
Taxes are not paid and
Temples and palaces are being insulted.

Lower Egypt weeps;
the king's storehouse is the common property of
  everyone,
and the entire palace is without its revenues.
To it belong emmer and barley, fowl and fish;
to it belong white cloth and fine linen, copper and oil;
to it belong carpet and mat, [. . .]
flowers and wheat-sheaf and all good revenues[902]

The first two lines refer to the eventual return of the Nile floodwaters after the end of the drought, but by then all the food, including seed grain, had been consumed, leaving nothing for the starving people to plant as their next crop. The second stanza is preoccupied with the pillaging of the king's granaries and the loss of 'all good revenues'. This poem has interesting parallels with the *Lamentation over the Destruction of Ur*, as quoted in

Chapter 10. Although the two works were almost certainly written independently, they each describe the unexpected demise of a highly centralized cereals-based civilization and the ensuing social chaos. Thanks to written records, we know that the earliest signs of the arid period in Egypt were in *4184 BP*, when the Nile floods were much lower than usual. This ushered in a 150-year drought that resulted in a failure of the irrigation system and the societal breakdown described above.

When Pharaoh Mentuhotep I eventually reunified Egypt in *4046 BP*, he greatly elaborated the irrigation and grain-storage infrastructure to ward off any recurrence such a disaster. A further interesting development at this time was the portrayal of the ruler as guarantor of agricultural bounty and shepherd of his people.[903] As in much of post-Ubaid Mesopotamia, Egypt was an intensely bureaucratic state, based on meticulous record keeping by a cadre of privileged officials of whom the Pharaoh was the head. For more than a millennium after its invention in about *5000 BP*, writing in Egypt was used almost exclusively for administrative purposes and no literary texts (such as the *Admonitions* above) were produced until about *3950 BP*. For much of the next two millennia, agriculture continued in more-or-less the same vein in the Nile Valley, until the arrival of the Ptolemies, as we will see in the next chapter.

The rest of Africa

There were probably several additional centres of agriculture in remaining parts of north and west Africa. However, the kinds of agriculture developed there were less intensive than in the contemporary Near East or Nile Valley societies. For example for many millennia to come, most of these other African farmers continued to use hoes and digging sticks, instead of ploughs, draught animals, and irrigation. Although most of the crops grown in the Nile Valley were introduced from the Near East, cultures further upstream, along the Upper Nile as far as Ethiopia, were already using specialized grindstones to process indigenous small-seeded wild grasses into flour by *c. 14,000 BP*. By *8000 BP*, these people had developed a broad-based cereal economy using local and imported crops, including sorghum and millets from Saharan agriculturalists and wheat and barley from the Lower Nile and Near East.[904] By *7000 BP*, there is linguistic evidence that Ethiopia was an independent centre of domestication of numerous crops including species such as tef, noog, and ensete, that never became staples outside the immediate region.

Further south, in the more forested, equatorial regions, yams and oil palm were possibly being cultivated by *6000 BP*. But, in general, hunter gathering and nomadic pastoralism were predominant lifestyles for subequatorial African cultures, especially in the far south of the continent. The decision not to adopt agriculture by numerous cultures in regions as far apart as Africa, Australia, and the Americas was due neither to ignorance nor any lack of sophistication. In fact, quite the reverse was true. In the case of much of southern Africa, the available plant domesticants could not compete with other food sources, such as wild flora and fauna, either in terms of its productivity or the comparative ease of plant management and collection. As we saw in Chapter 1, African hunter gatherer societies such as the !Kung have been aware of agriculture for many millennia but have opted instead for a non-farming lifestyle that, while rigorous and even at times precarious, has successfully sustained them and their culture to this day.[905]

Europe

The development of agricultural societies in Europe was influenced by several factors. First, much of the region was virtually bereft of indigenous food plants suitable for domestication into crops. Second, as the north-westerly projection of the main Eurasian landmass, Europe was adjacent to the major Near Eastern cradles of cereal and legume agriculture. Third, the climatic conditions of continental Europe were sufficiently different from the Near East to preclude the rapid dissemination of most early crops, without the selection of hardier varieties and changes in agronomic practice, such as spring-sowing of grains. There has been a tendency to regard the spread of agriculture to Europe as an almost passive process of technological diffusion that seeped from the Near East, westwards along the Mediterranean Basin to Iberia;

and north-westwards via Anatolia and the Balkans, to the Central European Plains and beyond to the Atlantic fringe (see Figure 8.2). This depiction of the trajectory of European agricultural expansion, which was already underway in Italy and the Balkans by *8500 BP*, is superficially accurate, but in reality it was a much more complex and dynamic process than is often stated (see Table 12.2 for chronology of major events).[906]

Crops such as emmer wheat and barley fared reasonably well in the warm Mediterranean climate, and by *8200 BP* farming villages had already been established in southern Italy and Sicily. Further west along the Mediterranean coastline, farming spread much more slowly, because it could not compete with well-established and successful fishing and hunter–gathering lifestyles. Gradually, a mixed farming/fishing/hunter–gathering economy developed in this region, with some societies becoming sedentary and developing distinctive styles of pottery, without completely relinquishing their traditional marine and terrestrial food resources. These communities used a new pottery style that was decorated with impressions of one of their principal forms of seafood, namely cockles, *Cardium edule*. This so-called 'Cardial' pottery style spread to almost all the Neolithic cultures of the western Mediterranean and, by *7000 BP*, the first inland villages based on cereal farming had been established in southern France.[907] Meanwhile, in central and north-west Europe, the uptake of crop-based agriculture followed an entirely different course, as we will now discuss.

Linearbandkeramik cultures: *7500 BP* and beyond

The springboard for the north-westerly expansion of Near Eastern farming was Anatolia. Large agrarian and/or trading towns such as Çatalhöyük, which had as many as 10,000 inhabitants, were already flourishing in Central Anatolia by *9500 BP*, as relatively warm and moist conditions prevailed across much of Eurasia. These cultures were in contact with their westerly neighbours in the Balkans, with whom they shared some aspects of material culture. This region was to provide the conduit for the introduction of domesticated emmer wheat and

barley into the Balkan Peninsula by *8500 BP*. However, further progress north was initially restricted by the cooler, damper climates that reduced yields and facilitated diseases in the new crops. It is still a matter of controversy whether cereal farming was brought to the Balkans and beyond to Central Europe by settlers from Anatolia, or was spread by cultural diffusion. The reality is almost certainly a combination of the two processes, although recent genetic data strongly suggest that immigration into Europe has probably been quantitatively much less important in determining the makeup of the present day population than previously believed.[908]

Analysis of male lineages (Y chromosomal DNA) suggests that indigenous European populations of today are overwhelmingly descended from two waves of Palaeolithic migrants who arrived from the Near East between *40,000 to 35,000 BP* and *25,000 to 20,000 BP*, that is many millennia before the dissemination of agriculture.[909] A complementary study of female lineages (mitochondrial DNA), published in 2005, reveals a fascinating story of limited migration by Near-Eastern agriculturalists at around *7500 BP*, followed by an eventual swamping of the newcomers by the indigenous population.[910] In this study, DNA from female skeletons from several sites occupied by the incoming, agrarian *Linearbandkeramik* culture was found to have similar genetic markers to other Neolithic farming sites across Central Europe. However, these markers are 150-times less frequent in modern Central European females. The data are consistent with the migration of small groups of pioneers originating from the Near East. These migrants initially introduced cereal farming into Central Europe via the Balkans by following the rich loess soil deposits and by moving along the alluvial plains of the major river valleys of the Danube, Elbe, Rhine, and Oder. However, the migrants were always in a minority and once they intermarried with the indigenous population, many aspects of their culture were incorporated into this larger group. The end result, which has been seen in many other places and times in human history, was the dissemination of many new aspects of material and farming culture by small groups of incomers without radically changing the genetic makeup of the resident population.[911]

Table 12.2 Chronology of European farming cultures, 40,000 to 1000 BP

Date BP	Location	Climate	Major food plants	Agricultural technologies	Other technologies	Social systems
40,000–35,000 and 25,000–20,000	Europe	Hypervariable, Mainly cold and dry	Seasonal fruits, acorns to supplement game diet	None	Larger stone tools	Two waves of highly nomadic hunter gatherers enter from Africa, diminishing megafauna
20,000–8500	Europe	Hypervariable, then moist/mild after 11,500	Seasonal fruits, acorns to supplement small game diet	None	Larger and some smaller stone tools game	Highly nomadic hunter gatherers, switch from megafauna to small game
8500	Balkans	Moist and mild	Emmer, einkorn, barley, lentils	Dispersed extensive farming	Full range of stone tools, grinding, some pottery	Migration or cultural exchange results in Anatolian-like farming cultures in Balkans
8200	Mediterranean littoral	Moist and mild	Emmer wheat, barley to supplement fishing	Dispersed extensive farming	Cardial pottery, fishing	Seaborne migration of coastal cultures across Mediterranean
7600	Euxine Lake Anatolia/Balkans	Euxine Lake inundated to form Black Sea	Emmer, einkorn, barley, lentils	Dispersed extensive farming	Full range of stone tools, grinding, some pottery	Displaced farming communities migrate through Balkans into Central Europe
7500	Southeast–Central Europe	Moist and mild	Emmer, einkorn, barley lentils	Dispersed extensive farming	*Linearbandkeramik*	Farming groups spread north and west along river valleys, small villages with longhouses
7300	Eastern France, Northern Poland	Moist and mild	Emmer, einkorn, barley, lentils, peas, flax	Dispersed extensive farming	*Linearbandkeramik*	Farming groups spread north and west along river valleys, small villages with longhouses
7000	Mediterranean, Southern France	Moist and mild	Emmer, einkorn, barley, pulses	Dispersed extensive farming	Cardial pottery, some trade	Dispersed village communities with wider trading and cultural links
7000	Southeast–Central Europe	Moist and mild	Free-threshing wheat, naked barley, pulses	Localized intensive garden-farming, ox-drawn and ploughs	*Linearbandkeramik*	Mixed livestock, cultural/genetic exchange with indigenous peoples
7400–6400	Europe	Moist and mild	New varieties for dry upland soils and cool, damp climates	Localized intensive garden-farming, and ploughs and manure	*Linearbandkeramik*	Abandonment of hunter gathering by much of indigenous population, mixed cropping/pastoral economy

Date (BP)	Region	Climate	Crops	Farming	Technology	Social/cultural
6400	Central and Northern Europe	Moist and mild	Emmer, barley, pulses	Localized intensive garden-farming, ard ploughs and manure	Increasingly diverse localized pottery and housing styles	Relatively egalitarian, possibly matriarchal, communities
6000	Central and Northern Europe	Moist and mild/cool	New hardier, cold-adapted cereals allow farming to move North	Localized intensive garden-farming, ard ploughs and manure	Localized pottery and housing styles	Most people live in farming-based villages ruled by chiefs with regional links, some game hunting
4800	Central and Northern Europe	Moist and mild/cool	Emmer, barley, pulses	Larger scale intensive cereal/livestock farming	Localized pottery and housing styles	Emergence of social stratification, elite houses and graves
4500–3700	Central and Northern Europe (e.g. Únětice culture)	Moist and mild/cool	Emmer, barley, pulses	Larger scale intensive cereal/livestock farming	Bronze tools and weapons	Spread of social stratification, elite houses and 'princely' graves, megaliths in Northwest
4000–2500	Europe	Moist and mild/cool	Emmer, barley, pulses Domestication of wild oats	Enclosed fields, intensification of cereal farming	Bronze tools and weapons	Large population increase, brief settlement of marginal uplands, regional hierarchical cultures
3200–2000	Europe	Moist and mild/cool	Emmer, barley, pulses Domestication of wild oats	Enclosed fields intensification of cereal farming	Spread of iron tools and weapons	Stable population, retreat from marginal uplands, regional states, sporadic urbanization
2000	South and Western Europe	Moist and mild	Emmer, barley, pulses, vines	Enclosed fields, huge intensification of cereal farming	Sophisticated iron tools and weapons, complex pottery	Roman imperium, social stability, increasing urbanization, population increase
1500–1000	Western Europe	Moist and mild	Emmer, barley, pulses, breadwheat	Sharp decline in productivity, land abandoned	Technological simplification, some cities abandoned	Widespread deurbanization, social upheaval, population decline, cultural simplification
1500–1000	Southeast Europe and Mediterranean Basin	Moist and mild	Emmer, barley, pulses, breadwheat, vines or stasis in productivity	Much slower decline	Continuity and innovation after Arab conquest	Some urban decline, especially in northwest, but less so in east, stable population numbers

Agrosocial development in Europe was dominated by the gradual dissemination of crops, technologies, and cultural trends, which had originated in the Near East, although they were frequently modified on their way northwest towards the Atlantic littoral. Farming was brought by small groups of migrants to Central and northwest Europe from Anatolia, via the Balkans, or directly across the Mediterranean to Italy, southern France, and Spain. The migrants eventually intermarried with indigenous hunter–gatherers who increasingly adopted farming themselves, so that by *3000 BP* various forms of agropastoralism had spread across the continent as far as Scandinavia and Ireland.

These migrant farmers may have been instrumental in establishing the distinctive *Linearbandkeramik* (linear pottery) culture that began in Hungary and swept through Central Europe to become the dominant ceramic motif between *7400 BP* and *6400 BP*.[912] There is an intriguing possibility that some of the migrants into Central Europe may have been propelled hence by a hitherto unremarked environmental event, namely the draining of the Mediterranean Sea into the Euxine Lake.[913] The Euxine was a huge freshwater lake occupying part of the area now known as the Black Sea. It was separated from the Sea of Marmara and the Mediterranean by an immense dam-like earthen berm that lay athwart the Bosporus Straits. Prior to the Younger Dryas, the Euxine Lake had flowed over this berm into the Sea of Marmara. But in the newly arid climate during and immediately after the Younger Dryas, the lake retreated and its level fell to 150 metres below that of the Mediterranean. By *7800 BP*, farming cultures were established on the rich, recently-exposed soils all the way around the Euxine lakeshore from Anatolia in the south, to the Ukraine in the north, and the Balkans in the west. However, the return of a wetter climate after the Younger Dryas also meant that sea levels were rising around the world, eventually threatening the berm that divided the Euxine Lake from the Mediterranean Sea.

In about *7600 BP* the berm was finally breached and the freshwater Euxine Lake was gradually (or rapidly—depending on who one consults) inundated by a huge inflow of seawater to form the much larger Black Sea, as it exists today. The Euxine littoral communities would have seen their croplands flooded and rendered useless by the incoming salt water. Those living to the north-west, in modern Bulgaria and Romania, would have been forced to move inland across the Bulgarian Plain and along the Danube Valley towards the Hungarian Plain. Along these river valleys, the soils would have been not too dissimilar to those of the former Euxine littoral, and the migrants may have established small farming communities to maintain their previous lifestyle. Such communities might have initially coexisted with indigenous groups of more mobile hunter–gatherers, who exploited the floral and faunal resources of the woodland habitats in the interior highlands, rather than the riverine strips favoured by the agriculturalist newcomers. The settled farming communities formed a ribbon-like development along the heavy soils of the major and minor river valleys that extended deep into the heart of Europe.

By *7300 BP*, the *Linearbandkeramik* culture had reached eastern France and northern Poland. The advance of farming then paused again for several centuries until new crop varieties arose (almost certainly via spontaneous mutation), which were better adapted for the cooler maritime climates of north-west Europe. Although they came from the Balkans, a region that was now growing a wide variety of cereals and pulses, the newcomers in Central Europe relied mainly on emmer and einkorn wheat, which they probably grew together as a single staple crop. Possibly, these were the only species in their crop portfolio that grew well and reliably enough in the cooler climate. It was to be another 500 years before the crop spectrum was widened to include peas, lentils, flax, and a little barley. The latter crop became more common after *7000 BP*, when naked barley varieties and free-threshing forms of wheat were developed.[914] Unlike the situation in the Near East, adverse climate and soil conditions precluded intensive farming in early Neolithic Europe and communities remained small. It is calculated that a family of six on a typical *Linearbandkeramik* farm would have needed at least 2.5 ha of cereals for its annual sustenance. To achieve this, most farms employed ox-drawn ard ploughs.[915] This strategy would have reduced the yield of cereals per hectare, but would have enabled much larger areas to be worked by fewer people.

An alternative agronomic strategy was a small-scale but intensive form of garden cultivation with mixed livestock/crop areas and high labour inputs.[916] In such a model, it is calculated that a family of five could live off as little as 1 ha.[917] Intensive garden cultivation would have been suitable for poorer soils, once farming spread beyond the fertile valley bottoms. Meanwhile, the process of cultural exchange and intermarriage between the farmer-migrants and the indigenous hunter–gatherers had already started.[918] A study of two *Linearbandkeramik* cemeteries in the upper Rhine

Valley, dating from *7300–7150 BP*, provides clear evidence of such cross-cultural intercourse.[919] This mainly involved women from the adjoining highlands on either side of the valley joining farming communities on the valley floor.

Eventually, farming spread up to the lighter soils above the valleys, perhaps facilitated by development of new crop landraces more suited to the lighter, drier soils.[920] As we saw above, such a move would have been facilitated by the development of intensive garden cultivation. Soils would have been enriched by manure from the household and from domestic livestock.[921] Because many valley farmers were now related to the highland occupants, the colonization of these new soils may even have been a joint venture between these now-converging social groups. In some of the new farmlands, a few of the weedy grasses that had accompanied the original cereal crops may have started to fare better than the crops themselves. This may have been how, during the later Neolithic, rye and oats eventually came to be used in preference to wheat and barley in some cooler areas with poorer soils. With its continual replenishment of the soil, the practice of intensive garden cultivation can be sustainable over relatively long timescales.[922]

There is little evidence of social stratification in early *Linearbandkeramik* societies. Most settlements consisted of collections of similar-sized longhouses that often occupied the same site for many centuries (Figure 12.3).[923] During much of the Neolithic, constraints on crop yields due to factors such as climate, soil type, and crop genotype, precluded sufficient surpluses to allow for evolution of larger social units. The vast majority of the population worked in a very basic subsistence economy of relatively low-yield farming (even in areas of intensive garden cultivation), or pastoral/hunter–gathering lifestyles. Although the cooler, cloudier, damper climate was the proximate cause of low yields in Central European farming, it was not an absolute limitation, as it could eventually be overcome by judicious manipulation of crop genetics and soil fertility. Hence, new varieties and new crop species better adapted to the climate gradually enabled yields to be increased. Another important contribution to increased yield was the refinement of physical (rather than biological) technologies involved in agriculture. Here the early development of the wooden ard plough, and the later invention of metal ploughheads, enabled an individual farmer to work more land and hence generate higher overall yields (although not necessarily on a per hectare basis).

New evidence suggests that *Linearbandkeramik* agricultural strategies did not necessarily force farmers to keep moving due to soil exhaustion, as often assumed hitherto.[924] The extensive ard plough method and intensive garden cultivation were both relatively sustainable strategies that may have performed well in particular areas. It now appears that farming in Central Europe eventually spread during the middle and late Neolithic periods of *7000* to *6000 BP* more due to uptake by indigenous hunter–gatherers, than to migration of farmers away from depleted soils. Previously, this process had been regarded as involving a gradual supplementation of cattle herding and wild plant collection with produce from cultivated crops.[925] In contrast, Bogaard and others have now suggested that the forms of hunter–gathering prevalent in late Neolithic Europe could not have satisfactorily coexisted with a little farming 'on the side'. The reasons for this are mainly logistical, such as conflicting periods when either one activity or the other would have to be favoured.

The implication is that, once selected, agriculture would have rapidly displaced hunter–gathering as the major form of food production. In other words, migrants from the south-east probably introduced farming into Central Europe along the major arterial routes, especially river valleys. Over a longer period, these incoming migrant groups intermarried with the much larger, indigenous hunter–gatherer communities that occupied most of the remaining land in the region. The adoption of farming by the indigenous community in a specific location may have been a sudden decision because simultaneous adherence to the two lifestyles may not have been possible. This process led to the steady expansion of farming whenever it proved to be a more adaptive method of food production than the alternatives.

By *6000 BP*, the development of hardier cereals, most notably new cold-adapted varieties of wheat and barley, allowed the resumption of agricultural

Figure 12.3 Central European *Linearbandkeramik* community, *7500* to *6400 BP*. The dispersed farming-based *Linearbandkeramik* (linear pottery) culture flourished in Central Europe during the early- to mid-Neolithic introduction of cereal/legume cultivation from the Near East. The drawing shows a group of communal longhouses and adjacent arable and livestock areas. Farming was locally intensive with crop yields increased by manuring and weeding. The major staples were emmer wheat and barley, which were stored within the huts after harvest, or ground to flour for breadmaking. Such farming communities were descended from immigrants who initially occupied river valleys throughout central and northern Europe and freely intermarried with neighbouring groups of indigenous hunter–gatherers. Eventually farming spread across the continent as the immigrants and many of their social systems became absorbed into the wider population. Unlike the Near East or Indus Valley of the mid-Holocene, European rainfed agriculture of this period was unable to provide large enough crop surpluses to support urbanization and cities did not develop until the Iron Age, some four millennia later.

expansion as far as north-west Europe. As we have seen in many other places, farming was not normally adopted by indigenous communities unless they faced problems with their existing food economy. Sometime the trigger was climate change, but it could also be due to much more localized, small-scale phenomena. Hence, in coastal regions of Denmark, disappearance of oysters as an

Date	Location	Climate	Crops/plants	Farming system	Technology	Society
1400	Colorado Plateau Mogollon Highlands	Warm and seasonally moist	*Maiz de ocho* irrigation farming	Locally intensive	Improved grinding systems	Larger villages size, more prominent elites
1250–1100	Colorado Plateau Mogollon Highlands	**Drought**	Partial switch from maize/squash/beans staple to wild plants	Locally intensive irrigation farming and renewed gathering	Stasis or regression	Social stasis, halt in innovation, some depopulation but eventual recovery
1200–1100	Yucatán Peninsula	**Drought**	Dependence on cultivated maize, squash, beans	Overexploitation of fragile soil system	Stasis or regression	Many cities in central Yucatán suddenly abandoned but others survive in North and South
1100–900	Colorado Plateau Mogollon Highlands	Warm and seasonally moist	Dependence on cultivated maize, squash, beans	Locally intensive irrigation farming	Road network of >100 km, complex canal system	Large Anasazi towns, Pueblo Bonito had 3–4000 people, rule by elites
1000–500	Central Mexico	Warm and moist	Maize, squash, beans cacao, peppers amaranth, chillies *etc*	Intensive *milpa* system and vast maize monocultures	Large range of technologies but no wheeled vehicles	Large cities, highly organized regional empires Toltecs then Aztecs
850–750	Colorado Plateau Mogollon Highlands	**Drought**	No cultivated plants and few wild plants	Irreversible collapse of irrigation systems, loss of vegetation	Stasis or regression	Sudden abandonment of towns, depopulation of region,
800	Amazon Basin	Hot and wet	Manioc, maize, yams, numerous wild plants	Locally intensive supplementation of poor soil with charcoal	Monumental buildings	Small and large villages, possibly towns of 1–5000 people
500	Central Mexico	Warm and moist	Maize, squash, beans, cacao, amaranth, chillies	Intensive *milpa* system *chinampas* and vast maize monocultures	Large range of technologies but no wheeled vehicles	Aztec Empire based in huge metropolis of Tenochtitlán, 500,000 people
500	Peru and region	Warm and moist	Maize, squash, potatoes, beans, chillies	Highly intensive terrace and irrigation farming	40,000 km road network, *quipus* recording system	Inca Empire based on efficient, highly centralized exploitation of farming

Three of the most interesting aspects of agrosocial development in the Americas are: (i) the delay (relative to China, Africa, Near East, and India) in the emergence of intensive farming and complex civilizations, which was probably due at least partially due to the gradual increase in the size of maize cobs from *7000–3200 BP*; (ii) the virtual absence of domesticated livestock (apart from llamas and alpacas in some parts of the Andes); and (iii) the development of exceptionally large, bureaucratically managed cities and empires with impressive road networks, despite the complete absence of either wheeled vehicles or advanced writing systems. The vast majority of American civilizations were underpinned by the uniquely nutritious and productive maize/beans/squash trinity of crops, often grown together in the *milpa* system.

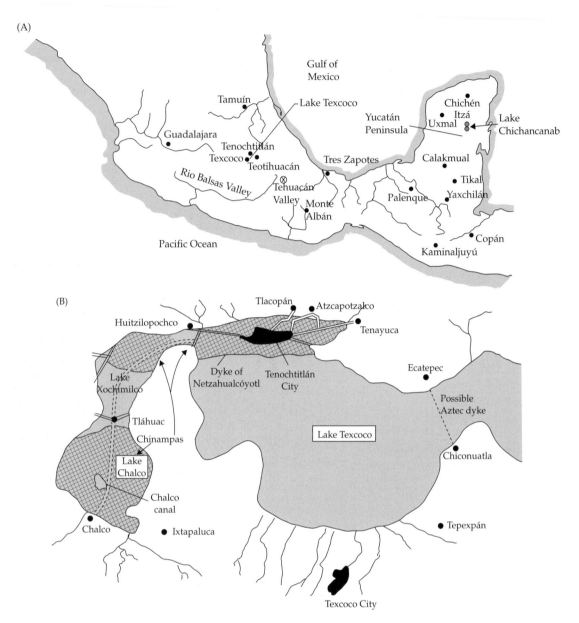

Figure 12.4 Mesoamerica: home of maize farming and the *milpa* system. (A) The key sites of Mesoamerican agriculture, which was based on squash, maize, and later beans. It originated in the modern Mexican states of Puebla, Guerro and Oaxaca, especially in the Tehuacán and Rio Balsas Valleys, where squash cultivation dates from before *10,000 BP*. After *7000 BP*. After *2000 BP*, the cultivation of large-cobbed maize with squash and beans in the highly productive *milpa* system coincided with the rise of the Olmec, Zatopec, and later the Mayan civilizations of south and east Mexico (around Monte Albán and Tres Zapotes) and the Yucatán Peninsula. These highly urbanized societies practised intensive agriculture for more than a millennium, using an innovative mixture of techniques including drainage, irrigation, and *chinampas*. (B) Lake Texcoco area in the late fifteenth century CE. This lake was the cradle of several imperial city-states, including Teotihuacán, Texcoco, and Tenochtitlán. By the time of the Aztecs, a vast system of intensive agriculture had developed whereby *milpa* crops were grown on a network of unique floating islands called *chinampas* (see Figure 12.6), which dominated the entire northwestern part of Lake Texcoco and the adjacent Lake Chalco. This system was so productive that the *chinampas* supplied almost half of the food requirements of the vast Aztec metropolis of Tenochtitlán, the largest city in the world at the time. Tenochtitlán was located on an island in the lake, surrounded by *chinampas*, and was linked to the outside world by causeways and deepwater canals (Scarre, 1989).

As maize cultivation gradually spread across the Americas, farmers selected and grew new variants with ever-increasing cob sizes, and hence higher yields, as well as improved meal quality. These empirical breeders were largely selecting for mutations in recently identified domestication trait genes such as *tb1*, *su1*, and *pbf* (see Chapter 6). However, although some of the earlier teosinte/maize variants had undergone relatively rapid selection during the major teosinte/maize wild plant/crop transition, these newer changes were much more subtle and complex. Characters such as cob size and meal quality are quantitative traits that are controlled by numerous genes, so the selection process is much more lengthy and problematic.[939] Nevertheless, from the remains of cobs in continuously inhabited sites such as the Tehuacán Valley of Mexico, we can find evidence of a steady improvement in maize yields over the millennia. Here, the earliest cobs, from *7200* to *5400 BP*, were less than 2 cm long, but by *5400* to *4300 BP*, cob length had doubled to over 4 cm.[940]

Despite this increase in crop yield, it was to be more than another millennium before the earliest (Olmec and Zapotec) civilizations arose in the wider region.[941] Hence, maize was cultivated as a low-intensity, subsistence crop by village-based farmers for as long as five millennia before more complex urban centres evolved. While there may have been social and/or environmental reasons for this delay, it is likely that one of the major causes was biological. Not only was maize a relatively low-yielding crop during the preurban period, the simultaneous cultivation of beans and squash in the highly productive *milpa* system (see below) did not start until after *2300 BP*. Meanwhile, the process of cob lengthening in maize continued with an increase to 6.5 cm by *2200* to *1300 BP*, a period that coincides with the apogees of the Mayan and Teotihuacán civilizations.[942] Genetic evidence shows that these farmers were still selecting for meal quality traits regulated by genes such as *su1*, while other quality-related genes, such as *pbf*, were already in their modern forms.[943] By the Toltec and Aztec periods, about one millennium ago, maize cob lengths had grown to over 10 cm, with a commensurate increase in overall grain yield. This crop, grown intensively in the *milpa* system, was

the staple food supply for the series of complex and dynamic Mesoamerican civilizations that flourished until the European incursions of the early sixteenth century CE.

One of the earliest of these civilizations was the ancient Maya, who originated in the Yucatán Peninsula, and gradually spread across a vast area of Mesoamerica over the four millennia between *4600 BP* and *800 BP* (Figure 12.4).[944] The Mayans were skilled architects with a highly sophisticated knowledge of astronomy and mathematics. Their cultural achievements peaked between the years of *1400* to *1200 BP*. Shortly after *1200 BP*, however, much of the Mayan civilization underwent a rapid decline that has long been a mystery to archaeologists. The decline was most marked in Petén, which was the core region of Classical Mayan civilization. A whole string of cities was abandoned for good, including important centres such as Calakmul, Copán, Palenque, and Yaxchilán. Not even the pre-eminent metropolis of Tikal was spared. With a population of over 50,000 people, Tikal was the most powerful Mayan city-state, with close links to the Toltec capital at Teotihuacán. Despite their power and sophistication, Tikal and many surrounding cites were abandoned abruptly at around *1100 BP*. The suddenness of the urban collapse in Petén has been illustrated by recent excavations at the site of La Milpa, a city of about 46,000 people, some 90 km north-east of Tikal. The city was in the midst of a rebuilding programme when it was abruptly deserted, with major public works, including a ceremonial pyramid, left unfinished. In the words of the authors of the study: 'there was no slow dusk and drawing down of blinds at La Milpa: the Maya collapse here came quickly'.[945]

The Mayan collapse was probably due to a combination of interacting factors, none of which would necessarily have been sufficient to precipitate such a drastic response on its own (Box 12.1 and Box 12.1).[946] However, there is compelling evidence that there was at least some climatic component in the Mayan disaster. Geoclimatic studies of lake sediment cores in the Yucatán Peninsula have revealed a strong correlation between times of drought and major cultural discontinuities in Classical Maya civilization.[947] In one of the latest analyses, sediment core data from Lake

Box 12.1 Documenting the effects of climate change on Mayan civilization

The effect of climate change on the trajectory of human societal development remains a controversial topic, especially for many historians and social scientists. One of the challenges in attributing an effect to a specific climatic event has been the difficulty in estimating its precise date, location, and magnitude in relation to the society that is allegedly subject to such an effect. Another challenge is the lack of historical records of climatic events, either because they occurred before the invention of record keeping, or because climate itself was rarely mentioned and can only be inferred via references to disasters such as crop failure or drought, that may in some cases have had non-climatic causes. The recent reconstruction of the past climate of the Yucatán Peninsula and its correlation with Mayan historical records has enabled palaeoclimatologists to address some of these concerns by establishing robust correlations between episodic aridity and major cultural discontinuities in Classic Maya civilization. While one or two conjunctions of climate change with social upheaval might be dismissed as coincidence, it is much more difficult to ignore five such occasions, as we will now see.

The past climate was reconstructed by studying lake sediment cores at various locations in the Yucatán Peninsula. In closed basin lakes, such as Lake Chichancanab, the ratio of oxygen isotopes (^{18}O:^{16}O) in lake water is mainly controlled by the balance between evaporation and precipitation (rainfall). The isotope ratio of lake water is recorded by aquatic organisms, such as gastropods and ostracods, that precipitate shells of calcium carbonate ($CaCO_3$). The isotopic ratio in fossil shells in sediment cores can be used to reconstruct changes in evaporation/precipitation through time, hence inferring the relative wetness or aridity of the climate over a period of more than 3500 years. For example ostracods of the genus *Candona* were obtained from lake sediments at Punta Laguna and their age and oxygen isotope ratios were compared with historical records of Mayan cultural periods.

The data show oxygen isotope ratios varying around a value of -0.5, which is indicative of a relatively moist climate, until about 550 CE. After this, there were five well-defined peaks where the ratio increased to over 1.0, which indicates relatively arid conditions. These peaks occur at 585 CE, 862 CE, 986 CE, 1051 CE, and 1391 CE, with an error of about +/−50 years. The first peak, at 585 CE, coincides with the early/late Classic boundary, which was marked by a sharp decline in monument carving, abandonment of cities in some areas, and widespread social upheaval. This event may have been drought-related. During the next 200 years, from 600 to 800 CE, the late Classic Maya flourished and reached their cultural and artistic apex. The next peak in ^{18}O:^{16}O occurs at 862 CE and coincides with the collapse of Classic Maya civilization in the Central Yucatán between 800 and 900 CE. The earliest Postclassical Period, between 986 and 1051 CE, was also relatively dry. At about 1000 CE, there was a return to more humid conditions. Although a Postclassical resurgence occurred in the northern Yucatán, city-states in the southern lowlands remained sparsely occupied. These findings indicate a strong correlation between periods of drought and major cultural upheavals in Mayan civilization. Data from Curtis *et al.* (1996), reproduced from the US National Oceanic and Atmospheric Administration (NOAA), http://www.ncdc.noaa.gov/.

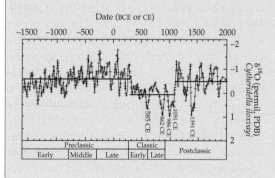

Chichancanab demonstrate that drought has been a recurring climatic feature in the Yucatán Peninsula over the past 2600 years.[948] The data rule out a role for increasing CO_2 concentration in these climatic fluctuations, as the CO_2 content has not varied in an oscillatory manner over the period of study. As with the abrupt termination of the African Humid Period (see above), the Yucatán data suggest that variable solar activity has been the major player in orchestrating the cyclical history of drought in this region. The Saharan and Yucatán episodes of abrupt climate change may both be due to relatively small oscillations in the orbit of the earth. Such oscillations can lead to changes in solar intensity, known as insolation or solar forcing, that can trigger sudden, localized alterations in weather patterns, often with drastic ecological and human consequences.

The consequences of renewed drought in the Yucatán might not have been so severe for the Maya if they had not developed a form of highly intensive agriculture that was particularly vulnerable to such perturbation. Much of the Yucatán is swathed in tropical rainforest which, with its thin soils constantly leached of nutrients by rainfall, is suitable only for low-intensity 'slash and burn' farming. The Maya settled instead in sediment-rich flood plains and favoured the cultivation of levees and raised fields created by draining and canalizing swampland.[949] These areas, now called *bajo* wetlands, were farmed intensively by a vast peasant population living in adjacent villages and closely co-ordinated by overseers from urban centres built on higher ground. *Bajo* wetlands were immensely productive of *milpa* crops, while tree crops such as avocado and cocoa were grown wherever it was suitable. The vast surpluses produced by this complex, highly regulated system were transported to the nearby cities where they both supported and conferred power on the large urban populations of labourers, artisans, soldiers, priests, and nobility.

Such a system was delicately poised and immensely fragile in the face of any sort of perturbation, whether environmental or social. As well as the climatic changes that brought drought to the region, it is possible that the soils were becoming exhausted by intensive use that was not adequately complemented by replenishment of minerals or other nutrients.[950] These two environmental factors probably contributed to a sudden drop in crop yields that triggered or exacerbated a series of economic, religious, and political upheavals. Eventually, the physical and organizational basis of this farming system was compromised to the extent that in some areas the entire agrourban edifice collapsed. Interestingly, not all Mayan cities suffered the disasters visited on the central region of Petén. The cities of the southern highlands, which did not practice *bajo* wetland farming, and those of the far north of the Yucatán, around Chichén Itzá, continued to flourish and were still inhabited when

the first Europeans arrived in Mesoamerica several centuries later.

A particularly innovative form of intensive maize-based agriculture, called *chinampas*, was developed to the north of the Mayan region in the Toltec, and later Aztec, heartlands of Central Mexico.[951] *Chinampas* are sometimes referred to as 'floating gardens', although they were tethered to the bed of a lake or river. The *chinampas* were stationary, artificial islands built along many lakeshores and rivers in the Valley of Mexico. Each unit was an approximately 100×10-metre rectangle, made of canes held in place by stakes and trees planted around the edge (Figure 12.5). Soil was heaped onto the canes to create a fertile platform for crop cultivation. *Chinampas* were used for the intensive cultivation of *milpa* crops as well as amaranth, chillies, and various ornamental flowers (as tributes to the gods). The platforms were periodically replenished with rich alluvial soil and proved to be spectacularly productive, regularly generating food surpluses that were donated to the gods by a grateful populace. The apogee of the *chinampa* network occurred around the fifteenth century CE when several geographically constrained city-states arose around the shores of Lake Texcoco.

One of the most remarkable rulers of this period was Netzahualcóyotl, king of the city-state of Texcoco, who was not only a well-regarded ruler and successful military leader, but also an erudite scholar, philosopher, poet, and engineer.[952] Netzahualcóyotl designed his eponymous dyke to separate coastal fresh water from the brackish waters of Lake Texcoco, thus enabling a massive expansion of the *chinampa* crop-growing area (Figure 12.6B). A few decades later, following the rise of the neighbouring Aztecs, it is estimated that *chinampas* produced between half and two thirds of the food requirements of the mighty city of Tenochtitlán. This metropolis eventually had a population of half a million people, making it by far the greatest urban centre in the world at that time.[953] By the time the first Europeans arrived in Mesoamerica, maize was being cultivated on such an epic scale across the whole region that the newcomers found it astonishing. In the 1520s, Diego Columbus (brother of Christopher) noted with awe

Figure 12.5 *Chinampas*: the 'floating gardens' of central Mexico. The *chinampas* were artificial islands, tethered to lake and river beds, which were an especially effective form of sustainable intensive agriculture. The *chinampa* network on Lake Texcoco alone generated well over half the food needed by the half-million population of the neighbouring Aztec metropolis of Tenochtitlán. The figure shows an artist's reconstruction of a network of chinampas showing (foreground) the harvesting of a bean crop and replenishment of the floating reed platform with new soil. Other platforms in the background are carrying maize crops or have already been harvested.

in his journal that he had walked for almost 30 km through a single, unfenced, and seemingly never-ending field of maize, beans, and squash.[954]

These preconquest Mesoamerican civilizations practiced an immensely productive form of state-sponsored, intensive agriculture that was comparable with the imperial Sumerian barley monoculturalists prior to the collapse of *4200 BP*. So efficient was their farming that it is estimated that only 20% of the Aztec population was required for agriculture, freeing up the remainder for other state-enriching occupations ranging from craftsmanship to warfare.[955] By comparison, as recently as 1850 CE, over 50% of the United States population still worked on the land.[956] As with the first Mesopotamian empires, such as that of Akkad, more than three millennia previously, the Aztecs based their power on an elaborate system of agricultural tribute. Each year, over 7000 tonnes of maize, 4000 tonnes of beans, two million cotton

cloaks, and many tonnes of valuable cocoa beans poured into the city of Tenochtitlán from conquered regions throughout central Mexico. Sadly, most of the agricultural achievements of Mesoamerican civilizations were extinguished in the mayhem of disease and destruction that accompanied the Hispanic incursions of the sixteenth century. State-organized, collective farming on such a heroic scale was not seen again until its much more poorly organized, and thankfully ephemeral, imposition in the Soviet Union and China in the twentieth century.

South America

Farming-based civilizations in South America arose in several parts of the Andean region, but at a much later date than in most other agrourban regions of the world. It was previously thought that this delay was due to the relatively late dissemination of potentially high-yielding domesticants, such as maize, from their Mesoamerican centres of origin. However, this view has been challenged by recent evidence of maize cultivation in Columbia and Ecuador as early as *8000–7500 BP*, and of cassava/manioc farming in Columbia by *7500 BP*.[957] These early South American farmers appear to have followed the Levantine model of small village-based societies rather than the Sumerian/Harappan/late-Mesoamerican models of increasingly urbanized societies based on appropriation and control of cereal surpluses by small elites (see below). In this regard, South American societies behaved in very similar ways to others around the world. Indeed, as discussed in the final chapter, the entire Holocene era can be seen as a period of flux between societal models based on a range of smaller, less complex, relatively stable village-based units all the way up to the huge, highly complex, but less stable, hyperurbanized polities that so dominate the written versions of human history.

Agrourban development in much of South America was restricted in much of the vast rainforest region by the poor soils and the lack of high-yielding domesticants. Contrary to popular belief, however, a number of relatively dispersed and small-scale, but complex, hierarchical societies developed in the Amazon Basin. These people created localized, intensive farming systems based on the addition of vast quantities of charcoal and other organic matter to the poor topsoil, thus effectively forming a uniquely fertile compost that supported an abundance of crops. Genetic evidence suggests that the starch-rich root crop, cassava (now one of the major staples of sub-Saharan Africa), may have been domesticated by groups of lowland tropical farmers on the southern fringes of the Amazon basin.[958] Unfortunately, very little is known about these societies, which collapsed in the sixteenth century mainly due to the massive disease epidemics introduced by Europeans, possibly exacerbated by localized overexploitation of the relatively meagre soil.[959]

Early farming of tuber crops in the high Andes did not generate sufficient yield to sustain large populations and, unusually, the first complex societies in South America were coastal fishing communities living in an arid region with neither crops nor pottery. Instead, the rich marine resources provided sufficiently reliable food surpluses to sustain the growth of well-organized, sedentary societies capable of ambitious construction projects, such as temples and ceremonial centres.[960] An example of such a centre is Aspero, which was built about *4600 BP*. At about the same time, Andean farming communities based on potatoes and quinoa (*Chenopodium quinoa*) crops, and herding of llama and alpaca, were establishing large permanent villages that by *4000 BP* included elaborate ceremonial centres. Pottery arrived in the Andes about *3600 BP*, having spread slowly after its initial development in Guyana and Columbia *c. 5500 BP*. Many of the indigenous crops, such as potatoes and oca, were restricted to the Andean highlands and were impractical to transport in the same way as the more compact grain crops.[961]

The arrival in the region of the new high-yield, large-cob maize after *3000 BP*, combined with indigenous crops such as potato and quinoa, gave the opportunity for much greater agricultural and urban intensification and the first true Andean civilization, the Moche state, emerged by *2100 BP*.[962] From *3800–2800 BP*, agriculture and its associated technologies and social systems spread to the coastal lowlands, where the use of irrigation greatly increased the area of potential arable land and hence the population that could be sustained. Farmers raised a wide range of crops, such as peppers, sweet potatoes, and peanuts, in addition to their staples of maize and potatoes. Between *2800* and *2200 BP*, long-distance trading developed between agrarian and fishing communities, as did monumental architecture and larger regional alliances with common religio-technological cultural attributes. It was from such regional groupings that the Moche state developed in *2100 BP*, with its immense adobe pyramids, skilled textile workers, potters, and metal workers. Instead of writing, the Moche used a less efficient recording system of marked beans, although this did not stop them establishing a large empire that endured for over 600 years.

Over the next millennium, several empires emerged in the greater Andean region, from the Atacama Desert in the south as far as present-day

Quito in the north. Little-known cultures and imperial states, such as the Nazca, Tiwanku, Chimú, and Wari ruled larger areas than most Old World empires.[963] These states relied for their cohesion on elaborate bureaucracies, good communication networks via roads reserved for state functionaries, shared religiocultural beliefs, and the immense bounty of a diverse but often locally-intensive agricultural system that extended from the irrigated coastal lowlands to elaborate terraced field as high as 4000 m in the Andes. In the past few years, thanks to a new synthesis of archaeological, climatological, and palaeobotanical studies, we have gleaned some fascinating insights into the adaptive agrosocial changes exhibited by such societies in response to repeated environmental changes. As recently pointed out by Dillehay and Kolata, societies in the greater Andean region adapted in different ways to different agrosocial stresses.[964] As ever, a reliable supply of water was by far the most important factor for efficient agriculture. In order to deal with the often considerable variations in rainfall in the region, some groups such as the Late Moche culture opted for decentralized, less complex agrosocial models, while others such as the Chimú instead adopted a highly centralized, technology-based, agrourban solution á la Sumer of the Akkadian/Ur II period (see Chapter 10).

The 'low-complexity' model of agrosocial adaptation was seen during the Late and Post Moche periods, *c. 1200–1000 BP*, when the onset of a series of prolonged droughts precipitated a series of social changes including decentralization, deintensification, and diversification of their agricultural and political systems. The population became more mobile, more spread out, more politically fragmented, and probably less prone to resource-wasting, large-scale conflicts. Such adaptations are reminiscent of the changes in the agrosocial landscape of northern Mesopotamia after the drought episodes of *8200, 5200,* and *4200 BP* (see Chapter 10). However, such adaptations also rendered these dispersed societies less technologically advanced and less able to defend themselves against threats from more centralized cultures (*cf.* the Akkadian invasions of northern Mesopotamia around *4300 BP*, as discussed in Chapter 10). As we will now see, while the fragmented Post Moche

societies coped well with environmental stresses such as drought, they eventually fell under the domination of a new and more urbanized culture, the Chimú.[965]

The adaptive response of the Chimú culture to environmental fluctuations was a Sumerian-like focus on centrally managed agriculture. This society was based on large urban areas where populations had restricted mobility and were closely tied to highly elaborate, labour-intensive, technology-driven farming systems. The winning of adequate crop yields to sustain such concentrated populations depended on the massive harnessing of water resources via canals and aqueducts for the high-yield production of maize, squash, beans, potatoes, and other locally important crops such as peppers, cassava, and quinoa. In some cases, the Chimú anticipated fluctuations in water availability by 'over-engineering' their agricultural infrastructure. Hence, numerous redundant canals were dug in areas such as the Jequetapeque Valley (about 700 km north of Lima), whereby either scarce or excess water supplies could be accommodated and continue to supply crops in both flood and drought conditions. In the large Chimú urban centres of Farfán, Cañconcillo, and Talambo, the focus was more on massive flood defences that were apparently effective for centuries of urban occupation until these cities were eventually overwhelmed by an especially prolonged drought *c. 600 BP*.

The last and most successful Andean civilization was that of the Incas, who united the entire region from *700–500 BP*. The Incas built an extensive road network, extending to more than 40,000 km, that linked the major farming regions. After the harvest, grain was transported along these roads, either by llamas or people, to huge regional stores such as the granaries of Huánuco Pampa, which could hold 36 million litres (one million tonnes) of grain. Instead of using writing to store information, this complex civilization depended on the *quipus* system, which involved lengths of string knotted at specific intervals. Specialized administrators, the *quipucamayoc* who must have had formidable memories, oversaw this sophisticated system that also used decimal reckoning and was able to record and predict astronomical events such as comets and eclipses.[966]

North America

The first complex societies in North America were farming-based cultures, such as the Anasazi, who flourished on the Colorado Plateau and surrounding areas from about *1200 BP*. However, farming had already been practiced for almost three millennia before this time (Table 12.3). Hence the Anasazi were descendents of an older, hunter–gatherer culture, called the Basketmakers, who started to grow maize as a minor food source and only later turned to fulltime farming and a sedentary lifestyle.[967] Some of the earliest cultivated plants in the Colorado Plateau and nearby Mogollon Highlands include maize from about *4100 BP*, squash from *3200 BP*, and common beans from *2500 BP*.[968] By *2300 BP*, a form of maize called *chapelote* had spread to the region from Mesoamerica via the Sonora Desert, closely followed by beans and bottle gourd. It is likely that these domesticants were initially grown as supplements to an already-rich diet that included indigenous grains as well as acorns, yucca, and cacti, all of which would have been gathered before it was time to harvest the maize crop. After the maize harvest, the people collected autumn-ripening species such as piñon seeds and juniper berries. The Colorado Plateau cultures of this period had an extremely varied diet, depending on the ever-changing plant and animal resources in their respective home ranges. So why did these adaptable and resourceful folk eventually opt for a highly risky commitment to the much more restricted fare offered by an agricultural lifestyle?

Part of the answer may be access to improved technologies for exploiting crop resources. By about *2000 BP*, ceramic vessels had come into widespread use, enabling maize grains to be presoaked and facilitating the strenuous and time-consuming task of grinding. Grinding implements were also continually being improved. The first grindstones had small grinding surfaces for crushing the small seeds of wild plants in a circular motion. Following the introduction of larger-seeded domesticates, large two-hand *manos* and basin-like *metates*, which employed a more efficient back-and-forth motion, were developed.[969] Eventually, much larger, trough-shaped *metates* were produced to facilitate the

processing of large batches of hardened, dried maize grains.[970] At this point, the further development of an agricultural lifestyle was temporarily checked by almost two centuries of low rainfall that would have especially impacted on crops such as maize. With the return of more favourable climatic conditions after *1600 BP*, the opportunities were again available for a greater emphasis on cultivated, rather then collected, plant resources.[971]

One of the key factors in tipping the scales towards farming and eventual urbanization was the appearance soon after *1400 BP* of higher yielding, earlier flowering varieties of maize, called *maiz de ocho*. These varieties had larger kernels plus a single deep and more drought-resistant taproot. Although it was originally believed that *maiz de ocho* was introduced from Mesoamerica, it now appears that the variety was selected locally by the Anasazi themselves. Even more impressively, the Anasazi went on to hybridize *maiz de ocho* with the previous *chapelote* variety to produce new varieties with a combination of improved flour texture and greater drought resistance that was much better suited to cultivation under local conditions. Most of the new phenotypically diverse varieties cultivated by the Anasazi differed from *maiz de ocho* by the presence of just one genetic mutation. This demonstrates the proficiency of these early farmer-breeders of the South-west, and was an achievement of which any modern scientific plant breeder would be justifiably proud. Thanks to their breeding and technological innovations, the Anasazi were able to focus on high-productivity maize-based farming (supplemented by smaller amounts of crops like beans, squash, and cotton), to the virtual exclusion of their previous hunter–gathering lifestyle.[972] This required an ever more sedentary lifestyle and a greater focus on the construction of more permanent habitations.

Throughout the Anasazi homeland, populations increased and so did surplus production of grain, which required storage and protection from the elements, from vermin, and from other people. To effect this, the Anasazi became increasingly urbanized in large stone or adobe settlements. Not even the onset of yet another dry period, from *1250 BP* to *1100 BP*, could dissuade the Anasazi from agriculture, although they became more conservative and did not either produce or import any new cultivars during this demanding time. The reappearance of wetter conditions after *1100 BP* witnessed a great deal of agricultural innovation, with the appearance of many new maize varieties, and several additional new crops. Anasazi settlements grew to small cities capable of housing several thousand people. The greatest of these urban centres was Pueblo Bonito in Chaco Canyon, with over 650 substantial, family apartment units. The construction of dams, canals, and other water features attests to the importance of being able to control and regulate the water supply. The settlements at Chaco were probably the administrative and religious focus of a much larger political entity, or protostate, that may have extended for as much as 100 km in every direction from the main urban centre. Secondary settlements were linked to Chaco by a network of well-constructed roads that must have been built and maintained by organized labour under the direction of a central authority.

The Chaco state reached its height at *900 BP*, and in the larger settlements such as Pueblo Bonito there were the beginnings of overt signs of economic inequality and possibly more autocratic leadership. Such signs include the presence of

Figure 12.6 Hohokam village and agricultural hinterland in Colorado. The Hohokam, Anasazi, and related agrourban cultures occupied parts of the Colorado Plateau and Mogollon Highlands from *4100 BP* until their sudden demise about 750 years ago. Many Anasazi towns were made up of highly clustered, interlinked dwellings hewn out of solid rock in places such as Pueblo Bonito in the Chaco Canyon. As shown here, Hohokam communities tended to have more dispersed adobe buildings, adjacent to intensively managed farmland. The major staple of this relatively arid region was maize, and the people depended on wells and small canals to irrigate their crops. At its height around *900 BP*, Hohokam and Anasazi agriculture supported cites with many thousands of people, but a series of droughts and possible resource depletion a few centuries later made it impossible for farming to continue. Within a short time, these unique societies disappeared for good from landscape and memory alike.

differentiated burial patterns, with some people being interred with large quantities of valuables, and the appearance of a new type of larger habitations, called 'great houses'. The precise function of Anasazi 'great houses' is still unresolved, and some of them may have had ceremonial or other functions. There is some evidence of conflict between different Anasazi groups that led to the loss of the pre-eminent position of the Chaco Canyon pueblos by *850 BP*, although the canyon was still occupied for another century. However, by *750 BP*, the entire series of cultures, of which the Anasazi were just one example, had disappeared and their cities lay abandoned, as they remain to this day. The abandonment of the Pueblo cities and towns appears to have been an organized process, at least in some centres, such as that of Sand Canyon. Here, the main buildings were not damaged, but the ceremonial burial chambers, called 'kivas', were burned. The lack of more general damage argues against the town being abandoned due to warfare or civil strife. Rather, it is suggestive of a deliberate act to close off or seal the resting place of their ancestors before the people of Sand Canyon left their centuries-old homes and irrigated crop fields in the sure knowledge that they would never return.

The exact cause(s) of the collapse of the South-western cultures is not known for sure. But it is clear that there was a series of severe droughts in the region during this period.[973] This was coupled with a decline in agricultural productivity that made it unsustainable to maintain such large and concentrated population centres. It is possible that, as with the Maya in Yucatán, the South-west farming cultures contributed to their own downfall. For example they may have overexploited the relatively poor soil and extracted too much groundwater, leading to a drop in the water table that would have been fatal for the livelihood of such high-altitude Mesa communities. Climatic change was also a major contributory factor, as suggested by tree ring data that reveal a series of prolonged droughts and an extensive loss of vegetation after about *825 BP*. The first drought lasted over 50 years, followed a few decades later by another drought of 14 years. When this drought cycle started, the Anasazi were living in dense, albeit somewhat scattered, communities each of which was intensively cultivating maize, squash, and other highly water-requiring crops.

Although the Anasazi had successfully withstood previous cycles of aridity, including the prolonged dry period from *1250* to *1100 BP*, this new drought proved to be beyond even their impressive powers of environmental resilience. Not only was agriculture now impossible in the Colorado plateau, even hunter gathering would have been difficult as the local flora and fauna were decimated. Hunger stalked the land and the elaborate Anasazi culture became increasingly unsustainable, even in their smaller village centres. Eventually, and doubtless with heavy hearts, the famine-wracked survivors trekked away from their stricken homeland in search of better prospects elsewhere. South-west of the Anasazi, a related culture called the Hohokam lived in similar villages based around a series of maize and bean fields (Figure 12.6).[974] Their architecture and customs show that the Hohokam were much influenced by Mesoamericans to the south but, as with the Anasazi, their culture had completely died out by *600 BP*.[975] The mysterious fate of the Hohokam is still reflected in the very name by which we know them today; it is from a Pima word meaning 'the vanished ones'.[976]

that carried out many functions of a modern civil service, as well as running their own commercial agricultural and industrial enterprises. These functions were in addition to, but informed by, the religious and ideological roles played by temples and their elites. As in Sumerian times, much of the urban populace still worked for the Babylonian and Assyrian states according to a corvée system, and were generally paid in kind with rations of grain, wool, wine, or oil.[985] Early Sumerian astronomers, who were employed by the temple priesthood, were meticulous observers of the cosmos, and many of their institutions, such as the seven-day week, the lunar month, and the solar year, are still in use today. These protoscientists were able to predict astronomical events, such as eclipses and comets, with great accuracy. However, the conflation of the science of astronomy with the superstitions of astrology eventually undermined further progress in the entire field, and is a useful metaphor for the lack of subsequent scientific development in Mesopotamian society in general.

In the area of crop management, the Babylonians made several significant technical innovations, including the invention of the seed plough, and the increasing use of draught animals such as oxen and horses to increase crop productivity per worker. Ploughs were made more effective by the use of iron, which became increasingly available during the Iron Age of 1200–300 BCE. The use of a seeder as part of the ploughing process was an important advance, ensuring a more even and efficient broadcasting of seed. It took almost 1000 years before this technology spread to Egypt, where it was introduced by the Ptolemies during their ambitious agricultural improvement programme after 300 BCE (see below). As we will see in the next chapter, the refinement of the Babylonian seeder into an automatic drill did not occur for another two millennia, and Jethro Tull's famous invention in eighteenth century England.

The Neo-Babylonians

Several centuries after the Old Babylonian Era, in about 700 BCE, we find the Neo-Babylonian king, Merodach-Baladan, proudly proclaiming on inscriptions how he had acclimatized horticultural plants such as lettuce (*Lactuca sativa*), leeks and garlic (*Allium* spp.), and cress (various Brassicaceae), as well as species such as coriander (*Coriandrum sativum*), hyssop (*Hyssopus* spp.), and cardamom (*Elettaria cardamomum*).[986] In a foretaste of the great botanizing ventures of the eighteenth century CE and the establishment of commercial gardens by European colonizers, Merodach-Baladan also noted that while his exotic transplants were growing well in small-scale trial plots in his gardens, they had yet to be developed for mass cultivation in the field.[987] Several years later, Merodach-Baladan was overthrown by the even more powerful king, Sennacherib, who ruled from the rich city of Nineveh in the Neo-Assyrian heartland of northern Mesopotamia. Sennacherib celebrated his many military conquests by bringing botanical, as well as human and material, booty back to adorn his palace at Nineveh, as he recorded in an inscription:[988]

A park, the image of Mount Amanus, in which all kinds of spices, fruit trees and timber trees, . . . I had collected and I planted them next to my palace.[989]

Sennacherib went on to relate how he brought prosperity to Nineveh by constructing an irrigation system that enabled new orchards to be planted. One of his many new canals was carried across a valley at Jerwan by a massive aqueduct made from over two million blocks of limestone, parts of which can still be seen today.[990] In the King's words:

In order to plant orchards I gave to the inhabitants of Nineveh, two panu of land . . . In order to make fields flourish, I tore open mountain and valley with iron picks to dig a canal. From the Hosr I caused a ceaseless stream of water to flow.

By this time, the power of the temples was somewhat on the wane, and there is evidence that the state had retreated from the kinds of all-pervasive micromanagement of agriculture that had characterized much of the previous four to five millennia of Sumerian and Old Babylonian rule. In the city of Sippar, the temple still owned a great deal of farmland, but it now leased most of it on a long-term basis to private individuals or institutions. In a foretaste of what also occurred in late-medieval England (see below), a new class of farmer–entrepreneurs emerged in some areas of Mesopotamia. Such people invested their own wealth in agricultural enterprises, acquired their own seed and implements, and were responsible for infrastructure such as irrigation systems.[991] One of the greatest rulers of this era was Ashurbanipal of Assyria (669–623 BCE), who established the first systematically collected library at Nineveh.[992] A broader interest in matters botanical, rather than merely agronomical, by the time of Ashurbanipal's reign is hinted at by an Assyrian herbal of *c.* 650 BCE, which lists almost 1000 plants.[993] Ashurbanipal himself issued the following dire warning to potential abusers of the library:

May the gods curse anyone who breaks, defaces, or removes this tablet with a curse that cannot be relieved, terrible and merciless as long as he lives, may they let his name, his seed be carried off from the land, and may they put his flesh in a dog's mouth.[994]

The Neo-Babylonian Empire, and with it 2000 years of Mesopotamian imperial tradition, was eventually extinguished by the Persian king, Cyrus II, who occupied the city of Babylon in 539 BCE. Anxious to demonstrate its sophistication, his parvenu Achaemenid dynasty soon adopted the botanical interests of its Babylonian and Assyrian forebears.[995] The later usurper, Darius I, in addition to conquering much of the known world, found time to study recent horticultural innovations, and constantly harangued his satraps to collect rare seedlings and to experiment with new plant cuttings. His illustrious son, and successor as 'Great King, King of Kings', was Xerxes who was not just a formidable warrior monarch, scourge of Egypt and Greece, but also a passionate gardener and practical botanist. Even as a young man, Xerxes would take time from military exercises or academic studies to work in his gardens, as Strabo relates: 'planting trees and collecting medicinal roots'.[996] At the height of an important military campaign against the Greeks of Ionia, Xerxes was so struck with the beauty of a single plane tree that he ordered it to be strictly guarded lest it be damaged during the passage of his army.

The Hellenistic Era

Almost exactly two centuries after the Persian overthrow of Babylonia, Hellenistic culture arrived in the region when Alexander of Macedon and his army crossed the Hellespont in 334 BCE. Within 3 years, Alexander had seized the Persian Empire, and Greek influence spread from Spain to India. In some areas the Greek incursion was a cultural disaster as cities and irrigation systems were badly damaged or destroyed.[997] However, the synthesis of Greek and Near Eastern knowledge was to have important consequences for the study of botany, and eventually for the improvement of agriculture. The major Greek innovation was a more systematic approach to both scholarship and experimentation.[998] And perhaps just as important, the Greeks recorded their findings as written texts that were not mere eulogies of the achievements of mighty kings. Alexander's tutor was the philosopher and naturalist Aristotle, one of many Greek thinkers with an interest in agriculture. One of the noteworthy characteristics of texts by Aristotle, Theophrastus, and many other classical Greek authors, is that for the first time there are frequent references, not just to agriculture itself, but also to underlying biological processes such as plant reproduction and propagation. Theophrastus, a pupil of Aristotle who ran a school in Athens from 317 until 307 BCE, was one of the most important early writers on botanical subjects. He based his work upon keen observation rather than mere received wisdom, and collected much data on the changes in plant morphology caused by their cultivation and propagation; he even noted several instances of what we now know as mutations. However, Theophrastus was seemingly less interested in more practical matters of botany or agronomy and he never produced a systematic herbal.[999]

Indeed the later Roman author, Varro, made a rather jaundiced remark in his *De re rustica* to the effect that the writings of Theophrastus were not so much for people who really wish to cultivate land as for those who desire erudite philosophical learning.[1000] The unprecedented extent of Greek interest in, and writing about, nature and agriculture is shown by Varro's passing observation that, in addition to the texts of Theophrastus, there were at least 50 similar works in Greek by other authors. Unfortunately, although Varro named them, none of these works have survived to the present day. Most of the later Roman and medieval texts were directly based on these lost Greek writings, and the

Hellenistic period is beginning to be regarded as witnessing an early flowering of a more rigorous and quasiscientific attitude to natural phenomena (see Box 13.1). Early Greek writers observed that

farmers did not confine themselves to manipulating crops, but also applied similar principles of empirical breeding to their animals. Hence, writers on matters botanical, such as Homer (800 BCE),

---

Box 13.1 Was there any real science or crop breeding before the eighteenth century?

It is widely assumed that 'real' science did not start until the sixteenth century and that scientific crop breeding was not practiced until the twentieth century. The reality is more complex, with intriguing evidence of a brief flowering of an early form of science in the Hellenistic Era, from about 400–0 BCE. In contrast, crop breeding seems to have been an almost exclusively empirical process until the eighteenth century, after which it gradually benefited from newly emerging botanical knowledge, especially concerning plant reproduction and genetics. The original crops started out as wild plants that became adapted to the new environment created by human cultivators. There seems little doubt that the early stages of this so-called 'domestication' process mostly occurred spontaneously as new varieties appeared without the conscious intervention of people. By only choosing the best performing plants, the first farmers acted as agents of selection but were not deliberately manipulating the genetics of their new crops. One could argue, therefore, that these farmers were no more acting as true breeders than those attine ants that have maintained pure clonal lines of selected fungi in their underground 'gardens' for the past 50 million years (Chapter 3).

Throughout the first ten millennia of Near Eastern farming, lack of understanding of the basis of plant reproduction precluded anything but the most rudimentary forms of selection and propagation of improved crop cultigens. During this period, the ability to manipulate the biology of plants was largely restricted to selection of those fortuitously occurring mutations that happened to be spotted by observant farmers. Instead of breeding, there was an overwhelming focus on maximizing the efficiency of crop growing, harvesting, and distribution by means of improved forms of bureaucratic organization by the state. Further benefits were accrued by the use of new forms of technology to improve the sowing, harvesting, transport, and processing of crops. These developments were complemented by the evolution of those associated social structures that were required to maintain the newly complex agrourban landscape.

As early as *7500 BP*, the Samarrans probably recognized and selected new barley genotypes that better adapted to

irrigation but, as discussed above, by far the most important factors behind the success of Sumerian farming were unrelated to plant biology. In post-Sumerian times, Babylonian, Assyrian, and Persian elites were much concerned with the collection of plants, including possible new crops, and experiments were sometimes carried out to acclimatize potentially useful new cultigens obtained from further afield. However, there was little systematic study of plants and virtually no record keeping until the Hellenistic Era. Indeed there was still much basic ignorance about biology in the time of Theophrastus and onwards into the medieval period. Different types of grain were often confused, and most barley or wheat crops probably contained substantial admixtures of oats and rye; these contaminating grains were sometimes regarded as 'degraded' forms of the main crop (Chapter 6).

The nature and extent of 'real' science in the Hellenistic Era is a matter of some controversy, but Greek writers such as Theophrastus certainly introduced new ways of thinking about the natural world and carried out some of the first recorded, systematic botanical studies. This knowledge was limited in scope and quantity, but still had a huge influence on succeeding generations of scholars and practitioners as Greek texts and their derivatives remained virtually the only source of information until the sixteenth century. The enduring scientific revolution that started in the sixteenth and seventeenth centuries CE (and continues to this day) is based on experimentation and the widespread dissemination of empirically based knowledge. In terms of crop breeding, one of the major keys to success is an understanding of plant reproduction and how it can be manipulated. This knowledge was developed from the eighteenth century, as plant sexuality was elucidated by Camerarius and others; and as Dutch and English breeders created the first deliberately produced hybrid forms. The power of modern breeding is epitomized by the major cereal crops where relatively low-yielding, tall genotypes predominated from Neolithic times. Following the so-called 'Green Revolution' of the 1960s–1970s, these traditional varieties were replaced by the much higher yielding semidwarf forms that are the mainstays of today's global agriculture (see Figure 16.2).

Democritus (420 BCE), and Herodotus (300 BCE), also noted how the sexual crossing of horses and donkeys had been used to produce those sterile, but extremely useful interspecific hybrids, known as mules.

It is now increasingly appreciated that, as part of the broader international diffusion of Greek culture in the Hellenistic Period (300–100 BCE), there was an early application of evidence-based knowledge to aspects of agriculture in many areas of the Mediterranean and Near East. Much of this region was ruled by the Hellenistic successors of Alexander, such as the Ptolemies in Egypt and the Seleucids in Mesopotamia and Persia. One such dynasty was the Attalids of Pergamum in northwest Anatolia, who created a library second only to that of Alexandria.[1001] Attalus III Philometor (138–133 BCE) was especially interested in practical agriculture and botany. He wrote books on botany and horticulture, as well as an agricultural treatise.[1002] Like other members of the dynasty, Attalus III sponsored and sometimes participated in experimental work on botanical subjects, even allowing the royal gardens to be used as laboratories for field studies of plants.[1003] In Egypt, the Ptolemeic dynasty greatly improved agriculture by introducing new crops, extending the irrigation system, and reclaiming huge tracts of land for farming.[1004] Much of the most fertile agricultural land of Pharaonic Egypt had been owned either by the state or by temples, but the Ptolemies extended ownership by granting such land to civil servants.[1005] Such was the success of Ptolemeic agriculture that, when Augustus conquered Egypt after the death of Cleopatra VII in 30 BCE, the Nile Valley and North Africa became the major granary of the Roman Empire. In particular, the million-strong population of the imperial metropolis of Rome was fed for almost five centuries on a diet based largely on the annual 400,000 tonnes of Nile Valley wheat shipped across the Mediterranean from Alexandria to Ostia.

The Romans

With the advent of the Roman imperium, the short-lived Greek venture into science came to a standstill that was to last for a further 1500 years.[1006] The Romans tended to reinterpret and republish Greek texts rather than uncover new knowledge about agricultural and crop processes. However, they were also more systematic and rigorous than the Greeks in their focus on practical agronomy. The Romans regarded the farmer as an independent, free citizen of similar importance to the soldier. Indeed, retired legionaries were normally given a parcel of land to cultivate as yeoman farmers. It is during this time that we get some of the earliest surviving references to the process of crop breeding itself. For example, authors such as Virgil noted the need for the continual selection of elite seed varieties; otherwise, he noted, the diligent labour of many years would be wasted as the seed reverted to the wild type.[1007] Virgil's sentiments on the need for selection of cereal varieties were echoed by Columella and Varro, while other writers such as Pliny the Elder discussed the phenomenon of sex in plants, but did not mention any experimental studies.

As already noted, much of this knowledge was taken from earlier Greek texts and was rarely, if ever, checked for its veracity by the Roman copyists. Hence, we find that Pliny the Elder was far from accurate in many of his writings on various aspects of crop cultivation and behaviour. Indeed, in his notorious diatribe against oats, Pliny echoed and reinforced the prejudices of Theophrastus, as published in the *Historiae Plantarum* some three centuries previously. Here, Pliny seems to be merely reflecting the opinions of an established authority, rather than basing his writing on reliably observed facts about plant behaviour. Such willingness to accept the writ of authoritative texts became even more pronounced after the fall of the Roman Empire in the fifth century CE, and would bedevil progress in the natural sciences until the seventeenth century and beyond.[1008]

In the second century CE, Galen and other Roman authors wrote about the numerous, and obviously distinctive, local varieties, or land races, of crops with which they were familiar. Later authors, most notably Palladius in the fourth and fifth century CE, presented several useful descriptions of cultivation methods and agronomic techniques in general, without dealing in any detail with the plants themselves. This is a little frustrating for cereal specialists because it was during the first few centuries of the Roman imperium that breadwheat started to the transformation of sugar-rich plant extracts, with the assistance of brewer's yeast, into the vast range of alcoholic liquors that are prepared by many human cultures around the world. Hence we can change barley to beer, grapes to wine, rice to saki, or milk to chang (a fermented mare's milk that is a delicacy for certain nomads in Central Asia).

There is evidence of beer making and other forms of human-induced microbial fermentation dating as least as far as the beginnings of plant cultivation, well before *10,000 BP*. Indeed, since hunter–gatherers often ferment many plant products that they collect, biotechnology might predate agriculture. Once large-scale organized agriculture was underway, fermentation became a more organized, industrial process, for example the ancient Egyptians and Sumerians used much of their staple barley crop to brew beer in huge, state-run breweries. This beer served to sterilize the frequently polluted drinking water in urban centres, and perhaps also helped to wash down an otherwise monotonous diet of barley cakes and hard bread.[1056] Like these forms of traditional biotechnology, empirical plant breeding was very much a 'low-tech' process. But it still enabled farmers around the world to feed a global population that expanded several hundred-fold, from under ten million people at the Neolithic dawn of agriculture to more than 600 million in 1700 CE.[1057]

We know from ancient engravings that the early Babylonian and Assyrian civilizations, of more than 5000 years ago, knew about the bisexual nature of the date palm and that they practiced artificial pollination.[1058] One suspects that many of these ancient farmers would have been constantly experimenting with new varieties of potentially useful crops. As we have seen, some of the ancients, such as the Neo-Babylonian kings, were also ever on the lookout for new types of wild plant or exotic crop that could be brought into cultivation. Hence, farmers in the Near East began, over thirteen millennia ago, with diploid wheats such as einkorn but then recognized the advantages of new, spontaneously produced tetraploid species such as durum. A little later they discovered a newly created hexaploid species, namely breadwheat. Following a long period of experimentation with breadwheat cultivation and processing, farmers produced the easily milled forms of wheat that we still grow today.

One of the major factors that held back progress in the ancient and medieval worlds was the lack of methods for systematic recording and disseminating knowledge of new discoveries. Those few written records were often produced more for the glory of elite groups or the fame of the writer, than to convey accurate knowledge to posterity. Unfortunately, such knowledge as existed was all too vulnerable to the depredations of vandals who destroyed cities such as Ur and Persepolis, or burned the irreplaceable libraries of Alexandria and Granada, and ruined the irrigation systems of Al-Andalus and Sumer.[1059] The repeated loss of knowledge and infrastructure was partially responsible for the lack of progress in many aspects of agriculture during the almost 4000-year period between the fall of Ur and the 'English revolution' of 1600 to 1800 CE. The so-called scientific revolution of the post-Renaissance period was not just a change in our world view; after all, curious and inventive people had been making discoveries since Palaeolithic times. Perhaps even more important than the creation of knowledge itself, was the way it was disseminated more widely between scholars and practitioners of science and technology. It was this new access to information, coupled with a willingness to share the fruits of one's knowledge, that really stimulated the study and manipulation of the biological world and underpinned the agricultural achievements of succeeding centuries.

Evolution of modern agricultural economies

Renaissance and neonaissance

As we discussed in the previous chapter, agriculture has always had both pragmatic and mystic dimensions.[1060] Until the era of printing; the wider penetration of literacy in the population; and the wider emergence of ideas of 'progress' and 'improvement'; there was virtually no systematic corpus of knowledge in Western Europe that related to the practical cultivation and breeding of crops. This situation began to change dramatically towards the end of the fifteenth century. During the first few decades following the invention of the Gutenberg Press in 1450, many existing works of the ancients were printed and widely disseminated for the first time in recorded history. At this time, many of the original handwritten texts had only recently come to the attention of Western scholars, either via copies imported from the Islamic world or in the form of documents rescued from the débâcle that followed the fall of Constantinople to the Ottoman Turks in 1453.[1061]

Among the first ancient texts relating to plants, medicine, and agriculture to appear in print were the classical works of Theophrastus and Galen, which were published in the late fifteenth century. The first printed work on agriculture was the late medieval vernacular text, *Ruralia Commoda*, written by the Bolognese lawyer and writer, Pietro de' Crescenzi.[1062] But this was very much the exception, and more typical was Galen's *De Materia Medica*, which was published in Venice, first in Latin in 1478 and then in Greek in 1499. The major agricultural works of Cato, Palladius, Columella, and Varro were frequently published together during the Renaissance as the *Scriptores Rei Rusticae*.[1063] This period witnessed increasing advances in botanical knowledge and its application. Beginning in Italy, practical and ornamental botanical gardens were established, and descriptive herbals and treatises published.[1064] As the Renaissance progressed, it took more than a century for the relatively small and scattered communities of Western European scholars and others to collate and digest the newly reprinted texts of the ancients. However, it was soon realized that these texts could be improved upon in various respects, not least by making them more practically focussed and accessible to the increasing numbers of non-Latinate readers who were actually involved at the 'cutting edge' of crop cultivation.

What I shall refer to as the 'neonaissance' occurred after the Renaissance period. Whereas the Renaissance was indeed a kind of re-birth, with much rediscovery and dissemination of ancient knowledge, the neonaissance was an altogether more profound phenomenon. The neonaissance was all about creation *de novo* of new and reliable knowledge based on impartial observation, rather than either written or oral revelation and tradition. As we saw in the previous chapter, there was a fore-taste of the potential power of scientific inquiry in the ancient Greek and Hellenistic periods, from about 500 to 100 BCE, but this early bloom soon withered in the chill winds of the Roman imperium. In contrast, the postmedieval neonaissance was a gradual process that began in the late fifteenth century and slowly gained momentum until it flowered in the scientific revolution of the seventeenth century and beyond.

The more practical reader of early Renaissance texts would have soon realized that many writings by venerable Classical authors were either outdated or incorrect. After all, although there had not been much progress in plant breeding per se in medieval Europe, the types of crop varieties grown and the prevailing agronomic practices had evolved considerably from the situation during the Classical period, well over a millennium previously. The older texts would also have been less useful as sources of practical farming advice in the much-changed climatic and social environments of the Renaissance compared to the Roman period. For example, since the fourteenth century, Europe had been in the midst of a 'Little Ice Age' that had greatly altered the profile of crop cultivation across the continent. Furthermore, some of the more populous regions, most notably England and the Dutch provinces, were on the brink of a revolution in the organization and management of farming, with the rise of the farmer–entrepreneur and the enclosure movement. This produced a growing market for practical, up-to-date manuals aimed at the educated and enlightened yeoman–farmer who wished to improve both himself and his crops.

It is important to stress here the surprisingly high degree of literacy in England at this time. It is estimated that from 1551 to 1651, between half a million and over one million adult males could read to some extent.[1065] This is a very large number, given that the total population of England was just three million in 1551, growing to five million in 1651, and that fewer than 35% of these would have been adult males. Hence, between one-third and a half of male adults would have had at least a degree of familiarity with the written word. This literacy was linked with what has been termed an 'educational revolution' in England during this period.[1066] Ironically, the use of vernacular texts aimed at a relatively literate, local population was a double-edged sword

CHAPTER 15

Imperial botany and the early scientific breeders

Le destineé des nations dépend de la manière don't elles se nourissent

[The destiny of countries depends on how they feed themselves]

Anthelme Brillant-Savarin. 1755–1825, *Physiologie du Goût*

Introduction

The gradual awakening of a new spirit of enquiry during and after the European Renaissance was considerably enhanced by the immense broadening of physical and mental horizons engendered by the voyages of discovery and conquest of the period. Such developments were especially marked in the emerging naval, scientific, and mercantile powers of Anglo-Dutch private adventurers and official voyagers. It was in the Netherlands and England that many of the most significant developments in agronomy and plant breeding came about during the seventeenth and eighteenth centuries. In this chapter, we will see how a little-known agricultural revolution in England laid the foundations for the later Industrial Revolution of the eighteenth and nineteenth centuries. We will also examine the phenomenon of 'imperial botany' that took much of Western Europe by storm in the eighteenth century. Finally, we will discuss how experiments in plant biology led to rational manipulation of crops by using hybrids and mutants to enhance variation in breeding programmes.

The English revolution

The timing of the 'English agricultural revolution' has been much debated by historians and economists. As we saw in the previous chapter, there was a something of a paradigm shift during the sixteenth century as entrepreneurs sought to improve output using enclosure, reclamation, new crops, and better management methods. By the eighteenth century, these individual, small-scale farmers had been joined by the 'gentlemen improvers', who were often effective publicists, able to bring agricultural innovations to a much wider audience.[1092] The result was a significant increase in the intensity and yields of arable crops, and especially of the major commercial cereal, breadwheat. During the eighteenth and nineteenth centuries, two of the most important topics that engaged the interest of those involved in applied plant science were: (i) collection, cataloguing, and evaluation of new varieties and new species of crops; and (ii) elucidation of reproductive mechanisms and their manipulation for crop improvement. From the earliest days of scientific breeding, progress was ensured by a mixed economy of private and public initiatives. The first private agricultural society in Britain was probably the *Honourable Society of Improvers in the Knowledge of Agriculture in Scotland*, founded in Edinburgh in 1723. Some of the better-known gentlemen innovators of this period are Charles 'Turnip' Townshend and Thomas Coke, both of Norfolk. Townshend was a successful politician who served as Secretary of State to George I, and Lord Lieutenant of Ireland. Following his enforced retirement from politics in 1730, he set up a profitable farming business in which he made effective use of turnip rotations to provide winter fodder for his animals. Although he did not personally establish the four-course turnips–barley–clover–wheat rotation to which his name was subsequently attached, Townshend certainly brought these innovations to wider public notice.

In much the same vein as Townshend, Thomas Coke, Earl of Leicester, was not so much a hands-on innovator as an extremely effective publicist and an eloquent proponent of scientific agriculture. From 1776 onwards, Coke held many demonstrations of improved cropping techniques on his impressive seaside estate at Holkham, on the north Norfolk coast. Enthusiasts flocked to Holkham from around the country to learn about the new farming methods. These two gentlemen, who were widely admired by their peers, contributed greatly to the betterment of eighteenth century English agriculture. Most entrepreneurs of the ilk of Coke and Townshend were interested, not only in the betterment, for profit, of their own farming businesses, but also in making known their discoveries and inventions to the public at large. In other words, they combined self interest with a very real desire to contribute to the public weal—and doubtless they also wished to be recognized for such philanthropy. The majority of scientists of this period were either themselves well-heeled gentlemen, or were financed by such men. It should be noted, however, that by the eighteenth century many practical farmer–entrepreneurs, most of whom were yeomen of modest means rather than wealthy gentlemen, were also making effective contributions to agricultural improvement.[1093]

Although the nineteenth century is often regarded as the key epoch in the scientific and industrial 'revolutions' that were pioneered so prominently in the UK, recent historical and economic research suggests that many major developments that underpinned these 'revolutions' had already occurred during the eighteenth century or earlier.[1094] Two of the more important developments of the time were the huge expansion in the use of coal, and the pressures and opportunities created by a steadily rising, and increasingly urban, population. During the eighteenth century there was a massive switch from wood to the vastly more efficient coal-based fuels. By 1800, Britain was already producing the equivalent amount of energy from coal that would have required the annual harvesting of 6 million hectares of forest. This amount of forest simply was not available in such a small country, where the total area of woodland is just 2.8 million hectares. It should also be borne in mind that the total area of arable crop cultivation in Britain today is only 4.5 million hectares.[1095] Therefore, the development of coal as an alternative

to wood was clearly crucial in allowing the increasingly power-hungry, new industries to thrive, while at the same time enabling the relatively small area of really productive land to be used for the improvement and cultivation of the food crops needed to sustain the rapidly increasing populations.

During the eighteenth century, an increasing proportion of the British population was urban and required ever-more efficient agricultural production for its sustenance. Already, by 1700, the population of England was 13.4% urban, in contrast to an average of 9.2% for the rest of Western Europe. By 1800, the English urban population had almost doubled to 24.0%, while that of the rest of Europe remained virtually constant at 9.5%.[1096] It is now becoming clear that this industrial revolution was largely based on a previous, and hitherto largely ignored, agricultural revolution.[1097] Key elements of increased agricultural productivity during the eighteenth century were the use of new crop types and varieties; the development of new and more effective crop rotational systems; and the recruitment of more productive land by drainage and woodland felling. The latter would not have been pristine woodland, of which there was virtually none left in Britain. A great deal of the woodland in post-Roman Britain was regrowth from abandoned arable land, or was actively coppiced to provide charcoal for the smelting of metals. With the advent of coal, much of the coppiced woodland became available for other purposes. If the land was potentially arable, it was felled, but in regions with poorer soils, some of the old medieval coppiced woods were left intact for exploitation as timber or for shipbuilding. Much of the extant woodland of southern England, such as the Weald, is therefore merely the remains of a medieval agroindustrial economy, rather than the primeval forest of popular imagination.

Many innovative farming techniques in eighteenth century Britain were imported from the Netherlands, which was probably the most agriculturally advanced region in Europe from c. 1650 to 1750. However, while the Dutch agricultural economy made only modest gains later in the eighteenth century, British productivity and innovation continued to advance steadily until, by the late eighteenth century, it had decisively eclipsed its maritime and agricultural rival across the North Sea. Two of the most notable English innovations were Charles Towshend's four-field crop rotation system (see above) and Jethro Tull's seed drill. Both were widely adopted and contributed greatly to yield improvements.[1098] Jethro Tull was a typical eighteenth century English innovator. He received the usual Classical education and went on to study law at Oxford and Gray's Inn (one of the four Inns of Court in London). As his extensive writings on agronomy reveal, he was profoundly conversant with the ancient texts but only used then to support his own evidence-based scientific approach to crop improvement.[1099] During his lifetime, Tull published full technical details of his revolutionary machines, making them freely available to all. However, such public-spiritedness was not to last and after 1750, all new inventions relating to agricultural machinery were jealously guarded by patents.[1100] Tull's best-known machine was an easily workable horse-drawn seed drill, which dropped seeds in rows, making seed setting much more precise and efficient.

This innovation, which we have already seen presaged in the ancient Sumerian and Babylonian seeders (Chapter 13, Agriculture during the Classical period: 2000 BCE–500 CE), made an enormous difference to seed yields and farming efficiency by ensuring optimal seed distances. It was also the start of ever greater mechanization culminating in today's industrial-scale, technology-dominated agribusiness. The next such advance occurred in the 1780s when Scottish millwright, Andrew Meikle, developed a human or animal-powered threshing machine for removing husks from grain. This device later became steam-powered and was the direct precursor of today's combine harvesters. Thereafter, following many centuries of very slow change in farming techniques, the pace of mechanization increased dramatically with the introduction of steam power during the nineteenth century, and the massive displacement of the rural population of England into the new urban manufactories (as bemoaned in the verses of many romantic poets of the era and in the early novels of Thomas Hardy[1101]). To this day, the average rural population of the island of Britain remains at pre-Elizabethan levels. Several new vegetable and grain crops were also introduced into England at this time, including turnips and other brassicas, for both fodder and oilseed use.

Botany in the ascendant: the seventeenth and eighteenth centuries

An early strategy for systematic crop improvement was to mount both private and state-sponsored expeditions to the newly discovered islands and continents that were opening up to European explorers after the fifteenth and sixteenth centuries.[1102] One of the aims of such ventures was to acquire new and better varieties of existing crops, as well as to collect some of the astonishing range of entirely new crops that was being discovered. From the earliest days of the emerging British Empire, such botanizing had started to emerge as an overt aspect of official naval exploration. The noted naturalist, John Tradescant, had been seeking new fruits since 1610. Best known today to gardeners for the eponymous herbaceous perennials (*Tradescantia* spp.), John Tradescant was the scion of a noted family with botanical and other interests.[1103] Tradescant himself was an influential naturalist and inveterate plant improver who was in turn cultivated by powerful political figures of the early seventeenth century, including Lord Salisbury and the Duke of Buckingham. He began his career by travelling to Holland in 1610, under the sponsorship of Lord Cecil, to collect new European varieties of orchard trees and ornamental flowers. In 1620, he accompanied a punitive military expedition to confront the pirates of the North African Barbary Coast, and managed to bring back several new varieties of apricot for the benefit of British horticulture.

In 1625, following the approval of senior government officials, Tradescant exhorted all British merchants to 'procure all manner of curiosities abroad', stating that: 'All Marchants from All Places But Espetially the Virgine and Bermewde and Newfound Land . . . will take Care to furnish his Grace With All manner of Beasts and fowells . . . seeds Plants trees or shrubs'.[1104] In other words, Tradescant was asking all merchant seamen sailing from Britain, but especially those visiting the eastern seaboard of present-day Canada and the USA, plus the islands of the West Indies, to collect or otherwise obtain any potentially useful specimens of plant or animal life. His son, John Tradescant the younger, was also a noted botanist who introduced many American plants into Europe, including magnolias, bald cypresses, and the asters, eventually becoming head-gardener to Charles I.[1105] However, despite the great talents of the likes of the Tradescants or John Parkinson, most seventeenth century enterprises were essentially dilettante affairs, largely reliant upon the whims of aristocratic backers and the ephemeral fashions of European high society.[1106] It was not until the eighteenth century that truly systematic and scientifically-based state botanical ventures really got underway. The success of these expeditions was largely due to the reinvention of that ancient Babylonian concept, the utilitarian botanical garden (see Chapter 10, Agriculture during the Classical period: 2000 BCE–500 CE, and Box 15.1).

Role of the botanical garden

Private botanical gardens emerged across northern Europe in the seventeenth century and there was much intercourse between their various patrons and curators across the continent. For example, in 1611, John Tradescant travelled to gardens in Holland and France to acquire material for the garden of his patron, Lord Salisbury, at Hatfield. A few years later, Jacob Bobart, curator of the Botanic Garden of Oxford, was associated with the gardens at Blois. Such links were encouraged by the Anglo-French alliance forged by Henrietta Maria, queen to Charles I and sister of the French king, Louis XIII. These gardens were not true commercial ventures, although they were useful in terms of seed multiplication and the trial cultivation of potentially valuable medicinal crops that had hitherto been imported from overseas. The impetus for more organized, state-sponsored botanical ventures came about in the eighteenth century, with the development of the global, trade-based maritime empires of the British and Dutch. The earliest, truly systematic collection and cataloguing endeavours were undertaken in the mid-eighteenth century by the Dutch East India and Dutch West India Companies, both of which established a series of

formal botanic gardens throughout their respective colonies in the tropics. The British soon emulated their North Sea neighbours as they became involved in their own burgeoning imperial project around the globe.

The process in Britain started in earnest with the establishment, in 1759, of the Royal Botanical Gardens at Kew.[1107] Earlier versions of the gardens at Kew had been set up by previous Hanoverian monarchs in the 1720s and 1740s, but these were nothing like the enduring establishment that dates from 1759. This latter venture was successful largely due to the efforts of Augusta, Princess of Wales, and John Stuart, 3rd Earl of Bute. This gentleman was descended from the Stuarts of the Scottish, and later British, royal family and briefly served as Prime Minister from 1762 to 1763. Notwithstanding his aristocratic pedigree, the Earl of Bute travelled to Leiden in Holland in order to study botany as a student of Linnaeus. By the 1750s, he was widely regarded as the finest botanist in England and from 1754 he based himself, and his extensive botanical library, in a house on Kew Green that opened out directly onto Kew Gardens itself.[1108] The angiosperm genus, *Stuartia* a relative of the camellias, is named after him. Following a fortunate marriage, the 3rd Earl's son acquired land in south Wales that later became the most productive coalfield in Europe, making the Bute family into the equivalent of present day multibillionaires. Their botanical legacy endures today in the capacious grounds of Bute Park, which has been hailed as the 'tree capital of the UK'.[1109] Bute Park extends for almost 55 ha in the centre of the Welsh capital city of Cardiff and contains more than 2200 types of trees from all over the world.

Part of the mission of Kew Gardens was to study the plethora of plant samples that was flooding into the country as the government dispatched more and more botanical expeditions throughout the world. Most of the great naval voyages of discovery during this period included botanists as a matter of course. On Captain James Cook's famous first circumnavigation expedition of 1768 to 1771, there were two former students of Linnaeus, Daniel Solander and Herman Spöring, not to mention the redoubtable Joseph Banks, then a wealthy 25-year old landowner from Lincolnshire. Thanks to the

Box 15.1 Botanical Gardens and paradise: from Nineveh to Svalbard

Formal botanical gardens date back well over 3000 years and probably much further. Part of the age-old appeal of these unique foundations is their intrinsic combination of beauty and functionality. Many of the most notorious and fearsome Mesopotamian rulers had strong botanical leanings. The first empire builder of the ancient world, Sargon of Akkad (2334–2279 BCE), was the son of a gardener, while the later Assyrian king, Tiglath-Pileser III (745–727 BCE), started his career as a gardener and ended as one of the greatest rulers of the empire, inaugurating the last and greatest phase of Assyrian expansion. This monarch was named after that mightiest of the Assyrian kings, the ferocious Tiglath-Pileser I (1114–1076 BCE), who laid waste lands far and wide, but in his own country paid particular attention to agriculture and fruit growing. Tiglath-Pileser I was one of the first monarchs recorded as establishing a botanical garden, in his capital city of Nineveh, where he planted hundreds of plant specimens collected during his many military campaigns.

The brutal Neo-Assyrian ruler, Sennacherib (705–681 BCE), immortalized in Byron's line: 'The Assyrian came down like the wolf on the fold' (Byron, 1815), rebuilt the Nineveh gardens and watered them with over 10 km of irrigation canals. The gardens of Nineveh were the direct precursor of perhaps the most famous of botanical edifices, the Hanging Gardens of Babylon, one of the Seven Wonders of the Ancient World. The latter were built as roof gardens within the walls of the royal palace at Babylon and were probably named for their profusely overhanging vegetation. Their provenance remains enshrouded in myth, but King Nebuchadnezzar II (605–561 BCE) may have built them to console his Median wife, Amytis, who yearned for her verdant montane homeland. In first century BCE, the Greek geographer Strabo described the garden thus:

It consists of vaulted terraces raised one above another, and resting upon cube-shaped pillars. These are hollow and filled with earth to allow trees of the largest size to be planted. The pillars, the vaults, and terraces are constructed of baked brick and asphalt . . . The ascent to the highest story is by stairs, and at their side are water engines, by means of which persons, appointed expressly for the purpose, are continually employed in raising water from the Euphrates into the garden.

One of the reasons that this structure was so remarkable to the ancients was that it was a garden filled with soil that brought forth a bounty of plants, and yet it was suspended high in the air on the rooftops of the palace. Moreover, its prolific greenery was in contrast with the arid climate of Babylon, and was only sustained by gravity-defying engines (probably chain pumps) that continually raised the water from ground level. The tradition of the botanical garden was continued by the Persians after their conquest of Babylon in 539 BCE. King Xerxes was an especially keen botanist who ordered his satraps cross the Persian Empire to establish such gardens to serve both as practical centres of plant cultivation and as areas of beauty and recreation for the populace. The Persians called their gardens, *paradaida* (Briant, 2002, pp. 442–443). A few centuries later, the incoming Greeks were so impressed by these botanical gardens that they used the same word (*paradeisos* in Greek) to refer to that otherworldly place of eternal bliss that we now call 'paradise'.

Botanical and experimental gardens became widespread in the Islamic world from the eighth century CE, often sponsored by local rulers and staffed by professional botanists and agronomists. They collected exotic plants, studied acclimatization, and occasionally developed new crop varieties (Watson, 1994). The earliest botanical gardens in Europe were probably established in Italy (in the south of which Muslim rulers had held sway for many years), in Salerno by Sylaticus in 1310 and in Venice by Gualterius in 1330. It was not until after the Renaissance that similar gardens were set up in other cites and universities: Pisa in 1543, Padua, Parma, and Florence in 1545, Bologna in 1568, Leyden in 1577, Leipzig in 1580, Königsberg in 1581, Paris in 1590, and Oxford in 1621 (Chiarugi, 1953; Hill, 1915; Watson, 1995).

As discussed in this chapter, botanical gardens became emblems of state power and scientific advance during the heyday of post-eighteenth century European imperial expansion. Nowadays, such gardens serve a mixed role as public parks, research establishments, and conservatories of useful plant germplasm. Thanks to modern preservation methods, it is no longer necessary to keep plants as growing specimens, and they are often preserved instead as seeds, cuttings, or even DNA. One of the most recent such ventures is the Svalbard Global Seed Vault, on the remote Arctic island of Spitsbergen (Svalbard), which was announced in 2006 as a secure repository of frozen seeds for the benefit of future generations. It is difficult to imagine a greater contrast between this hypermodern, aseptic, and utterly utilitarian Nordic seed vault, sealed forever from the public gaze in the icy wastes of the far Arctic, and those luxuriant oases of verdancy that once soothed the minds of kings and populace alike in the pitiless dry heat of an ancient Mesopotamian summer.

efforts of Banks and Solander (the unfortunate Spöring having died from a disease contracted in Batavia), the expedition returned to Britain with no fewer than 1300 new species of plant. Banks and Cook were welcomed back as heroes, and matters botanical seized the public imagination across Europe. The tradition of British naval botanizing continued for well over a century, up to and including Charles Darwin's momentous voyage of 1831 to 1836. Darwin travelled on the naval warship, HMS Beagle, captained by Robert FitzRoy, where the young scientist spent much of his time on botanical studies.[1110]

During the mid-eighteenth century, there was a veritable explosion of botanical interest in the highest social circles of Europe. In 1751, Louis XV of France, who already rejoiced in the title of *Roi-géographe*, began to take lessons in botany. The botanical pretensions of Louis *le Bien-Aimé* were encouraged by his mistress, the formidable Madame Pompadour, and further facilitated by the eminent botanist, Louis Guillaume Lemonnier, head of the splendid botanical garden of the Trianon at Versailles.[1111] Meanwhile, not to be outdone by their French rivals, the Holy Roman Emperor, Francis I, set up the great Hapsburg botanical collection at the palace of Schönbrunn in 1753, while in 1755 the Bourbon King Fernando VI of Spain created the Royal Botanic Garden of Madrid.[1112] It also seems that Madame Pompadour was not the only French royal bedfellow with an interest in botany and crop improvement. Just 35 years later, in 1786, the unfortunate and much-maligned Marie Antoinette, queen to Louis XVI, played a key role in the introduction of potatoes into Europe. At that time, most people had an almost superstitious aversion to potatoes, which were still a rather novel crop in Europe although, as we saw in Chapter 8, they had already been a staple food in the South America for more than ten millennia. Potatoes were regarded as fit only for animals and as positively bad for the health of the soil. By publicly sporting potato flowers in her posies, Marie Antoinette did much to improve the reputation of this important crop amongst the general populace.[1113] It is perhaps more fitting to remember this lady for her very real contribution to European agriculture rather than for the falsely attributed (and mistranslated)

remark after the failure of the grain harvest in 1789; namely, 'let them eat cake'.[1114]

With the added impetus of the overseas plant discoveries flooding into London, the British Royal Family became especially enthusiastic botanizers. George III may have notoriously lost the war to keep the American colonies, but he was much more effective as a vigorous champion of science. As he himself remarked: 'I spend money on war because it is necessary, but to spend it on science, that is pleasant to me.'[1115] The royal interest in botany was famously celebrated by Erasmus Darwin (Box 15.2), grandfather of Charles Darwin. Erasmus was one of the leading intellectuals of eighteenth century England, with a remarkable array of interests and pursuits. He led a stellar group of likeminded amateur experimentalists and thinkers who termed themselves the 'Lunar Society of Birmingham'. Other members of this elite club included Joseph Priestly, James Watt, Josiah Wedgwood, and Samuel Galton.[1116]

Darwin was a respected physician, a noted poet, philosopher, botanist, and naturalist. His writings influenced a later generation of Romantic writers and poets, including Coleridge, Wordsworth, and the Shelleys. Wordsworth cited Erasmus Darwin as a source for 'Goody Blake and Harry Gill' in *Lyrical Ballads* (1798), while Coleridge averred that Darwin possessed 'perhaps, a greater range of knowledge than any other man in Europe, and is the most inventive of philosophical men'. Half a century before his more famous grandson became interested in the issue, Erasmus Darwin was already much concerned with the conundrum of how plant and animal species evolve and change their form. Erasmus Darwin was remarkably modern in adopting an integrative approach to the study of this and other important scientific questions. For example he used his observations of livestock and wildlife behaviour, and his knowledge of fields such as palaeontology, biogeography, botany, systematics, embryology, and comparative anatomy to elucidate such important, multifaceted topics as evolution and speciation.

One of the key English figures in the early stages of the practical application of botanical knowledge to agriculture and commerce is Joseph Banks, President of the Royal Society from 1788 to 1820.

Box 15.2 *The Botanical Garden*—a poem by Erasmus Darwin

The botanical garden is far more than a place of beauty and utility, as demonstrated in this allegorical work by one of the most celebrated scientists of the late eighteenth century. In 1784, the scientific studies of George III and Queen Charlotte were eulogized by Erasmus Darwin in a poem entitled *The Botanical Garden*. In his rather over-lengthy paean, Darwin memorably uses the royal botanical work at Kew Gardens as an explicit emblem of scientific advancement, imperial growth, and national prosperity, as exemplified in the following extract:

> So Sits enthron'd in vegetable pride
> Imperial Kew by Thames's Glittering Side
> Obedient sails from realms unfurrow'd bring
> For her the unnam'd progeny of spring . . .
>
> Attendant Nymphs her dulcet mandates hear,
> And nurse in fostering arms the tender year,
> Plant the young bulb, inhume the living seed,
> Prop the weak stem, the erring tendril lead;
> Or fan in glass-built fanes the stranger flowers
> With milder gales, and steep with warmer showers.
>
> Delighted Thames through tropic umbrage glides,
> And flowers antarctic, bending o'er his tides;
> Drinks the new tints, the sweets unknown inhales,
> And calls the sons of science to his vales.
>
> In one bright point admiring Nature eyes
> The fruits and foliage of discordant skies,
> Twines the gay floret with the fragrant bough,
> And bends the wreath round GEORGE'S royal brow.
> – Sometimes retiring, from the public weal

> One tranquil hour the ROYAL PARTNERS steal;
> Through glades exotic pass with step sublime,
> Or mark the growths of Britain's happier clime;
> With beauty blossom'd, and with virtue blaz'd,
> Mark the fair Scions, that themselves have rais'd;
> Sweet blooms the Rose, the towering Oak expands,
> The Grace and Guard of Britain's golden land

(Darwin, 1784)

Putting aside its seeming sycophancy and bombast to the reader of today, this poem is interesting in the context of its period. In the above extract, Darwin weaves together classical imagery with a sensitive evocation of the delicate new plant life that has been recruited from the 'antarctic' ends of the earth. Thanks to the efforts of the 'sons of science' (not to mention the 'ROYAL PARTNERS'), these frail seedlings of 'weak stem' and 'erring tendril' will burgeon into mighty plants such as 'the towering Oak' to augment further the emerging imperial might of 'Britain's golden land'. Whatever their poetic merits, these verses demonstrate the self-aware marriage of botanical science with its practical employment for both material well-being and the advancement of state power. Darwin is saying that knowledge was not so much for knowing as for applying, and applying in a very specific direction. His poem also anticipates the more systematic deployment of botanical knowledge for the benefit of the state that was pioneered so effectively in the USA after the mid-nineteenth century.

Banks was a scientific imperialist *par excellence*, who took over as advisor at the Royal Botanical Gardens at Kew in 1772, soon after returning from the famous circumnavigation voyage with Captain Cook (see above). Once established at Kew, Banks sought to outdo the rival *Jardin du Roi* in Paris, which had been turned into a noted research centre by the Comte de Buffon during his directorship from 1739 to 1788.[1117] Banks also benefited greatly from his long-standing interest in agriculture, which not only supplied him with much of his considerable wealth, but was also at the core of his intellectual life.[1118] Although a supreme British nationalist, Banks was also a true scientist in his internationalist leanings. Hence, he arranged with the Admiralty for all scientific collections captured during the wars of the early nineteenth century to

be returned to their French and American owners.[1119] No other form of property was accorded such a privilege. This illustrates how scientific knowledge was deemed to transcend petty national rivalries. Throughout the medieval period, the Enlightenment and beyond to today, scientists in different countries have generally maintained strong links with each other, despite warfare, blockade and repression.[1120]

Economic and political botany

Sugar, tea, coffee, and chocolate

During the sixteenth to the late eighteenth centuries, dozens of crops were transplanted from one continent to another for a wide range of purposes.[1121] Various species of palm were taken from the jungles of Africa and Malaya to Bengal in India, to stave off famine in this new British colonial possession. Breadfruit was brought from Tahiti to the West Indies to be grown as a cheap staple to feed the recently imported African slaves. Many of the latter unfortunates had themselves been brought over to work in the new and immensely profitable Caribbean sugar cane plantations.[1122] One of the motivations for the British conquest of the Caribbean island of Jamaica in 1655 was its agricultural potential. Unlike the previous Spanish rulers, the British soon established a vast and immensely lucrative export industry based on sugar cane and cocoa. During this period, the boundaries between the public and private sectors were rather blurred in European societies. Hence, many colonial ventures that ultimately involved the expansion of the state originally arose from trading enterprises led by aggressive, private companies, such as the Dutch West and East Indies Companies or the British East India Company. In the case of the conquest of Jamaica, the sugar cane industry was funded by a series of entrepreneurs headed by the King and his relatives, as well as many private investors, including noted scientists such as John Locke. Once they realized that a new supply of labour was needed for sugar cultivation, these investors founded the Company of Royal Adventurers to pursue a trade in African slaves. In 1672, the name was changed to the Royal Africa Company and the transatlantic slave trade began in earnest.

Agriculture, trade, botany, and colonial power were now enmeshed in a nexus involving merchants, financiers, landowners, politicians, soldiers, and royalty; all seeking to further the collective enterprise of the exploitation of overseas plant resources for personal profit. Sugar cane, originally a Southeast Asian plant was transplanted across the world for cultivation in the Caribbean, while the sugar itself was shipped across the Atlantic in order to sweeten the palates of the newly emerging tea and coffee drinking classes of the European Enlightenment. Sugar cane refers to six species of perennial grasses of the genus *Saccharum*, two of which are wild, *S. spontaneum* L. and *S. robustum*, while four species are cultivated, that is *S. officinarum*, *S. barberi*, *S. sinense*, and *S. edule*.[1123] The four

cultivated species are complex hybrids, and all intercross readily with each other. All commercial canes grown today are interspecific hybrids.[1124] Although the centre of origin of sugar cane is probably in northern India, the cultivated varieties of today are derived primarily from New Guinea and the Indonesian Archipelago. Tea (*Camellia sinensis*) and coffee (*Coffea* spp.) were, respectively, exotic east Asian and Ethiopian crops that were transplanted to new centres of cultivation in various European colonies in the tropics to meet the seemingly insatiable demand from the proliferating teahouses and coffeehouses of seventeenth and eighteenth century Europe.

Cocoa (*Theobroma cacao*) originated in South America and was brought to Mesoamerica by the Mayans. The beans were used to make a powerful drink, variously called *chicolatl*, *xocalatl*, or *cacahuatl* (meaning 'bitter water') by the Aztecs in their Nahuatl tongue.[1125] The paste of cocoa beans was mixed with spices, vanilla (*Vanilla plantifolia*), and a small amount of honey to make a highly prized beverage. This somewhat acrid concoction, often drunk out of pure gold goblets, was a luxury item, largely reserved for ceremonial use by the aristocracy and royalty (Figure 8.3). In an echo of the regal treasure-house/granaries of the ancient Mesopotamian kings, the major item of value stored in Aztec treasuries was hundreds of thousands of cocoa beans. This was much to the chagrin of the Spanish conquistadors who, in 1519, had stormed Montezuma's main treasury in Tenochtitlán at great cost to life and limb, in the expectation that it would be full of gold, only to find heaped piles of cocoa beans. In the seventeenth century, cocoa seeds were taken from the Americas for replanting in West Africa and the East Indies by Dutch and British traders to supply an increasingly lucrative demand for chocolate in fashionable salons across Europe. From the mid-seventeenth century, chocolate houses were all the rage with the elites of Vienna, London, Paris, and Amsterdam. In 1657, the first chocolate house in London opened and soon became at least as popular as the coffeehouses of the period. Chocolate contains theobromine, plus small amounts of caffeine and anandamide, an endogenous cannabinoid also found in the brain. The presence of these and other psychoactive compounds, and the rich creamy texture of chocolate, make it a soothing and satisfying drink. In contrast, coffee contains much higher quantities of the alkaloid stimulant, caffeine, and has a more rapid and acute physiological effect on the drinker.

The palatability of chocolate was greatly enhanced by the addition of generous measures of milk and sugar, which moderated its otherwise acerbic taste. This innovation was first described by the eminent physician and botanist, Sir Hans Sloane, who found the ordinary bitter chocolate of the time 'nauseous'. Sloane devised a secret recipe for milk chocolate that was later acquired by the Cadbury family and is still used today in the many products sold under this well-known, global brand. Sloane was a typical polymath of the time. During a brief visit to Jamaica, he catalogued over 800 new plant species, publishing the work in a two-folio volume in 1696.[1126] Returning to Britain, he resumed his medical work and pioneered an early version of smallpox immunization before becoming personal physician to Queen Anne. He went on to succeed Isaac Newton as President of the Royal Society in 1727, a post that he held until his death in 1753. Soon after his death, Sloane's vast collection of over 50,000 books and 200,000 specimens, many of them botanical, were purchased by the nation and formed the nucleus of the British Museum. Thanks to Sloane's innovation, chocolate houses reigned supreme in England for well over a century before becoming eclipsed by coffee and teahouses in the 1790s.

Botany, commerce, and power
The mid-eighteenth century can be said to mark the beginning of economic botany, not only as a scientific discipline, but also as an often lucrative commercial opportunity. The promotion of economic and applied botany soon became an integral aspect of government policy, most notably in Britain and the Netherlands.[1127] As illustrated in the poem, *The Botanical Garden* by Erasmus Darwin (Box 15.2), plants could serve as very effective instruments of statecraft. In particular, the broader power of food as a political tool that could be wielded by governments with access to food surpluses, became increasingly apparent during this period. As with so many developments in agriculture, this harks back to the agroimperialism of ancient cultures such as the Akkadians who explicitly used crop surpluses as a weapon of statecraft. An interesting illustration of British policy is provided by the events that immediately preceded the French Revolution of 1789. The spectre of famine was continually present at this time, and the capacity of food shortages to fuel social unrest was well known and feared in all the major countries of Europe. Over the preceding century, France had fallen far behind Britain in its agricultural efficiency and suffered chronic grain shortages, while British farmers produced a reliable surplus. The year of 1789 was especially bad in France, with widespread crop failures and food shortages.[1128] Most French farmers continued to grow wheat during this period, despite its poor response to heavy rainfall. In contrast, Dutch and English farmers diversified their crop base by growing more barley, oats, legumes, and brassicas.

By June 1789, wheat shortages in France were getting increasingly desperate and the beleaguered government appealed to their British neighbours for a shipment of much needed flour.[1129] Despite having grain stocks in abundance, British Prime Minister William Pitt at first procrastinated and finally refused the French request. The famished French populace exploded into rebellion soon afterwards. Within a month, the Parisian *sans culottes* had stormed the Bastille and the Revolution had started. A few weeks later, the French government and monarchy, hitherto the strongest power in Europe, had been overthrown and a republic declared. There is little doubt that, while it in no way directly caused the revolution, the withholding of food shipments by the British probably played a part in accelerating its progress. Although he could not have foreseen the momentous consequences of his actions, Pitt would have been duly impressed by such a demonstration of the potential power of food as a tool of statecraft. Pitt and his contemporaries doubtless hoped that, with the help of science, the food weapon might be rendered even more potent in the future. This lesson was duly learned by the British government, which was especially diligent in its deployment of the food weapon as an instrument of state policy from that time onwards. Indeed, only few years later, the Royal Navy blockaded

Napoleonic Europe as part of the British strategy to reverse French hegemony on the Continent.[1130]

As the nineteenth century dawned, plants of all descriptions were big business in Europe and immense fortunes could be made by botanical entrepreneurs. Entire cities and colonial regions were becoming increasingly reliant on the new trade in exotic crops. For example the great west English port cities of Bristol and Liverpool owed much of their wealth in the late eighteenth and early nineteenth centuries to the combined trade in Caribbean sugar and African slaves.[1131] Many merchants established huge slave-worked sugarcane plantations in the West Indies, becoming fabulously wealthy on the proceeds of this squalid venture.[1132] By the beginning of the nineteenth century, the establishment of botanic gardens and the auditing of the indigenous crops were becoming a routine accompaniment to the seizure of new colonial possessions, especially by the British. In 1802, a garden was set up in newly-conquered Trinidad; in 1810 a botanic collection was ordered in Ceylon;[1133] in 1814 it was the turn of the former French possession of Mauritius; and in the same year Stanford Raffles established the botanic garden at Buitenzorg (now known as Bogor) on Java, during a brief interlude of British control of the island. The Bogor gardens were kept on by the Dutch when they recaptured Java soon afterwards, whereupon Raffles established a new botanic garden in the nearby British entrepôt of Singapore. Meanwhile, now back under Dutch management, the Bogor gardens expanded to include some especially impressive collections of ornamental and practical plants. To house the latter, a separate Economic Garden was established at Bogor. By the early the twentieth century, the Bogor gardens were the most important centre for tropical botany and agriculture in the world.[1134]

Some of the older British colonies also received similar botanical beneficence, as exemplified by the new Glasnevin Botanic Garden that was launched with great fanfare and celebration in the city of Dublin in 1800. The good gentlemen of the 'Dublin Society for Promoting Husbandry and Other Useful Arts' made clear their agenda in their newly launched *Transactions*. The Garden was to be: 'a complete repository, for practical knowledge in every thing which respects vegetation, agriculture, trees, farming and all uses of the surface of the land

and its produce . . . and to be a school for instructing all persons concerned in the produce of the land.' Alas, these worthy sentiments only applied to 'improved' private land and to the wealthier, anglicized, landowning classes. They most assuredly did not apply to the poverty-stricken tenant farmers working 'unimproved' land in Munster and Connaught in the south and west of Ireland. Here the Dublin gentlemen looked with indifference upon the ever-proliferating multitudes of impoverished Gaelic natives and their ill-fated dependence on a potato monoculture.

It was this reliance on single clonal variety of Andean potato that led a massive outbreak of fungal infection in 1845. The resulting catastrophic famine, known locally as *An Gorta Mór* (the great hunger), killed over a million people and depopulated much of western Ireland during the 1840s.[1135] Only the potato crop was affected by the fungal blight and other crops fared well. However, many folk in the subsistence farming communities starved simply because they could not afford to buy food. Little useful purpose did the Glasnevin Botanic Garden serve for these poor wretches. Notwithstanding this unfortunate episode in colonial Ireland, the dissemination of plant-related knowledge was greatly improved during from the period from the mid-eighteenth century Enlightenment until the early nineteenth century. This was largely due to the activities by European (and later by American) botanists and agronomists who were able to take advantage of the new samples and other information that became available from the botanical explorers of the time, as well as the emerging knowledge of plant reproduction and its potential for manipulation for crop improvement.

Beginnings of scientific breeding

Plant reproduction and systematic botany

Modern scientific crop breeding relies on the systematic manipulation of plant reproduction, which in turn requires a detailed knowledge of the wider biology of plants in general. Beginning in the early eighteenth century, investigators such as Camerarius and Linnaeus established the framework for later generations of botanists and practical breeders up to the present day. Rudolph Jacob Camerarius (1665–1721) was director of the Tübingen Botanic Garden in Germany. He was the first to demonstrate that sexual reproduction occurred in plants and proposed the role of pollen as the equivalent of animal sperm in plant fertilization.[1136] Camerarius also suggested that crossbreeding of different varieties, or even different species, could be used to create new and potentially more useful types of plant. Another key achievement of this period was the creation, in 1718, of the first interspecific hybrid (see below). With these successes, the increasingly intrusive intervention of humans into the mechanisms of plant reproduction was already well underway by the early eighteenth century.

Another landmark in the advancement of botanical knowledge was the publication, in 1753, of *Species Plantarum*.[1137] In this book, Swedish naturalist Karl Linnaeus produced an extensive catalogue of plants and first used the system of binomial nomenclature that is now universally applied to all species from *Escherichia coli* to *Homo sapiens*. The works of Linnaeus are especially interesting in the way that they show how his ideas on the formation of species gradually altered as his research matured, and he learned more about the behaviour and classification of plants. In some of his earlier published work from 1735, Linnaeus still clung to the old dogma that the number of plant and animal species had remained the same since the supposed 'creation' of ancient Hebrew and later Christian myth.[1138] However, new data from experimental crossings of plants convinced him that hybridization produced new combinations of parental traits. While he did not completely abandon the notion of creationism, in his later work of the 1750s, Linnaeus proposed that the genus rather than the species was likely to be the basic unit of divine creation. At this point he even admitted, for the first time, the possibility that new species could appear in the world and existing species disappear from it. As we will now see, the deliberate production of new varieties and new species of plant by hybridization was set to become one of the key fields of scientific study in the eighteenth and nineteenth centuries.

Hybrids and their importance in crop improvement

The term hybrid is used in several different ways in biology, but it always denotes the progeny of two genetically dissimilar parents. Two major classes of hybrid are relevant to plant breeding, namely *intra*specific and *inter*specific hybrids. More commonly, a hybrid is intraspecific, that is the offspring of two members of the same species. Such intraspecific hybrids are very common in agriculture. Today, some of our most important crops and ornamental plants are hybrids. For example, since the 1930s almost all the major commercial varieties of maize have been intraspecific hybrids. A second type of hybrid is the interspecific hybrid, where the individuals are products of mating between parents of two different species. Although such hybrids are often sterile, fertile progeny are sometimes produced, especially in plants (see Chapter 4). The earliest recorded, manmade interspecific hybrid plant was called Fairchild's mule after its creator who produced the new ornamental species in 1718 by crossing the gillyflower, *Dianthus caryophyllus*, with sweet William, *Dianthus barbatus*. As its name implies, Fairchild's mule is a sterile interspecific hybrid. It was produced just after the phenomenon of the Dutch 'tulip mania',[1139] which was also the supreme age of the early-Georgian formal garden. The new plant was created as a commercial venture with its distinctive floral pattern that appealed to well-heeled gardeners.[1140]

Much money could be made by enterprising breeders of Fairchild's ilk who were able to satisfy the ever-increasing demand for floral novelties. Fairchild's special triumph was to show that a plant breeder could produce a novel variant by recombining existing species, rather than having to rely on discovery and importation of expensive exotics from abroad. Many hybrids cannot reproduce and must be recreated by the breeder each new generation. Even today, many annual ornamentals sold by garden centres are sterile hybrids, requiring gardeners to repurchase more seed each year, to the great profit of producers of such hybrids. Gardeners gladly repurchase sterile hybrid seeds because the resulting plants are often superior to their fertile, but unimproved, counterparts. We can therefore say that it was the likes of Fairchild and his ilk that invented this eminently lucrative business model of sterile plant procreation, almost three centuries ago.[1141]

We have seen previously that some of our commonest and most ancient crops are spontaneously occurring, interspecific hybrids that have been formed and reformed, without deliberate human intervention, since the dawn of agriculture. Hence, all modern forms of breadwheat and durum wheat are interspecific hybrids between various goat grass species and einkorn wheat. Oilseed rape (*Brassica napus*) is a spontaneous hybrid between two different species of the *Brassica* genus, namely cabbage (*Brassica oleracea*) and turnip (*Brassica rapa*). Both intraspecific and interspecific hybrids were gradually recognized for what they were and then manipulated by plant breeders from the mid-eighteenth century onwards. Intraspecific hybrids are commonly used in crop cultivation because of the phenomenon of heterosis, or 'hybrid vigour' (see below). Inbreeding has long been appreciated by animal breeders, for example all pedigree dogs are extremely inbred and, while many of these breeds suffer from various types of congenital deformities, they are nevertheless prized as 'pure breeds'.

Despite the success of inbreeding as a strategy for producing specialized varieties of animal or plant, there was a general feeling, often with good reason, that it was somehow 'wrong'. However, after 1760, Robert Bakewell demonstrated that inbreeding was not necessarily a bad thing in an agricultural context. For example inbreeding ensures uniformity in a crop, which is a trait that is greatly prized by farmers, and is also the quickest and most expedient way to fix a new genetic character in a population. From the eighteenth century onwards, crops manipulated by the new methods of early scientific breeding became increasingly inbred and lacking in variation. This may not necessarily be problematic in species that rarely outbreed in their normal environment. However, in species that are normally accustomed to outbreeding, enforced inbreeding often leads to so-called 'inbreeding depression'. This is marked by deleterious traits such as poor seed germination, slower growth rate, and reduced disease resistance. Inbreeding causes a reduction in heterozygosity, as most alleles tend to become homozygous. As a result, many deleterious recessive genes that are normally masked become expressed, leading to a marked reduction in fitness and vigour. A similar phenomenon occurs in inbred groups of humans, a good example being the high incidence of haemophilia in the descendants of Queen Victoria. The remedy, in both plants and humans, is to set up an outbreeding cross with a non-relative so as to restore heterozygosity and hence suppress the deleterious alleles in future generations.[1142]

Inbred crop varieties often carry many useful agronomic traits that may have been selected over several centuries or more, so breeders wish to maintain the inbred lines, but they still need to somehow get around the problem of inbreeding depression. Fortuitously, crops can be rescued from inbreeding depression by crossing two different inbred lines together to produce an intraspecific hybrid. The results are often dramatic, with the new hybrids producing as much as 25 to 50% higher yields than the inbred parental lines. Originally called 'hybrid vigour', this phenomenon was first described by the German botanist, Josef Kölreuter, in 1761. This pioneering scientific breeder demonstrated that hybrid offspring received traits from both plant parents and were intermediate in most traits; he also produced the first hybrid crop variety.[1143] The first generation of such hybrids is called the F_1.

In 1823, Thomas Knight in England demonstrated, by using several species including peas and wheat, that male and female parents contribute equally to the F_1 generation but that the next generation, or F_2, is very different.[1144] The reason for the extreme variation observed in the F_2 generation is genetic segregation, which produces a wide range of progeny, many of them agronomically useless. The result is that farmers who grow an F_1 hybrid crop cannot grow a useful crop from their own saved seed, which will comprise the useless F_2 generation. Knight had previously published an influential paper on the practical application of hybridization in agriculture.[1145] He was also the first person to suggest the use of a convenient model system for future experimental studies, namely the garden pea. Knight made this suggestion many decades before the same plant was used with such effect by Gregor Mendel to establish the principles of genetic inheritance. In Knight's words:

... none appeared so well calculated to answer my purpose as the common pea, not only because I could obtain many varieties of this plant, of different forms, sizes, and colours, but because the structure of its blossom, by preventing the ingress of adventitious farina, has rendered its varieties remarkably permanent.

As Knight also noted, the pea is a conveniently sized, rapidly growing annual that is easily cultivated either indoors or outdoors. This combination of useful characteristics led to the use of peas as perhaps the most important plant model system of nineteenth century genetic research. Fifty years after Knight had originally suggested their use as a model system, Mendel used the same model system to work out his laws of inheritance. Thomas Knight deserves credit for first suggesting the use of a convenient experimental model from which results could be extrapolated to other plants, including the major crops many of which were not so easy to work with experimentally. The potential of using hybrids for crop improvement soon piqued the interest of several of the new scientific academies that had been established in Europe, following the earlier example of the Royal Society in England, and several prizes were offered for practical suggestions that could be applied in agriculture.[1146]

The hybrid crops discussed above are the result of crossing different varieties of plants from the same species, but it is also possible to produce hybrids from plants of different species, and even from different genera. During the twentieth century, these so-called wide hybrids were increasingly used to create ever more radical and useful forms of variation in many of our most important crops. Interspecific hybrids are normally the progeny of a cross between two different species from the same genus and many crops will occasionally hybridize spontaneously with other species in the same genus to produce fertile progeny. Such an event occurred about 2000 years ago to create oilseed rape, and similar, spontaneously produced wide hybrids have often been of great utility to crop breeders. However, such chance events are extremely rare and in some species do not occur at all. Therefore it is often the case that agronomically useful, interspecific hybrids can only be created by breeders after a great deal of technological manipulation, involving methods such as embryo removal, *in vitro* culture, chemical alteration of the genome, and regeneration using plant growth regulators.[1147]

Intergenus hybrids are often bizarre organisms because the genetic differences between such species are much greater than between species belonging to the same genus. Therefore, successful intergenus

hybridization, whether spontaneous or manmade, is comparatively rare. The best-known examples of spontaneous intergenus hybridizations in agriculture are those between the goat grass genus, *Aegilops*, and the wheat genus, *Triticum*, which produced the polyploid species of durum wheat and breadwheat (Chapter 6). In the nineteenth century, French botanist Esprit Fabre observed a similar spontaneous hybridization between wild species of *Aegilops* and *Triticum* that had been collected in Provence.[1148] When Fabre communicated his results to the French Academy in 1852, the data were questioned by established scientists such as Jordan, who still held to the dogma of the immutability of species.[1149] It was only in 1854 that Fabre's botanical colleague, Dominique Alexandre Godron, conclusively demonstrated by direct experimentation that, not only could *Aegilops* and *Triticum* produce viable intergenus hybrids, but that this phenomenon could be reproducibly recreated by any careful plant breeder.[1150]

Mutations and their uses

Another way of introducing variation into a population is via mutation. A mutation is caused by a sudden change in chromosomal DNA that often results in a change in gene function and a consequent alteration in the appearance, or phenotype, of an organism. The existence of spontaneous mutants or 'sports' was known by farmers and naturalists for millennia, but their significance for generating new biological variation was not realized. This was largely because most people still believed that all the original variations in living organisms had been produced by an external deity in a single act of creation. As a result, mutations were regarded as rare and aberrant monsters that were generally considered to be abominations of the naturally created order of life. An early clue about the biological significance of mutations, that also shed light on plant development and evolution, came from the careful series of observations by Johann Wolfgang von Goethe, the German poet, philosopher, and botanist. In 1798, Goethe published his seminal essay on the *Metamorphosis of Plants*. In this study, he described the phenomenon of mutated flowers in which one floral organ is replaced by a different one. Goethe realized that mutations can be regularly seen in some species of plant, especially when they result in a gross change in appearance. He found that sometimes the different organs of the flower change, or mutate, from one form to another, for example petals may develop instead of anthers. Goethe also discovered that leaves and flowers are obviously related when he observed that some of his mutant plants produced intermediate forms of these two organs. To quote from his essay:

Anyone who has paid even a little attention to plant growth will readily see that certain external parts of the plant undergo frequent change and take on the shape of the adjacent parts—sometimes fully, sometimes more and sometimes less. Thus, for example, the single flower often turns in to a double one when petals develop instead of stamens and anthers; these petals are either identical in form and colour to the other petals of the corolla, or still bear visible signs of their origin.[1151]

Given his lack of modern botanical knowledge, Goethe's detailed description of metamorphosis, that is mutagenesis, in flowers is remarkably prescient. It is also the first description of what we now know as homeotic mutations, although that term was only coined by William Bateson over a century later, in 1894.[1152] We now know that all of the floral organs, including petals, sepals, stamens, and carpels, originally arose from modified leaves via a series of spontaneous mutations, similar to those studied by Goethe in the eighteenth century. The discovery of the principles of heredity by the Austrian monk Gregor Mendel and of evolution by Charles Darwin in the mid-nineteenth century were two more key building blocks for the development of modern forms of scientific crop breeding, as we will discuss in the next chapter.

CHAPTER 16

Agricultural improvement in modern times

Genetic changes underlie the evolution of organisms; mutations are the ultimate source of the genetic variation that makes possible the evolutionary process.

Francisco Ayala and G. Ledyard Stebbins, 1981,
Science

Introduction

After more than ten millennia of rather slow, fitful progress via empirical crop improvement, the period from the Renaissance to the Enlightenment witnessed a much faster transition to a radically different, more scientifically informed process. This transition was fuelled, in part, by the conjunction of new forms of commercial opportunism that were relatively unfettered from state control, and by the related process of imperial expansion that opened up new vistas for crop exchange across oceans and continents. It was also driven by the identification of new or improved techniques of cultivation from other regions of the world. These developments occurred in parallel with the emergence of new science-based methods of knowledge creation. Perhaps more important than the mere creation of additional knowledge, however, was its improved dissemination, thanks to innovations such as printing. An international community of scholars, researchers, and technologists emerged, who communicated more freely and effectively with one other as part of a new tradition of the sharing, rather than the hiding, of knowledge. It was this more ready availability of new evidence-based knowledge, coupled with greatly improved opportunities to harness such knowledge, that launched a quantum leap in our ability to both understand and manipulate the

biological world, including how plants develop and reproduce.

At the same time as the burgeoning of new and more reliable knowledge about how plants worked and how they might be manipulated, voyages of discovery and colonization resulted in the (re)discovery of hundreds of new, potentially exploitable plants.[1153] As we saw in the previous chapter, the synergy of economic opportunity, improved knowledge, and confidence in the ability to improve and exploit these resources led to the blossoming of the European project of imperial botany, which was largely led by the maritime powers of the Netherlands and Britain. This was the age of the globally minded botanists; men such as Tradescant, Banks, and Darwin; just as much as that of more home grown agrarian improvers such as Townshend, Tull, and Coke. By the late nineteenth century, the more ready availability of an ever-growing corpus of new knowledge, and the realization that it could be exploited for the radical improvement of agriculture, led to the establishment of a professional cadre of trained plant breeders in new forms of public sector institutions. These developments constituted something of a break with the previous paradigm of crop improvement, which in Europe had been mainly carried out by individual husbandmen and private entrepreneurs, rather than public functionaries.[1154] Over the past 150 years, this public-sector paradigm has served agriculture and society very well indeed. Virtually all the new plant breeding technologies of the twentieth century, from mutagenesis to mass propagation, were developed within the public sector milieu. Later in the twentieth century, public sector plant breeders were responsible for the Green Revolution of the

1960s and 1970s. This was an especially momentous achievement that enabled local farmers to feed billions of people in developing countries, who otherwise faced the very real prospect of widespread hunger or starvation.

The achievements of modern agriculture

The success of modern agriculture in feeding the ever-increasing population of the world is often underestimated. The human population has increased ten-fold over the past three centuries, almost quadrupling during the past century alone.[1155] This unprecedentedly rapid population growth could not have occurred were it not for matching increases in food production from agriculture. Although the recent population explosion is now levelling off, it is likely that an additional 2.6 billion people will need to be fed during the next 50 years.[1156] Again, this will be a task for agriculture but, as we have seen repeatedly in the past, food production is susceptible to environmental factors, including, this time, some climatic changes of possible anthropogenic origin.[1157] The near trebling of global food production in the two centuries from 1700 to 1900 was sustained largely by the cultivation of new land that was won either by reclamation or the establishment of overseas colonies. As we saw in Chapter 14, agricultural improvements, such as better crop rotation systems and use of new varieties, also played their part in enabling higher yields, but expansion of arable cultivation was undoubtedly a key factor. This situation continued into the early twentieth century as pioneers and migrants continued to win new arable land, especially in the Americas. From 1860 to 1920, more than 440 million hectares of land was brought under the plough for the first time. This increased the global area of arable land by about 70%, and the world population also increased by about 70% over the same period.

After the mid-twentieth century, the even more dramatic population increases of the next five decades were mainly sustained by increased plant productivity on existing land, rather than by expansion into new arable cultivation. This improvement in crop output was due to a combination of the better use of inputs (e.g. fertilizers, herbicides, and

insecticides), and the breeding of higher yielding crops in a process that was increasingly informed by the science of genetics. A few more statistics may demonstrate the significant achievements of crop scientists over the past half-century. We have continued to win more arable land since the 1950s, but this has slowed to a growth rate of less than four million hectares of new land per year, and we are now beginning to approach the global limit of suitable soils and climatic regions.[1158] Over the past 50 years, we have only expanded the area of global crop cultivation by about 10 to 15%. During the same period, the population has more than doubled. But what is most impressive is that food production has actually *trebled* since 1950. This means that not only have farmers and breeders kept pace with the unprecedented population increase of recent times, they have been able to produce an impressive 40% more food *per capita* for every person on earth compared with 50 years ago.[1159] How was this done and what are the implications for our future relationship with crops? We will now examine these topics by looking at the combination of improved management of crops, and increasingly sophisticated genetic manipulation of plants by breeders, that enables us to feed more people today than the combined total number of humans that has lived on the earth over the past million years.

Improving crop management

Scientifically informed crop management, or agronomy, has made a decisive contribution to agricultural production over the past century or so. As we saw in previous chapters, many incremental improvements were made in crop management over the past 8000 years. These include technological innovations such as stone and metal tools, the seed broadcaster, and the use of crop rotation systems. The still unparalleled success of Mesopotamian civilization over so many millennia was largely based on careful management of complex networks of irrigated crops, plus the exquisitely controlled distribution of grain to urban populations. During the post-Renaissance period in northwest Europe, the first steps were taken in a more rational, evidence-based approach to using inputs such as fertilizers and mechanization of tasks such as harvesting and grain processing. Rather than consulting ancient texts, experiments were performed to determine optimal fertilizer and crop rotation regimes, and the results were published and/or disseminated at agricultural fairs or other demonstration venues. Combined with the winning of new lands via drainage or expansion, these early protoscientific attempts to increase crop productivity by improved management were surprisingly successful in underpinning the tripling of global food production from 1700 to 1900.

During the nineteenth century, several new strands emerged in the practice of agronomy that were to make a key contribution to our current high levels of food output, despite the even greater increases in population since 1900. On the scientific side, new or improved inputs were developed, including chemical fertilizers, and a variety of crop protection agents ranging from herbicides and pesticides to fungicides and antiviral formulations. On the engineering side, on-farm mechanization was accelerated thanks to powered tractors and combine harvesters. Grain-processing units also became larger and more efficient and storage conditions were greatly improved. Finally, agriculture moved from being a relatively small-scale, family or community-centred, operation to the large-scale agribusiness venture prevalent today in many of the most productive crop-growing areas of the world.

To a great extent, the advances in agriculture over the past century or so have been due to the establishment of effective public-sector networks of research and dissemination of knowledge to farmers. Probably the most influential of these developments occurred in the USA in the late nineteenth and early twentieth centuries.[1160] Some idea of the success of US agriculture between 1867 and 1901 can be gleaned from the unprecedented 12-fold increase in maize and five to six-fold increase in wheat exports (to 5.4 and 4.9 million tonnes respectively).[1161] These achievements were due, in part, to the establishment of crop-improving institutions in the USA, as informed by two separate but linked strands in wider US society. First, there was the desire to extend 'useful and relevant scientific education' to the agricultural

and artisan classes that in those days formed the backbone of the nation's energies and future prospects. Secondly, there was pressure from private groups, including bankers, wealthy farmers, and editors of agricultural journals, for a more systematic use of scientific knowledge to improve agricultural productivity. These US groups were both stimulated by, and worried about, advances in applied agricultural research in Europe, especially regarding the use of inputs such as fertilizers. The more global impact of such knowledge can be seen in the case of Asian rice, which yielded a meagre average of about 1 to 2 tonnes per hectare during the millennium up to the twentieth century.[1162] The eventual use of fertilizers enabled yields to be doubled by 1950, while a combination of breeding advances and improved crop management enabled Chinese yields to exceed 6 tonnes per hectare by 2000, with maximum potential yields of well over 12 tonnes per hectare.

Inputs

One of the main examples of European research that concerned would-be agricultural improvers in the USA was the work of German chemist, Justus von Liebig, which appeared there in the 1840s.[1163] Liebig and his followers worked out how to improve soil fertility with inorganic fertilizers, which not only vastly increased crop yields but also spawned a massive new chemical industry that directly benefited the newly unifying German state. Liebig began the scientific study of soil fertility, and especially the effects of nitrogenous compounds on the yield of crops and other plants. In addition to inspiring farmers and scientists in Europe and the Americas, his work led directly to the search for a way to manufacture nitrogenous fertilizers that culminated in the invention, in 1908, of the Haber–Bosch process for industrial-scale fixation of nitrogen gas into ammonia. Liebig was also one of the first people to establish a professionally organized, university-based research laboratory, first at the University of Giessen, and later Munich, in Germany.[1164] This development marked the beginnings of a shift in the research paradigm from a focus on amateur individuals or small groups of self-funded gentlemen scientists to more organized state-funded networks of professional scientist-educators.[1165] Thanks to Liebig and colleagues, chemical fertilizers are now an indispensable part of high-yield cropping systems, although in some places their misuse has caused environmental problems, such as the adverse effects on watercourses of excess nitrate runoff from fields.[1166]

Nitrogen availability is one of the most important limitations on crop growth, ranking in importance with such key inputs as sunlight and water. Prior to the use of chemical fertilizers, farmers had been forced to rely on biological sources, such as farmyard manure or bird guano. Plants cannot assimilate the organic (carbon-linked) forms of nitrogen in these biologically derived fertilizers, and can only use the inorganic breakdown products, such as nitrates, which are gradually released from the organic compounds. Due to their high cost, limited availability, and their slow rate of nitrogen release, use of organic fertilizers on a large scale can significantly limit food yield from crops. Second only to nitrate as a yield-limiting mineral for crops is phosphate. Success in the search for an efficient inorganic form of phosphate fertilizer came in 1842 when James Murray, an Irish doctor who dabbled in chemistry as a hobby, discovered that acid converts calcium phosphate into a soluble mixture of calcium hydrogen phosphate and calcium dihydrogen phosphate.[1167] This is the basis of superphosphate, an extremely effective, slow-release fertilizer, which was soon in widespread use in Britain.

During the nineteenth century, the restricted availability of conventional fertilizers such as manure caused richer European countries to scour the world for guano as an alternative, causing great environmental damage in the process. For example the first cargo of Peruvian guano arrived at the port of Liverpool in 1835. By 1841, 1700 tonnes were being imported, rising to 220,000 tonnes by 1847. These unsustainable activities resulted in the destruction of bird colonies and the long-term impoverishment of soils in those regions that supplied guano.[1168] In Europe itself, horse manure, and even human waste, was collected from urban streets and transported (by horse-drawn wagons) to farms where it was applied, in often massive quantities, to the soil. For growing vegetables, as much as 1.2 tonnes of manure per hectare was

used.[1169] Such quantities of manure were often impractical and expensive to collect, and the supply of guano was fast running out.

Food production was rescued by the arrival of the inorganic fertilizers. In particular, the introduction of chemical forms of nitrate and phosphate fertilizers greatly improved crop yields after the late nineteenth century and these are still mainstays of conventional farming across the world today. Although these inorganic fertilizers have revolutionized food yields, there is a biological limit to their effectiveness, especially in the case of cereals. This is because, at high doses, such fertilizers can cause plants to produce tall, spindly stems that are more easily blown over or flattened by rain, a phenomenon called lodging. As we will see below, this limitation was eventually removed by the biological manipulation of cereals by breeders in the mid-twentieth century in a development that ushered in the Green Revolution and successfully averted the threat of starvation in much of the developing world.

After fertilizers, the other major class of inputs contributing to high yields are the crop protection agents. Every year, between one-third and a half of most crops used to be lost due to competition from weeds, to diseases caused by viruses, bacteria, and fungi and from damage caused by insects, rodents, and other pests. In bad years, an entire crop could be wiped out by a sudden outbreak of disease or pest infestation. For many centuries, and with mixed success, farmers experimented informally with hundreds of treatments against the pests and diseases that regularly ravaged their crops. For example, the use of chalk and alum as pesticides are mentioned respectively by Varro in the first century BCE and by Bassus in the seventh century CE, while the efficacy of sulphur was discussed in the fifth century BCE in Homer's *Odyssey*.[1170] During the medieval period, the twelfth century Moorish writer from Al-Andalus, Ibn Al-Awwam, reported the use of arsenical sulphides to control pests.[1171]

It was only after the sixteenth century that the use of chemical crop protection agents began to be investigated, and the results published, in a systematic and widespread manner. These studies led to reports, among many others, of the antiworm properties of potash (1631);[1172] the fungicidal and herbicidal effects of copper sulphate (1761);[1173] and the antifungal effects of sulphuric acid (1799).[1174] Two of the most significant scientific pioneers of work on crop protection were Mathieu Tillet in the 1750s and Bénéndict Prévost in the early 1800s, who laid the foundations of our understanding of fungal diseases such as cereal blasts and their control by chemical agents such as copper/lime mixtures.[1175] Already, by 1850, several dozen crop-protection chemicals were in widespread use in European agriculture.[1176] The modern era of chemical control in agriculture is often deemed to have started in the late nineteenth century with the use of Bordeàux mixture (copper sulphate and hydrated lime) to suppress powdery mildew and weeds in French vineyards. Bordeaux mixture is still used around the world today.

Many of these early formulations were highly toxic to humans and animals, as well as to their target organisms, but were used nevertheless because of their effectiveness in boosting crop yields. The development of cheap and effective chemical insecticides and fungicides did not really take off until the mid-twentieth century. As with the early herbicides, early insecticides and fungicides sometimes had adverse environmental side-effects, one of the best known of which is DDT. Interestingly, although the pesticide DDT has been greatly maligned in recent decades, new evidence on its efficacy prompted the World Health Organization (WHO) in 2006 to recommend resumption of its use for the control of endemic malaria.[1177] Nowadays, crop protection agents are much more selective and have far lower environmental impacts than their predecessors.[1178] Breeders have even developed crop varieties that produce their own protective agents, rather than relying on external inputs.[1179]

Intensification

The intensity of crop cultivation has varied enormously at different times and in different places over the past 12,000 years. For example, as discussed in earlier chapters, barley cultivation was much more intensive in the irrigated river valleys of southern Mesopotamia than in the rainfed farmlands of the north. However, the extent to which these intensification strategies were applied has waxed and waned during the many climatic and political upheavals of the Holocene period. In ancient China, early millet and rice cultivation only gradually intensified over a period of several millennia. In Mesoamerica, low yields of early maize varieties precluded intensive farming until about 1500 BCE. In much of Europe, farming intensity declined markedly after the fall of the Roman Empire and during parts of the 'Little Ice Age'. Agricultural intensification depends on an increased level of inputs (these can include a variety of external agents, such as fertilizers, human and animal labour, and technological aids), coupled with more active management of the farming system. The latter might include frequent weeding, guarding against pest and disease outbreaks, and even multi-cropping in the same field, as in the ancient *milpa* system in the Americas.

The archetypical intensive cultivation system is the small home plot for domestic consumption or local sale. As we saw in Chapter 12, some of the early European *Linearbandkeramik* cultures probably practiced this kind of highly intensive, but small-scale farming. However, we have also seen that several forms of large-scale intensive farming evolved in different parts of the ancient world. In the Near East, this was very much dependent on state-sponsored irrigation projects and the drafting *en masse* of both urban and rural human labour. In Mesoamerica, the Mayan cities of the Yucatán organized large rural peasant populations to tend the levees and raised cultivated fields of the immensely productive *bajo* wetlands. And in Central Mexico, the carefully tended *chinampas*, and vast *milpa* fields that so impressed the conquistadores, were intensively farmed under the firm direction of Toltec and Aztec rulers.

Despite their often epic scale, such state-organized ventures into intensive farming have tended to be the exception rather than the rule over the past ten millennia or so of human history. Apart from Pharaonic Egypt, there were few examples of large-scale intensive farming in Africa or Europe. Even in Asia, such experiments were rarely of long duration. As a rule, states have found it more expedient to leave the actual practice of farming to the private sector, whether in the form of a large landowner with his retinue of tied peasantry and considerable economies of scale, or the more highly motivated, independent, but small-scale yeoman farmer. Most large landowners lacked the drive to invest in agricultural improvement and such innovations might also be beyond the more modest economic means of the yeoman farmer. Until the eighteenth century, the demand for and the wherewithal to achieve a serious commitment to agricultural intensification were both largely absent. All of this changed as technological innovations, such as mechanization, joined forces with economic stimuli, such as the industrial revolution, and increased demand due to the growth in population, to create new conditions that were at last conducive to large-scale investment in more intensive systems of crop production. At first, this was very much a piecemeal process that was patchily applied across a few areas of Europe, and later in North America, over the course of the nineteenth century.

Intensification received a huge boost in the early twentieth century as mechanized devices, such as tractors and harvesters, and the fuel to power them fuel, became cheaper and more efficient. Plant breeders assisted the process by developing new crop varieties that were higher yielding and more amenable to mechanized harvesting and long-distance transport. The newly established agrochemical industry produced cheaper and more effective inputs, ranging from slow-release fertilizers to targeted pesticides and herbicides. In the UK, this led to consolidation of many smaller farms and removal of field boundaries to facilitate mechanized sowing and harvesting, for example in the intensively farmed 'prairies' of East Anglia. In the real Prairies of North America, a uniquely intensive form of agroindustry now produces food on an unprecedented scale and currently supplies a large proportion of the most important traded crops including maize, wheat, and soybeans. As we will see in the next chapter, this new form of large-scale, private-sector, intensive farming is heavily reliant on cheap inputs manufactured using non-renewable energy sources, and therefore may not be sustainable in the long term. Meanwhile, as we will now discuss, the biological manipulation of crops using genetics and breeding has assisted both large-scale agroindustry and smaller-scale farming, especially in developing countries.

Genetic variation and its manipulation for crop improvement

For most of the millennia-long history of agricultural improvement, increased crop yields have been won mainly by technical and management innovations. Biological improvements were a matter of happenstance that relied on random appearance of favourable mutants and the ability of farmers to recognize and successfully propagate such rarities. During the first two centuries of more scientifically informed plant breeding, the main stress was on the discovery and recording of new forms of existing variation, for example by means of botanical expeditions and the establishment of illustrated and annotated catalogues. The second phase of scientific breeding, which began in the twentieth century, involved the deliberate creation of additional forms of plant variation using newly invented technologies ranging from induced mutagenesis to transgenesis (Figure 16.1). Variation is arguably the most important resource for the breeder. As Charles Darwin observed in *The Origin of Species*, the inherited variation between individuals of a species is the major raw material for evolutionary change and development of new species. Likewise, all forms of variation, whether manmade or not, are also the raw materials for the deliberate selection that is exercised by plant breeders. After all, breeding is simply another form of evolution, in which the selection now occurs by considered human choice rather than via stochastic, environmental processes. Darwin himself was an enthusiastic and innovative plant breeder, who emphasized the importance of meticulous study and observation as follows:

Not one man in a thousand has accuracy of eye or judgment to become an eminent breeder. If gifted with these qualities, and he studies his subject for years, and devotes his lifetime to it with indomitable perseverance, he will succeed, and may make great improvements: if he wants any one of these qualities he will assuredly fail.[1180]

For the past three centuries, breeders have used the immense, but steadily diminishing, pool of existing variation amongst the various crops and their wild relatives.[1181] From this raw material, they have produced hundreds of improved varieties and created new, manmade species.[1182] The mechanisms

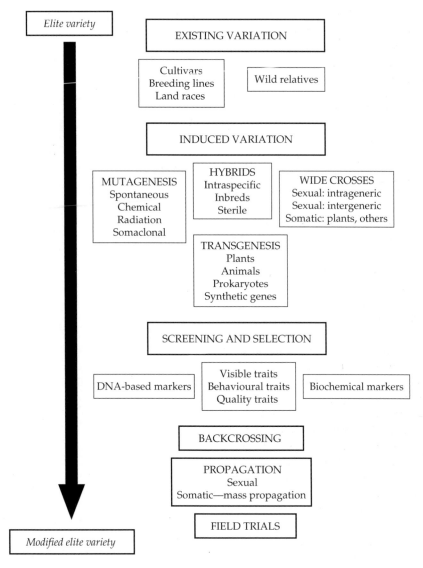

Figure 16.1 The mechanism of modern crop breeding. Plant breeding depends on *variation* and *selection*. Most crop varieties (and domesticated animals) tend to be relatively inbred, and new genetic characters such as disease resistance must be acquired by bringing in variation from other varieties or wild relatives. Before the eighteenth century, breeders relied for variation on sexual reproduction or the chance appearance of mutations in a population. Nowadays, additional variation can be created by deliberate mutagenesis or transgenesis and, with the assistance of tissue culture methods, even unrelated organisms can often be hybridized to generate fertile offspring. Following the creation of new variants, the population (possible numbering many thousands of plants) must be screened to identify suitable individuals. Selected plants are then backcrossed with the original elite parental line to generate a new elite variety that can be propagated and tested in the field. This entire process can take a decade or more but has nevertheless been the cornerstone of recent, immensely successful efforts to increase global food production, most notably the Green Revolution of the 1960s and 1970s that saved millions of famine-threatened lives in developing countries (Murphy, 2007).

responsible for creation of new variations in organisms were still an enigma in Darwin's time. However, it was obvious to these observers that, once we have a series of variations in a population, these can be reshuffled by the process of sexual reproduction to form thousands of new combinations. This gives rise to a series of new individuals each of which is slightly different, both from each other and from their parents. To give a very simple example, it was clear that a red rose could be crossed with a white rose to produce a new form of pink rose. But it was not at all clear how the original red and white colours had come about in roses in the first place. Perhaps all the original variants were present from the moment of creation, as many people of the time believed to be a literal truth and, astonishingly, even today still believe? Or maybe these variants had arisen subsequently via the still mysterious and somewhat feared process of mutation?

Later in the nineteenth century, it became apparent that there were two major sources of existing variation that could be manipulated by plant breeders, namely long-standing genetic variants of the crop (or their sexually compatible relatives), and those much rarer exotic freaks or sports that we call mutations.[1183] Traditional land races of crops provided a rich source of variation and these were avidly collected by early botanists. Many crop-breeding centres still contain collections of traditional land races collected during this and later periods. But most crop land races disappeared from large-scale cultivation during the twentieth century and those not already stored in collections have been lost for good.[1184] As well as the collection of existing crop varieties, the main impact of botany during the seventeenth and eighteenth centuries was in discovering, transporting, cataloguing, and propagating the many new crops and other potentially useful plants that were emerging from the newly discovered continents across the oceans. With the rediscovery of Mendel's work on plant inheritance, and the beginning of the new science of genetics at the start of the twentieth century, the stage was set for much more precise and radical manipulations of crop performance.

Quantitative genetics

The foundation of modern genetics was Mendel's principles of heredity, which deal with individual traits that segregate simply and are controlled by a small number of genes. Several early scientific breeders, such as Rowland Biffen at Cambridge, who invented the term 'genetics', used a Mendelian approach with great effect in selecting disease resistant varieties of wheat. However, breeders soon realized that the vast majority of agronomically useful characters in crops was not regulated by simple Mendelian inheritance. Key traits such as overall yield, grain quality, and stress tolerance often behave in a complex way that indicates they are regulated by many different genes.[1185] To make things even more complicated, these genes sometimes interact with environmental factors to determine the final phenotype of the crop plant.[1186] Many breeders believed the behaviour of such complex characters was beyond the power of science to explain or manipulate. However, a statistical, approach known as quantitative genetics, enabled the limited existing knowledge to be harnessed for crop breeding, even if the underlying molecular principles remained a mystery.

It is difficult to overestimate the importance of quantitative genetics for practical breeding, and it has been described as 'the intellectual cornerstone of plant breeding for close to 100 years.'[1187] The foundations of quantitative genetics were laid by British geneticist, Ronald Fisher in his seminal paper of 1918, in which he showed that continuous variation between members of a population could be as a result of Mendelian inheritance, albeit involving many genes, plus an environmental component.[1188] Previous work by geneticists, such as Bateson at Cambridge, had suggested that Mendelian mechanisms only gave rise to large and discrete, or quantum, changes in phenotype. Fisher and others established a statistical framework allowing crop breeders to apply quantitative genetics in a practical manner.[1189] In particular, quantitative genetics allows breeders to develop robust mechanisms to predict phenotypic performance from their knowledge of a given genotype. It also enabled them to design manageable and affordable field trials that had a good likelihood of detecting useful characters without being too large or unwieldy.[1190]

Creating new variation

The greater understanding of how plants reproduced and how their agronomic characters are determined was coupled with techniques such as sexual crossing and recurrent backcrossing, allowing breeders to produce dozens of new crop varieties. By the early twentieth century, other new technologies were developed that would enable breeders to create completely different types of variation without relying on the vagaries of random mutation. Mutations could now be caused deliberately using chemical or radiation treatments and new combinations of sexually incompatible parent species could be produced thanks to advances in techniques such as tissue culture.[1191]

Hybrids and wide crosses

Most crop plants are derived from related species that still grow in their centres of origin, for example there are many wild relatives of wheat in southwest and central Asia. These wild relatives often carry useful traits, such as disease resistance, that are absent in the more inbred crop species.[1192] As discussed in Part II and Chapter 15, some crops can be readily hybridized with a wild relative to which it is closely related. However, there are many potentially useful wild relatives that only yield sterile hybrids or cannot form hybrids at all. Two of the major keys that have allowed breeders to unlock the genetic potential of wild relatives are tissue culture and the use of growth regulators, both of which were developed in the mid-twentieth century.[1193] Attempts to produce wide crosses between relatively distant species often fail due to the incompatibility of their genomes, which leads to an inability of chromosomes to pair at meiosis, resulting in sterility.

One of the technical advances that helped breeders to surmount this challenge was the development of chemically induced chromosome doubling, which has been the key to the success of many crop-breeding programmes. As well as making possible much wider genetic crosses, chromosome doubling has enabled the use of powerful methods such as somatic hybridization and haploid breeding, which have been especially useful in developing countries. In the past few decades, the technique of mass propagation has also been of considerable benefit in breeding programmes for tree crops, most of which are too long lived to be accessible to the sorts of approaches

developed for the much shorter-lived, annual crops. The development of methods to prevent seed propagation is another important target for many commercial breeding programmes. Over the past century, new techniques have been devised either to induce fruit or seed sterility, or to prevent seed saving by using hybrid varieties.[1194] All of these methods are considered to be part of 'conventional' plant breeding, although in reality they involve direct, scientifically informed human manipulation of crop genomes in ways that are often less predictable and less precise that so-called 'genetic engineering'.

Mutagenesis

Mutations occur all the time, albeit with low frequencies, and there are many agents in the physical environment that cause mutations. All living organisms on earth are constantly exposed to such mutagenic agents, many of which originate from the sun, including neutrons, UV radiation, and more powerful forms of high-energy electromagnetic radiation, such as X-rays, and γ-rays. These forms of radiation can sometime penetrate the atmosphere as far as ground level and may then be absorbed by terrestrial organisms, resulting in possible DNA damage and cellular mutation. The other important category of mutagenic agent is the vast group of DNA-reactive chemicals that occur in both biotic and abiotic environments. Some of these chemicals, such as benzene and mustard gas,[1195] are manmade but many others are by-products of environmental processes such as volcanism, or may result from normal cellular activity, for example during stress or ageing.[1196] Exposure to environmental mutagens, including solar radiation and many chemicals, can increase the incidence of mutation, depending on the type of mutagen and the duration of exposure.[1197] However, most crops have relatively massive genomes and numerous duplicated genes, giving them an impressive degree of genetic redundancy. This means that the likelihood of a useful, spontaneous mutation suddenly appearing in any given crop plant is extremely low.

The work of Dutch botanist Hugo de Vries and others highlighted the distinction between mutations and environmental variations.[1198] They showed that germ-line genes could be directly changed by mutation and any offspring from such a 'mutated' organism would also carry the altered genes.[1199] This also revealed the importance of mutation as a key provider of those genetic variations that provide the raw material of evolution. Since anything that increases variation in a population is obviously also of great interest to the crop breeder, the ways in which mutations might be exploited in agriculture became a major topic of research in the twentieth century. Mutations only occur in the existing DNA of an organism, and normally reduce or eliminate a particular gene function. If only a single gene is affected by a mutation, the effects may be quite minor, unless the gene is involved in a particularly important process. Although many mutations have little or no effect, some minor mutations can have spectacular consequences that seem out of all proportion to their effects on the genomic DNA. For example the alteration of a single nucleotide in a genome of over 600 million nucleotides changed the small, bushy teosinte plant into the tall, erect crop that we now know as maize. Moreover, this tiny change in the maize genome has transformed the lives of millions of people in Mesoamerica and beyond.

The reason for the huge effects of such minor mutations is that they sometimes disrupt expression of key regulatory genes that in turn control many additional genes. One well-studied group of such mutations is the homeotic genes that were the subject of Goethe's famous study on the floral organs of plants (see Chapter 15). Homeotic genes are commonly involved in determining the nature of a particular organ during animal or plant development. In animals, homeotic genes cause limbs to develop in the correct location and with the correct structure. Mistakes in the expression of homeotic genes in animals can lead to truly monstrous abnormalities, such as legs or wings growing out of the head. In plants, a particularly important class of homeotic genes plays a key role in regulating the development of reproductive structures, including flowers. Some of the first homeotic genes were discovered in the early twentieth century by studying a series of curious mutations that were found in the fruit fly, *Drosophila melanogaster*. These mutant insects had additional organs that had developed in

the wrong location, for example extra limbs emerged from the head and eyes were produced on the thorax.[1200] In plants, floral development and seed development are similarly regulated by a small number of homeotic genes.[1201] A single DNA base change in such a gene can have considerable effects on the appearance of the plant that might, in turn, have huge consequences for its use as a crop. For example several strikingly different looking vegetables such as cabbage, broccoli, and cauliflower are due to homeotic mutations in genes determining floral or meristem development (Figure 5.1, Box 5.3). As we saw in Chapter 7, these diverse vegetables are members of the same brassica species and vary from another by just a few tiny changes in their homeotic genes.

On occasion, spontaneously occurring mutations in crops can have huge implications for agriculture. Two recent examples are the semidwarf forms of rice, discovered in the 1950s, and of wheat in the 1960s (Figure 16.2). These mutations were largely responsible for the Green Revolution of the 1960s and 1970s that dramatically improved crop global yields and averted a threatened famine in southern Asia. We now know that the dwarfing mutations affected one gene in rice and one in wheat that regulate the action of the hormone, gibberellin. The gibberellins play important roles in many developmental processes in plants, but in these two cases the only visible result was a disruption in the elongation of the stem so that semidwarf plants were produced. Plants carrying the mutation had shorter, thicker stems than usual but compensated by making more grain. Their reduced height made the plants much less likely to topple over due to lodging, for example during heavy rain or strong winds, and the plants also responded exceptionally well to fertilizers.[1202] The end result was huge yield increases in these new rice and wheat varieties, and a massive boost in food production wherever they were grown. For example, between 1967 and 2000, wheat yields increased more than four-fold in India and Pakistan and as much as six-fold in Mexico (Figure 16.2F).

At the same time as de Vries and others were demonstrating the important role of mutations in producing new types of variation, other scientists realized that mutations could be deliberately

Figure 16.2 Dwarf cereal crops and the Green Revolution. For over 10,000 years, the major food staples, wheat, barley, and rice, were grown as relatively low-yielding, tall varieties prone to lodging and a restricted response to fertilizers. The introduction of new semidwarf varieties after the mid-twentieth century revolutionized cereal yields and averted mass famine, especially in parts of Asia. (A) Traditional forms of wheat were almost as tall as the people harvesting the crop, as shown in this illustration of a medieval harvesting scene; and (B) in Peter Breughel's painting, *The Harvesters*. (C) Modern dwarf wheat (left) is less than half the height of older varieties (right), and has a far higher yield of grain. (D) Norman Borlaug, was one of the pioneers of the disease-resistant, semidwarf wheat that led to the Green Revolution of the 1960s and 1970s (photo courtesy of Texas Agricultural Experiment Station). (E) Conventional (left) and dwarf (right) varieties of rice. The dwarf phenotype in all the major cereal crops is due to a mutation in a gene regulating responses to the hormone gibberellin. In the future, recreation of this mutation by breeders could produce dwarf varieties of any cereal crop, with prospects of further yield improvements for important local staples such as sorghum and the millets (photo courtesy of International Rice Research Institue, Philippines). (F) Over the last few decades, access to the new wheat short varieties has transformed wheat farming in developing countries across the world, with yield increases from three- to six-fold in India, Pakistan, and Mexico (data from FAO).

manufactured by human intervention. This was a radical step in the application of scientific knowledge and the use of newly invented technologies, such as X-ray sources. The ability to deliberately manufacture mutations in plants and animals enabled breeders to cease their reliance on spontaneous genetic variation. These methods enabled the creation of much wider forms of variation and selection for crop breeding. Use of such intrusive approaches to plant breeding has allowed our food production to more than keep pace with the rapid growth of human populations. By the early twentieth century, the combination of knowledge about the roles of both hybrids and mutations in increasing variation, together with the rediscovery of Mendel's work on inheritance, set the scene for a dramatic leap forward in plant breeding. This occurred within a research paradigm dominated by large, publicly funded institutions staffed by professional scientists and breeders. These workers used sophisticated methods of genetic manipulation in order to enhance variation in crops for much of the past century. It was only towards the end of the twentieth century that there was a resurgence of private sector interest and investment in crop breeding. One of the best-known breeding technologies to emerge in this recent period is genetic engineering, or transgenesis.

Transgenesis

We can define transgenesis as the addition of small segments of externally derived DNA sequences to the genome of a recipient organism, such as a plant or animal. In the case of plants, DNA is normally added to cells using either of two techniques. First, the DNA can be added directly by propelling tiny DNA-coated gold particles into a plant tissue. This technique, called biolistics, can be used for any plant, crop or otherwise, but is relatively inefficient and does not always result in the incorporation of the DNA into the plant genome.[1203] Alternatively, the DNA can be added in a more controlled fashion by means of bacterial vectors, such as *Agrobacterium tumefaciens* or several *Rhizobium* species.[1204] These bacteria can insert a specific region of transgene-containing DNA into the genome of the plant.[1205] Despite their limitations, these methods of DNA

transfer, or transgenesis, are often more efficient in delivering desired genes into crops than alternative methods of crop genetic manipulation, such as induced mutation or wide crosses. At present, and despite the considerable hype surrounding this technology, transgenesis has only resulted in the manipulation of a few simple input traits, although much more is promised in the coming decades (see Box 16.1).[1206]

Screening and selection

New variants of crops may be generated from germplasm collections, or they can be deliberately manufactured by breeding techniques such as wide crossing, mutagenesis, or transgenesis. In all cases, however, the population of variants, which can number in the tens of thousands or more, must be screened for the presence of the desired phenotype(s). Sometimes, breeders can select suitable plants and varieties using highly visible traits such as height, branching, seed size, tuber shape, etc. But many of the most important attributes of a crop, such as the quality traits that determine taste and nutritional content, are often invisible and can only be determined in the seed or tuber after harvest. In the case of wheat, an important criterion is the bread-making ability of the flour. This character depends on the presence of a particular ratio of gliadin and glutenin storage proteins in the seed. The presence or absence of this kind of quality trait would not become apparent until well after harvest. One can imagine the difficulty of attempting, on an empirical basis, to select for any useful variation in such traits. Not only are such traits invisible in the growing crop; they are also frequently regulated by numerous, unlinked genes. This made for exceedingly slow progress in the selection of many useful quality traits before the advent of more recent methods of screening and analysis.

Phenotypic and chemical markers

Since the late 1980s, researchers have developed a host of ever more accurate and rapid techniques for the simultaneous screening and selection of thousands of different compounds in plants. These technologies, often referred to collectively as metabolomics,

Box 16.1 Genetic manipulation in agriculture—ancient art or modern science?

To what extent have people been practicing genetic manipulation since the first plants and animals were domesticated? We have already seen that many of the earlier stages of crop and livestock domestication were largely unconscious processes of coevolutionary development. The word 'manipulation' is derived from the Latin for 'handful', as in a handful of grain, and implies a conscious procedure of handling or exploitation. Therefore we can only use it to describe the later stages of deliberate and purposeful changes wrought by people on other living organisms.

Humans have always been skilled manipulators of their environment and early farmers and pastoralists soon realized that their new semitamed crops and livestock could be modified further by judicious selection of favoured traits. In the case of animals, the easily recognizable similarities between their reproductive mechanisms and those of humans made it possible for farmers to exercise direct control over mate choice and the survival of offspring. Hence, favoured animals were mated with similarly favoured close relatives to reinforce traits such as tractability, meat production, and milk yield. Unsuitable males were castrated to prevent breeding, and many litters were ruthlessly culled to remove unfavourable traits. Excessive inbreeding was mitigated by regularly mating selected females with unrelated tame or wild males to replenish genetic diversity. Thanks to the genetic linkage between many of the most desirable traits, animal genomes have been drastically altered by relatively clear-cut processes of selective mating and culling over the past ten millennia, without the need to understand anything of the biological processes involved. In this manner, the fierce, wild aurochs was transformed into the placid cow and the aggressive wolf became the friendly and faithful dog.

In the case of plants, matters were less straightforward. Because they are not motile in the same way as animals, plants were thought to belong to a completely different category of life. However, early farmers soon realized that 'like begets like' in the plant world just as much as with animals. They learned to keep back some of their best seed to sow as the next season's crop, rather than eating it all after harvest. However, for most crops, there was little understanding of how to manipulate the process of reproduction itself. One notable exception was in the case of horticulture, where the ancient Egyptians and Assyrians learned how to fertilize female flowers with pollen from selected male flowers. This was a rare example, however, and the phenomenon of sexual reproduction in plants was not discovered until the eighteenth century CE. In the meantime, although crops were certainly improved by an empirical process of selection by farmers, in many cases their genomes were relatively unaltered compared with their wild ancestors for many millennia after domestication.

During the nineteenth and early twentieth centuries, more radical types of genetic manipulation of crops became possible when people were able to introduce a much greater degree of variation into their genomes by means of hybridization, both within and between species, and by deliberately creating new mutations instead of relying on the much slower 'natural' process. Greater scientific understanding also allowed for a more rational approach to the breeding and selection of genetically different crop variants that were able to grow in new climates or were resistant to new diseases never encountered by their ancestors. Throughout the twentieth century, the manipulation of plant genomes became progressively more intrusive and precise, although in many respects it has always retained many attributes of an artisan's craft rather than a truly precise, scientific process. Since the 1930s, chemical agents such as colchicine have been used to cause genome duplications and, after the 1940s, irradiation by cobolt-60 was used to create hitherto unknown mutations that have led to over 3000 new varieties of crop plant. All this happened many decades before the latest technology, known as transgenesis or genetic engineering, whereby completely new genes could be added to a plant or animal.

take advantage of robotic systems to automate sample collection and processing for analysis. Hence, it is possible to run round-the-clock screening pro-grammes, for example for the presence or absence of a particular compound. Chemical analysis has also been revolutionized in the past 20 years by the development of techniques that are more accurate, faster, cheaper, and require much smaller equipment than previously. Examples include spectroscopic and chromatographic methods, as also used in bio-medical research and hospital practice. Useful spectroscopic techniques include mass spectroscopy (MS); plus the various forms of infra-red (IR) and nuclear magnetic resonance (NMR) spectroscopy.

Two of the most powerful methods of metabolite screening are gas–liquid chromatography (GLC) and high-performance liquid chromatography (HPLC).

DNA-based markers

Another relatively new 'high-tech' screening method that is proving to be extremely beneficial to plant breeders is the use of DNA-based molecular markers. This involves the same basic technology as that used in DNA fingerprinting for forensic analysis in criminology, and for genetic profiling in medicine.[1207] Molecular markers can save much time and money in crop improvement programmes because breeders can select plants that are likely to express traits of interest while they are still at the early seedling stage. Molecular markers have now been developed for many major commercial crops, including several tree species. These markers can be used to track the presence of useful characters in crop-breeding programmes. For example, in a recent review of the use of DNA markers in oilseed rape breeding, the following marker-linked agronomic traits were identified: resistance to five major diseases, seed fatty acid content, glucosinolate content, cold tolerance, flowering time, and plant height.[1208]

At present, the use of DNA-based molecular markers is largely limited to the major economically important, annual, temperate crops. The main factors delaying their more widespread use include high upfront costs and the technical sophistication needed to produce the markers and interpret data from large populations. These and other limitations on the use of marker technology are gradually being overcome as costs come down and the requisite technical expertise becomes more widely available, especially in developing countries. A recent study from the International Maize and Wheat Improvement Center (CIMMYT) in Mexico has looked at the cost/benefit considerations of using marker-assisted selection in resource-limited public breeding programmes.[1209] The conclusion of this and related studies is that justification for developing marker-assisted breeding depends critically on the nature of the crop, including its genomic organization; availability of the requisite

technical infrastructure; and of external capital to meet the set-up costs.[1210] The bottom line is that for many developing country crops, such high-tech methods of breeding may not be appropriate for the foreseeable future.

Modern molecular breeding methods can also be combined with earlier methods like mutation breeding to generate powerful new hybrid technologies, such as TILLING.[1211] This method is based on molecular genetics—the acronym stands for: 'Targeting Induced Local Lesions IN Genomes'. In a TILLING programme, mutagenic agents, such as alkylating agents or various forms of radiation, are used to create a large, genetically diverse population consisting of thousands of mutagenized plants. The mutants are then screened by a semiautomated, high throughput, DNA-based method to detect mutations in genes of interest. The third step is to evaluate the phenotype of mutants identified during the screen. The screening stage involves the polymerase chain reaction (PCR) to amplify gene fragments of interest. Finally, any mutation-induced lesions in the genome are identified by looking for mismatches in duplexes with non-mutagenized DNA sequences. TILLING can detect mutations without the need to grow up the plant and screen it for an observable phenotype, such as plant height or disease resistance. It can also be automated using high-throughput screening systems, making it suitable for some of the large polyploid genomes of major crops such as wheat. In a short time, large pools of genetic variation can be produced for introduction into breeding programmes. For the first time, in 2005, TILLING was used to identify variants in a gene, known as *Waxy*, which plays an important role in determining flour and bread quality, and the method is also being used for soybean improvement.[1212] In the future, TILLING and other high-tech breeding methods will both speed up and broaden the scope of crop improvement programmes around the world (Figure 16.3).

Domesticating new crops—a new vision for agriculture

At the beginning of this book, I made the point that over the past twelve millennia of agriculture around the world, human societies have domesticated only a

CHAPTER 17

The future of agriculture and humanity

The corn was orient and immortal wheat, which never should be reaped, nor was ever sown. I thought it had stood from everlasting to everlasting.

Thomas Traherne, *c.* 1637–1674, *Centuries of Meditations*

Introduction

In previous chapters, we have seen that cognitively advanced humans, possibly capable of developing agriculture, and with it all the trappings of complex societies and urban cultures, might have been on earth for as much as 50 to 100 millennia. For the vast majority of that time, however, their most adaptive way to survive was to live in small bands that exploited a diverse range of faunal and floral resources. Such strategies were especially suited to the turbulent conditions of the Middle Palaeolithic Era, with its sudden and drastic changes in the biological and physical environments. We have also seen that, as the climate became more stable in the Holocene Era, new methods of resource exploitation sometimes became more adaptive. In localized regions of Africa, Asia, and the Americas, people cultivated a small number of domestication-friendly plant species. Over several millennia, as these lifestyles became increasingly successful in the new socioenvironmental conditions of the mid to late Holocene, agriculture and complex agrourban societies gradually spread around the world. In this final chapter, we will examine human population numbers and the impact of agriculture. We will then speculate about future prospects for agrourban societal models in the context of a likely reversion to more variable climatic conditions in the next few millennia. Given the importance of genome architecture in determining agricultural utility (or not) of plants, we will also examine the biological manipulation of crops as a possible way of adapting to future environmental changes, such as aridity or temperature fluctuations.

Agriculture and human population fluctuations

Over the past two million years, human numbers have oscillated considerably, with often dramatic, local booms, extinctions, and migrations. Population tends to be a function of available resources and, like any other species, humans tend to respond to changing resource levels by adjusting their numbers.[1221] At a global level, there have been at least three large-scale, incremental increases in population that coincide with: (i) late-Pleistocene migrations from Africa; (ii) early to mid-Holocene development of agriculture;[1222] and (iii) post-1700 CE agroscientific developments (Figure 17.1).[1223] Estimates of human numbers during the Palaeolithic Era are necessarily inexact, but before one million years ago the total population of African hominids is unlikely to have exceeded 100,000.[1224] Later migrations from Africa by *H. erectus*, *H. neanderthalis*, and *H. sapiens*, may have brought human numbers up to about 0.5 to 1.0 million by *c.* 150,000 BP, and the population may have fluctuated around this level until the end of the Middle Palaeolithic *c.* 50,000 BP.[1225] It has been suggested that the latest global spread of *H. sapiens*, and the increase in hunting efficiency due to inventions such as the harpoon, bow and arrow, and spear thrower, may have led to an increase in the population to about six million by the time of the final Neanderthal extinctions shortly after *30,000 BP*.[1226] This level was probably more or less

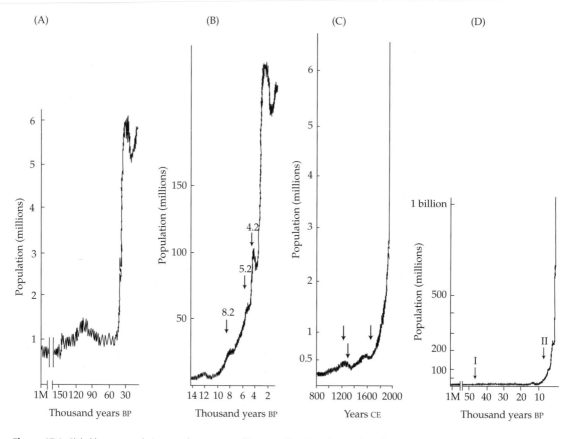

Figure 17.1 Global human population over the past two million years. There have been at least three major increases in human population since the speciation of *Homo sapiens*, as follows. (A) The series of mid-Pleistocene global expansions during a milder period after *50,000 BP* that took *H. sapiens* across the world. (B) The development of agriculture after the Younger Dryas at *c. 11,500 BP*, which led to several millennia of dramatic but punctuated growth until populations stabilized or fell at around *2000 BP*. Reverses in population growth are correlated with the climatic events around *8200 BP* (8.2), *5200 BP* (5.2), and *4200 BP* (4.2). (C) The past 1200 years have been dominated by fluctuating population numbers during the Medieval Warm Period (left arrow) and Little Ice Ages (right arrows). This was followed by the post-1700 CE agroscientific revolutions, which led to several centuries of near-exponential growth that started to level off after 1970 CE. If present trends continue, the global population will probably eventually stabilize at about 10–12 billion in the twenty-second century. While the first major increase in human population was due to the occupation of new habitats, the second and third increases were associated with the development of agriculture and its subsequent intensification and industrialization. (D) Population increases over the past million years, shown on the same scale as the Pleistocene expansion (I) to emphasize the relatively recent expansions of the Neolithic (II) and modern periods in the context of the history of our species.

maintained until the end of the last Ice Age and the start of the relatively settled Holocene climatic period at about *12,000 BP*.

The global human population increased significantly during the first five millennia after the earliest adoption of agriculture, from about six million in early Neolithic times to about 100 to 300 million by *4500 BP*.[1227] This rise, which was the largest relative increase in human numbers in our history, is directly attributable to the far greater useful productivity of agricultural versus uncultivated ecosystems. Thanks to this improved strategy for the exploitation of new plant and animal domesticants, human populations increased between 20 and 50-fold in about five millennia. However, after *c. 4500 BP*, there was a period of retrenchment. During the next four millennia, population numbers fluctuated up and down in response to often-interrelated sociobiotic factors such as epidemics, famine, and wars; coupled with abiotic factors such as environmental degradation and climate change.[1228] By the Medieval Climatic Optimum, the world population was still only about 300 to 400 million, and this then fell significantly after 1300 CE due to lower crop yields in the Little Ice Age, and to new epidemic diseases, such as the Black Death.

It was only after the sixteenth century that there was a more dramatic and sustained population increase; up to 700 million by 1700, and one billion soon after 1800. To begin with, this more recent population surge, which coincided with the early Industrial Revolution in Britain, was largely fuelled by more efficient exploitation of existing land, thanks to post-Enlightenment advances in agroomy and technology, plus the exploitation of new land for agriculture, especially in the Americas (Figure 17.2A). As we saw in the last chapter, the considerable yield gains of modern agriculture have enabled the global population to reach well over six billion today, with a projected total of 8 to 9 billion by 2050. *Per capita* crop production has increased by

40% since the 1950s, although much of this additional food is used for animal rather than for human consumption. Nevertheless, since 1950, average *per capita* calorie intake has increased 25%, from 2250 to 2800 kilocalories per day, so despite localized areas of hunger the average person is much better fed today than at almost any time in our history (Figure 17.2B). Indeed, in 2006 it was reported that, for the first time in human history, the number of morbidly overfed people exceeded those who were underfed.[1229]

In looking towards the possible future of our agrourban cultures, we should bear in mind the very short timescale of the most recent population increase, which has occurred over a mere three to four centuries. Our current mode of agrosocietal development is dependent on: (i) highly complex, but inherently fragile, scientifically informed global information networks; (ii) an atypical interlude of stable and benign climates (in the context of the last few million years); and (iii) an unprecedented and unsustainable rate of exploitation of non-renewable energy resources that is still accelerating. As we

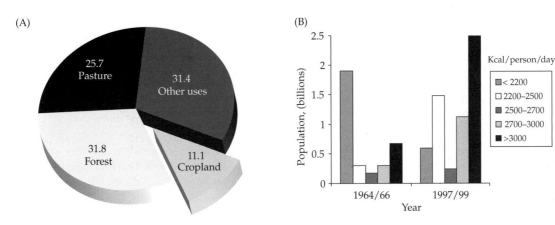

Figure 17.2 Global land use and food consumption. (A) Global land use in 1996, showing percentages of land covered respectively by crops (11.1%), pasture (25.7%), forest (31.8%), and others (31.4%). Much of the cropland is still underutilized and some pasture could be converted to arable use if there were a need to increase food production in the future. Note: data do not include Antarctica (Source: World Resources Institute, 1996). (B) Changing patterns of global food consumption from 1964/66 to 1997/99. From these data it is evident that food production has more than kept pace with population growth. Two trends are especially noteworthy: firstly, despite an almost doubling in population from 3.3 to 6 billion, the number of underfed people (<2200 Kcal/day) has fallen by almost 70%, from 1.90 to 0.59 billion and, secondly, the number of well fed (or overfed) people has increased almost four-fold from 0.67 to 2.49 billion. Indeed, in 2006, for the first time in human history, the number of people who are seriously overweight and obese exceeded the number that is malnourished. The take-home message is that, notwithstanding the growing global population, there is no overall shortage of food production in the world, but there remains a (diminishing) problem in distributing this food to those that truly need it. Data from FAO (http://www.fao.org/docrep/004/y3557e/y3557e06.htm).

will discuss below, these conditions are most unlikely to be maintained in the longer term, but what is the prognosis in the short term and especially over the next few decades, where the actions of the current generation could possibly make a real difference to our immediate descendents?

The short- to medium-term future

Can we continue to feed the world over the next dew decades? Global agricultural production provides a significant overall food surplus. But, for a variety of mainly economic and/or political reasons, this food does not always reach the most disadvantaged groups. Given the political will to distribute it, there is already sufficient food to cope with the seemingly unavoidable, but thankfully localized, periodic shortages caused by factors such as drought, disease, flooding, or human conflict. Moreover, the Food and Agriculture Organization (FAO) predicts that, due to improved agricultural and economic development, the number of hungry people will actually fall from 770 million to 440 million by 2030, despite a predicted population increase of over one billion people during the same period.[1230] In the past, the one region that seemingly failed to benefit from agricultural advances was sub-Saharan Africa. According to a series of FAO reports, the number of hungry people will hardly fall in sub-Saharan Africa between now and 2030.[1231] In contrast, as discussed above, hunger in the rest of the world is predicted to fall sharply over the same period.[1232] Therefore, one effective way of tackling lack of access to food, would be to improve the agricultural and politicoeconomic conditions in sub-Saharan Africa.[1233] As discussed in the previous chapter, a useful option might be to introduce new, protein-rich crops into African farming systems.

During the next few decades, population growth will probably continue to level off, possibly reaching about nine billion by 2050. As discussed above, the prognosis for the capacity of agriculture to feed this additional 40% of people (compared to mid-2006) is cautiously optimistic. The major unknown is the magnitude and effect on crop productivity of short- to medium-term climate change, associated mainly with increased atmospheric CO_2 levels, which are predicted to increase global temperatures

and alter rainfall patterns.[1234] Warmer weather and higher CO_2 levels might even favour higher crop yields (but see discussion below), although localized aridification due to reduced rainfall could severely affect output in affected regions. Therefore, the overall effect of short-term, CO_2-related climate change on global agriculture will largely depend on whether any of the key producer regions, such as Chinese rice-growing areas or the US Midwest, suffers serious and sustained drought.

Providing crop global productivity can be maintained at or slightly above current levels, we can probably maintain as many as ten billion people on earth indefinitely—at least until the supply of fossil fuels runs out. It is even possible that by maximizing the biological potential of our present crops, and by optimizing agronomy and management/economic systems, a maximum carrying capacity of as many as 20 billion people could be achieved. After all, given that the population has more than trebled from two billion in 1930 to about 6.5 billion in 2006, it cannot be ruled out that a further trebling to 20 billion could occur at some time in the next two centuries. As shown in Figure 17.2, only 11% of the global area is used to raise crops, while just over 25% is animal pasture. Some of this pastureland could be converted to arable use, and there is still immense untapped potential for increased yields of many traditional subsistence crops, some of which have yet to receive the kinds of attention from scientific breeders that has led to as much as ten-fold yield gains in some of the major commercial cereal crops. We can also generate more food for people by adjusting our use of some of the edible crops, such as soybeans and maize, that are currently grown more for livestock than for human nutrition.

In the absence of any drastic climatic shifts or seismic political upheavals, the major factors that will probably limit human population growth in the medium term (i.e. the next few centuries) are likely to be the availability of water and energy. Increased pressure is being put on water resources by urbanization, industrialization, irrigation, and physical factors such as erosion, pollution, salinization, nutrient depletion, and the intrusion of seawater. The misuse of scarce water resources is often exacerbated by inappropriate subsidies that distort markets and amplify inequalities to the overall detriment of agricultural production. In Chapter 10, we saw how the well-organized use of raw human power on a massive scale enabled the Sumerians to run the intensive irrigation farming system that underpinned their agrourban culture for several millennia. Later economies successfully used a combination of human and animal power, increasingly supplemented by the ever more efficient mechanical devices that have now largely replaced human and animal labour. Nowadays, agriculture is critically dependent on fossil fuel to power the machinery and to manufacture the chemical inputs upon which its productivity largely rests.

While oil and gas will become a lot scarcer during the present century, there is probably enough coal for several hundred years.[1235] However, long before coal runs out, the adverse economics of its use in farming will almost certainly drastically reduce our ability to manufacture inputs and to power machinery to work the land. As a result of more expensive energy inputs, crop productivity, and hence population, seem bound to fall within a century or two. Such a reduction might destabilize, or even destroy, some societies, but this will not necessarily be the case. Drastic declines in agricultural productivity over large parts of mid-Holocene Mesopotamia and China led to centuries of cultural simplification, but the societies eventually recovered. More recently, the medieval European boom period was abruptly halted by the Little Ice Age and Black Death, which together killed one-third to half of the population. But, despite severe economic decline and agricultural retrenchment, plus a degree of social upheaval, the prevailing elites survived and these complex societies rapidly rebounded to recover within a century or two. It is, perhaps, a moot point whether today's highly complex, dynamic, supposedly enlightened, and technologically 'advanced' civilization will prove as resilient to such a shock as the relatively 'backward' medieval Europeans, with their gloomy religiosity and stagnant agricultural systems (see Chapter 13).

The far future—an uncertain environment

Climate, whether local, regional, or global, is a complex phenomenon that we are only now beginning to understand. The more we discover about past climatic events, the less confidence we have in our ability to predict future trends, except at the broadest level. It is clear that the global climate has undergone dramatic fluctuations over the past few million years, and that we are currently living in an unusually benign and stable period (Figure 17.3). However, we have seen that even this relatively stable climatic regime has been punctuated by several episodes that were sufficiently drastic to extinguish civilizations, such as that of the Indus Valley, and to cause the temporary abandonment of agrourban culture in many other parts of the world. There are too many variables to make precise predictions about climate over the coming few millennia. But using extrapolations from previous data, it seems likely that we will eventually experience greater deviations from the present stable, mild, moist regime, as climatic systems revert to a cycle of more frequent and extreme cool/dry and warm/wet periods, as occurred before the Holocene.

On previous experience, it is likely that climatic oscillations will be more pronounced in temperate latitudes, where most of the human population now lives. This means that, as has happened so often over the past million years of our history, some temperate regions might once again become uninhabitable. For example people have colonized the present-day island of Britain no fewer than eight times since *700,000 BP*.[1236] However, due to climatic vagaries, all but one of these occupations, some lasting for tens of millennia, ended in failure, with the retreat or extinction of the resident human population. The last occupation of Britain only dates from *12,000 BP* and occurred during the favourable Holocene period. Given the known climatic variability of the last 700,000 years, it would therefore seem at least plausible that this eighth attempt to colonize Britain might also, eventually, end in failure.

Although it is difficult to predict the future climate with any reliability, there are generalized trends that may allow us to make some tentative forecasts. Firstly, current concerns about anthropogenic climate change relate mainly to increased greenhouse gas emissions, especially CO_2 from fossil hydrocarbon combustion. However, since

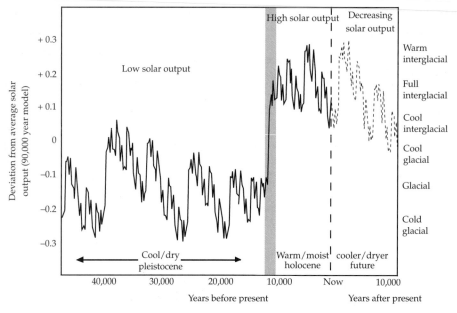

Figure 17.3 Global climate change: past, present, and future. Variations in solar luminosity from 40,000 years before present to 10,000 years after present together with the timing of selected environmental and human-related events). The overall picture is of a lower than average solar output during the latter part of the Pleistocene Era, and a significantly higher than average output during the current Holocene Era. Note that short-term fluctuations in solar activity shown here are obviously related to, but do not exactly coincide with, terrestrial climatic data such as proxy temperature records from ice cores, as shown in Figures 1.1, 1.2, 1.3, and 10.3. Several low points of solar activity coincide with especially cold periods on earth, such as the coldest recent glacial episode *c. 18,000 BP*. Extrapolation over the next 10,000 years is based on the solar output model as discussed by Perry and Hsu (2000) and the present figure is based on Figure 1 from the same report. The conclusion is that, over the next few millennia, solar output is predicted to fall to levels previously associated with much cooler glacial/interglacial terrestrial climates. Such conditions may pose serious challenges for maintenance of the current agrourban strategies of food winning and societal management.

these fuels are non-renewable, such emissions are self limiting and will decline and then cease in the next few centuries. During this period, warmer weather and higher CO_2 levels might even enhance crop yields. However, several studies suggest that increased yields due to elevated atmospheric CO_2 may only be half the previously accepted values, and may be insufficient to offset yield decreases due to lower amounts of soil moisture.[1237] The key issue of water scarcity in agriculture has been highlighted in a recent report from the International Water Management Institute.[1238] So it seems that in both the short to medium term and the long term, by far the most serious threat to food production will come from aridification, rather than temperature change (Box 2.2).

Further into the future, climate change will tend to be dominated by solar activity and other non-anthropogenic cycles, occasionally modulated by large-scale, terrestrial events such as massive volcanic eruptions or meteorite strikes.[1239] Models based on solar activity cycles can explain many large-scale fluctuations in global climate over the past 40,000 years, for which relatively robust data are available.[1240] Extrapolations of such models into the future suggest a possible transient recurrence of Little Ice Age conditions over the next few centuries. This will be followed by warmer, mid-Holocene-like, conditions for the next millennium or so, which will then give way to a long-term cooling trend towards a more glacial climate three to four millennia from now (Figure 17.3). Of course, these and other predictions about long-term climatic change are necessarily speculative, but they are still useful in stimulating us to think about future climates, their possible impact on agriculture and human societal development, and ways in which we might cope with them.

Can we ensure that agriculture survives in the long term?

Our entire technoscientific civilization is underpinned by agriculture and it seems likely that one of the major reasons for its relatively recent emergence is the exceptional stability of the Holocene climate; a stability that could well come to an end as we enter a new and less certain climatic period. But given our new knowledge of crops and genetics, will it be possible to devise mechanisms whereby agriculture, and hence our complex societies, could survive a return to the unstable conditions that our remote ancestors endured during the Palaeolithic Era? In the previous chapter, we surveyed some of the remarkable advances made by plant breeders over the past few centuries. We also looked at the possibility of domesticating new plants to supplement the restricted portfolio of tried and tested ancient species that have served us so well during the current period of relative climatic stability. In the past, cultures such as the Mesopotamians adapted to cool/arid spells by using crops such as barley that were better adapted than wheat to such harsher conditions. Also, the pioneer farmers in Central Europe selected cold-tolerant crop genotypes, enabling them to spread into the cooler maritime regions of the northwest of the continent. On the other hand, the Indus Valley and some Mayan societies succumbed to climatic shifts, which either occurred too quickly for them to adapt, and/or were exacerbated by social factors such as warfare or internal disorder.

It would therefore seem prudent for us to make every possible effort to prepare ourselves for more arid climates in the future. We can do this by developing new food resources, including newly domesticated crops and genetically improved cultivars of existing crops that can tolerate a wider range of climates than we have experienced hitherto. Our present methods of genetic engineering are still too primitive to contribute significantly in this direction. This is mainly because traits such as drought tolerance are complex and regulated by many genes, which makes it difficult to manipulate them by the kinds of simple transgenic approaches in use today. However, other high-tech methods, such as marker-assisted selection, may enable us to develop new crops in the next few decades.[1241] Moreover, given the rapid progress of research over the past few years, it cannot be ruled out that radical forms of genetic engineering will be invented during the present century. Therefore it is entirely possible that, given time, hard work, and a little luck, we could produce some of the crops and technologies needed to sustain some form of agriculture in much harsher climates than we have experienced over the past 12,000 years. But will this be enough for our complex cultures to survive?

People, plants, and genes in the next 100,000 years

Given the changing nature of the terrestrial climate, both locally and globally, it is a moot point as to whether large-scale agriculture, and the associated technological, urban-based cultures in which most of us now live, are truly adaptive strategies in the very long term. My suspicion is that they are not. Should the climate return to its more typical changeable form, it seems likely that most of our current crops will quite rapidly become extinct. For example even with our current knowledge of crop science, it is unlikely that we could adapt our agricultural systems in time to cope with an environmental transition of the scale and rapidity of the Younger Dryas, in which the cooling and rewarming episodes may have occurred in as little as a decade or two. Even if we were able to develop a new and more diverse portfolio of crops to suit a wider range of climates and soil conditions, some agrourban societies might struggle on, but the probably much-reduced crop yields would necessitate a far lower human population and probably a reduced level of social complexity.

We might also face the spectre of sudden and irreversible social collapse, as happened to the Harappans and some Mayans. However, these previous collapses were invariably localized, and this would probably also be the case for future societal failures. Moreover, it is probably equally legitimate to view such events from a less urban-centric perspective and to regard them as adaptive changes in social structure (from more to less complex), which may be more suitable strategies for survival under harsher and more rapidly changing environmental conditions.[1242] We have seen numerous examples of such social downshifting, both short term (e.g. northern Mesopotamia after *8200, 5200,* and *4200 BP*, see Chapter 10) and longer term (e.g. post-Harappan in India and post-Anasazi in the US southwest) in our survey of human development in the Holocene.[1243] Of course, such considerations will do little to sweeten the bitter pill that is the likely future demise of our contemporary urban-based society. In 2007, for the first time in history, the number of city dwellers exceeded the rural population so urban societies are now both dominant and ubiquitous around the world.[1244] Unfortunately for the majority of urban humanity, cities would probably be one of the first casualties of sustained and largely unpredictable (on a local level) environmental change. It seems unlikely that our global economic and scientific networks, and the complex technological cultures that they sustain (including intensive agriculture), could be maintained under multiple stresses of climatic change, resource depletion, and consequent social upheaval.

But is this really such a 'bad thing' for the future of our species? As we have seen, there is a case for regarding technourban societies as relatively unstable and even aberrant entities that are only evolutionarily adaptive (in both cultural and biological respects) in the rather exceptional current circumstances of environmental stability and widespread availability of genetically suitable agricultural domesticants. Our perspectives on many aspects of alternative modes of social organization are coloured by deep-seated, cultural assumptions and by our restricted access to reliable evidence. As we saw in Chapter 2, this may have led us to underestimate the prevalence and importance of non-agricultural forms of plant management from pre-Neolithic times almost up to the present day. Similar arguments may apply to our views on the importance and adaptive fitness of complex urban-based societies *versus* simpler village-based units in post-Neolithic times. Here, many of our views are seen through the prism of an historical (i.e. written) record that has overwhelmingly been laid down in a very selective manner by members of urban elites, who often have quite open agendas in advancing the interests their own society and class. This history has been written by a tiny, and arguably unrepresentative, fraction of humanity. Moreover, such histories are focused on (sometimes inaccurate) accounts of their own societies; they have largely ignored the vast majority of less visible, but vastly more diverse, social models that have enabled people to survive the repeated collapse of unstable urban cultures, or even to expand over entire continents and live successfully for tens of millennia (as in Australia) without ever building a single town.

In our future scenario of increased environmental instability, it seems likely that, as has happened often over the past ten millennia, a few smaller groups of people in some favoured regions would doubtless maintain somewhat simplified, but still precarious, farming-based cultures. In the worst-case scenario, however, a series of severe Heinrich events could well extinguish agriculture altogether as the climate oscillated in an even more drastic and unpredictable fashion than we have so far experienced in the Holocene. Being such a resourceful species, however, it seems most unlikely that humankind would become extinct, even after such a catastrophe. The survivors would probably simply revert once again to the kind of hunter–gatherer lifestyle that has sustained us for most of our million-odd year existence as a species. Who knows, they might even go on to rediscover agriculture at some time in the distant future.

After all, at the beginning of the Neolithic Era, some of the major cereal crops developed domesticated forms with a few decades of the first experiments at tilling by nascent farming cultures. Also, it only took about four millennia from the earliest appearance of domesticated rye at Abu Hureyra to the emergence of complex urban cultures in Sumer. Providing humans can survive future ice ages and droughts as they did in the Pleistocene, there is no reason to suppose that they could not go through the whole process of agrosocial development once again in the more distant future. But any such karmic reinvention of agriculture might well depend more on the genetic constitutions of available edible plants than on the expertise of our descendents. Therefore, any prospect for a renewal of agrourban culture in the far future might lie, not so much in the hands of our descendents, but more in the survival of plant species that have domestication-friendly genomic architectures similar to those found in our current handful of global crop staples. Whatever the future holds, there seems little doubt that both the fates and the genomes of people, and their plant and animal domesticants, will remain inextricably linked for a very long time to come.

Notes

1. Early human societies and their plants

1. For example, a recent popular book on human evolution refers to agriculture as 'the second Big Bang'. This book, by Spencer Wells (2003), provides a good, readable introduction to some of the latest genetic evidence that has shed light on the course of human evolution over the past million years. In addition to debunking the myth of biologically based racial differences, the book has a wonderful series of pictures of different types of people, taken by photographer Mark Read. I would, however, counsel caution about some of the interpretations of human development in the book. In particular, Wells refers to the 'great leap forward' (the arrival of modern humans) and the 'second big bang' (agriculture), whereas I shall be arguing for a much more gradualist interpretation of these developments, albeit punctuated and diverted in different directions by sudden, periodic changes in climate.

2. The term 'Neolithic Revolution' was coined by Australian archaeologist, W.G. Childe, in 1941 (Childe, 1941). According to this thesis, the near-simultaneous development in the Near East of sedentism, agriculture, and improved technology in the Neolithic period at about *10,000 BP* was an epochal event. Most authorities today would argue for a much more extended and gradual series of complex, interlinked developments, over different timescales (tens of millennia) in different parts of the world (from Asia to the Americas), that gave rise to some very diverse forms of agriculture, technology, and social organization in these different regions.

3. For example the following quote from Smith (2001a) 'Not surprisingly, this 'Neolithic Revolution' has attracted increasing attention from both biologists and archaeologists. . . . No longer open to easy and universal explanation as a rapid and straightforward transition between adaptational steady states, the developmental shift from hunting and gathering to agriculture has in the past several decades blossomed out into a set of long-unfolding and fascinatingly complex, regional scale developmental puzzles. The most dramatic recent advances in understanding these diverse and extended regional transformations center on documenting the domestication of individual species and involve a consilience and cross-illumination of biological and archaeological approaches.'

4. See Watson (1995) for an account of the changing ideas about agricultural transitions.

5. The impact of recent genomic studies on our understanding of the origins and consequences of agriculture on human societies has been reviewed by Armelagos and Harper (2005a).

6. The definition of 'domestication' has changed over the past 50 years with scientists and scholars in different fields sometimes having widely varying notions of what constitutes a domesticated plant or animal, as discussed by Hayden (1995).

7. One popular interpretation of what is referred to as 'protofarming' is presented in the brief text by Tudge (1998). Although interesting, provocative, and worth reading, this text contains some over-simplifications and inaccuracies and should therefore be treated with caution.

8. Moore (1978).

9. Caldwell and Caldwell (2003).

10. Surovell (2000).

11. Williams (1957) and Hawkes *et al.* (1998).

12. Austad (1997) and Washburn (1981).

13. Blurton Jones *et al.* (2002) and O'Connell (2002).

14. In this study by Richard Lee (1968), he showed that, even though Bushmen live in one of the most inhospitable climates in the world, they can collect enough food in 2.3 days to provide their family group with an average of 2140 Kcal per day. This is in excess of the USDA recommended daily caloric intake (1975 Kcal) for vigorously active people of their stature. It is important to note, however, that these !Kung people live in the modern period and their behaviour as studied now should not be taken too literally as reflecting exactly how hunter–gatherers may have behaved over 10,000 years ago. See Cohen (1997) for counterarguments on the interpretation of the !Kung studies.

15. Modern !Kung have repeatedly made the transition both into and out of agriculture, as dictated by its varying cost-benefit ratio compared to hunter gathering (Wilmsen, 1989).

16. Such a conclusion is consistent with the Human Behavioural Ecology approach to the analysis of subsistence strategies, as reviewed by Kennett and Winterhalder (2006).

17. This process still goes on today whenever people run short of food. There is a marvellous website hosted by Purdue University, called 'Famine Foods', listing some of the thousands of plants that people have experimented with through the ages (see: http://www.hort.purdue.edu/newcrop/faminefoods/ff_home.html). As discussed in Chapter 16, some of these plants have the potential to be cultivated as useful new crops.

18. Obviously it is only possible to collect cereal grains during the summer season, but these grains also readily lend themselves to storage and can then be used later in the year when other food sources are no longer available (Harlan, 1992b).

19. For a discussion of sedentism and the origins of farming, the 'Neareast Historians' website is a good starting point at: http://www.neareast.historians.co.uk/html/sedentism.html.

20. The story of the interplay between plants, animals, and climate over the history of the earth is engagingly documented by Beerling (2007).

21. For example the study of Weiss et al. (2004) recounts the finding of 23,000-year-old cereal grains in the Jordan Valley. About 50,000 years ago, there seems to have been a gradual shift in the diet of Near Eastern peoples from reliance on larger prey species to smaller animals such as rabbits and birds (Flannery, 1998). This was accompanied by a growing reliance on plant foods (Stiner, 2001), and by 20,000 years ago, wild cereals were becoming major dietary components.

22. For a brief summary of this concept, see the useful recent article by Doebley (2006).

23. Smith (2001a, 2001b).

24. Harlan (1992b).

25. Mooney (1983) and Fowler and Mooney (1990).

26. Pedersen et al. (1989). Three of these crops account for about 80% of today's total world cereal production. These 'big three' cereal crops are wheat (28%), maize (27%), and rice (25%). A fourth crop, barley, accounts for a respectable, but gradually diminishing, total of 10% and all the remaining cereals together account for the final 10% of world production.

27. Waldey and Martin (1993) and Vasal (2002). Obviously this proportion of cereal-derived nutrients will be lower in those regions where meat is readily available. However, across much of the world, plant-derived foodstuffs still overwhelmingly predominate in the human diet. This is especially true for the most populous and fastest growing regions of Asia, Africa, and South America, where the 'big three' cereal crops (rice, maize, and wheat) are especially dominant. Dietary meat may be largely precluded in such regions for a variety of reasons that are principally economic but may also include cultural and religious prohibitions.

28. There is also an increasingly robust emerging literature on the economics of the transition from hunter gathering to agriculture, of which a useful and relatively accessible introduction is the article by Olsson (2001).

29. Useful reviews of the origins of modern human behaviour can be found in Henshilwood and Marean (2003) and Wong (2005b). The possibility of advanced aspects of cognitive modernity among Neanderthals is entertainingly explored in a fictional context by Arsuaga (2003). See also, Wells (2003) and Zilhão et al. (2006).

30. Henshilwood et al. (2001, 2002); d'Errico et al., (2005); Henshilwood (2006, 2007).

31. For example Steele (2003) argues that human hunting behaviour demonstrated evidence of advanced cognition well before 50,000 BP. On the other hand, others such as Richard Klein et al. have argued persuasively for a sudden cognitive leap forward at about 50,000 BP, possibly involving a 'fortuitous mutation' or 'big bang' that altered human brain function uniquely in H. sapiens (Klein and Edgar, 2002). Personally, I am cautious about accepting such an unverifiable (at present) hypothesis and instead prefer the gradualist argument of slow, incremental changes in cognitive capacity and behaviour over the past 250,000 years.

32. This view was cogently challenged by McBrearty and Brooks (2000) and by Deacon (2001) who provide a good case for the gradual evolution of more complex behaviours in African human populations over the past 250,000 years, rather than the commonly accepted view of a sudden 'human revolution' after 50,000 BP. The more orthodox view of a recent transition to cognitive modernity is lucidly explained in Klein and Edgar (2002) and in a more spiritual context by Lewis-Williams (2002).

33. Vanhaeren et al. (2006).

34. See Lewis-Williams (2002) for an interesting and imaginative account of the possible origins of European cave paintings, although his case for a recent

(i.e. late Palaeolithic) neurologically-based transition to what is termed 'higher-order consciousness' in *H. sapiens* (Chapter 7 *et seq.*) remains controversial.

35. Ambrose (1998); Ingman *et al.* (2000); Forster (2004).

36. *Homo erectus* originated in Africa but, about 1–2 million years ago, the species spread in a series of migrations via Eurasia and as far as Southeast Asia. This is shown by finds of 1.7 million-year-old *Homo erectus* remains in Georgia (Gabunia *et al.*, 2000) and 1.6 million-year-old remains in Java (Swisher *et al.*, 1994). *Homo erectus* probably died out in Africa and the Asian mainland by 400,000 years ago, but there is evidence (Swisher *et al.*, 1996) that some populations may have persisted much later in Java and may even have coexisted with migrating *Homo sapiens* as recently as 30–50,000 years ago.

37. The discovery of this diminutive species of human was only reported in late 2004. It is suggested that *Homo floresiensis* may have diverged from populations of *Homo erectus* who had previously migrated from Africa to the Indonesian archipelago (Brown *et al.*, 2004). The small human species on the island of Flores may have subsisted partially on another diminutive mammal, the pygmy stegodon (*Stegodon sondaari*), which is a type of dwarf elephant (Morwood *et al.*, 1999; 2004; 2005). For a popular account of the discovery and significance of *Homo floresiensis*, see Wong (2005a).

38. The origins of Neanderthal humans can be traced back to their divergence from the *H. sapiens* between 370,000 and 500,000 years ago (Green *at al.*, 2006; Noonan *et al.*, 2006). Neanderthals, as characterized by their robust anatomy and heavy skeleton, became firmly established about 125,000 years ago in Eurasia between Western Europe and the Near East. Many anthropologists have regarded Neanderthals as a closely related subspecies of *Homo sapiens*, called *Homo sapiens* ssp. *neanderthalis*. It has even been claimed that modern humans may have interbred with some Neanderthal populations in regions where the two groups overlapped. A very readable account of Neanderthals and their image is the book by Trinkaus and Shipmen (1993), although certain aspects are now a little out of date. For example more recent evidence suggests that Neanderthals have not contributed significantly to the modern human gene pool and therefore did not interbreed to any great extent with *Homo sapiens* even when the two types of hominid coexisted in the same region (see Rak *et al.*, 2002; Caramelli *et al.*, 2003; Serre *et al.*, 2004; Currat and Excoffier, 2004; Noonan *et al.*, 2006; Mellars *et al.*, 2007). For a popular treatment of Neanderthal

evolution, see Arsuaga (2003). Significant morphological differences between *H. sapiens* and the Neanderthals suggest that the latter should be reclassified as a separate species of the genus *Homo*, that is they should be called *Homo neanderthalis* (see Harvati *et al.*, 2004). For a variety of reasons, including competition from *Homo sapiens*, Neanderthal populations declined until they finally became extinct about 28,000 years ago (Finlayson *et al.*, 2006).

39. So-called archaic, or premodern, populations of *Homo sapiens* have been found throughout Eurasia over the period 250,000 to 600,000 years ago. The premodern *Homo sapiens* appear to be virtually identical to modern humans in terms of body structure and brain size, but are regarded as not 'cognitively modern', as discussed previously. Genetic evidence suggests that these early humans Eurasian populations were supplanted by more modern African migrants, without significant interbreeding, after about 70,000 years ago (see Cavalli-Sforza and Cavalli-Sforza, 1995; Cavalli-Sforza, 2001).

40. Recent DNA evidence suggests that a single, rapid human migration from Africa involving relatively small numbers of people may have given rise to modern populations on the other continents (Macaulay *et al.*, 2005; Forster and Matsumura, 2005), although this is disputed by others who espouse the 'Multiregional Continuity Model' (Wolpoff and Caspari, 1996).

41. Reviewed in Forster (2004).

42. Macaulay *et al.* (2005) estimate that nearly all of the non-African peoples of the world are descended from just a few hundred women who left Africa about 70,000 years ago.

43. Tishkoff *et al.* (2003).

44. Obviously, racial issues are still very important in most human societies. My point is that what we call a race, that is a group of people who share some common superficial physical or behavioural attributes, is almost entirely a cultural artefact that has no biological meaning. This means that people of all races are biologically equivalent in terms of their humanity. For a useful recent review on genetics and race, see Bamshad *et al.* (2004).

45. This includes ethnicity, race, nationality, regional identity, or any other arbitrary division of present-day *Homo sapiens*.

46. An excellent introduction to the latest research on genetics, archaeology, and linguistics and the scientific absurdity of racism is Cavalli-Sforza (2001). The book contains an especially telling image in Figure 3 (p. 89 of the 2001 edition), which depicts the genetic relatedness of 42 of the world's major population

groups. The Figure is dominated by a large, tightly packed cluster of genetically similar, non-African groups that includes Mongols, Danes, Indians, Koreans, and Amerinds. In contrast, the various African groups are much more scattered, which is indicative of a greater degree of genetic variability. This is to be expected because well over 90% of the human evolution that gave rise to all present-day people occurred within populations living in Africa.

47. For example, see Cavalli-Sforza and Feldman (2003).

48. The US National Academies report on prehistoric climate change, *Abrupt Climate Change: Inevitable Surprises*, is available online (Various, 2002). For an interesting popular account of the possible implications of such climate changes for human evolution, see Calvin (2002).

49. Despite its rather alarmist title, the book, *Climate Crash*, is a well-researched and informative account of recent research into climatic changes, and in particular the frequent occurrence of relatively rapid warming and cooling events over the past 100,000 years (Cox, 2005).

50. The term 'hominid' is used here because it will be most familiar to the vast majority of readers. More recently, evolutionary anthropologists have instead started to use the term 'hominin' to describe the group of anthropoids that diverged from the apes about 6–7 million years ago and gave rise to the various species of *Homo*. In this modern nomenclature, the older term 'hominid' now includes all of the anthropoids, including the extant species of great apes. Unfortunately, while recent scientific literature tends to use 'hominin' for the human lineage, all of the older literature, most textbooks, and most non-anthropologists still use 'hominid' in this context. See the following website for further discussion of this issue: http://www.madsci.org/posts/archives/Apr2003/1050350684.Ev.r.html.

51. The Eemian interglacial period was almost as mild as today across much of the world, but the longest sustained warm spell only lasted for 2900 years and the rest of the Eemian was punctuated by numerous cold, dry interludes (Field *et al.*, 1994; Müller *et al.*, 2005).

52. The dating of the great migration from Africa is based mainly on genetic evidence as reviewed by Macaulay *et al.* (2005) and Forster and Matsumura (2005). Hence genetic marker studies indicated that modern humans first appeared in the Chinese region of northeast Eurasia at about *50,000 BP* (Su *et al.*, 1999; Ke *et al.*, 2001). These data have yet to be reconciled with archaeological evidence that shows a smooth transition of lithic technologies (core and blade stone industries) that started before *70,000 BP* (Brantingham *et al.*,

2001; Gao and Norton, 2002). However, other recent archaeological findings are supportive of the genetic data. For example, Dolukhanov *et al.* (2002) have used radiocarbon dating to infer three waves of migration into northern Eurasia from further west. These putative migrations occurred at *40–30,000, 24–18,000,* and *17–1,000 BP*.

53. This period is known as the Lower Pleniglacial and lasted from *70,000* to *60,000 BP*.

54. Peteet (2000).

55. The classical Heinrich events, designated H6 (oldest) to H1 (most recent) are timed at about *60,000, 45,000, 38,000, 31,000, 24,000,* and *16,800 BP* (Hemming, 2004).

56. Cox (2005).

57. Rahmstorf (2003).

58. For a useful overview of climate change over the past 100,000 years, see Burroughs (2005).

59. These data have come from many parts of the world and involve several independent techniques of measurement of climate, geophysical parameters, and biotic factors, such as pollen records. One of the best summaries of the data is in the report of the Committee on Abrupt Climate Change, chaired by Richard Alley, and published by the US National Research Council; see Various (2002), while other useful accounts of past climates include Bradley (1999, 2000) and Macdougall (2004).

60. Useful primary sources include: Alley *et al.* (1993); Alley (2000); Cuffey and Clow (1997); Weaver and Hughes (1994) for Greenland and Arctic data. Complementary data from Antarctica are discussed by Petit *et al.* (1999) and Augustin *et al.* (2004). Most of the Greenland ice core data can be accessed online from the National Geophysical Data Center, Boulder, Colorado, USA at: http://www.ngdc.noaa.gov/paleo/icecore/greenland/summit.

61. Perry and Hsu (2000); Landscheidt (2003).

62. As argued by Perry and Hsu (2000).

63. For an informative popular account of the 'out of Africa' human migrations, see Oppenheimer (2004).

64. Shea (1998, 1999, 2001a, 2001b); Bar-Yosef (1998b); Finlayson *et al.* (2006); See also the special issue of *Athena Review* on the Neanderthals and modern humans, available online at: http://www.athenapub.com/index8.htm

65. The Bering land bridge extended from northeast Siberia to Alaska from *26,000–11,000 BP*. There is both DNA (Merriwether *et al.*, 1994; Mokrousov *et al.*, 2000, 2004) and linguistic (Greenberg *et al.*, 1986; Greenberg, 1987) evidence that there was a single migration of a group that gave rise to the majority of present day North and South American Indians at about *25,000 BP*.

These early migrants all spoke languages of the Amerind family, as derived from a common Asiatic ancestor. The land bridge was eventually inundated after *11,000 BP* as sea levels rose in response to a period of generalized global warming. A smaller second migration from northeast Asia seems to have occurred after *11,000 BP* and was almost certainly by boat as the Bering land bridge no longer existed. This brought a new and linguistically distinct group of people, called the Na-Dene, who settled in western Canada and the southwestern United States. The Na-Dene peoples include the Navajo and Apache. It should be noted that some of the interpretations of the DNA data in particular have been challenged more recently and that, although the broad picture presented above is probably correct, there are many details about these migrations that remain controversial. See Dillehay (2001) and Eshelman *et al.* (2003) for recent perspectives on the peopling of the Americas. In a recent twist, Albert Goodyear has published evidence of a possible culture dating back as early as *50,000 BP* (Goodyear, 2004; Anonymous, 2005). While such an early migration is entirely possible (people had travelled from Africa to Australia by then), it seems likely that these people either did not survive or were genetically and culturally swamped by later arrivals.

66. Elias *et al.* (1997); Lambeck *et al.* (2002).
67. Elias *et al.* (1997).
68. This refers to the series of immense migrations of mainly Germanic peoples, such as the Vandals, Alemanni, Burgundians, Ostrogoths, and Visigoths, into the Roman Empire during the third and fourth centuries CE. In 378 CE, one Visigoth camp alone reportedly contained over 200,000 people: the remnants of a whole nation that had been displaced by the invading Huns. Many modern authors have questioned the scale, and even the veracity, of some of these mass-migrations, *e.g.* the Anglo-Saxon invasion of Britain after 450 CE (for a popular account, see Pryor, 2004; for an interesting hypothesis on how a small number of Anglo-Saxon migrants may have contributed disproportionately to the gene pool of the modern English population, see Thomas *et al.*, 2006). However, there does seem to be robust evidence for at the Central European *Völkerwanderung* during the late-Roman period.
69. Forster and Matsumura (2005); Thangaraj *et al.*, (2005).
70. Hey (2005).
71. These data are especially interesting in view of the current debate on 'global warming'. Whether or not the recently observed climatic changes are primarily due to human activity is really something of a moot point. For example the often-acrimonious debate between the proponents (Mann *et al.*, 1998, see weblog: http://www.realclimate.org/) and opponents (McIntyre and McKitrick, 2005, see website: http://www.climate2003.com/) of the so-called 'hockey stick' model of recent anthropogenic climate change continues to rage unabated. Throughout human development, we have experienced similar, non-anthropogenic climatic oscillations and have learned to (sometimes literally) 'go with the flow', which is a colloquial description of what anthropologists call 'habitat tracking' (Weiss and Bradley, 2001). A sober introduction to the climate change debate, which cuts through much of the disinformation and special interest pleading found elsewhere, can be found in the book *Hard Choices* by Coward and Weaver (2004). Although written from a Canadian perspective, the book is universally relevant. An account of conventional scientific opinion on global warming can be found in Houghton (2004).
72. Archaic groups of *Homo sapiens* (sometimes classified as a separate species, *Homo heidelbergensis*) spread across much of Africa and Eurasia from 600,000 to 100,000 years ago. Some of these archaic populations in Africa gave rise to modern forms of *Homo sapiens* but other archaic humans remained relatively unchanged until they became extinct about 28,000 years ago (Oppenheimer, 2004; Stringer and Andrews, 2005).
73. The reasons behind the success of modern *H. sapiens* are still controversial. Some of the theories range from mutations resulting in improved brain function (Klein and Edgar, 2002) to a more adaptable gender-based division of labour compared to competing Neanderthals (Kuhn and Stiner, 2006).
74. Stiner *et al.* (1999).
75. Stiner *et al.* (1999).
76. Enloe (2001); Stiner (2001).
77. Ungar and Teaford (2002).
78. Stahl (1984); Toth (1985); Ulijaszek (1992); Stiner and Kuhn (1992).
79. Early humans were able to occupy the niches of climax carnivores mainly due to their newly developed stone tools. By two million years ago, hominids were using stone flakes to mimic the flesh-slicing incisors of big cats and other heavier stones to mimic the bone-cracking premolars of hyenas (Klein and Edgar, 2002).
80. Stiner *et al.*(1999).
81. 'Trophic' means relating to nutrition. A trophic level is the position of an organism in the ecological food chain or web. Hence, species like lions and tigers

occupy the highest trophic level at the top of the food chain as climax carnivores. Smaller carnivores like wild dogs occupy the next level. Then we have the many herbivorous animals that in turn live on the primary producers, i.e. the plants that occupy the lowest trophic level. There are other trophic levels occupied by microorganisms, for example the many fungi and bacteria that act as decomposers by living off dead organic matter. Most animals tend to specialize both physiologically and behaviourally to exploit one particular trophic level so humans are unusual, but not quite unique, in their great flexibility in this regard.

82. For an account of the volatile faunal and floral assemblages in the Middle Palaeolithic Levant, see Shea (2003).

83. Milton (1993); Eaton *et al.* (2002).

84. Madsen *et al.* (2003); Barton *et al.* (2006); Bettinger *et al.* (2006); Madsen *et al.* (2007); Madsen and Elston (2007).

85. Cellulose is the hard material from which plant cell walls are made. Cellulose is indigestible by most animals and can only be digested by specialized ruminants, such as sheep and cows, thanks to symbiotic, cellulolytic bacteria and fungi that these animals carry in their rumens. Lignin is the main component of wood and hardly any animals, except termites, can digest it. Chitin is the main component of the hard outer skeleton of insects and is only digestible by microbes, including many fungi.

86. Stahl (1989).

87. The period of the Last Glacial Maximum is what is commonly known as the last Ice Age. Although it was the last time there was a full glacial world, both cold and dry, there have been several sudden and relatively severe cold phases since the last glacial maximum. These include the Younger Dryas Interval and the cooling event at about *8200 BP*.

88. Cereals are members of the grass family (Poaceae) that are exploited by humans, normally by means of cultivation, although some cereals have been collected as wild plants by non-agricultural societies until very recently.

2. Plant management and agriculture

89. People may have also reached the Americas as early *50,000 BP*, but these claims have yet to be fully substantiated and are treated with caution by most investigators (Stone, 2003). It is quite possible that some groups of migrants crossed from Eastern Asia between *19,500* and *16,000 BP* (Sarnthein *et al.*, 2006), but genetic and linguistic data suggest that, even if

there were human migrants in the Americas before *14,000 BP*, their populations either died out or were totally swamped by the Amerinds who arrived subsequently (Merriwether *et al.*, 1994; Greenberg *et al.*, 1986; Greenberg, 1987). For a recent account of the ongoing investigations into the peopling of the Americas, see the book by Adovasio and Page (2002).

90. There is much evidence of the preagricultural use of grinding stones (Kraybill, 1977). Some of the earliest cave paintings were made using pigments extracted by grinding plant tissues or minerals (Chalmin *et al.*, 2003), and Australian foragers were already using stone grinding tools to process wild seeds by *30,000 BP* (Fullagar and Field, 1997).

91. Excavation of a 200,000-year-old site in Sudan revealed distinctive stone-tool kits (called Sangoan), ochre, sandstone slabs, and polished quartzite cobbles with starch granules (possibly from seeds), suggesting that humans were processing seeds before they dispersed out of Africa (van Peer *et al.*, 2003).

92. Jarmo is in the foothills of the Zagros Mountains to the east of the modern city of Kirkuk.

93. See Braidwood (1960) and Braidwood *et al.* (1983) for more details.

94. Mithen (2003) has written a quirkily imaginative and accessible account of human history over the period of *22,000–7000 BP*, using the device of a modern time traveller in the prehistoric past who acts as a bridge to the contemporary reader. Although primarily focused on the human aspects of societal development, the book also touches briefly on some aspects of crop domestication and usage.

95. Strictly speaking, what we think of as the seeds or grains of cereals are really fruits. This is because the seeds are enclosed by a pericarp that is fused to the seed coat, or testa. This structure, which in cereals is called a caryopsis, is therefore a fruit containing a single seed, but for all practical purposes it behaves like a seed and will be referred to as such here.

96. Weiss *et al.* (2004); Piperno *et al.* (2004).

97. Nadel and Werker (1999); Nadel *et al.* (2004). Toothwear patterns from human remains at Ohalo II suggest a mixed diet containing aquatic foods and ground cereals (Mahoney, 2007).

98. Colledge (2001); Weiss *et al.* (2004).

99. This is based on DNA analysis of wild and domesticated barleys, as reported by the Egyptian/German team of Badr *et al.* (2000).

100. Kislev *et al.* (2004).

101. Harlan (1967).

102. Morrison and Morrison (2000).

103. Cane (1989); Harlan (1995).

104. The most common species of pistachio in the Neolithic Levant was the terebinth, *Pistacia terebinthus*, which bears small turpentine-flavoured nuts that were a popular food in antiquity. The larger commercial pistachio nuts come from *Pistacia vera*, which grows further east in Iran and Turkmenistan.

105. This period is often called the Bølling–Allerød interstadial. It extended from about *14,600* until *13,000 BP*, although it was by no means a uniformly warm period and its effects were more marked in some regions than others (Yu and Eicher, 2001).

106. Sage (1995).

107. I use ppm (parts per million) here to describe CO_2 concentrations because this term will be almost universally familiar to readers. Strictly speaking, however, ppm is a mixing ratio rather than a concentration, and the more correct units would be micromoles per mole, or $\mu mol.mol^{-1}$.

108. This region corresponded to the belt where wild cereals were present during the Younger Dryas (see Figure 2.2). The 100 to 300-km-wide belt started about 50 km inland from the Mediterranean Sea and extended in an arc from the Dead Sea in the south right up through present-day Syria and Turkey until it reached the Aegean Sea near Izmir. For more details, see Hillman (1996) and Bar-Yosef (1998a, 1998b).

109. Turville-Petre (1932).

110. Henry (1989).

111. Moore (1978).

112. In Kebaran times, the Levantine coast lay about 15 km west of its present location.

113. See Bar-Yosef and Meadow (1995); Bar-Yosef (1998a); and Cauvin (2000) for more detailed accounts of the role of the Natufians in the origins of agriculture.

114. Bellwood (2005, p. 51).

115. Garrod (1928).

116. Bar-Yosef (1998a).

117. Bar-Yosef (1996).

118. The neurological context of Neolithic developments in art, religion, and social complexity, and their relationship to the uptake of agriculture and urban cultures are discussed in the book by Lewis-Williams and Pearce (2005).

119. Gopher *et al.* (2000).

120. Jones and Meehan (1989); Evans (1998, pp. 11–13).

121. Anderson (2004).

122. For a broader overview of what is termed 'protoagricultural practices', and the dangers of extrapolating from a few selected examples, see Keeley (1995).

123. For more detailed accounts of the now vanished Kumeyaay lifestyle and social organization, the works of Florence Shipek are an invaluable resource (Shipek, 1981, 1982, 1991), while Melicent Lee has written an evocative account of her association with the Kumeyaay Indians of more recent times (Lee, 1989).

124. The Kumeyaay were aware of agriculture, as practiced by neighbouring Mesoamerican cultures, and sometimes grew small plots of maize, beans, and squash in favoured locations. Due to the aridity and mountainous nature of much of their home range, however, such agricultural ventures were very much the exception in their plant exploitation strategy.

125. Ladastida and Caldeira (1995).

126. See Jackson and Castillo (1995) for an account of the impact of the Hispanic Mission System on California Indians.

127. During Spanish–Mexican times, California was split into Upper or Alta California and Lower, or Baja, California. Following a disastrous war with the USA, Mexico was forced to sign the Treaty of Guadalupe Hidalgo in 1848, in which it was obliged to cede the States of New Mexico, Upper California, Texas, and the part of Tamaulipas lying between the Nueces and Bravo rivers. This opened up the southwest for colonization by Anglo settlers and the bringing of agriculture to many areas of California, including parts of the Kumeyaay home range.

128. See Fagan (2005) for an account of the human societies of pre-European California.

129. The wider role of acorns in human subsistence is discussed by Mason (1995).

130. Baumhoff (1963).

131. Koenig (1994).

132. Goldschmidt (1951).

133. Pringle (1998b).

134. Ewers (1959, pp.122–125).

135. The most common wild rose in the Owens Valley is still Wood's Rose, *Rosa woodsii*, named after the nineteenth century English botanist, Joseph Woods. The Wood's Rose grows profusely along the meadowland of the Valley floor and up the slopes of the Sierra foothills. The rose hips are one of the richest sources of vitamin C and can also be stewed to make a refreshing and fragrant tea; both properties would have greatly endeared these wild plants to the Paiute people. For further information, see USDA National Resources Conservation Service website at: http://Plant-Materials.nrcs.usda.gov.

136. Pesto sauce is based on the herb, basil, but pine nuts are the main bulking and textural component. Pesto also contains garlic, salt, olive oil, and Parmigiano Reggiano cheese.

137. A classical account of white–Indian relations and history in California and Nevada can be found in Forbes (1969) and a more recent account is in Costo and Costo (1995).
138. Cain (1961, pp. 29, 88–90); Cragen (1975, pp. 11–13, 21–36, 46–62); Wilek and Lawton (1976).
139. Anderson (2004).
140. Frederickson *et al.* (2005).

3. How some people became farmers

141. For many years, the Younger Dryas was regarded as a solely European phenomenon but improved analytical tools and dating techniques now suggest that it was a global event as discussed by Various (2002). Some of the main hypotheses that seek to explain the causes of the Younger Dryas are reviewed by Broecker (2003).
142. As originally reported in Jansen (1938).
143. Peteet (2000).
144. Data are from several ice core samples from Greenland, Canada, Peru and Bolivia and from sediment cores from Venezuela, as reviewed by Various, 2002. See especially Figs 2.1—2.4 on pp. 26–32.
145. Peteet *et al.* (1993).
146. Meltzer and Mead (1983).
147. Trueman *et al.* (2005); Wroe *et al.* (2005).
148. Martin (1984).
149. Trueman (2005) and Wroe (2005) argue for a considerable overlap between Australian human populations and the resident megafauna that may have extended for as long as 15,000 years in some parts of the continent. These authors posit a more important role for climate change, rather than human agency, in the eventual extinction of the megafauna. In contrast, Miller *et al.* (1999, 2005) infer a rapid megafaunal extinction soon after the initial human colonization at *50,000–45,000 BP*. They propose that the primary cause of the sudden mass extinctions may have been habitat destruction due to the burning of large areas of vegetation by humans as part of their plant management strategy. As shown by several recent publications (Wroe and Field, 2006; Brook *et al.*, 2007; Wroe and Field, 2007), this issue is still highly controversial.
150. For a more general discussion of some of the factors, including climate, that have driven previous societal collapses, see Renfrew (1979); Yoffee and Cowgill (1988); and Weiss and Bradley (2001).
151. Johnsen *et al.* (1997); Grootes and Stuiver (1997).
152. Bar-Yosef (1998a).
153. Anderson (1994) and Willcox (1998).
154. Anderson (2004).
155. For example, see Trut (1999) and the earlier discussion by Morey (1994) about the issue of intentionality in animal, and especially canine, domestication by humans. Other similar evolutionary arrangements have been achieved by non-intentional processes. For example nobody would suggest that any intentionality could be ascribed to either partner of the immensely successful and long-lasting ant–aphid or ant–fungus domestications, as described by Shingleton and Stern (2003) and Chapela *et al.* (1994). The question of intentionality can be regarded as irrelevant in purely biological terms, especially given the lengthy, gradual, and incremental nature of the familiarization/ domestication process. If one adopts the relativistic viewpoint of the non-human partner, one can regard feline or canine 'domestication' as the selection of a protective human host by the animal concerned, rather than the selection by humans of an animal for domestication. Their associations with humans have led to the successful proliferation of these animals across the world, to the great advantage of their respective species. One could plausibly argue that cats and dogs now extract much more benefit from their association with people that do their human 'domesticators'. Consider an alien observer of human/dog interactions in a modern Western city: the dog is carefully fed, housed, and protected by its human guardian; the dog leads the human on periodic forays ('walks'), it excretes openly and the excreta are diligently collected by the faithful human; if the dog is ill, the human takes care of it; and none of this nurturing behaviour is reciprocated by the dog. Some dogs are even allowed to sleep in the vicinity of a human, which would normally be an unthinkable privilege for other humans, even from the same family. Who would the alien believe to be the most favoured partner in this association? Admittedly this is an extreme case, but the general point is valid, i.e. domestication involves reciprocal benefits to all partners and can be viewed as a classical biological process of coevolution.
156. Rindos (1980, 1984).
157. Colledge (1998).
158. There is also an analogous 'chicken and egg' argument about the interaction of population numbers and agriculture. Here, Gordon Childe *et al.* (Childe, 1928, 1934, 1936) have asserted that farming made possible the Neolithic population surge, while Ester Boserup *et al.* (Boserup, 1965; Cohen, 1977) posit a pre-Neolithic build-up in numbers that forced

hunter–gatherers to turn to agriculture. My own view is that agriculture was initially a very localized phenomenon that may have been sparked by a range of factors including: the availability of domestication-friendly plants; the fortuitous appearance of favourable mutations; depletion of existing floral and faunal resources (with climate change playing a large role); cultural adaptations to a farming lifestyle; and possible local population pressures (but this would be a minor factor).

159. Henry (1989); Smith (1994).

160. McCorriston and Hole (1991).

161. In the early 1990s, geneticist Jack Harlan was so fed up with the various prescriptive models for the development of agriculture that he proposed, with some irony, a 'no-model model' (see Harlan, 1992a, 1992b and 1995). His basic thesis is that both humans and their potential crop plants underwent an extended period of preadaptation, over several millennia, whereby it only required a minor event, such as a local climatic change, to provide enough impetus to drive them along an increasingly narrow trajectory towards full-scale agriculture. Once people had made such commitments, the option of abandoning the crops on which they increasingly relied, not to mention their home villages with their associated familiar social structures, became ever more difficult (but, as we now know, not impossible).

162. In general, the simpler a scientific explanation of a set of phenomena, the more likely it is to be both intellectually satisfying and a closer approximation to the truth. Therefore, when we are challenged by several explanations or theories, the principle of parsimony entails acceptance of the least complex explanation, providing of course that it gives a satisfactory account of the phenomena in question. The principle of parsimony has assumed even greater importance in recent years as huge amounts of information from molecular biology are used in fields such as medicine, population studies, and taxonomy. For example when constructing an evolutionary tree based on such information, the scientist will automatically select the most parsimonious model. For more on the modern uses of parsimony, especially in genetics research, see Sober (1988), Kitching *et al.* (1998), or Albert (2005).

The concept of parsimony has a venerable tradition that extends as least as far back as the Franciscan friar, William of Ockham (*c.*1287–1347), after whom 'Ockham's razor' (often spelt Occam's razor) is named. This rule states that, in science and philosophy, simpler theories are always to be preferred. Or to put it more pithily: '*Entia non sunt multiplicanda praetor necessitatem*', i.e. 'It is vain to do with more what can be done with less'. William of Ockham was a remarkable thinker for his time, initiating the school of scholastic nominalism, which provided a more reality-based approach to enquiry that validates science as objective knowledge. He drew a clear distinction between the empirically-based natural sciences and the mass of theological speculation that is based only on revealed premises. Perhaps unsurprisingly, he was excommunicated in 1328 and fled to Bavaria, where he was able to continue his work. See McGrade (1974) or Adams (1987) for more on this innovative medieval English thinker.

163. See discussions in Harris (1989), Damania *et al.* (1998), and Anderson (2004).

164. Richerson *et al.* (2001).

165. Kislev *et al.* (2004).

166. These issues are discussed in depth in several chapters in a forthcoming book (Barton *et al.*, 2006; Bettinger *et al.*, 2006; Madsen and Elston, 2007; Madsen *et al.*, 2007).

167. Munro (2003).

168. Goring-Morris (1991).

169. This evidence suggests that human population densities in the Levant decreased during the late Natufian period of *12,900–11,600 BP*, which coincided with the most acute phase of Younger Dryas Interval. This was partially due to a series of migrations, including the successful venture to the cereal-rich north and the failed attempt to revert to hunter–gathering in the more arid south. The reduction in population density allowed the successful late Natufian groups to maintain levels of food availability as measured by the index of small game (e.g. waterfowl, rabbits, partridge, turtles etc) in the study by Munro (2003).

170. These 'weeds' would have included plants that would have grown in the newly established fields in the absence of human intervention. A weed is simply a plant growing in the wrong place and/or at the wrong time. Crops can sometimes act as weeds when they grow in the field of another crop. Hence, wheat plants growing in a barley field are weeds. Such weed crops are known as volunteers.

171. Kislev *et al.* (2006).

172. Davis and Valla (1978). Note, however, the genetic evidence for the origin of dogs is less clear. Mitochondrial DNA data suggest a single site of origin in a population of wolves from east Asia at least *15,000 BP*, although the authors suggest that there were probably numerous separate instances of domestication by people who independently

sampled members of this wolf population (Savaolainen *et al.*, 2002). By, *14,000* to *12,000 BP* domesticated dogs were in western Eurasia (Levant and Germany) and by *10,000* to *9000 BP* they were in North America (Utah) (Leonard *et al.*, 2002). Hence, human migrants who crossed from Asia into America before the flooding of Beringia at *11,000* to *10,500 BP* probably brought domesticated dogs with them. More controversially, there is also some genetic evidence of a possible, more ancient domestication of dogs dating from as early as *40,000 BP* (Vilà *et al.*, 1997). This could indicate that the earliest colonizers of the Siberian steppes may already have had dogs with them as long ago as *26,000* to *19,000 BP* (Goebel, 1999; Savaolainen *et al.*, 2002). Although there is no physical evidence for any form of association between people and dogs before about *12,000 BP*, there is a suggestion from DNA analyses that domestication-related changes in dogs may date back as much as 100,000 years (Vilà *et al.*, 1997). For more background on the origins of domesticated dogs, see Morey (1994) and Clutton-Brock (1995).

173. Even compared with our closest human relatives, the Neanderthals, adults of our own species have several distinctly paedomorphic traits. These include small, minimally protruding faces, the lack of a supraorbital (brow) ridge, and prominent, high foreheads. Such traits are universally associated with immaturity among other mammals but persist into adulthood in humans. As pointed out by Arsuaga (2003) and in the novel by Björn Kurtén (1995), our Cro-Magnon ancestors would have reminded Neanderthals of their own children and might have looked disarmingly appealing as a result. Unfortunately, cute and sweet as the Cro-Magnon adults may have seemed, the Neanderthals would have soon discovered, doubtless to their great dismay, the sort of behaviour of which our species is capable.

174. Analysis of gene expression profiles using microarrays (http://en.wikipedia.org/wiki/DNA_microarray) has revealed that several genes are expressed differently in the brains of domesticated and wild foxes, which may underlie the different behaviours of the two groups of animals (Lindberg *et al.*, 2005).

175. See Trut (1999) and Hare *et al.* (2005) for more detailed accounts of this fascinating experiment in animal domestication.

176. Bar-Yosef and Valla (1991).

177. This elderly person, who is of indeterminate sex, is holding a canid puppy in their left hand (Davis and Valla, 1978). Because the puppy is very young, it is difficult to determine whether it is a wolf pup or a dog. Whatever is the case, the animal was been taken from its conspecifics to live in a human society and we can therefore say that it has been domesticated. As shown by Trut (1999), and discussed above, many of the biological adaptations to domestication might have occurred in as little as a single human generation.

178. There are several non-human examples of domestication-like associations involving both animals and plants. Perhaps the best-known exemplars of such domestications are attine ants. Some ants manage herd-like groups of aphids, which they 'milk' to obtain a highly nutritious sugary secretion called honeydew. The ants protect the aphids from predation and will even carry them to more suitable feeding locations to encourage them to increase their honeydew production. This mutualistic association has evolved and then been lost again many times during ant/aphid coevolution, as shown by recent DNA evidence, e.g. Shingleton and Stern (2003). Other ants cultivate Lepiotacean fungi in elaborate underground gardens. The transition from a 'hunter-gatherer' (of arthropod prey, nectar, and other plant juices) to 'farmer' (cultivating and feeding off fungal gardens) lifestyle originated 45–65 million years ago (see Mueller *et al.*, 2001 for more detailed discussion). These ants do not allow the fungi to mature and produce mushrooms. Therefore, the fungi must rely solely upon the ants for reproduction. The founding queen ant (the mated female) carries fungal spores from the original nest to restart a genetically identical fungus garden in the new nest. By comparing the phylogenies of both the ant and the fungus, mycologists have discovered that some of the more developed attines have been clonally propagating fungi from the same Basidiomycete family lineage for over 23 million years, as described by Ignacio Chapela and colleagues (see Chapela *et al.*, 1994 for the original scientific paper and Wade, 1999, for a newspaper article).

179. For a recent review on canine genetics, see Sutter and Ostrander (2004).

180. The story of the Abu Hureyra excavations is well told in the classical account by Moore *et al.* (2000) and other articles by the same research team, such as Hillman *et al.* (2001), from which much of the information presented here is gleaned.

181. Moore *et al.* (2000, Chapter 3).

182. Moore *et al.* (2000)

183. Cauvin (2000); Akkermans and Schwartz (2003).

184. Among the heavy grinding tools used by such semi-sedentary groups were boulder mortars that weighed as much as 150 kg (Ames, 1999).

185. Nowadays the nearest wild cereal stands to Abu Hureyra are more than 100 km away.

186. Leinonen *et al.* (1999).

187. Miller *et al.* (1999).

188. Blumler (1996).

189. The ergot fungus, *Claviceps purpurea*, is a parasitic organism commonly found on rye, but not on other cereal crops. The fungus spreads to the grains of the rye, which reduces crop yield, but by far its most serious effect is in causing the disease ergotism. Ergot-contaminated rye bread often contains toxic alkaloids produced by the fungus, including several peptide alkaloids of the ergotamine group (e.g. ergotamine, ergosine and ergocristine) that cause vomiting, diarrhoea, hallucinations, and may lead to gangrene in serious cases. People eating rye bread have always been bedevilled by the risk of ergotism, which was referred to as St Anthony's Fire in the Middle Ages. Tens of thousands of people in France were killed by ergotism during this period. There was an outbreak of rye-ergotism as recently as 1951, in the French town of Pont-St. Esprit on the River Rhône. There is even well researched, albeit circumstantial, evidence that ergotism caused some of the behaviour that led to the Salem Witch Trials of 1692, as documented by Caporael (1976). Despite its toxicity, the ergot fungus is also a source of several beneficial alkaloids, such as ergonovine, which is used to induce labour in pregnancy and to control haemorrhaging in surgical operations; and ergotamine, which is used extensively to relieve migraine headaches through the constriction of blood vessels. Finally, the ergot fungus was the original source of lysergic acid. In 1943, Swiss chemist Albert Hoffman added diethylamide to lysergic acid to produce the psychotropic drug, better known today as LSD.

190. Feil and Schmid (2002).

191. This proposal by Hillman *et al.* (2001) is based on the known effects of temperature on the fertility mechanisms of rye, as detailed by Gertz and Wricke (1991).

192. Hillman *et al.* (2001).

193. Hancock (2004).

194. For example some groups of Amerindian hunter–gatherers have strong ethical, ecological, and spiritual perceptions of their relationship with their environment and their creator that precludes them from wantonly disturbing the earth (MacLuhan, 1972). Other cultures regard the earth as the primeval mother figure and the penetration of her skin (the soil) as a sacrilegious act.

195. This perspective is close to the human behavioural ecology (HBE) approach to resource exploitation in human populations, as expounded in the recent informative volume edited by Kennett and Winterhalder (2006). One of the early examples of this socioecological approach is the influential paper by Layton *et al.* (1991) proposing that agriculture might be used either alone or in combination with other strategies, any of which could be discarded as dictated by social and environmental conditions. One of the problems with the human behavioural ecology approach is its focus on individual motivations in the tradition of Adam Smith. However, strong social networks can create an emergent 'super-organism', e.g. a city-state with additional population-level properties that can override individual needs and even adaptive fitness as long as this enhances its own fitness. I would assert that the imperial agroeconomies of Sumer and Akkad in the Uruk period are early examples of such 'super-organisms', most of whose members had considerably reduced adaptive fitness at the individual level but collectively were more adaptively successful than alternative societal organizations in the region.

196. Callen (1967).

197. Oota *et al.* (2005).

4. Plant genomes

198. This value of 400,000 plant species is taken from Govaerts (2001) and Bramwell (2002). However, this value may well be a substantial underestimate of the true total. With the application of recent advances in molecular genetics and bioinformatics to the study plant populations, especially in the tropics, it is likely that the total number of plant species will continue to rise, possibly to as many as several million.

199. Tudge (1988) and Cordain (1999).

200. Tudge (1988).

201. Darwin (1868)—this book is a compendium of interesting information about plant domestication, as it was understood in the nineteenth century. Most of Darwin's works are now available on the internet at http://darwin-online.org.uk The continuing importance of studies of plant domestication for our understanding of more general evolutionary processes has been reviewed by Hancock (2005).

202. Buckmann (1857).

203. de Candolle (1883).

204. Hawkes (1998).

205. Roll-Hansen (2004).

206. For more on the life of Vavilov, see Popovsky (1984).

207. The Lysenko saga has been far from an isolated case. The sciences of evolution and genetics seem to have

attracted more than their fair share of pseudoscientific adherents, often with tragic consequences. Examples include the eugenics movement that blighted the lives of thousands in the USA and Scandinavia until the 1970s, the sinister racial lunacy of the Third Reich, and the puzzling rise of contemporary Christian creationism.

208. Vavilov's observations were soon picked up and reinforced by other scientists. For example while he was living in a suburb of the Mexican city of Guadalajara, US botanist Edgar Anderson noted that he found 'more variation in the corn of this one township than in all of the maize in the United States' (quoted in Fowler and Mooney, 1990).

209. The link between wild relatives of crops and their origins of domestication was proposed in 1927 by Peake and Fleure, who also suggested several other characteristics that would have contributed to domestication, including an impediment to migration and the absence of heavy forest cover (Peake and Fleure, 1927), as also discussed in Gepts (2004), which is available online.

210. Vavilov (1926, 1935).

211. Jack Harlan died in 1998 and in the same year a UN-sponsored symposium on *The Origins of Agriculture and Crop Domestication* was named *The Harlan Symposium* in his honour (see Damania *et al.*, 1998). Several of Harlan's works are cited in the bibliography; see Harlan (1981, 1992a, 1992b, 1995). Other notable contributions came from the likes of Sauer (1952); Hutchinson (1965); and Harris (1967, 1989, 1996), to mention but a few.

212. The importance of knowledge of crop origins for the understanding of their evolution and domestication is discussed in the review by Matsuoka (2005).

213. Anderson (2004).

214. One of the earliest authors to propose that agriculture could be described as a coevolutionary process was David Rindos (Rindos, 1980, 1984), a genuinely creative thinker, influential scholar, and popular teacher, whose life was tragically cut short by illness during a prolonged struggle for academic tenure against the University of Western Australia (see: http://www.uow.edu.au/arts/sts/bmartin/ dissent/documents/sau/sau06.html). For a useful review of Rindos' ideas on agriculture, see Zubrow (1986), and for another perspective on agriculture as a process of human–plant coevolution, see Clement (1999a, 1999b). In a lighter vein, Michael Pollan examines aspects of plant–human coevolution from the plant perspective in *The Botany of Desire* (Pollan, 2001).

215. Diamond (1997, 2003).

216. It used to be thought that no mammalian herbivores were domesticated in the entire continent of Africa (Diamond, 1997, pp. 162–164, 389). However, more recent evident suggests that the donkey, *Equus asinus*, was domesticated in two separate locations in north and east Africa about *5000* BP (Beja-Pereira *et al.*, 2004).

217. Comai (2005).

218. The process of polyploidy in plants and its role in crop genetics and evolution is described in detail in the book *Plant Evolution* (Hancock, 2004). The importance of polyploidy is underscored by the fact that the author devotes an entire chapter to it. Although aimed primarily at postgraduates, some of the chapters on farming and domestication may be of more general interest. A second book *Plants, Genes and Plant Biotechnology* (Chrispeels and Sadava, 2003) covers some of the same ground (albeit less comprehensively) in a simpler, more undergraduate-orientated format. For recent reviews of the research literature, see Wendel (2000), Liu and Wendel (2002), and Adams and Wendel (2005).

219. This is probably a conservative estimate. Some workers have calculated that as many as 95% of all angiosperms have experienced significant chromosome doubling at some stage in their evolutionary history, even if some of them now behave as diploids (Grant, 1981; Leitch and Bennett, 1997). This is consistent with recent estimates that polyploids account for 97% of species in lineages of non-flowering plants, such as the pteridophytes (e.g. ferns), that are much older than the angiosperms (Stebbins, 1950).

220. Whereas, 35–50% of flowering plant species currently behave as polyploids (Grant, 1971), the proportion rises to 78% in crop species (Hancock, 2004). Not even the plant geneticist's favourite model species, *Arabidopsis thaliana*, has been exempt from this relentless polyploidization. Although it has only five chromosomes and possesses one of the smallest of all plant genomes, Arabidopsis now appears to be an ancient tetraploid (Blanc *et al.*, 2000) that underwent a whole-genome duplication to form an autotetraploid, or possibly an allotetraploid, about 38 million years ago (Ermolaeva *et al.*, 2003).

221. This was only discovered recently, following detailed analysis of the rice genome (Paterson *et al.*, 2004; Yu *et al.*, 2005). The ancestor of the rice genome underwent a complete duplication about 70 million years ago, i.e. well before the divergence of the various grass species of the Poaceae that started about 50 million years ago.

222. Muller (1925).

223. In a revealing quotation, LG Stebbins, one of the leading researchers on polyploidy, asserted that it 'retards rather than promotes progressive evolution' (Stebbins, 1950, 1971). Quite apart from the fact that few scientists today would agree that evolution is 'progressive', polyploidy is now recognized as being one of the key drivers of increased variation, thereby making a huge contribution to the evolution of both animals and plants (Hancock, 2004; Mable, 2004).

224. We now know that all vertebrates are ancient polyploids. Recent genome analysis supports earlier hypotheses that there were several rounds of genome duplication during the evolution of fish into the land vertebrates and eventually mammals (Furlong and Holland, 2004). This means that our ancestors would have been tetraploid on several occasions. Gradually, the gene sequences and order on the duplicated chromosomes would have diverged from one another so that they were no longer copies and the organism effectively reverted to being a functional diploid, albeit with twice the original number of chromosomes of its distant diploid ancestor. This is why humans have a relatively large number of chromosomes (2n = 46) for a 'diploid' organism that has a much smaller genome than many plants.

225. Ohno (1970).

226. For recent studies and reviews of ancient vertebrate polyploidy, see Pébusque et al. (1998); Spring (1997); and Furlong and Holland (2005).

227. It is now known that many amphibians and fish are polyploids, although unlike plant polyploids, animal polyploids are almost invariably autopolyploids, i.e. they condition has arisen from an endogenous genome duplication, rather than an interspecific hybridization (Gregory and Mable, 2005). The recently discovered, and hitherto unique, tetraploid mammal is a desert-dwelling, octodontid rodent called the red viscacha rat, *Tympanoctomys barrerae*, which is endemic to arid regions of central-western Argentina (Gallardo et al., 1999, 2004).

228. Carothers and Beatty (1975).

229. Polyploid humans are almost invariably unviable and do not survive gestation. However, there are rare cases of chimeric individuals, so called 'mixoploids' who are made up of a mixture of cells exhibiting different levels of ploidy. For example, in a study from Australia, two cases of diploid/tetraploid (2n/4n) mixoploidy in surviving females were described (Edwards et al., 1994). Both individuals manifested severe mental retardation and other physical abnormalities but one of them survived into her twenties.

230. Modern commercial export bananas all belong to the Cavendish subgroup of triploid clones (from the family name of the Duke of Devonshire in England who, in 1836, managed to get this clone to flower in his greenhouse). This clonal line has a monospecific *Musa acuminata* origin. Triploid plants were formed when normal haploid gametes fused with aberrant, non-reduced diploid gametes (Raboin et al., 2005).

231. It would take just a single opportunistic pathogen to decimate the world's supply of commercial bananas, as happened in the 1930s with the previous clonal variety, Gros Michel (originally propagated in the Far East during the early nineteenth century), after it was attacked by the fungus, *Fusarium oxysporum*.

232. As noted in the original literature, estimates of the proportion of angiosperm species undergoing some form of polyploidization have tended to increase as knowledge of their genomic organization has improved, particularly over the past decade (Stebbins, 1950; Stebbins, 1971; Lewis, 1980; Grant, 1981; Masterson, 1994; Wendel, 2000).

233. Murray et al. (2005).

234. De Pamphilis et al. (2006).

235. A tetraploid would be expected to have a genome that is twice as large as each of its diploid progenitors, but this is normally only found with newly formed polyploids. Following their formation, most polyploids seem to undergo a process of selective 'genome downsizing' that is presumably of adaptive significance for the plant (Leitch and Bennett, 2004).

236. Caetano-Anollés (2005).

237. Polyploidy is also a key mechanism for the evolution of new plant species following interspecific hybridization events (Darlington, 1963; Rieseberg and Wendel, 2004). New molecular approaches are now allowing us to investigate in detail the ways in which such hybrids sometimes become reproductively isolated from their parents and hence can be regarded as new species (Hegarty et al., 2006). For a review, see Hegarty and Hiscock (2005).

238. Unless stated otherwise, the term 'genome' is used here to refer to the nuclear genome, where the vast majority of genes of a eukaryotic cell are located. In addition to their nuclear gnome, all plants contain two much smaller genomes, namely those of the mitochondrion and plastid. These two groups of organelles contain their own DNA because they originate from once free-living bacteria that were captured by a larger cell in a process called endosymbiosis (Alberts et al., 2002). Human cells

also contain mitochondria and therefore we also have a second genome, of ancient bacterial origin. Unlike nuclear genes, of which we inherit two sets, one set from each parent, we only have a single set of mitochondrial genes and these are all inherited maternally. The analysis of mitochondrial DNA has been of immense value in tracing inheritance patterns during human evolution, such as the extent of our relationship with Neanderthals (Serre *et al.*, 2004).

239. For a more detailed, but accessibly non-technical explanation of polyploidy and its role in crops, see Tudge (1988).

240. Essentially, an autotetraploid organism has a full genome available for all of its 'normal' functions plus an entirely new genome that is surplus to requirements and therefore available for evolutionary 'experimentation'. Mutations are generally deleterious as they affect gene function. However, genes on the second 'extra' genome of a tetraploid can mutate without affecting the function of the equivalent copies on the first genome. This is one way in which new metabolic pathways can arise. Even relatively simple organisms may be descended from ancient autotetraploids. For example it is now believed that the humble Brewer's yeast, *Saccharomyces cerevisiae*, is the result of a genome duplication that allowed it to develop the metabolic pathways responsible for the fermentation of sugars to ethanol (Kellis *et al.*, 2004). Therefore, one could justifiably say that were it not for polyploidy we might have no wines, beers, or spirits, which is a somewhat sobering prospect.

241. The concept of evolution via the duplication of both individual genes and entire genomes was most powerfully stated by Ohno, in the classic study from 1970 (Ohno, 1970). We now know that the duplication rate of individual genes in most higher plants is of the same order as that of yeast, at 0.002 duplicates per gene per million years (Moore and Purugganan, 2005). In the case of a species such as *Arabidopsis*, which contains about 25,000 genes, this would mean one gene duplication every 20,000 years—a rapid rate on an evolutionary timescale.

242. This remarkable phenomenon has only recently been recognized, as reported by Adams *et al.* (2003). These authors report a series of gene expression changes that occur after allotetraploid formation, most of which involve reciprocal organ-specific, gene silencing. In some cases the silencing is reciprocally absolute, i.e. one homoeologue is completely suppressed in some organs but fully expressed in others,

while the other homoeologue exhibits the exact reverse pattern of expression. Some of these silencing events occur immediately after polyploidization while others occur over evolutionary timescales of thousands of generations. These and similar studies (e.g. Chen *et al.*, 1998; Pikaard, 2001; and the reviews by Comai *et al.*, 2003; Rapp and Wendel (2005); and Veitia, 2005) highlight the dynamic nature of polyploid genomes and their ability to express variation and hence to evolve by such epigenetic (non-Mendelian), as well as by conventional Mendelian, mechanisms.

243. Zhao *et al.* (1998) showed that, after polyploid formation in *Gossypium* spp., repetitive DNA sequences from one genome colonized the alternative genome, possibly leading to alterations in gene expression patterns. This can occur in either direction; hence the two genomes are effectively engaging in a kind of dialogue that involves mutual exchange of information, hence 'cross-talk'.

244. Wendel (2000).

245. Saltational variation (from the Latin *saltus*, 'leap') is the occurrence of a sudden inherited change from one generation to the next. Such non-Darwinian changes can arise as a consequence of various forms of polyploidization or from the mutation of a key regulatory gene. Intergenomic invasion involves the colonization of one genome in a polyploid organism by DNA sequences from the other genome. This has been studied in detail in cotton, where a great deal of asymmetric colonization of the D genome by repetitive elements from the A genome has occurred (Hanson *et al.*, 1998; Zhao *et al.*, 1998). Cytonuclear interactions are especially important in plants because they have two semiautonomous extranuclear cytosolic genomes, namely those of their plastids and mitochondria (Song *et al.*, 1995). These cytosolic genomes must be co-ordinately regulated with the nuclear genome, e.g. during the expression of key photosynthetic enzymes such as rubisco where the different subunits are encoded respectively on the nuclear and plastidial genomes (Rodermel, 1999). These mechanisms of variation enhancement following polyploidization have been reviewed by Wendel (2000) and Liu and Wendel (2002).

246. This mechanism was originally proposed by Feldman *et al.* (1997), and has been more recently reviewed by Özkan *et al.* (2001).

247. Shaked *et al.* (2001).

248. Much of the early work on characterizing the control of chromosome pairing in polyploid wheats was

carried out by Ralph Riley and colleagues at the Plant Breeding Institute in Cambridge, UK (Riley and Chapman, 1958; Riley *et al.*, 1959; Chapman and Riley, 1970).

249. For a recent comprehensive review of the evolutionary significance of polyploidy in plants, see Wendel (2000).

250. Polyploids are not always larger than diploid forms of the same species. For example when an interspecific hybrid was created between foxtail millet and giant green foxtail, the tetraploids were smaller, flowered later, and had a two-fold reduced fertility (grain number per cm of spike) compared with their diploid parents (Ahanchede *et al.*, 2004). Although grain weight increased by 20% with polyploidy, total grain yield decreased by 46%. This study illustrates the difficulty of breeding polyploid lines of foxtail millet that are of agronomic use and perhaps explains why ancient farmers only selected diploid varieties for cultivation.

251. Thompson *et al.* (2004).

252. For example, autopolyploidy in the saxifrage, *Heuchera grossulariifolia*, may have arisen up to seven times during its evolutionary history (Segraves *et al.*, 1999) and this is by no means an isolated or extreme case.

253. Hegde *et al.* (2006).

254. Radically new genetic combinations can also be introduced into isolated populations of the same species when new forms are introduced from outside. This is now recognized as a key factor in the creation of invasive new forms of hitherto quiescent species (Novak, 2005; Lavergne and Molofsky, 2007) and may also be relevant to the improvement of crops via interbreeding with wild relatives (see Chapter 6).

255. Sauer (1957); Levin *et al.* (1996); Martin and Cruzan (1999).

256. There are yet more extreme types of polyploidy where some organisms have eight, ten, or even twelve sets of chromosomes. For example there are members of the wheat group that are octoploid, *Elytricum fertile*, and decaploid, *Agropyron elongatum*, while members of the birch genus, *Betula*, can be dodecaploid (Särkilahti and Valanne, 1990).

257. See Forbes and Watson (1992), pp. 234–235 for a useful discussion on crop polyploidy.

258. Özkan *et al.* (2001); Feldman and Levy (2005); Griffiths *et al.* (2006).

259. To quote Feldman and Levy (2005): 'The revolutionary changes comprise (1) non-random elimination of coding and non-coding DNA sequences, (2) epigenetic changes such as DNA methylation of coding and non-coding DNA leading, among others, to gene silencing, (3) activation of genes and retroelements which in turn alters the expression of adjacent genes. These highly reproducible changes occur in the F_1 hybrids or in the first generation(s) of the nascent allopolyploids and were similar to those that occurred twice in nature: first in the formation of allotetraploid wheat (~0.5 million years ago) and second in the formation of hexaploid wheat (~10,000 years ago). Elimination of non-coding sequences from one of the two homoeologous pairs in tetraploids and from two homoeologous pairs in hexaploids, augments the differentiation of homoeologous chromosomes at the polyploid level, thus providing the physical basis for the diploid-like meiotic behavior of allopolyploid wheat. Regulation of gene expression may lead to improved intergenomic interactions. Gene inactivation brings about rapid diploidization while activation of genes through demethylation or through transcriptional activation of retroelements altering the expression of adjacent genes, leads to novel expression patterns.' For a discussion of non-Mendelian phenomena in the evolution of allopolyploid genomes, see Liu and Wendel (2002) and Levy and Feldman (2005).

260. To quote Feldman and Levy (2005) again: 'The evolutionary changes comprise (1) horizontal intergenomic transfer of chromosome segments between the constituent genomes, (2) production of recombinant genomes through hybridization and introgression between different allopolyploid species or, more seldom, between allopolyploids and diploids, and (3) mutations.'

261. Özkan *et al.* (2001); Griffiths *et al.* (2006).

262. Griffiths *et al.* (2006) localized *Ph1* to a 2.5-Mb region of wheat chromosome 5B, which contains a heterochromatin segment that became inserted into a cluster of *cdc2*-related genes after polyploidization. This insertion event, and its consequences for chromosome pairing, differentiates many of the allopolyploid wheats from their diploid counterparts.

263. Graham Moore and colleagues from the John Innes Centre, UK, have recently published cytogenetic evidence for the role of *Ph1* in chromosome pairing (Martinez-Perez *et al.*, 2001, 2003); and the first molecular description of the wheat *Ph1* locus (Griffiths *et al.*, 2006).

264. The two major competing theories are that: (1) *Ph1* promotes premeiotic association of homologous chromosomes (Feldman, 1993; Martinez-Perez *et al.*, 2003; Naranjo and Corredor, 2005; Feldman and Levy, 2005); and (2) *Ph1* regulates synapsis (homologous

chromosome alignment at meiosis) and crossing-over at meiotic prophase (Luo *et al.*, 1996).

5. Fluid genomes, uncertain species, and the genetics of crop domestication

265. As with the vertebrate animals discussed in Chapter 4, many apparently diploid crops, such as rice and maize, are almost certainly descended from ancient polyploid ancestors and there is still considerable evidence of duplicated chromosome regions in both maize (Gaut and Doebley, 1997) and rice (Ge *et al.*, 1999).

266. A megabase is one million bases. DNA is made up of a series of paired nucleotides, called bases. A single gene is normally in the region of a few thousand bases. The human genome contains 3300 million bases, of which 90% is made up of repetitive DNA. Recent results from the human genome sequencing project suggest that we may have as few as 20–25,000 genes, which is even fewer than the tiny model plant *Arabidopsis thaliana*. (International Human Genome Sequencing Consortium, 2004). Estimates of genome size can vary according to the technique employed and values given here are normally at the midpoint of the range of published values (Greilhuber *et al.*, 2005).

267. Rayburn (1990); Rayburn and Auger (1990); Rayburn *et al.* (1985).

268. Ma *et al.* (2004).

269. One recent study showed that repetitive, non-coding DNA accounted for over 90% of a sample of the genome of a diploid member of the wheat family, *Aegilops tauschii* (Li *et al.*, 2004).

270. Leitch *et al.* (1998) .

271. Bennett and Smith (1976).

272. Knight *et al.* (2005).

273. Ma *et al.* (2004). This study shows that the genome of Asian rice has lost two-thirds of the huge 300 Mb injection of repetitive DNA that it has acquired over the past eight million years. This ability to lose such non-coding DNA may be one of the main reasons that rice continues to have such a small genome compared to other cereals. It also implies that much of the repetitive DNA may indeed be non-useful for the rice plant in evolutionary terms and therefore its loss is either neutral or may even be adaptive. In contrast, most of the other cereals appear to be either less able to lose their massive loads of repetitive DNA or are better able to tolerate such an apparent burden.

274. This phenomenon of DNA removal has been well studied in some insects, where the rate of DNA removal seems to be correlated with the size of the genome. Hence, the fruit fly, *Drosophila melanogaster*, has been able to lose exogenous DNA at 40-times the rate of the Hawaiian cricket, *Laupala cerasina*, with the result that the cricket now has an 11-fold larger genome than the fruit fly (Petrov *et al.*, 2000). So far we do not know whether a rapid rate of DNA loss is more adaptive than a slower loss, but the existence of such variations in the rate of DNA loss in both plants and animals suggests that rapid loss may not always be advantageous.

275. SanMiguel *et al.* (1998).

276. The term 'non-coding DNA' is used here to refer to DNA sequences that are not involved in encoding or regulating the transcription of DNA into RNA. In addition to their coding regions, or 'open reading frames', eukaryotic genes contain regulatory elements, such as promoters, 3' and 5' enhancers, and introns. It is the interaction of these regulatory DNA sequences with other genetic or environmental factors that determines when, where, and to what extent a particular gene is expressed. Note also that not all genes ultimately encode proteins, as was stated by the now-outdated 'central dogma' of molecular biology. Some genes simply encode a type of RNA, called micro-RNA, that is involved in regulating several important classes of developmental genes in both plants and animals (Eddy, 2001; Yelin *et al.*, 2003; Gibbs, 2003).

277. In a study with mice, Nobrega *et al.* (2004) have shown that it is possible to remove about 1.4 Mb of non-coding DNA from the genome with no apparent effect on factors including morphology, reproductive fitness, growth, and longevity. However, it is also becoming evident that some of this non-coding DNA (sometimes inaccurately called 'junk' DNA) is not only useful but may be essential for animal development. In another study, Peaston *et al.* (2004) have demonstrated that retrotransposon elements from this repetitive DNA play an important role in the early stages of embryogenesis.

278. The key roles of LTR retrotransposons in angiosperm, and especially crop, evolution are discussed by Vitte and Bennetzen (2006).

279. Retroviruses are members of a class of RNA viruses that transcribe their RNA into DNA, once they have infected a host cell. Many retroviruses infect vertebrates and cause a range of diseases including tumours, leukaemia, and immunodeficiency. Retroviruses express a gene product, the integrase protein, which facilitates the integration of the viral genome into the host chromosomes where it can lie dormant for a considerable time. Related viruses in

plants are the pararetroviruses, some of which are pathogens of important food crops. Until recently, it was thought that plants did not contain true retroviruses but data from genome sequencing projects show that plants do indeed harbour retroviral-like sequences (Peterson-Burch et al., 2000; Zaki, 2003). Some plant retroviruses are already being used by biotech companies to facilitate the process of gene transfer (for example, see: http://www.isupark.org/news/index.cfm?step = 1&id = 277).

280. Many retrotransposons in plants are made up of so-called gypsy-like or copia-like elements. These DNA elements are now known to be ubiquitous in plants. Very similar DNA elements are found in other groups of organisms, including the endogenous retroviruses in the genomes of mammals. Many of these elements encode very similar gene products and their organization and mechanisms of mobility are also remarkably conserved in plants and animals. It seems unlikely that such distinctive and highly similar structures would have evolved more than once, i.e. they are probably homologous rather than analogous. This implies that they were already in existence in the last common ancestor of the fungi, plants, and animals. The antiquity, ubiquitous distribution, and dynamic nature of retrotransposons seems to belie the concept that they are mere parasitic or 'junk DNA' and raises many interesting questions about their role and impact on organisms and their genomes.

281. Almost 10% of the human genome consists of LTR sequences derived from retroviruses that have colonized the genome; some of these exviruses may play a role in processes such as reproductive isolation and speciation (Yohn et al., 2005).

282. Schulman et al. (2002).

283. See Biemont and Vieira (2006) for a recent review.

284. Morrish et al. (2002).

285. Vicient et al. (2001).

286. Zhao et al. (1998) and Wendel et al. (1995).

287. Chantret et al. (2005).

288. For example it has been reported that very rare and endangered plant species, i.e. those that are now on the brink of extinction, have on average larger genomes and more 'extra' DNA than their more secure relatives, indicating that the 'extra' DNA might increase the likelihood of extinction. In contrast, polyploidy was not associated with the increased risk of extinction (Vinogradov, 2003).

289. Vitte and Bennetzen (2006) discuss the ways in which the differing nature and behaviour of LTR retrotransposon structures in a wide range of crop and other plant species has dramatically affected their evolution. In the same journal issue, Wang and Dooner (2006) describe how transposons have contributed to the unique genetic and phenotypic diversity of Zea mays.

290. Evidence is also growing about the role of transposable elements in human genomic evolution (Johnson et al., 2006).

291. Genes are also being exchanged between bacteria, such as E. coli and Synechocystis, at the relatively rapid rate of 16 kb per million years (Martin, 1999). It is also estimated that no less than 18% of the entire E. coli genome may be of relatively recent foreign origin. Horizontal gene transfer is reviewed in the book edited by Syvanen and Kado (2002).

292. Davis et al. (2004, 2005) have described the transfer of genes to an endophytic parasite in the Rafflesiaceae from its obligate hosts in the genus Tetrastigma; as well as gene transfer in the opposite direction, from members of the parasitic angiosperm order, the Santalales (which includes the sandalwoods and mistletoes) to one of their non-angiosperm host plants, the rattlesnake fern (Botrychium virginianum). Meanwhile, Mower et al. (2004) described evidence for gene transfer from parasitic members of the Bartsia and Cuscuta genera to their host species in the unrelated genus, Plantago. Hence, gene transfer can occur either from host to parasite, or vice versa, among plants and can cross the boundaries of genus, order, family, and even kingdom.

293. In one case, the genome of the non-parasitic, tropical shrub Amborella trichopoda was found to contain no less than 26 foreign genes (Bergthorsson et al., 2004). For more about virus-mediated gene transfer, and other mechanisms, see Bergthorsson et al. (2003, 2004) and Martin (2005).

294. Richards et al. (2006).

295. Fungal–plant gene transfer in the context of mycorrhizal associations (which affect as much as 90% of soil-growing plants) is potentially a very powerful and uniquely pervasive mechanism for horizontal gene flow between almost all plant species (Davis et al., 2004).

296. As discussed in Syvanen and Kado (2002).

297. Ghatnekar et al. (2005).

298. The gene transferred from Poa to sheep's fescue encoded the enzyme phosphoglucose isomerase; in this case an entire functional gene was transferred between these divergent plant lineages (Ghatnekar et al., 2006).

299. The gene transferred from rice to Setaria spp encoded a Mu-like transposable element, or MULE (Diao et al., 2006).

300. Won and Renner (2003).

301. Richardson and Palmer (2007) report numerous instances of horizontal gene transfer between plants, almost all of which involve mitochondrial genes, with parasitic plants frequently acting as either donors or recipients of such exogenous genes. In one unusual case, the New Caledonian shrubby angiosperm, *Amborella trichopoda*, has acquired up to several hundred foreign genes from a wide range of species including mosses and other angiosperms (Bergthorsson *et al.*, 2004).

302. This report was from Habetha and Bosch (2005), who discovered that the nuclear genome of *Hydra viridis* contained a plant ascorbate peroxidase gene. This hydra is a member of the phylum Cnidaria, which includes the hydras, jellies (or jellyfish), sea anemones, and corals. The gene probably originated from the alga, *Chlorella vulgaris*, which can associate symbiotically with animals such as hydra. The plant ascorbate peroxidase is fully functional in the modern hydra where it is expressed specifically during oogenesis. Lateral gene transfer is also commonly found between prokaryotes and single-celled eukaryotes (Andersson *et al.*, 2003).

303. The size of mitochondrial genome varies from 200 kb in *Brassica rapa* to 2500 kb in *Zea mays*, while plastid genomes are much less variable, mostly in the range 130–150 kb, although there are a few instances of genomes as small as 35 kb and as large as 217 kb.

304. This process is known as endosymbiosis. There were several endosymbiotic processes, involving different types of bacteria, that gave rise to the red, brown, and green algae. Green algae were formed by the entry into a heterotrophic eukaryotic cell of a cyanobacterium. The cyanobacteria are one of the most ancient forms of life, dating back at least 3.5 billion years and responsible for the creation of our current oxygen-rich atmosphere—a task which took them over a billion years. It was the green algae that went on to evolve into the land plants that dominate today's vegetation.

305. This process of endosymbiosis has recently been seen in action with a vestigial green alga called *Nephroselmis* being engulfed by a flagellate protist called *Hatena* (Okamoto and Inouye, 2005).

306. An interesting twist to the story of algal evolution has come with the recent discovery that the water moulds, or oomycetes (including the organism responsible for potato blight and many other crop diseases), are probably descended from algae that lost their ability to photosynthesize, and instead became exclusively heterotrophic (Randall *et al.*, 2005).

307. Matsuo *et al.* (2005).

308. We know that most of the original plastid genes were successfully transferred to the nucleus because the nuclear genome contains genes encoding many proteins involved in photosynthesis, which could only have come from the plastid genome. For some reason that we are still unsure of, the successful transfer of genes from plastids to nucleus was then selected against. The actual transfer of genes still occurs but, in the words of a recent study 'once plastid DNAs are integrated into the nuclear genome, they are rapidly fragmented and vigorously shuffled, and surprisingly, 80% of them are eliminated from the nuclear genome within a million years.' (Matsuo *et al.*, 2005).

309. As happens so often in biology, we now know that there are exceptions to the 'rule' that many plastid genes, especially those encoding some photosynthetically related proteins, always remain within the organelle. The exceptions seem to be the plastids of dinoflagellate algae, most of which have lost almost all of their genes apart from a few single-gene DNA minicircles (Hackett *et al.*, 2004).

310. For some idea of the complexities of the species problem, especially for geneticists, see the paper by Hay (1997).

311. A useful chapter on species and their formation can be found in Futuyma (1998).

312. Lions (*Panthera leo*) and tigers (*Panthera tigris*) have a high similarity at the DNA level and can mate with each other, although the offspring are sterile. However, the two species of big cat are significantly different in both appearance and social behaviour. Whereas lions live in large matriarchal prides, tigers are normally solitary predators. Their divergent behaviour patterns and distinct geographical ranges mean that tigers and lions rarely encounter one another in the wild.

313. The closest living relatives of humans are the two chimpanzee species, *Pan troglodytes* (common chimpanzee) and *Pan paniscus* (pygmy chimpanzee). Although these apes share many morphological features and 95% of their genes with humans, they are obviously in a quite different category and constitute separate species (Britten, 2002).

314. Ants as a group date back over 90 million years (Grimaldi and Agosti, 2000).

315. A useful discussion of the species concept, and especially its relevance to plants and agriculture, can be found in Chapter 4 of the book, *Mendel in the Kitchen: Scientist's View of Genetically Modified Food*, by Fedoroff and Brown (2004). See also the review by Hancock (2005).

316. Bessey (1908).

317. For example, as noted in Box 5.2, thanks to DNA sequence analysis, large numbers of new and hitherto unrecognized (so-called 'cryptic') species are now being identified, as reviewed by Bickford *et al.* (2007).

318. According to Thomas Aquinas, the thirteenth century Italian scholastic philosopher, everything on Earth is but a reflection (albeit a poor reflection) of what exists in Heaven. Each species represents an idea in the mind of God. Such concepts as 'intrinsic value' and 'species integrity' have been elaborated with respect to plants and crop breeding by Heap and Wirz (2002).

319. Quoted in Fedoroff and Brown (2004), p 62.

320. The brief article by Gepts and Papa (2002) is a useful introduction to domestication.

321. Harlan (1975); Harlan (1992).

322. For example Hammer (1984). Some recent examples include Xiong *et al.* (1999), Gepts (2004), and Hancock (2004).

323. For example, in the cereals, independent mutations in a small number of corresponding genes in each crop species have been selected for during domestication in a process that is closely related to convergent evolution by natural selection (Paterson *et al.*, 1995a).

324. Poncet *et al.*, (2004).

325. Ladizinsky (1998).

326. Doebley *et al.* (1990).

327. The *tga* gene has recently been characterized as a putative transcription factor that may in turn regulate the expression of several other genes (Wang *et al.*, 2005).

328. The most common class of such 'master genes' encodes transcription factor proteins (Doebley *et al.*, 1997; Frary *et al.*, 2000; Liu *et al.*, 2002; Wang *et al.*, 2005; Li *et al.*, 2006).

329. Koinange *et al.* (1996).

330. An account of the identification of domestication-related genes that regulate complex traits (so-called quantitative trait loci, or QTL) can be found on pp. 163–166 of Hancock (2004) and reviews by Paterson (2002) and Poncet *et al.* (2004).

331. Spooner *et al.* (2005).

332. Badr *et al.* (2005). However, as discussed in Chapter 11, it is also possible that barley was independently domesticated a second time at a more easterly location such as Mehrgarh in modern Pakistan, as suggested by recent genetic evidence (Morrell and Clegg, 2007).

333. Özkan *et al.* (2005).

334. Heun *et al.* (1998).

335. Olsen and Schall (2001); Olsen and Schall (2006).

336. Matsuoka *et al.* (2002).

337. Erickson *et al.* (2005).

338. Decker (1988).

339. Wendel (1995).

340. Yabuno (1962)

341. Sonnate *et al.* (1994).

6. The domestication of cereal crops

342. Buckler *et al.* (2001) The reduction in genetic diversity due to 'bottlenecks' during crop domestication is discussed at greater length in Gepts (2004, pp. 29–31).

343. Hey (2005).

344. Vilà *et al.* (2005).

345. For a simple introduction to the use of wild relatives for the improvement of rice crops, see Barclay (2004, available online).

346. For a recent wide-ranging review see Motley *et al.* (2006).

347. We can define the broader wheat group as members of the tribe, Triticeae. The front page of the Wheat Genetics Resource Center website (Kansas State University) notes pithily: 'As is the case with *Triticum* and *Aegilops*, the taxonomy of other genera in the Tribe Triticeae is complex and controversial.' (http://www.k-state.edu/wgrc/Taxonomy/triticeaetax.html). For more information on taxonomy and Triticeae research, see the website of the International Triticeae Consortium (based at Utah State) at: http://herbarium.usu.edu/Triticeae/default.htm.

348. More than 40 species of *Triticum* and *Aegilops* are listed in the International Triticeae Consortium website entry entitled *Genomes in Aegilops, Amblyopyrum, and Triticum* (http://herbarium.usu.edu/Triticeae/genomesaegilops.htm). For a recent account of the evolution of the wheat group, see Mac Key (2005).

349. This feat was first achieved by McFadden and Sears (1946), as also described by Anderson (1967).

350. In case you were wondering, there are members of the wheat family with a CC genome, including *Aegilops cylindrica*, but they do not contribute to any of the crop genomes.

351. It has recently been proposed that individuals in the hominid line may have occasionally interbred with members of the chimpanzee line after their divergence over five million years ago (Patterson *et al.*, 2006). Although several hyped up versions of these claims appeared in the popular media, considerable caution should be exercised in considering this

small weedy cress species will have direct application for improvements in the breeding of brassica crops in the future.

505. Murphy (1998).

506. U (1935).

507. Seyis *et al.* (2003).

508. The evidence that brassicas are ancient hexaploids has been challenged by other data that suggests that there is only good evidence for one genome duplication, which would make them tetraploids (Quiros and Paterson, 2004; Leukens *et al.*, 2004). However, the same authors also acknowledge that there may have been as many as three earlier genome duplication events prior to the divergence of the *Arabidopsis* and *Brassica* lineages. Moreover, recent analyses of 34 species of the Brassicaceae led to the conclusion that there had been two ancient ploidy events, leading to a tetraploid and then a hexaploid genome in the supposedly diploid Brassicaceae of today (Johnston *et al.*, 2005). Whatever version is eventually verified, the fact remains that the present day brassica crops are the end result of an extraordinary series of genome duplications, the most recent of which may have occurred only a few thousand years ago.

509. Blanc *et al.* (2000).

510. One could be pedantic here and stick to the usual convention of assigning a name based on the Latin version of the number of genomes, but I thought that tetravigintiploid might not trip off the tongue quite as freely as 24-ploid.

511. This age estimate depends on whether the genome duplication involved autotetraploidy, in which case the best estimate is 38 million years ago, or allotetraploidy, in which case the duplication may be more recent (Ermolaeva *et al.*, 2003).

512. A great deal of evidence for rapid genome evolution in newly formed polyploid plants had come from studies of resynthesized *Brassica* allopolyploids (Song *et al.*, 1995; Schranz and Osborn, 2000; Osborn *et al.*, 2003; Osborn, 2004).

513. Rana *et al.* (2004).

514. Precise values for *Brassica* genome sizes vary according to the technique used; smaller values are generally provided by flow cytometry while the Feulgen method gives values that can be 20–40% higher. Flow cytometry data from several labs gives the following average values for *Brassica* genome sizes: *B. rapa* 505 Mb; *B. oleracea* 623 Mb; *B. nigra* 468 Mb; and *B. napus* 1160 Mb (Arumuganathan and Earl, 1991). For data from the Feulgen method, see Bennett and Smith (1976). For more information on *Brassica* genomes, see Schranz *et al.* (2006), and the

Multinational Brassica Genome Project (http://www.brassica.info/).

515. The contribution of polyploidy to creating some of the agronomically useful variants of brassica crops is explored in more detail by Osborn (2004); while some of the technical details of the genetic changes and genomic rearrangements are outlined by Rana *et al.* (2004).

516. Bradburne and Mithen (2000).

517. One of the glucosinolates is even called 'progoitrin' in recognition of its goitrogenic activity. Progoitrin, and other glucosinolates, can reduce the palatability of the meal produced from oilseed rape and hence restrict its use as an animal feed. Thanks to the efforts of breeders, many of the antifeeding glucosinolates have been reduced or removed from modern varieties of oilseed rape so that the crop can be used both as a source of a healthy, edible oil and a nutritious feed for farm stock (Bones and Rossiter, 1995).

518. Several US Presidents, including Bill Clinton and George Bush I, have been famously averse to brassica vegetables such as broccoli. In view of the formidable nutritional benefits of vegetables like broccoli, one can only say that such aversions are their loss.

519. See Mithen (2001) for a general review of this topic and Mithen *et al.* (2003) for a more technical account of progress in breeding anticarcinogenic varieties of broccoli. Although definitive proof has yet to be published, the National Cancer Institute in the USA currently advises that broccoli, along with its cruciferous family members, may be important in the prevention of some types of cancer (http://www.nci.nih.gov/).

520. Thompson (1979).

521. Helm (1963).

522. Pod shattering and other seed-loss traits are part of the 'domestication syndrome' and are usually among the first traits to be selected against, as a plant is cultivated. Oilseed rape is unusual in the extent of its pod shattering, despite several centuries of rigorous selection. In a bad year, as much as 20–30% of the seed can be lost in this way. Farmers are forced to spray chemical desiccants to prevent pod shattering, despite the cost and loss of yield entailed by driving tractors through crops that are fully grown and almost ready for harvest.

523. Morgan *et al.* (2003).

524. Hancock (2004). Radish was grown in Pharaonic Egypt both for its sharp-tasting bulbs and its oil-rich seeds. According to Pliny: 'In Egypt the radish is held in very high esteem, on account of the abundance of oil that is extracted from the seed.

Indeed, the people of that country sow this plant in preference to any other, whenever they can get the opportunity, the profits derived from it being larger than those obtained from the cultivation of corn, and the imposts levied upon it considerably less: there is no grain known that yields a larger quantity of oil.' (Pliny, 23–79 CE).

8. People and the emergence of crops

525. This quotation reminds us of the amount of forward planning, sometimes requiring several years, before the crop bore fruit, which would have required a radical conceptual shift for human groups who previously had much shorter planning horizons. The development or nurturing of this mental ability in early agrarian societies would have facilitated the conception and undertaking of other long-term logistical undertakings, such as the construction of permanent habitations, irrigation systems, and cultural-religious edifices such as tombs and temples. Similar sentiments regarding the future vision of the agrarian are expressed in Cicero's aphorism: '*Arbores serit diligens agricola, quarum adspiciet baccam ipse numquam.*' [The diligent farmer plants trees, of which he himself will never see the fruit] (Cicero MT, *Tusculanarum Disputationum* I, 14)

526. For an interesting perspective of neurological aspects of the Neolithic human transitions, see Lewis-Williams and Pearce (2005).

527. Examples include erstwhile starch-rich staples such as club-rush (*Scirpus maritimus/ tuberosus*), Euphrates knotgrass (*Polygonum corrigioloides*), and feather grass (*Stipa* spp.) (De Moulins, 2000).

528. While not wishing to sound overly alarmist, it is the case that the survival of many of the temperate wild cereals is under increasing threat, principally due to habitat destruction (Valkoun *et al.*, 1998). Some of these cereals have already been, and may well continue to be, invaluable sources of useful genetic material for future breeding of the major cereal crops. Efforts are now underway in places such as Turkey, Armenia, and Syria to preserve wild stands of cereals *in situ*, in addition to the less satisfactory but necessary *ex situ* conservation of the cereals in gardens and gene banks. See articles by Damania (1998) and Waines (1998).

529. This region of early crop domestication may have extended from the Jordan Valley in the southwest to eastern Anatolia and the Upper Tigris/Euphrates Valleys of modern Turkey and Syria in the northeast (Lev-Yadun *et al.*, 2000).

530. van Zeist and Bakker-Heeres (1982, 1985).

531. Hillman (1984).

532. Harris (1996) estimates that the average rate of spread of cereal farming from the Levant to Anatolia, and thence through the Balkans and beyond, was about 1 km per year, although this would have been an intermittent process with many pauses or even localized reversals.

533. Zohary (1986); Zohary and Hopf (2000).

534. Ho (1969).

535. This has been a controversial topic over the last few decades. From the 1970s, the so-called 'wave of advance' model, involving agricultural diffusion via mass migration, was widely, but not universally, accepted. Since 2000, more detailed studies of several DNA markers suggest that very few people migrated from the Near East during the expansion of agriculture. The topic is discussed in more detail by Wells (2003) and some of the original scientific evidence can be found in Semino *et al.* (2000).

536. See Figure 6.7 in Hancock (2004).

537. Hancock (2004).

538. Normile (1997); Zhao (1998).

539. Higham (1984); Maloney *et al.* (1989).

540. *Indica* and *japonica* are the main races of cultivated rice, but there is also a third race called *javanica*. Present-day rice cultivars are also categorized as long-, medium-, or short-grained. In addition, there are *upland* and *paddy* rice cultivars; the former being more typical of hilly locations and cooler regions such as Japan and Korea, while the latter are grown in tropical lowland regions.

541. This refers to the debate over whether *indica* and *japonica* are monophyletic (single origin) or diphyletic (two origins). The evidence tends to be more supportive of the diphyletic hypothesis, with a diffuse series of multiple origins of cultivation, both in space and in time, as discussed by Morishima (2001).

542. Vitte *et al.* (2004).

543. Further evidence for the dual origin of cultivated Asian rice comes from molecular marker analysis (using the retrotransposon, p-SINE1-r2) suggesting that the *indica* line of rice varieties evolved from annual wild rice with an AA genome, while the *japonica* varieties evolved from a separate series of perennial forms of wild rice, that also had an AA genome (Yamanaka *et al.*, 2003).

544. The earliest carbon-dated samples of domesticated rice date back to *9000–8000 BP* (Pringle, 1998a). However, in 2003, a team of archaeologists reported (albeit in a press release) the discovery, in Korea, of a handful of burnt rice grains with an estimated age of

15,000 BP. This challenges the widely accepted view that rice cultivation originated in China or India no earlier than about *12,000–10,000 BP*. However, DNA analysis showed that the rice was genetically different from the modern food crop so the significance of this as-yet unconfirmed finding remains to be established. See http://news.bbc.co.uk/1/hi/sci/tech/3207552.stm.

545. In much the same way, some Amerindian groups collected a distantly related edible species of American wild rice, called *Zizania aquatica*, until relatively recently.

546. Various (2002).

547. See Oka and Morishima (1971, 1982). Of course, this ready propensity of wild rice to evolve into domesticated-like forms is mirrored both by the temperate cereals of the Near East and by maize in Mesoamerica.

548. Morishima (2001).

549. This phenomenon was reported by the Chinese groups of Xiong *et al.* (1999) and Ji *et al.* (2006). Similar clusters of major genes regulating domestication-related traits have been reported in the case of the common bean, *Phaseolus vulgaris*, by Koinange *et al.* (1996).

550. Toyota and Honda are relatively common Japanese surnames. Therefore the eponymous car manufacturers are actually named after their founders, not because they have any present-day connection with rice agriculture.

551. In classical Chinese, the same term refers to both 'rice' and 'agriculture'. In many official languages (e.g. Lao and Thai) the verb 'to eat' means 'to eat rice'. Indeed, the words 'rice' and 'food' are sometimes one and the same in eastern semantics. For more information see the UNCTAD (United Nations Conference on Trade and Development) website at: http://r0.unctad.org/infocomm/anglais/rice/characteristics.htm .

552. Huke and Huke (1990). *Rice: Then and Now*, International Rice Research Institute, Los Baños, Philippines. An abridged version of the Chinese rice myth can be obtained online at: http://beaumont.tamu.edu/eLibrary/Newsletter/2001_May_Newsletter.pdf.

553. deWet (1995); Bettinger *et al.* (2006).

554. Bettinger *et al.* (2006).

555. Yan (1999) and Cohen (1998).

556. Werner (1998).

557. Gasohol, derived from maize and sugar cane, is used on a particularly wide scale in Brazil. The plant material is fermented to ethanol, which is then added to gasoline to form a 10% alcohol plus 90% gasoline mixture.

558. For example, *Zea diploperennis*, *Zea perennis*, and *Zea luxurians*.

559. Jaenicke-Després *et al.* (2003).

560. Eyre-Walker *et al.* (1998).

561. See Matsuoka *et al.* (2002) for original data, and Fedoroff (2003) or Matsuoka (2005) for wider discussion.

562. Hillman and Davies (1990a, 1999).

563. Beadle (1978).

564. Flannery (1973a, 1973b).

565. An example of one of the new methods is the analysis of phytoliths. As implied by their name, phytoliths are stony structures that are formed when plant cells take up silica from ground water until the hard silica assumes the shape of the plant cell. When the organic matter in the cell is broken down, the silaceous microfossil remains. Investigators including Dolores Piperno in Panama (Piperno and Pearsall, 1998; Piperno, 2001) and Deborah Pearsall in the USA (Pearsall, 1995; Buckler *et al.*, 1998) have pioneered the use of phytoliths to identify crops and their wild precursors in Mesoamerican strata, particularly from humid regions where plant preservation is normally very poor. However, the phytolith technique is still controversial in some quarters because of problems in the accurate dating of the phytoliths themselves. See Pringle (1998a) and Harvey and Fuller (2005) for more discussion of the potential value of phytolith data.

566. These three Mexican states are still rich centres of maize diversity. To some extent, it was the discovery of GM maize in this particularly sensitive region of Mexico in 2001 that fuelled so much of the outrage that led to the bitter controversy, involving Ignacio Chapela (Quist and Chapela, 2001; Mann, 2002). Another nearby region that has been suggested as an early site of maize cultivation is the Tehuacán Valley, which is immediately north of Oaxaca. This is where the thus-far oldest fossilised samples of domesticated maize have been found, as detailed by Piperno and Flannery (2001).

567. Benz (2001); Piperno and Flannery (2001).

568. Matsuoka *et al.* (2002). Note that this estimated date of *9200 BP* should be regarded as a maximum value. In the same way, the oldest maize fossils at about *7000 BP* give us a minimum value, so the date range extends from about *9000–7000 BP*. As some of these authors note in a subsequent paper (Vigouroux *et al.*, 2003), the use of certain types of DNA markers called microsatellites must be hedged with caution in the case of maize. Therefore these dates are provisional

but, taken in conjunction with the available fossil evidence, they do raise two important points: (1) the relative antiquity and single origin of maize domestication; and (2) its non-coincidence with the start of wheat or rice cultivation.

569. Note that this interpretation is the opposite of that suggested by data from phytoliths, which imply that it was the Oaxacan lowlands that were the initial domestication site (Piperno and Pearsall, 1998). I have opted for the 'highlands first' hypothesis here because it is supported by two independent and relatively uncontroversial lines of evidence.

570. Szabo and Burr (1996) and Jaenicke-Després *et al.* (2003).

571. Buckler *et al.* (1998).

572. Pearsall (1995) and Pope *et al.* (2001). Piperno and Pearsall (1998); Piperno *et al.* (2001). As discussed in Chapter 12, Dickau *et al.* (2007) report the cultivation of manioc and arrowroot, alongside maize crops, in Panama by *7400 BP*.

573. McClung de Tapia (1992).

574. Werner (1998).

575. Anthropologist, Mary Helms, has suggested that the shift from animal gods to the worship of ancestors and human-like gods may have accompanied the transition from hunter–gathering to farming and a more settled village life (Helms, 2003).

576. For a review of the changing views on Mayan maize gods, see Taube (1985).

577. For more about maize myths, see Taube (1985, 1993) and Coe (1999).

578. Matsuoka *et al.* (2002); Dickau *et al.* (2007).

579. This dispersal route is supported by the well-established morphological evidence of Galinat and Campbell RG (1967) as well as the more recent genetic data from Matsuoka *et al.* (2002).

580. Many of the Amerindians of North America took up maize cultivation, but this was largely as a supplement to a vigorous hunter–gathering lifestyle. Many villages in the northeast had maize gardens that were harvested in the autumn, while the rest of the year was spent in more mobile foraging for animal and plant foodstuffs. For a description of such cultures in the Hudson Valley, see Shorto (2005).

581. Huckell (1996).

582. Smith (1989) and Mann (2006). By the 17th century, maize was being grown in many places in the north and east of North America. Hence, when the 'Pilgrim Fathers' landed in 1620 in what is now the state of Massachusetts, and their European crops failed, they were saved from starvation by the local Iroquois Indians who taught them how to grow maize.

583. Lindsay (1986). Many other cultures in northeast America temporarily adopted and then rejected maize farming, most notably the Moundbuilder peoples such as the Adena, Hopwell, and Mississipians (Fagan, 2001b; Mann, 2006; Benson *et al.* 2007)

584. Hancock (2004).

585. For more comprehensive accounts of crop evolution and domestication, see the reviews and monographs by Chapman (1992); Pearsall (1995); Smartt and Simmonds (1995); Zohary and Hopf (2001); and Hancock (2004).

586. Edible cucurbits include the bottle gourd, *Lagenaria siceraria*; melon, *Cucumis melo*; watermelon, *Citrullus lanatus*; loofah, *Luffa* spp.; cucumber, *Cucumis sativus*; plus many species of squash itself such as pumpkin, *Cucurbita pepo* and winter squash, *C. maxima* (Bisognin, 2002).

587. Smith (1997).

588. Smith (2005).

589. This finding was made at two archaeological sites in southwest Ecuador although the plant remains were phytoliths, rather than conventional fossils. See Piperno and Stothert (2003).

590. Sanjur *et al.* (2002).

591. Bisognin (2002).

592. Nee (1990).

593. Smith (2006b).

594. See Spooner *et al.* (2005) and Spooner and Hetterscheid (2006).

595. Patel *et al.* (2002).

596. The bitter taste of alkaloids is a product of evolution that arose independently in humans and chimpanzees (Wooding *et al.*, 2006). Alkaloids are one of the major classes of life-threatening toxin occurring in plants that are a potential part of the human diet. Via our taste buds, we became highly sensitive to the presence of alkaloids in anything that we eat. Just to reinforce their toxicity, their taste is so unpleasant that it often triggers a reflex-like spitting out of the offending food. And even if they are inadvertently swallowed, some alkaloids will trigger a regurgitation reflex to ensure their prompt removal from the body.

597. Beier (1990).

598. Remains of potato tubers from at least *13,000 BP* have been fond in the Casma Valley, about 300 km north of Lima, in Peru, as reported by Ugent *et al.* (1982).

599. Harlan (1992).

600. Potatoes also seem to have attracted a great deal of prejudice both from traditional grain farmers and from consumers more used to cereal-based breads as their dietary staples (Chapman, 2000).

690. In recent years, much has been learned about the evolution of agriculture in several other regions, including Arabia, Oceania, Madagascar, and New Guinea, as reviewed in Kennett and Winterhalder (2006).

691. Despite dating to 1990, the splendid atlas by Michael Roaf is still an invaluable aid to understanding the geographical context of the ancient Near East (Roaf, 1990). A more recent, useful reference is the atlas by Hunt (2004).

692. Andie Byrnes of University College, London, has compiled a useful website on the development of agriculture in the Near East at: http://www.near-east.historians.co.uk/index.htmll.

693. van Loon (1988a).

694. Kirkbride (1982).

695. Gibson (2000).

696. Nissen (1988); Algaze (1989); Fagan (2001b, 2004).

697. For an early, but groundbreaking account of the rise of Levantine Neolithic cultures, see Moore (1978), while the detailed account of Akkermans and Schwartz (2003) is a classical study of Neolithic Syria, and that of Yoffee (1995) does the same for Mesopotamia. More recent, general accounts of the evolution Near Eastern farming and civilizations can be found in Fagan (2001b, 2004) and Yoffee (2005).

698. Perry and Hsu (2000). For a discussion of the interactions between climate change and technological/economic change, see Dow et al. (2006).

699. An interesting perspective on the effect of environmental change on societal development is in Coombes and Barber (2005).

700. Brooks (2006).

701. This topic has been examined at length by Issar and Zohar (2004).

702. Morrison (2006).

703. The extent and magnitude of the *8200 BP* climatic event have yet to be resolved conclusively, as discussed by Alley and Agustsdottir (2005); Rohling and Palike (2005); and Thomas et al. (2007). While the *8200 BP* event may have been comparable in its temperature effects to the Younger Dryas, it may have been less widespread in its effects (Thomas et al., 2007).

704. Alley et al. (1997).

705. Barber et al. (1999); Baldini et al. (2002); Teller and Leverington (2004).

706. Ellison et al (2006); Kerr (2006).

707. Rossignol-Strick (1999); Weiss (2000).

708. Kozlowski (1994).

709. Goring-Morris and Belfer-Cohen (1997).

710. Magny and Haas (2004); Thompson et al. (2006).

711. Weiss (2000); Thompson et al. (2006).

712. For an account of the interactions between environmental instabilities and urban origins, see the chapter by Hole (1994).

713. Adams (1981).

714. Weiss (1997).

715. Johnson (1988); Weiss (2003); Wetterstrom (2003).

716. Calderoni et al. (1994); Dyson (1987).

717. Weiss (2003).

718. Weiss (2000).

719. Glasson (2003); Brooks (2004).

720. There was an abrupt aridification event at the Great Salt Lake and at Elk Lake at *c. 4200 BP* (McKenzie and Eberli, 1987); and in the midcontinental region of North America (Booth et al., 2005); aridification and dust storms in Belize (Alcala-Herrera et al., 1994); and a huge dust spike in the Huascarán glacier core in Peru (Thompson, 2000), plus a sudden reduction in the level of Lake Titicaca (Cross et al., 2000).

721. Weiss (2000, p. 83).

722. Mayewski (1993); DeMenocal et al. (2000).

723. Lemcke and Sturm (1997).

724. Bar-Matthews et al. (1997); Bottema (1997).

725. The social effects of the *4200 BP* event in the Near East have been reviewed by Cullen et al. (2000), Weiss (2000, pp. 84–92), and DeMenocal (2001).

726. Weiss (1983).

727. Adams (1981).

728. It took about 6000 years from the domestication of the first crop at about *11,000 BP* until the emergence in several locations of such trappings of civilization as writing, complex cities, and empires. We are still living in this second 'urban' phase of agriculture-based society, which has so far lasted about 5000 years.

729. Molleson and Jones (1991).

730. Molleson et al. (1993).

731. Moore et al. (2000, Chapter 11). The late age of wearing at Abu Hureyra occurred despite the onset of lactose intolerance in older sucklings and was not an ideal situation for all concerned.

732. For more detailed accounts of these cultural phases and their material manifestations, and the ways in which views have changed over the years, the following literature is a useful start: Redman (1978); Nissen (1988); Bader (1993a, b, c); Merpert (1993); Pollock (1999, 2001). For a detailed review of the Neolithic period in the southern Levant, see Twiss (2007).

733. For an account of the Sumerians and their civilization, see Crawford (2004).

734. As stated by Mike Shupp of California State University, Northridge: 'It appears to me that larger Mesopotamian communities are not homogeneous

but contain people with different traditions who continued to maintain seperate (sic) social institutions until rather late in history. The mystery is when such combinations occurred, and the last possible date which suggests itself is about 3500 BC, when some sort of catastrophic realignment of population patterns began.' (see online article at: http://www.sumerian.org/Mesopotamian%20Archaeology.htm)

735. Tsuneki *et al.* (2000).

736. The latest discoveries in the remains of the mysterious and seemingly aberrant urban settlement of Çatalhöyük are described by Hodder (2006).

737. Haywood (2005).

738. Nissen (1988).

739. Haywood (2005).

740. Potts (1997); Yoffee (1997).

741. Adams (1966, 1970); Fernea (1970); Goldsmith and Hildyard (1984); Postgate (1992).

742. See also, Nissen (1988, pp/ 59–60).

743. For a more detailed account of the sociopolitical context of the Ubaid period and of Mesopotamia in general from 7000–4100 BP, see Susan Pollock's wide-ranging account: *Ancient Mesopotamia* (Pollock, 1999) and her more recent review chapter (Pollock, 2001).

744. The argument that early kings served desirable ends is explored by Elman Service (1962, 1975) and mirrors the concept of the king as manager/administrator, or at least as the chief executive of a group of managers in such societies, as discussed by the likes of Childe (1928, 1936), Wittfogel (1957), White (1959), and Johnson (1982). Others have argued more in terms of exploitation and coercion as being dominant themes (Fried, 1967; Diakonoff, 1991). As ever, the truth is probably a combination of all of these factors, and maybe others.

745. Liverani (1993).

746. For an authoritative account of the Uruk period, see Rothman (2001).

747. In medieval times, 'Iraq' was used as a geographical term for the area in the south and centre of the modern republic of Iraq.

748. As reviewed by Schmandt-Besserat (1992, 1997, pp. 166–170 and references therein).

749. Kohler-Rollefson (1988); Rollefson and Kohler-Rollefson (1993).

750. Englund (1991, 1993).

751. According to a review of the evidence by Glassner (2003, pp. 29–46), the first Mesopotamian written texts date from between 5400 and 5300 BP during the Uruk period.

752. Nissen (1986); Nissen *et al.* (1993); Hansen (1965). See also Walker (1987) for a useful brief introduction to cuneiform.

753. The Uruk colonization of the north is discussed in depth by Akkermans and Schwartz (2003, Chapter 6) and in the volume edited by Rothman (2001).

754. van Loon (1988b); Gibson (2000); Akkermans and Schwartz (2003).

755. The outlines of this controversy are discussed by Lawler (2006).

756. Weiss (2000).

757. Akkermans and Schwartz (2003, pp. 207–208).

758. Some archaeologists and pre-historians classify the Uruk period as lasting until about 5,100 BP, after which there is a Jamdet Nassr Period until about 4800 BP and an Early Dynastic Period until the rise of Akkad about 4300 BP. For the sake of simplicity, I have treated these latter divisions as extensions of the Later Uruk Period. For more detail on these chronologies, see Nissen (1988) and Hansen (1965).

759. Finkbeiner (1991).

760. Johnson (1973); Weiss (1977).

761. Leick (2001, p. xvi).

762. Adams (1972, Fig. 8) has shown how, between 5100 and 4600 BP, smaller settlements up to the size of modest towns were progressively abandoned as their populations moved to emerging dominant cities such as Uruk and Shuruppak. Hansen (1965) argues that the rural population probably moved into these cities due to a combination of voluntary (e.g. desire for greater security) and coercive (e.g. conscription, or corvée, for large state projects such as temple construction) influences. Such workers would have been dispatched to the fields during harvest period, possible living in temporary shelters or barracks during this time.

763. Adams and Nissen (1972).

764. Powell (1985; 1999, pp. 94–96).

765. The diet of the more fortunate was heavily based on barley, but included many other ingredients as shown in this passage from Andrew Dalby's interesting account of food in the ancient world: 'Mesopotamian food is known from archaeology and written records on cuneiform tablets, including bilingual Sumerian–Akkadian word lists. These sources indicate the importance of barley bread, of which many kinds are named, and barley and wheat cakes, and grain and legume soups; of onions, leeks and garlic; of vegetables including chate melon, and of fruits including apple, fig and grape; of honey and cheese; of several culinary herbs; and of butter and vegetable oil. Sumerians drank beer often, wine seldom if at all.' (Dalby, 2003).

766. From *The Electronic Text Corpus of Sumerian Literature*, available online at: http://www-etcsl.orient.ox.ac.uk/proverbs/t.6.2.3.html.

767. Jacobsen (1982).

768. Brewing of fermented beverages such as beer originated as a technique for ensuring that water was disease-free and fit to drink. Ancient Sumerian beer was made from a thick mash of dates and malted barley. It was often drunk through a straw (to filter out the solids) from a communal vat, rather like the hubble-bubble pipes still used in the same region.

769. Adams (1962).

770. Jacobsen and Adams (1958).

771. Walters (1970, p. 160).

772. *Atrahasis*, II, 4; 7/8, as quoted by Walters (1970, p. 160); see also Lambert and Millard (1969, p. 79).

773. Weiss (2003).

774. Jasny (1944).

775. Weiss (1986).

776. Charles (1984).

777. Powell (1985); Wetterstrom (2003).

778. Weiss (1983); Cooper (2006b).

779. These features of the resurgent Leilan settlement are described in the volume edited by Rova and Weiss (2003).

780. Ancient Ebla, nowadays known as Tell Mardikh, was a modest-sized settlement of several thousand people, covering some 60 ha, that is located about 60 km south of Aleppo. One remarkable find has been about 17,000 clay cuneiform tablets and fragments in the acropolis that demonstrate the importance and widening use of written texts throughout the Near East by *4500 BP* (Matthiae, 1981; Matthiae *et al.*, 1997; Akkermans and Schwartz, 2003).

781. These communities lie along the mid-Habur Valley in what is now the border region between northern Syria and Turkey. The interpretation of grain stores as evidence of state control of agriculture is widely, but by no means universally, accepted (see discussion in Akkermans and Schwartz, 2003, pp. 221–224).

782. Sargon's daughter, Enheduanna, is notable for being the first named author in history. As well as being a scribe, she was high priestess of the moon god Nanna at Ur. This also shows the high status and degree of education of some women in ancient Mesopotamia (Walker, 1987).

783. Weiss and Courty (1993).

784. Weiss *et al.* (2002).

785. Excavations at Leilan show evidence of the distribution of cereal rations from the palace, in accordance with Akkadian customs and unlike previous practices in northern Mesopotamia (Wetterstrom, 2003, p. 393).

786. Senior and Weiss (1992).

787. Charles and Bogaart (2001).

788. Weiss *et al.* (1993).

789. Part of this extract from *The Curse of Akkad* is quoted from Weiss (2000). For the full text, see Cooper (1983).

790. Cullen and deMenocal (2000); Cullen *et al.* (2000).

791. Weiss (1997); Ristvet and Weiss (2005).

792. For more information on the Akkadians and early Mesopotamian civilization, see Postgate (1992). For a more general discussion of climate–societal interactions, see Weiss (2000).

793. One such political entity was the so-called 'Kingdom of Urkesh and Nagar', which probably encompassed the former cities of Brak and Mozan (Oates and Oates, 1991; Weiss, 2000). The phenomenon is reminiscent of the plethora of self-proclaimed 'kingdoms', many of them of exceedingly modest proportions, that sprang up in post-Roman Celtic Britain.

794. Adams (1981).

795. Leick (2001).

796. Both men and women toiled as forced labourers in the Ur III period. While men worked on heavy manual projects such as building and canal maintenance, thousands of women were conscripted into giant textile concerns. Here, the omnipresent scribes carefully noted their rations as supplied by the state and the raw materials allocated to each 'factory', but tell us nothing about the doubtless grim conditions of their enforced employment (Neumann, 1993).

797. Powell (1985) has argued that fallowing and leaching, as described in documents of the period, may have enabled Sumerian farmers to keep salinization under control in some areas, but the overall contribution of salinization to the crises and subsequent collapse of Sumerian civilization after *4100 BP* remains the subject of vigorous debate (Hansen, 1965).

798. Goldsmith and Hildyard (1984).

799. Adams (1981); Leick (2001).

800. Gomi (1979, 1984).

801. We know this because the two sets of data were recorded by different scribes using different parts of the same tablet (Walker, 1987).

802. Campbell (1997, p. 240–214).

803. Larsen (1976, p. 37).

804. Gasche (1990).

805. Chiera (1934).

806. See a translation of the text of the letter from governor Ishbi-Erra to King Ibbi-Sin of Ur, in which the governor explains: 'I entered with 72,000 gur of grain—the entire amount of grain—inside Isin. Now I have let the Martu, all of them, penetrate inside the Land . . . Because of the Martu, I am unable to hand over this grain for threshing. They are stronger than me, while I am condemned to sitting around.' Available online from The Electronic Text Corpus of

Sumerian Literature at: http://www-etcsl.orient.ox.
ac.uk/section3/tr3117.htm.

807. Roux (1993, p. 168).
808. Diakonoff (1991).
809. Michalowski (1989).
810. Sollberger and Kupper (1971, pp. 205–206).
811. Wilkinson (1990); Akkermans and Schwartz (2003).
812. Matthiae (1997).
813. Akkermans and Schwartz (2003).

11. Evolution of agrourban cultures: II South and east Asia

814. Agriculture also developed independently in southern India, where the millet grasses, *Brachiaria ramosa* and *Setaria verticulata*, and the pulses, *Vigna radiata* and *Macrotyloma uniflorum*, were probably domesticated locally *c. 4800 BP*, after which other crops were imported from outside the region (Fuller *et al.*, 2004).

815. For a useful, if at times somewhat speculative, review of early agriculture on the Indian subcontinent, see Mehta (2002).

816. Jarrige *et al.* (1995).

817. Meadow (1993). It seems likely that the region around Mehrgarh was one of the two ancient centres of cattle domestication, producing the ancestor of Indian cattle (*Bos indicus*), while the more westerly domestication centred on the Euphrates basin produced the ancestor of *Bos taurus* breeds (Bradley and Magee, 2006).

818. Shaffer and Lichtenstein (1995, 1999).

819. Morrell and Clegg (2007).

820. For a useful survey of Indian civilizations, see Kenoyer (1998), Tharpar (2002), and Possehl (2003).

821. Kimber (2000).

822. A useful volume, available online, is *Sarasvati: Civilization* by Srinivasan Kalyanaraman (Kalyanaraman, 2003).

823. At its height in the fifth millennium BP, the Indus Valley civilization extended over 1.3 million square kilometres, while the contemporary Akkadian Empire covered less than 0.4 million square kilometres, and the Old Kingdom of Egypt occupied a mere 17,100 square kilometres (Kalyanaraman, 2003).

824. Access to stored grain was not just a useful hedge against the odd crop failure. Harappan elites traded their surplus grain with mountain dwellers from the north-west in exchange for valuable resources such as metals, precious stones, and timber. In turn, commodities such as unworked copper were exported as far afield as Sumer in exchange for other desirable

items, such as highly crafted copper, and later bronze, tools (the early Bronze Age in Mesopotamia started about *4500 BP*) (Haywood, 2005).

825. For example, there is no evidence of institutions such as formal monarchies, slavery, or compulsory agricultural labour that were so common throughout the Near East after the early Uruk period of about *6000 BP*.

826. It is possible that the lack of evidence of coercive Mesopotamian-style public institutions in the ancient Indus Valley civilization is due to the disappearance of physical remains, such as canals, etc., but most scholars take the more optimistic view that a complex state might have evolved at this time in the absence of many of the social pathologies seen in contemporaneous Near Eastern cultures.

827. Haywood (2005).

828. The extent of agricultural intensification and diversification in late-Harappan agriculture is still controversial, as discussed by Weber (1999) and Fuller (2000).

829. Staubwasser *et al.* (2003).

830. Staubwasser *et al.* (2003).

831. Masson (1968); Gupta (1982); Ghosh (1982).

832. The site of Ganweriwala was only discovered in the 1970s and, partially due to its sensitive location near the modern Indo-Pakistan border, it has yet to be excavated. With an area of about 80 ha, Ganweriwala is comparable in size to the largest of the other known Harappan urban centres, such as Mohenjo-Daro.

833. Haywood (2005); Thapar (2002).

834. The modern name *Mohenjo-Daro* is Sindhi for 'Mound of the Dead': like so much about the Indus Valley civilization, the original name of the city is still unknown.

835. Shaffer (1993).

836. Shelach (2000).

837. For a flavour of recent progress in the study of China during the Palaeolithic/Neolithic transition, see the special issue of the *Review of Archaeology*, edited and introduced by Kuzmin (2003).

838. For an overview of recent progress in the elucidation of the origins of rice agriculture in China, see the review by Crawford and Shen (1998), and the subsequent four papers in the special issue of the journal, *Antiquity*.

839. An *et al.* (2005).

840. According to the latest genetic and archaeological evidence, the exact timing and mechanism of human migration across eastern Eurasia is both complex and still rather controversial. A good introduction can be found in Chapters 4 and 5 of Stephen Oppenheimer's excellent book, *Out of Eden*

(Oppenheimer, 2003). Although genetic data are consistent with the arrival of modern humans in the Chinese region of northeast Eurasia at about *50,000 BP* (Su *et al.*, 1999; Ke *et al.*, 2001), these data have yet to be reconciled with some archaeological evidence consistent with an initial earlier occupancy (Brantingham *et al.*, 2001; Gao and Norton, 2002) followed by a gap in the Ice Age (Bettinger *et al.*, 2006).

841. The people of the Shuidonggou period used a flat-faced core technology, similar to the Levallois technique, to produce stone blades (Brantingham *et al.*, 2001).

842. Bettinger *et al.* (2006).

843. See map (Fig. 6.3) on p. 254 of Oppenheimer (2003).

844. The severe reduction in Central Asian human populations in the last Ice Age is discussed by Davis (1990), and their putative 'bounce back' to repopulate much eastern Eurasia with Mongoloid peoples is covered in Chapter 5 of Oppenheimer (2003).

845. Bettinger *et al.* (2006); Barton *et al.* (2006).

846. Underhill (1997).

847. Madsen *et al.* (2007); Madsen and Elston (2007).

848. Methods of wild millet collection and processing are discussed by Lu (1998).

849. Madsen *et al.* (1996).

850. Bettinger *et al.* (2006).

851. Evidence of early millet cultivation has also been found in the Lower Yellow River Valley where the Peiligang culture may have been growing the crop by *8500–7500 BP* (Higham, 1995, p. 134; Lu, 1999).

852. This view is endorsed by Higham (1995) and Dow *et al.* (2005); see also Bettinger *et al.* (2006).

853. Bettinger *et al.* (2004).

854. Barton (2004).

855. There is a more controversial claim of a possible precursor of writing in China that dates back as far as *8600 BP* (Li *et al.*, 2003). This relates to a series of signs carved into tortoise shells found in early Neolithic graves at Jiahu in Henan Province. It is possible that some of these signs might anticipate later Chinese characters and could have been intended as words, although the authors of the study believe it is more likely that the signs were simply symbols connected with ritual practices. However, as with the situation in Mesopotamia where marked tokens gradually developed into a written script between *9000 BP* and *5400 BP* (Pollock, 1999, 2001), these early Chinese signs may well presage a long period of sign use, which led eventually to a writing system.

856. Lu (1999).

857. Shelach argues for the autonomous development of agriculture and societal complexity in northeast China, and this is supported by a lack of evidence for large-scale migration into the region (Shelach, 2000). But this does not rule out the import of domesticated crops from other regions, followed by their independent development in the local area.

858. Yan (1992).

859. Loewe and Shaughnessy (1999).

860. Madsen *et al.* (2003); Madsen and Elston (2007).

861. An *et al.* (2005, 2006).

862. Wu and Liu (2001); Zhao (1997).

863. The earliest substantial state in northern or central China was probably based on the urban centre of Erlitou in Henan province, where remains of palatial buildings and royal tombs dating from *3900–3500 BP* may be correlates of the semilegendary Xia dynasty that supposedly ruled from *4070–3600 BP* (Liu and Chen, 2003). The Erlitou state appears to have declined from *3600 BP* and collapsed about *3500 BP*, to be succeeded by a more extensive polity based on the nearby city of Zhengzhou (also in Henan province), which held sway from *3600–3400 BP* and was the beginning of the Shang dynasty (Yoffee, 2006).

864. In 1912, the last emperor of the Manchu Qing dynasty, Aixinjueluo Puy, was forced to abdicate when the Republic of China was declared in Nanjing.

865. People who carry latent forms of the *M. tuberculosis* pathogen are normally asymptomatic and can go their entire life without developing TB.

866. Mokrousov *et al.* (2006).

867. Su *et al.* (2000); Deng *et al.* (2004).

868. Lu *et al.* (2002).

869. Pringle (1998a).

870. Gupta (2004).

871. For a recent discussion on the use of phytoliths as quantitative indicators in the reconstruction of palaeoclimates in China, see Lu *et al.* (2007).

872. Hillman *et al.* (1989); Hillman *et al.* (2001); Moore *et al.* (2000).

873. Higham and Lu (1998); MacNeish and Libby (1995); Kuzmin (2006).

874. MacNeish and Libby (1995); Yuan (1996).

875. Higham (1995); Lu (1999); Pei (1998).

876. Pei (1998).

877. Higham and Lu (1998).

878. Xueqin *et al.* (2003).

879. For primary sources on modern studies of early Chinese rice cultivation, see Crawford and Shen (1992); Higham (1995); Higham and Lu (1998); Pei (1998); Zhang and Wang (1998); and Zhao (1998).

880. See Zhao (1998, pp. 894–895).

881. Bray (1984).

12. Evolution of agrourban cultures: III Africa, Europe, and the Americas

882. Sahara is the English version of an Arabic word that means desert. The redundant term 'Sahara Desert' is one of numerous examples of toponymic tautology that include River Avon (Avon is Welsh for river) and East Timor (Timor is Bhasa Indonesian for east). One of the best examples of this genre is Torpenhow Hill in England in which *tor*, *pen*, and *how* each mean hill in various Anglo-Celtic languages, giving the marvellous tautology: Hill-hill-hill Hill.

883. Gasse (2000).

884. Scarre (1989).

885. Sweet potatoes, *Ipomoea batatas*, are sometimes incorrectly referred to as yams but are actually unrelated plants of Mesoamerican origin.

886. Ehret (2002).

887. Ehret (2002).

888. Wasylikowa *et al.* (1993); Wendorf and Schild (1998).

889. Haywood (2005).

890. Scarre (1989).

891. Schneider (1967); Pias (1970).

892. With the continuing aridification of the Sahara, coupled with human activities (including farming) that deplete ground water in the region, the area of Lake Chad is still declining from over 25,000 square kilometres in 1963 to a mere 1350 square kilometres in 2006.

893. Renssen *et al.* (2006).

894. A similar kind of 'tipping point' between two metastable states is probably responsible for the abrupt transitions of the high altitude thermohaline circulation, as best exemplified by the periodic diversion of the Gulf Stream away from northwest Europe, with consequent rapid cooling effects on the adjacent parts of the continental landmass (Stocker *et al.*, 1992; Rahmstdorf, 1995).

895. See McIntosh and McIntosh (1983), for a discussion of the human societies involved in the post-Holocene collapse in Africa and deMenocal *et al.* (2000) for the climatological evidence.

896. Genetic evidence suggests that the donkey was domesticated between *7000* and *5000* BP (Marshall, 2000), possibly as one of the responses of pastoralists and other societies in northeast Africa to the desertification of the Sahara (Beja-Pereira *et al.*, 2004).

897. Kuper and Kröpelin (2006).

898. Kuper and Kröpelin (2006).

899. Obviously there had been informal contacts between Near Eastern and Nile cultures well before this time, including the introduction of most of the crops grown in the Nile Valley. But the adoption of similar pottery styles is suggestive of a more formal system of contact between the two regions.

900. Ancient Egyptian breweries and other urban industrial installations are discussed by Friedman (1997) and Haywood (2005). Beer making in ancient Egypt was probably mainly based on malted barley mixed with a small amount of emmer wheat, but without any flavouring agents such as the hops used in modern beer (Samuel, 1996). The wild yeasts used for fermentation contained many additional microbial impurities, which must have made for uncertain results in the final brew. However, by the New Kingdom at *3500* BP, much purer yeast cultures were being used and the Egyptians were probably the first real biotechnologists who were able to culture more efficient 'domesticated' genetic strains of yeast.

901. It has been suggested that the *5200* BP event may have indirectly stimulated Egyptian social differentiation and cultural complexity due to the arrival of migrants from the Nubian Desert who were forced by the failure of the summer rains to settle in the Nile Valley, hence increasing the tendency towards agricultural intensification and urbanization (Malville *et al.*, 1998; Brooks, 2004).

902. Aldred (1961). Although the *Admonitions* was originally believed to refer to the disasters of the First Intermediate Period after *4000* BP, the work is now more commonly held to be a more generalized threnodic commentary on wider theme of catastrophe. It is interesting in the present context in the strong coupling of agricultural failure with social dissolution and the loss of power by the elite. See van Seters (1964) and Lichtheim (2006, pp. 149–163) for discussions about the provenance of the *Admonitions*.

903. Bell (1971).

904. Scarre (1989).

905. The !Kung, who refer to themselves as the Zhun/twasi, 'the real people', have been a distinct cultural group in and beyond the Kalahari region for many thousands of years; a distinction that they maintained even after the migration of Bantu-speaking people into their territory over two millennia ago (Shostak, 1981). Prospects for the !Kung and other so-called 'bushmen' cultures are currently bleak, as their lands are repeatedly invaded by outsiders, and they are relocated to new settlements where they suffer from malign influences such as alcoholism and a spirit-sapping dependence on government welfare.

906. Price (2000).

907. Barnett (2000).

908. Bocquet-Appel (2002) has presented evidence for another mechanism to explain the Neolithic demographic transition in Europe, namely the outbreeding of the indigenous population by more fertile incoming farmers. See Alinei (2004) for a discussion of the 'continuity theory' of Neolithic Europe, and Armelagos and Harper (2005b) for a review of the genetic evidence.

909. This study also shows that there was a third wave of migration from the Near East during the Neolithic, possibly including agriculturalists, but that these people settled mainly around the Mediterranean and relatively few of them contributed to European populations in the main part of the continent (Semino *et al.*, 2000).

910. Haak *et al.* (2005). For subsequent comments on this paper, see Ammerman *et al.* (2006) and Burger *et al.* (2006).

911. This 'cultural diffusion' model of the spread of farming into Central Europe is not accepted by some scholars who prefer the 'demic diffusion' model of mass migration and population replacement (e.g. Cavalli-Sforza and Cavalli-Sforza, 1995; Cavalli-Sforza, 2001). At present, the genetic evidence remains equivocal (see Currat and Excoffier, 2005, for a discussion of some of the technical problems involved), but I have chosen to highlight the 'cultural diffusion' model in this instance both because of its parsimony and its successful application to other instances of population movement.

912. The 'core' *Linearbandkeramik* culture lasted from about *7500–7000 BP*, followed by a series of *Linearbandkeramik*-related cultures from *7000–6400 BP* (Bogaard, 2004).

913. Based on a variety of biotic and abiotic evidence, this hypothesis was advanced by Columbia University scientists, William Ryan and Walter Pitman (Ryan and Pitman, 1999). While it is generally agreed that the Black Sea used to be a freshwater lake and was considerably lower than today's level, the deluge theory of its transformation in Neolithic times remains controversial, with addition evidence appearing in recent years that both supports and challenges it.

914. Bakels (1991, 1997).

915. An ard plough uses a simple wooden blade, normally pulled by oxen, and was used from Neolithic times until development of the more efficient metal-bladed plough in the Iron Age.

916. This model was first suggested by Halstead (1989).

917. Bogaard (2004, p. 159).

918. Bentley *et al.* (2003).

919. Price *et al.* (1995, 2001).

920. For an overview of the expansion of agriculture beyond the loess belt and onto poorer, higher soils, see Bogaard (2004, p.16–20).

921. For a detailed archaeobotanical study of Neolithic farming in Europe, see Bogaard (2004).

922. Jones (2005).

923. Individual longhouses would be remodelled, or even rebuilt, on a periodic basis, but a family or clan might occupy the same site for more than 400 years (Lüning, 2000).

924. Bogaard (2004, p. 161–162).

925. Whittle (1996).

926. Scarre (1989). The importance of oysters in Neolithic Denmark is shown by their predominant presence in the huge rubbish tips, called 'kjökken-möddings' (meaning 'kitchen refuse'), that are found in the vicinity of settlements (Lubbock, 1861; Steenstrup, 1872).

927. Kubiak-Martens (2002).

928. For accounts of the uses and importance of hazelnuts, see Bokelmann (1981) and Huntley (1993). In much of northern Europe, wild garlic was eaten on an almost daily basis when it was in season until medieval times and beyond.

929. For example, see the discussion in Bogaard (2003, pp. 167–170).

930. Thomas (1987).

931. Shennan (1993).

932. For an up-to-date, finely researched, and well-written account of pre-Columbian American societies, the recent book by Charles Mann is recommended (Mann, 2006).

933. The possibility of the evolution of independent centres of crop domestication in the east of North America has been much debated. Four species have been identified as putative domesticates: marsh elder (*Iva annua*), chenopod (*Chenopodium berlandieri*), squash, and sunflower, but evidence remains controversial. For a recent perspective, see Smith (2006a).

934. Pearsall (1995) and Pope *et al.* (2001).

935. McClung de Tapia (1992); Betz (1999).

936. MacNeish (1992); Pearsall (1995).

937. Betz (1999).

938. Dickau *et al.* (2007).

939. Jaenicke-Després *et al.* (2003).

940. Scarre (1989).

941. The earliest Olmec ceremonial centre was built at Tres Zapores at about *3200 BP* (Bunson and Bunson, 1996; Coe and Koontz, 2002; Haywood, 2005), while the earliest Zatopec remains at Monte Albán date from *2500 BP* (Scarre, 1989). For an account of the Zatopec civilization, see Marcus and Flannery (1996).

942. Millon (1973); Bunson and Bunson (1996); Coe (1999); Braswell (2003); Demarest (2004).

943. Jaenicke-Després *et al.* (2003).

944. For further reading on the Mayans and the partial collapse of *1100 BP*, several volumes by Pat Culbert can be recommended, including Culbert (1973, 1988, 2002) and Culbert and Rice (1990), plus the recent books and articles by Andrews *et al.* (2003), Demarest (2004), Gill (2000), Haug *et al.* (2003), Lucero (2002), and Webster (2002)

945. Hammond *et al.* (1998).

946. Although most authors speak in terms of the Maya collapse, it is probably more accurate to regard the phenomenon not as a single event but rather as a series of processes that occurred throughout the southern Maya lowlands, as argued by Chase and Chase (2006).

947. For the original study on the effect of climate in the Mayan collapse, see Hodell *et al.* (1995).

948. As reported by Hodell *et al.* (2001).

949. Jacob (1995).

950. For example see Johnston (2003) for a discussion of the importance of cultivation lengthening.

951. Most of the *chinampas* were abandoned soon after the Hispanic conquest, but a few of them can still be seen today where they serve as tourist attractions along the canals of Xochimilco, near Mexico City (Sanders, 1957; Rojas *et al.*, 1983; 1995).

952. Netzahualcóyotl is still recognized in Mexico today and is depicted on the 100-peso note. His works and those of other Nahuatl poets are celebrated in an anthology by Miguel León-Portilla (2000) while his philosophical outlook is discussed by Maffie (2002).

953. Haywood (2005).

954. Visser (1986).

955. Noguera (1974).

956. Murphy (2007). By 1850, even the most advanced industrial and agricultural country of the day, namely Britain, had only just caught up with this aspect of Aztec farming efficiency with a little over 20% of the British population listed on the census as working on the land.

957. See discussion in Dickau *et al.* (2007) and Dillehay and Kolata (2004).

958. Olsen and Schall (2006).

959. See Heckenberger (1998) and Heckenberger *et al.* (2003) for the case in favour of successful Amazonian farmer-cultures until the European incursions. However, Meggers (2003) has also argued that the farmers may have been more ephemeral incomers who overexploited the Amazonian habitat in an unsustainable manner, forcing the abandonment of the land and eventual reversion to subsistence hunter–gathering. Meggers stresses the known sustainability of the four-millennium long hunter–gatherer lifestyle in the Amazon Basin and regards the farming communities as isolated exceptions. Others, especially Heckenberger, Petersen, *et al.* point out that the absence of stone necessitated use of highly perishable building materials such as wood, which may explain why previous investigations found little evidence of complex cultures. If such cultures did exist, it may restore the tarnished reputation of the conquistador, Francisco de Orellana, who returned to Europe in 1542 with incredible tales of cities, temples, roadways, and women warriors deep in the Amazonian interior. These ideas received a fillip in May 2006 with (as yet unconfirmed) press reports of a putative 2000-year old 'Stonehenge-like' observatory in a remote part of the northern Amazon at Calçoene (http://news.bbc.co.uk/1/hi/world/americas/4767717.stm).

960. Haywood (2005).

961. Oca, *Oxalis tuberosa*, is an ancient octoploid crop that helped sustain many pre-Incan cultures. It is still second only to the potato in its importance as an Andean food staple (Emshwiller, 2006a, 2006b).

962. Recent data from Perry *et al.* (2006, 2007) indicate that complex agriculture, including cultivation of maize, peppers, and tubers, may have already been in progress by *4000 BP* in some inland mid-altitude Andean sites, such as the preceramic town of Waynuna in the Cotahuasi Valley, about 300 km west of Lake Titicaca.

963. McEwan (2006).

964. Dillehay and Kolata (2004).

965. Conrad (1990).

966. Scarre (1989).

967. Anasazi is a Navajo word meaning 'the ancient ones'.

968. See Wills (1995, Table 8.1 and references therein).

969. Diehl (1993).

970. Much useful information on the ancient societies of the US Southwest can be readily accessed via John Kantner's website at Georgia State University, available online at: http://sipapu.gsu.edu/timeline/timeline1000.html.

971. Kantner (2004).

972. Although the Anasazi cultivated a diverse range of crops, their overwhelming dependence on maize is demonstrated by coprolite data showing that up to 85% of their diet was derived from this one source.

973. In a detailed study, Benson *et al.* (2007) present evidence of three severe, multidecade warm/drought episodes at roughly *900, 800,* and *660 BP*. Soon after

the last of these events, the Anasazi and neighbouring cultures collapsed for the final time.

974. Crown (1990).

975. For example, the Hohokam constructed large, oval arenas similar to the sacred bull courts of the Maya (Haywood, 2005). Although some archaeologists have argued that the Hohokam were migrants from Mesoamerica, it is now generally agreed that they were an indigenous people that adopted southern practices via cultural diffusion. For a non-specialist overview, see Andrews and Bostwick (2001).

976. The Hohokam did not vanish totally. Although all of their villages were abandoned, the remnants of the population reverted to hunter–gathering and are the ancestors of the modern Pima and O'Odham tribal groups.

13. Crop management in the classical and medieval periods

977. Hansen (1965).

978. Hansen (1965).

979. Here are some of the key passages from the Code of Hammurabi (for the complete text, see: http://www.wsu.edu/~dee/MESO/CODE.HTM):

> 53. If a man neglects to reinforce the embankment of the irrigation canal of his field and does not reinforce its embankment, and then a breach opens in its embankment and allows the water to carry away the common irrigated area, the man in whose embankment the breach opened shall replace the grain whose loss he caused.
> 54. If he cannot replace the grain, they shall sell him and his property, and the residents of the common irrigated area whose grain crops the water carried away shall divide the proceeds.
> 55. If a man open the branch of the canal of irrigation and negligently allows the water to carry away his neighbour's field, he shall measure and deliver grain in accordance with his neighbour's yield.
> 56. If a man opens an irrigation gate and releases waters and thereby he allows the water to carry away whatever work has been done in his neighbour's field he shall measure and deliver 3000 sila of grain per 18 iku of field.

980. Kramer (1956).

981. Hesiod's agricultural text is part of the *Homeric Hymns and Homerica* as translated by Evelyn-White (1914).

982. One *she* was about 2.8 millimetres.

983. Zohary and Spiegel-Roy (1975).

984. Temples often rented farmland to tenants who were responsible for its upkeep and productivity. Several laws in the Code of Hammurabi specified strict rules for the maintenance of irrigation systems, with drastic punishments for malefactors. For more on temples and land ownership, see Jursa (1995, p 196).

985. Saggs (1965).

986. Ambrosoli (1997, p. 2–3).

987. Meißner (1920, p. 210).

988. Frahm (1997).

989. The fertile slopes of Mount Amanus rise above the Gulf of Alexandretta (or Iskenderon) near the modern Turkish/Syrian border. The Pinarus River Valley on the lower slopes of Mount Amanus is the site of the battle of Issus, where Alexander of Macedon defeated Darius III Codomannus of Persia in 333 BCE.

990. Roaf (1990, pp. 182–188).

991. Leick (2001, pp. 188–189); Jursa (1995, p. 196).

992. Ashurbanipal was remarkable for his learning; not only did he claim to be fully literate (as shown in his signing of cuneiform tablets), his inscriptions record how he had mastered difficult texts in Akkadian and Sumerian and could solve complex mathematical problems. His library was one of the scholarly treasures of the ancient world, with copies of all known archives and important texts from across Mesopotamia, bilingual vocabularies and wordlists, works of literature, and lists of medical diagnoses. This invaluable collection, much of which survives to the present day, remains the basis of much of our knowledge of Mesopotamian writing and scribal traditions (Walker, 1987; Roaf, 1990).

993. Thompson (1924).

994. Bancroft-Hunt (2004, pp. 110–111).

995. Much of the information about the ancient Persian penchant for botany is from Holland (2006, pp. 212–213).

996. Holland (2006, p. 212).

997. The destruction of Persepolis in 330 BCE was an especially unnecessary piece of vandalism, and Alexander is still remembered in Zoroastrian texts (*Book of Arda Wiraz*) as 'Guzastag'—the accursed.

998. For more on the often-underrated contribution of the ancient Greeks to systematic scientific investigation and the generation of reliable knowledge, see Lloyd (1974, 1975) and Russo (2004).

999. The two main texts of Theophrastus that have come down to us as Latin translations are *De causis plantarum* and *Historiae plantarum*, both of which are now available as English translations (Theophrastus, ca 310 BCEa and 310 BCEb). Theophrastus also made a

series of striking woodcuts of plant, including many crops, as described by Raven (2000).

1000. In his well know treatise, *Rerum rusticarum de agric cultura*, Varro (34 CE) comments in vol. I, v, 1–2 as follows about Theophrastus' work: '*non tam idonei iis qui agrum colere volunt quam qui scholas philosophorum*'.

1001. Pergamum, also known as Pergamon, is close to the site of the modern Turkish town of Bergama.

1002. Dalby (2003).

1003. Russo (2004).

1004. Finley (1985).

1005. These grants were revocable but nevertheless extended land ownership to a new class; during their brief period of rule just before the Ptolemies, the Persians had also introduced the concept of outright private ownership to Egypt (Rostovtzeff, 1941).

1006. The evidence for a Greek scientific renaissance from about 300–50 BCE is discussed at length in the monograph by Russo (2004).

1007. 'I've seen the largest seeds, tho' view'd with care, Degenerate, unless th' industrious hand Did yearly cull the largest.' Virgil, *Georgics I*: 197, as quoted on p. 318 of Darwin (1868).

1008. An example is the creationist and 'intelligent design' debates of the present day, which threaten to introduce quasimystical beliefs, opinions, and other speculations into the scientific curriculum of educational establishments around the world.

1009. The is evidence of small scale cultivation of breadwheat in Britain before the Roman invasion but it became the dominant crop throughout the arable areas of England by the fourth century CE (Fowler, 2002, pp. 212–216) and by the Early Anglo-Saxon period it was virtually the only wheat crop grown in Britain (Murphy, 1994b).

1010. In much of the literature, the tenth century *Geoponics* is wrongly attributed to the seventh century writer, Cassianus Bassus who wrote another work with the same title. The tenth century version of *Geoponics* used Bassus' work as well as others but was written by an unknown author. This work is available online as the 1805 translation by Thomas Owen at: http://www.ancientlibrary.com/geoponica/ (Owen, 1805).

1011. Ambrosoli (1997).

1012. Norwich (1993).

1013. Gibbon (1776–1788, Book VII, p. 76).

1014. Rashed (1996).

1015. The key role of Christians in the Near East in translating Greek works into Arabic, from which they were, in many cases, translated into Latin is described by O'Leary (1980).

1016. Andrew Wilson has described the 'story of failure' whereby European Christendom in particular was 'extremely unreceptive' to new crops and farming techniques from the more advanced Islamic countries of the Mediterranean and Near East (Watson, 1995).

1017. Ambrosoli (1997).

1018. Gibbon (1776–1788, Book VII, p 50). Al Ma'mun was a most unusual Arab ruler in his embrace of an obscure theological offshoot of Islam called Mu'tazili (meaning to leave or desert), which rejected fundamentalist mysticism and was deeply influenced by Aristotelian thought and Greek rationalism.

1019. The immediate post-Roman period in much of Europe, especially in the north and west, is generally regarded as one of considerable decline in both agriculture and societal complexity (Ward-Perkins, 2005). The mainstream view that there was relatively little agricultural progress during the subsequent medieval period is summarized by Postan (1973). For alternative perspectives of this complex issue, see Astill and Langdon (1997), Gimpel (1988), Stone (2005), and Sweeney (1995). Adams (1996, pp. 46–53) revealingly contrasts the medieval coincidence of 'stagnancy of agricultural regimes' with 'technical. florescence and scholastic withdrawal'.

1020. There were some exceptions to this, especially in some of the larger monastic foundations, such as the Cistercians and Benedictines, which had both libraries and a keen commercial interest in agriculture (Brunner, 1995).

1021. An example of this is the recolonization of the Welsh and Scottish uplands between 1150 and 1250 (Parry, 1985). In Wales, this led to a rebalancing of rural populations with large numbers of people again farming the uplands (Davies, 1987; Leighton and Silvester, 2003).

1022. The most productive period for English vineyards was from 1100 to 1300, and represents a northerly extension of 500 km compared to commercial grape-growing regions of today (Lamb, 1965).

1023. Langdon (1997) For an account of medieval faming in Flanders, see Thoen (1997), and for the Netherlands, see Hoppenbrouwers (1997).

1024. Campbell (1997). It is interesting to reflect that there is less land area under arable production in Britain today (4.5 million hectares) than there was in the High Middle Ages (4.7 million hectares).

1025. There was much room for increased productivity in English farms where demesne farmers produced only about 0.7 tonnes/hectare of grain, compared to 1.1 tonnes/hectare on similar soils in Flanders (Langdon, 1997, pp. 278–281).

1026. See Campbell (1997, pp. 225–230) for a more detailed analysis of the argument for a series of small incremental improvements in medieval agriculture.

1027. de'Crescenzi (1300).

1028. Southern (1986).

1029. Crombie (1953).

1030. Walter of Henley (1286).

1031. Fitzherbert (1523).

1032. Frank (1995).

1033. Bradley (2000).

1034. In *The Little Ice Age: How Climate Made History*, Brian Fagan details some of the impacts of this series of climatic events on European history from late-medieval times until the nineteenth century (Fagan, 2001a). Despite the title of this book, Fagan does not fall into the trap of environmental determinism, instead describing climate as often being a 'subtle catalyst' in human affairs.

1035. Leighton and Silvester (2003, p. 34).

1036. The revival of upland arable farming in some areas of Wales after the abandonments of the 'Little Ice Age' was made possible by climatic amelioration, but was also influenced by social factors. In many areas of Wales, the move from partible inheritance to the English system of primogeniture also helped ensure the economic viability of upland farms, which needed to be larger than lowland units (Hook, 1997; Leighton and Silvester, 2003).

1037. The tree line in the Alps was lowered by 70–300 metres in the Alps, by 100–200 metres in northern Germany, and by as much as 300 metres in Iceland (Lamb, 1977).

1038. Kershaw (1973).

1039. Mazoyer and Roudart (2006).

1040. In Scotland, the capital was moved from Perth to Edinburgh, and in Norway, from Trondheim to Oslo. Obviously there were complex political and cultural reasons for these moves in addition to the climatic background, but nevertheless these dramatic shifts of capital serve as a metaphor for the generalized retreat southwards that occurred throughout northern Europe during the Little Ice Age.

1041. Björnsson (2004).

1042. Marcus (1980).

1043. Palladius (late 4th century CE).

1044. Two notable exceptions to the Latinate tradition were Grosseteste's Anglo-Norman treatise, as described in a previous note, and the near contemporary and much-copied *Ruralia Commoda* by Pietro de'Crescenzi, which was written in Italian (Crescenzi, 1300). For an exquisitely comprehensive treatment of agricultural texts from the late medieval and Renaissance periods, the first few chapters of the book, *The Wild and the Sown*, by Mauro Ambrosoli are simply unbeatable (Ambrosoli, 1997).

1045. For example the statement about transgenic crops by Prince Charles: 'this kind of genetic modification takes mankind into realms that belong to God' (quoted in the *Daily Telegraph*, 7 February, 2003, available online at: http://www.telegraph.co.uk/news/main.jhtml?xml = /news/2003/02/08/nprin08.xml).

1046. Postgate (1992) contains a comprehensive account of the rise and fall of the Akkadian civilization.

1047. Cook (1937) makes the point in his *Chronology of Genetics* that: 'monstrosities were ascribed to hybridization, as the ancients generally looked upon the process of hybridization with abhorrence.'

1048. Hippocrates, *Law, book IV*.

14. Agricultural improvement and the rise of crop breeding

1049. Some of the many recent texts on plant breeding include Kang (2002); Schlegel (2003); Thomas *et al.* (2003); Jain and Kharkwal (2004); Acquaah (2006); Brown and Caligari (2007).

1050. McCouch (2004).

1051. The term 'variation' is used here in the sense of genetic, i.e. inherited, variation. There is another equally important source of variation that is caused by environmental effects. Farmers seek to minimize those environmental effects that are under their control, for example by trying to ensure that all parts of a field are equally watered and fertilized. But some environmental effects, such as the weather, are more difficult to control. Even here, however, breeding can help by providing crop varieties that are genetically more resistant to environmental insults, be they biotic (e.g. pests or diseases) or abiotic (e.g. salinity or climate). For a more technical discussion of variation and crop selection, see Chapter 9 of Forbes and Watson (1992).

1052. In this respect, breeding could be said to be just a special case of the process of evolution, as indeed it is. Darwin (1859) stated in his book *Origin of Species* that: 'The preservation of favourable variations and the rejection of injurious variations, I call Natural Selection . . . '. The early events in crop domestication by humans were really an example of coevolution between the two partners, to their mutual benefit. It is only in the past few centuries that people have extended the possibilities by creating new forms of variation and using more rapid and effective tools for selection, but all we are doing here is speeding up a biological process that had been happening anyway for many millennia.

1053. For example it is estimated that annual yields of up to 2 tonnes/hectare of emmer and 1 tonne/hectare of breadwheat may have been possible under the optimal climatic and agronomic conditions in Mesopotamia *c. 4400 BP*, which is not far removed from average global yields in the late nineteenth century (Araus *et al.*, 2007).

1054. Ereky (1919).

1055. See Chaucer (1386) and Ripley (1471).

1056. The purity of drinking water has always been a serious issue for sedentary societies, and this was especially true in large urban centres where it was almost impossible to avoid potentially life-threatening microbial contamination. Alcoholic fluids such as beer and wine were effective in sterilizing drinking water and were even given in diluted form to children. Widespread provision of safe public drinking water did not occur until late in the nineteenth century in industrialized countries, and is still far from the norm in many parts of the world. Water-borne diseases remain one of the most important contributors to human mortality, infecting 2.5 billion people each year, and most particularly affecting young children.

1057. Desmond (1964).

1058. Morris and Bryce (2000).

1059. Raven (2004) gives a good account of the lost libraries of antiquity, while Kamen (1997) is a good source on the fate of learning in post-Islamic Al-Andalus. Here, in 1499, Cardinal Francisco Ximénez de Cisneros, Primate of All Spain and Confessor to Queen Isabella, '*La Reina Catholica*', ordered the entire library of the Arab University of Granada to be publicly burned in the central square of that loveliest of Moorish cities. All Arabic manuscripts in were consigned to the flames, with the exception of a few medical texts. Records state that at least 1,500,000 items were destroyed, including unique works of Moorish culture, copies of the holy Qur'an, and many thousands of agricultural and engineering texts. Following this catastrophic extirpation of knowledge, hard-won by skilful breeders and scientists over the centuries, agriculture in Al Andalus fell into a deep and enduring decline.

1060. For example, in the Judeo-Christian tradition, Adam, the first man, is expelled from the Garden of Eden to be a toiler of the soil, charged with improving the fallen earth with the 'sweat of his brow'.

1061. Immediately after the fall of Constantinople in 1453, it is estimated that over 120,000 books and manuscripts were destroyed. Fortunately, many other books had previously been stolen, bought or otherwise taken to Italy by the Genoese or Venetians. Hence, many of the classics were saved from the inferno, albeit not necessarily for the best of motives.

1062. de'Crescenzi (1300).

1063. Columella's work, *De Re Rustica*, is available in English translation (Columella, first century CE).

1064. Ambrosoli (1997).

1065. Ambrosoli (1997, pp. 262–263).

1066. Except for the highest social classes, women were almost entirely excluded from formal education at this time (Stone, 1964).

1067. Ambrosoli (1997, pp. 180–183 and 317–318). During this period, what we call French today was just one of several languages such as Breton, Occitan, and Catalan.

1068. For example, *Five Hundred Points of Good Husbandry*, by Thomas Tusser (1560).

1069. Examples include Henry Plat (Plat, 1600) with his title page legend *Adams Toole Revived* and Adolphus Speed's manual of husbandry entitled *Adam out of Eden* (Speed, 1659).

1070. The tension between use of the more nationally accessible vernacular versus a more internationally accessible, but much more exclusive *lingua franca* has always been an issue in the dissemination of all types of knowledge, but applies particularly to science with its universalist aspirations. During my training as a biologist, I was obliged to learn sufficient French and German in order to read the many important papers then written in those languages. It is only in the past 20 years that English has emerged decisively as the true *lingua franca* of scientific discourse across the world.

1071. Turner published the *Names of Herbs* in 1548, followed by the *Herball* in 1551.

1072. For an entertaining biography of Nicholas Culpeper, Benjamin Woolley's *The Herbalist* is definitely the book to read (Woolley, 2004).

1073. The term 'scientist' was not invented until as late as 1840 when William Whewell stated 'We need very much a name to describe a cultivator of science in general.. I should be inclined to call him a scientist.' (Whewell, 1840). Previously the term 'natural philosopher' was used to describe one who studied the natural world. The word 'science' only acquired its present meaning in the eighteenth and nineteenth centuries; before then, science simply meant knowledge (as opposed to mere opinion or belief). For example, in the 1421 Rolls of the English Parliament, we are told that: 'Thre Sciences that ben Divinite, Fisik and Lawe' The spurious nature of astrology was apparent even to medieval thinkers, such as the great Córdoban Jewish physician and philosopher, Maimonides (*c.* 1135–1204 CE), who stated: 'Astrology is a sickness, not a science . . . It is a tree under the shade of which all sorts of superstitions thrive'.

1074. From the Oxford English Dictionary (1971), 1523: '*the lorde may improve him self*'.

1075. The long, narrow strips of medieval fields are still the basis of the old English unit of land area, the acre, which is a chain (22 yards) by a furlong (10 chains or 220 yards), and was reckoned as the amount of land that a team could normally plough in one day. The modern metric measurement, the hectare, is more prosaically defined as 100 metres × 100 metres, or about 2.5 acres.

1076. The continued pejorative use of the word 'improve' can be seen in a quotation from 1655 in which Thomas Fuller bemoans the exploitation of the university academics by the townsfolk of Cambridge: 'The Townsmen . . . unconscionably improving themselves on the Scholars necessities extorted unreasonable rents from them' (Fuller, 1655).

1077. Henry VIII, State Papers, volume ii, p. 10.

1078. Lambarde (1581).

1079. Although Fitzherbert (1523) was strongly in favour of enclosure as a socially progressive measure, the resultant displacement of common-land farmers led directly to Jack Kett's rebellion in Norfolk in 1549. This was the occasion of a great deal of bloodshed that is still remembered in the area (Rutledge, 1993).

1080. Clark (2004). For a pictorial impression of the spread of enclosures, see Kain *et al.* (2004).

1081. Moore (1516).

1082. Lambarde included in his text, *Perambulation of Kent*, both Classical and European references, plus a section called '*vox cantianorum*' – the voice of the people of Kent (Lambarde, 1570; Ambrosoli, 1997, pp. 238–243).

1083. Quoted in Tawney (1912, p 17), the passage is from a work entitled: *The defence of John Hales agenst certeyn sclaundres and false reaportes made of hym*.

1084. Descartes was an early champion of reductionism in scientific enquiry and also had a keen interest in agricultural improvement, especially the use of mechanical devices (Descartes, 1637). William Gilbert, of whom it is said, ' [he] deserves the title of first scientist' (Gribbin, 2003, p. 68) is best known for his treatise *De Magnete* (On the Magnet), which was published in 1600 and had a great influence on Galileo (Gilbert, 1600).

1085. The Baconian project for the betterment of state power by the harnessing of knowledge is discussed at length in Drayton (2000).

1086. Bacon (1597).

1087. Bacon (1626).

1088. For a detailed account of John Ray plus much more about the history of botany, Anna Pavord's book, *The Naming of Names: The Search for Order in the World of Plants*, is worth a read (Pavord, 2005).

1089. Eston (1645).

1090. The structures observed by Hooke were the cellulose-enclosed remains of former cells in lignified plant tissues, and were not therefore living cells. Hook did not realize that cells were the basic components of all organisms, or indeed that they had once contained living matter—to him they were just empty chambers with walls. The concept of the biological cell was not to be enunciated in its present form until the mid-nineteenth century.

1091. Examples of late seventeenth century 'improvers' are legion, but they include: Thomas Locke in his Treatise (Locke, 1690), who implicitly advocated enclosure; John Worlidge's *Systema Agriculturae* (Worlidge, 1669); John Houghton's *Collection of Letters for the Improvement of Husbandry and Trade* (Houghton 1681–3, and 1692–1703), which is considered to be the first agricultural periodical; Richard Blome's *The Gentleman's Recreation* (Blome, 1686); Leonard Meager's *The Mystery of Husbandry* (Meager, 1697); and Timothy Nourse's *Campania Felix, or a Discourse of the Benefits and Improvements of Husbandry* (Nourse, 1700), which set out the new technologies and argued the need for enclosure if agricultural yields were to increase to the level perceived necessary for general economic growth (examples taken from: Clark, 2004).

15. Imperial botany and the early scientific breeders

1092. The importance of small-scale farmers in the early part of the English agricultural revolution is discussed by Allen (1999), who also reviews evidence that this process occurred between 1600 and 1750, rather than the later dates of 1750–1800, favoured

1092. by some previous historians, as recently restated by Overton (1996).

1093. Overton (1996).

1094. The thesis that Britain had already undergone an economic, industrial, and agricultural transformation during the eighteenth century is usefully summarized by Wrigley (2004). For example, although the Parliamentary Enclosure acts were not formally passed into law until after 1760, most of the productive land in England had already been enclosed in the seventeenth and early eighteenth centuries.

1095. The total land area of the UK is 24 million hectares, of which about 4.5 million hectares is classified as arable, or suitable for crop cultivation, while only 2.8 million hectares are forest (statistics from DEFRA, 2005, see www.defra.gov.uk).

1096. Wrigley (1987).

1097. The case for an eighteenth century British agrarian revolution is put by Szreter (2004). However, this view is by no means uncontested and some economists believe that, although there were modest gains in agricultural productivity (especially cereal yields), many of the yield gains date from the seventeenth century. See, for example, Clark (1999) and Ambrosoli (1997, pp. 362–368) for discussions of these contrasting views.

1098. Many of these developments are well described in the classic, early twentieth century text of Lord Ernle (Ernle, 1936).

1099. Ambrosoli (1997, pp. 338–348).

1100. Boehm (1967).

1101. The best-known Romantic poets were Blake, Wordsworth, Coleridge, Byron, Shelley, and Keats and their poetry was characterized in part by a reaction against the rapid and unprecedented industrialization of Britain and consequent changes in its countryside.

1102. The role of botany in the British imperial project is eloquently described in Richard Drayton's book: *Nature's Government: Science, Imperial. Britain and the 'Improvement' of the World* (Drayton, 2000).

1103. Alan (1964); Leith-Ross (1984).

1104. Tradescant (1625).

1105. John Tradescant the younger eventually built up a huge collection of plants and artefacts that became known as the *Musaeum Tradescantianum*. This collection, and an equally impressive library fell, into the unsavoury hand of Elias Ashmole who later bequeathed it and other items to Oxford University. The Tradescant collection is now housed in the Ashmolean Museum in Oxford, where its true origin is often overlooked.

1106. See pp. 34–36 of Drayton (2000) for further discussion of 'aristocratic and princely enthusiasms' and the emulation of continental fashions in sustaining such exploratory endeavours.

1107. For more on the 'imperial' role of British botanical gardens, see the classical study by Lucile Brockway (1979) and for a recent perspective, see Chaplin (2003).

1108. Collinson (1755).

1109. Blake (2005).

1110. HMS Beagle was originally a warship used for a range of duties including coastal defence, antipiracy duties, intelligence gathering, and communications work. Captain FitzRoy was an exceptionally talented individual who had passed his naval exams in 1824 with the highest score ever seen. His account of the 1831–36 voyage of the Beagle earned him a gold medal from the Royal Geographical Society. He was a Member of Parliament and Governor of New Zealand, being dismissed from the latter post in 1846, largely because he contended that Maori land claims were as valid as those of white settlers. He became a noted meteorologist, was head of the British Meteorological Department in 1854, and pioneered the new science of weather forecasting.

1111. Described by Broc (1974).

1112. Jacquin (1797). The Royal Botanic Garden of Madrid was originally established by Fernando VI in a grove just outside the city, called *Migas Calientes*, but in 1781 it was moved to its present location on Paseo del Prado by King Carlos III.

1113. Zuckerman (1998).

1114. The originally attributed remark 'let them eat cake' actually referred to the brioche, which is a rich egg-based pastry rather than a form of cake; and it was philosopher Jean-Jacques Rousseau in his *Confessions* who wrote 'S'ils n'ont pas de pain, qu'ils mangent de la brioche' many years before Marie Antoinette ever came to France.

1115. As said by George III (1738–1820) to French astronomer Jérôme Lalande (quoted in Gregory, 1916, p. 47).

1116. An engaging account of this group of eighteenth century British innovators can be found in Jenny Uglow's book, *The Lunar Men* (Uglow, 2003).

1117. The *Jardin du Roi* was originally established as a medicinal herb garden in 1626 by Guy de la Broesse, physician to Louis XIII. After the Revolution, it was renamed *Jardin des Plantes*. It is still the principal botanical garden in France, and is conveniently located near *Gare d'Austerlitz*, just a brief walk from the cathedral of *Notre Dame*.

1118. This paraphrases the comment of Drayton (2000) who, on p. 97, also notes Banks' publications on matters agricultural that range from diseases of wheat; to the cultivation of flax and hemp; and the drainage of the Fens.

1119. Drayton (2000).

1120. This tradition of scientific internationalism has always been under threat from jealous governments, a threat that is still manifest today in the deeply misguided attempts to boycott scientific contacts between US researchers and their colleagues in Cuba and other proscribed states. Incredibly, even some hitherto respected scientific journals have been pressurized to reject papers from such proscribed nations, as if their knowledge were somehow tainted because of its geographical provenance.

1121. One of the consequences of the transcontinental movement of crops such as cacao, coffee, and potatoes, was the global spread of crop diseases, especially fungal, as entertainingly explored by Nicholas Money (2007).

1122. Sugar cane had already been a successful crop in the Hispanic domains, beginning with its introduction to the Iberian Peninsula by the Muslims during the tenth century. The crop was grown more intensively in the fifteenth and sixteenth centuries in the 'Sugar Islands' of Madeira and the Canaries. Following the discovery of the New World, first Brazil and later the Caribbean became the centres of sugar production for export to Europe. During the seventeenth century, Dutch and English planters and traders wrested control of a newly-expanded sugar trade from the Spanish and Portuguese and to a great extent this created the demand for labour that led to the massive growth of the African slave trade. The early genesis of the Atlantic sugar trade is described in the book: *Tropical Babylons*, edited by Schwartz (2004).

1123. Purseglove (1979).

1124. Wrigley (1982).

1125. Rust (1999).

1126. Sloane (1696).

1127. This process is epitomized by creation under the British Raj of the splendidly titled position of Imperial Economic Botanist to the Government of India. In a link with more modern times, this post was held from 1905 to 1924 by Sir Albert Howard, who was, ironically, an early proponent of traditional agriculture whose writings, such as *The Soil and Health* (Howard, 1924) have strongly influenced the current heir to the British throne, Charles Prince of Wales, as noted in his most recent book on plant cultivation (Wales and Donaldson, 2007).

1128. Archaeologist Brian Fagan has suggested a possible link between French shortages of corn with the cooler and damper climate of the Little Ice Age (Fagan, 2001a). This may well be true, but as Fagan also acknowledges, the principal reason for the sustained poor grain yields in France was a failure to innovate as much as their neighbours in Britain and the Netherlands who increased their average grain yields despite experiencing similarly poor climatic conditions.

1129. The particular shortage of wheat in France in the 1789 season may not have been solely due to the conservatism of farmers in the face of climatic change. There is evidence that speculators deliberately shipped wheat to England in order to drive up prices, and possibly to destabilize the French government. For a graphic description of the effects of the wheat shortages in 1789, see the classical account of Hippolyte Taine (Taine, 1878, Chapter 1).

1130. The Royal Navy went on to use the weapon of a maritime blockade against the Central Powers (in World War I), and the Third Reich (in World War II) to put pressure on food supplies. The blockade of 1914–18 was especially effective and led to widespread famine in Germany that is officially estimated to have contributed to as many as 750,000 deaths (http://www.nationalarchives. gov.uk/pathways/firstworldwar/spotlights/ blockade.htm).

1131. Bristol was the most important centre of the early sugar trade but, by 1673, Liverpool already had its first 'sugar house' and by 1768 there were eight of them.

1132. By the late eighteenth century, the port of Liverpool had eclipsed Bristol and London as a result of its involvement in the transatlantic trade in slaves and cotton. Ships from Liverpool would sail to West Africa and take cargoes of slaves to America, returning to Liverpool with a lucrative shipment of cotton destined for the burgeoning mills of Lancashire and Yorkshire. By 1790, Liverpool controlled 80% of the British and over 40% of the European slave trade (Klein, 1999).

1133. This Garden was ordered by Sir Joseph Banks after the British takeover of Ceylon (Sri Lanka) and was initially named Kew after the London Gardens. The Gardens were originally located on Slave Island, Colombo, before moving to Peradeniya, near Kandy, as the Royal Botanical Gardens in 1821.

1134. Drayton (2000). The Bogor gardens are also famous as the original site for the introduction of the oil palm (*Elaeis guineensis*) to Southeast Asia. This West African tree crop was brought to Java as an ornamental plant by the Dutch in 1848. Large-scale oil palm cultivation was commercialized, at first by the Dutch in Sumatra, but then far more extensively by British planters in Malaya, during the twentieth century. Nowadays, oil palm plantations cover much of the Indonesian and Malaysian archipelagos and in 2006 the crop became the most important source of vegetable oil in the world (Murphy, 2006). Oil palm is also a major export earner for Indonesia and Malaysia (Corley and Tinker, 2003).

1135. Póirtéir (1995).

1136. Camerarius (1694).

1137. Linnaeus (1753).

1138. The Hebrew creation myth, as related in the biblical story called *Genesis*, bears many remarkable similarities to a much older Babylonian creation story, known as *Enuma Elish*, which dates from about 1120 BCE. For a more detailed account of the *Enuma Elish*, see King (2004).

1139. Dash (2001).

1140. Murray (2003).

1141. Curiously, one of the principal objections to the commercial release of transgenic crops is that some varieties might at some time in the future be modified to make them sterile, hence forcing farmers to repurchase seed from suppliers each year. However, as we have seen, many ornamental plants and sterile forms of numerous 'traditionally' bred crops, such as bananas and maize, have been widely grown for several centuries.

1142. A useful explanation of homozygosity, inbreeding depression, and heterosis is given in Chapter 4 of Tudge (1988).

1143. Kölreuter (1761).

1144. Knight was the first person in Europe to make a cross between two wheat varieties to produce a new variety (Knight, 1823).

1145. Knight (1799).

1146. Murphy (2007).

1147. Despite this battery of modifications and manipulations, including the chemically induced doubling of its genome and the possible presence of additional genes from other species, a crop variety made by wide-hybridization is not regarded by policymakers as GM (genetically modified).

1148. Fabre (1852).

1149. An account of this controversy is given in Leighty *et al.* (1926). The prejudice of Jordan against results that imply intergenus hybridization mirrors the view expressed by Lamarck in the early eighteenth

century when he reluctantly accepted that different species might indeed hybridize, but not beyond the confines of the genus (Orel, 1996).

1150. Godron (1854).

1151. Goethe (1790).

1152. Bateson (1894).

16. Agricultural improvement in modern times

1153. Almost all of the useful plants that came to the attention of European explorers, traders, and colonists after the late fifteenth century were already well known and in widespread use by the indigenous peoples of Africa, Asia, and the Americas. Examples include oil palm, sugar cane, cocoa, maize, potatoes, tomatoes, and breadfruit.

1154. The evolution of late Victorian public/private sector partnerships in plant breeding is discussed in Palladino (2002) and by Murphy (2007).

1155. The role of agriculture in sustaining population growth has been examined in a useful text by crop physiologist Lloyd Evans (1998).

1156. According to the UN Population Division, the global population was predicted to grow from 6.5 billion in early 2005 to 9.1 billion in 2050, which is a 40% increase (2004 Revision, http://esa.un.org/unpp/). However, it is also estimated that there is an 85% probability that the world population will stabilize, and may even begin to fall, by the end of the 21st century (Lutz *et al.*, 2001).

1157. As we have seen here, the magnitude of many past climatic changes has been greater than the better-informed predictions from our current very short-term climatic shift that only dates from the mid-twentieth century. The extent to which the recent events are anthropogenic is still a moot point but it seems likely that in the longer term, much more serious non-anthropogenic climatic perturbations will recur, as they have done repeatedly in the past.

1158. Data from UN FAO (http://www.fao.org). As I have recently argued (Murphy, 2007), possibly more than 100 million hectares is still available for the expansion of arable cultivation, especially in South America. But eventually a limit will be reached and a more realistic prospect for increasing future food yields is to use biological methods (i.e. genetics applied as breeding) to intensify cultivation, especially in regions of high demand for food.

1159. Not all of our increased food crop production is used directly as human foodstuffs. In the case of crops such as maize, the majority of US and

Bibliography

Abbo S, Lev-Yadun S and Ladizinsky G (2001). Tracing the wild genetic stocks of crop plants, *Genome* **44**, 309–310.

Abdel-Aal ES, Hucl P and Sosulski FW (1995). Compositional and nutritional characteristics of spring einkorn and spelt wheats, *Cereal Chemistry* **72**, 621–624, available online at: http://www.aaccnet.org/cerealchemistry/backissues/1995/72_621.pdf.

Abel C (1998). Non-GM future is mapped out, *Farmers Weekly*, 11 September 1998, available online at: http://www.btinternet.com/~nlpwessex/Documents/murphy.htm.

Acquaah G (2006). *Principles of Plant Genetics and Breeding*, Blackwell, Oxford, UK.

Adams KL, Cronn R, Percifield R and Wendel JF (2003). Genes duplicated by polyploidy show unequal contributions to the transcriptome and organ-specific reciprocal silencing, *Proceedings of the National Academy of Sciences USA* **100**, 4649–4654, available online at: www.pnas.org/cgi/reprint/100/ 8/4649.pdf.

Adams KL and Wendel JF (2005). Polyploidy and genome evolution in plants, *Current Opinion in Plant Biology* **8**, 135–141, available online at: www.eeob.iastate.edu/faculty/WendelJ/pdfs/Adams_and_Wende_COPB_2005.pdf.

Adams RMcC (1996). *Paths of Fire. An Anthropologist's Inquiry into Western Technology*, Princeton University Press, Princeton, New Jersey, USA.

Adams RMcC and Nissen HJ (1972). *The Uruk Countryside: The Natural Setting of Urban Societies*, University of Chicago Press, Chicago, USA.

Adams MM (1987). *William Ockham*, University of Notre Dame Press, Notre Dame, Indiana, USA.

Adams RM (1962). A Synopsis of the Historical Demography and Ecology of the Diyala River Basin, Central Iraq. In Richard B. Woodbury RB, editor, *Civilisation in Desert Lands*, Anthropological papers No. **62**, Department of Anthropology, University of Utah, USA.

Adams RM (1966). *The Evolution of Urban Society*, Aldine, Chicago, USA.

Adams RM (1970). The study of ancient Mesopotamian settlement patterns and the problem of urban origins, *Sumer* **25**, 111–123.

Adams RM (1981). *Heartland of Cities*, University of Chicago Press, Chicago, USA.

Adovasio J and Page J (2002). *The First Americans: In Pursuit of Archaeology's Greatest Mystery*, Random House, New York. USA.

AgCanada (2004). The United States Canola Industry: Situation and Outlook, *Agriculture and Agri-Food Canada Bi-weekly Bulletin*, 2004 **17** No. **4**, available online at: http://www.agr.gc.ca/mad-dam/e/bulletine/v17e/v17n04_e.htm.

Ahanchede A, Poirier-Hamon S and Darmency H (2004). Why no tetraploid cultivar of foxtail millet? *Genetic Resources and Crop Evolution* **51**, 227–230.

Akkermans PMMG and Schwartz GM (2003). *The Archaeology of Syria: From Complex Hunter-Gatherers to Early Urban Societies (c. 16,000–300 BC)*, Cambridge University Press, Cambridge, UK.

Alan M (1964). *The Tradescants: Their Plants, Gardens and Museums, 1570–1662*, Michael Joseph, London, UK.

Alberts B, Johnson A, Julian Lewis J, Raff M, Roberts K and Walter P (2002). *Molecular Biology of the Cell*, Garland Publishing, New York.

Albert V (2005). *Parsimony, Phylogeny, and Genomics*, Oxford University Press, Oxford, UK.

Alcala-Herrera JA, Jacob JS, Castillo MLM and Neck RW (1994). Holocene paleosalinity in a Maya wetland, Belize, inferred from the microfaunal assemblage, *Quaternary Research* **41**, 121–130.

Aldred C (1961). *The Egyptians*, Thames and Hudson, London, UK.

Algaze G (1989). The Uruk expansion, cross-cultural exchange in early Mesopotamian civilization, *Current Anthropology* **30**, 571–608.

Algaze G (2005). The Sumerian takeoff, structure and dynamics: *eJournal of Anthropological and Related Sciences* **1**, article **2**, available online at: http://repositories.cdlib.org/imbs/socdyn/sdeas/vol1/iss1/art2/.

Alinei M (2004). *The Paleolithic Continuity Theory of Indo-European Origins*: An Introduction, available online at www.continuitas.com/intro.pdf.

Allen RC (1999). Tracking the agricultural revolution in England, *Economic History Review* **52**, 209–235.

Alley RB (2000). The Younger Dryas cold interval as viewed from central Greenland, *Quarterly Science Reviews* **19**, 213–226.

Alley RB and Agustsdottir AM (2005). The 8 k event: cause and consequences of a major Holocene abrupt climate change, *Quaternary Science Reviews* **24**, 1123–1149.

Alley RB, Mayewski PA, Sowers T, Stuiver M, Taylor KC and Clark PU (1997). Holocene climatic instability: A prominent, widespread event 8,200 years ago, *Geology* **25**, 483–486.

Alley RB, Meese DA, Shuman CA *et al.* (1993). Abrupt increase in greenland snow accumulation at the end of the Younger Dryas event, *Nature*, **362**, 527–529.

Allport S (2000). *The Primal Feast: Food, Sex, Foraging, and Love*, Harmony Books, New York, USA.

Alvarez-Venegas R, Sadder M, Hlavacka A *et al.* (2006). The Arabidopsis homolog of trithorax, ATX1, binds phosphatidylinositol 5-phosphate, and the two regulate a common set of target genes, *Proceedings of the National Academy of Sciences USA* **103**, 6049–6054, available online at: http://www.pnas.org/cgi/content/figsonly/103/15/6049.

Ambrose SH (1998). Late Pleistocene human population bottlenecks, volcanic winter, and differentiation of modern humans, *Journal of Human Evolution* **34**, 623–51.

Ambrosoli M (1997). *The Wild and the Sown: Botany and Agriculture in Western Europe 1350–1850*, Cambridge University Press, Cambridge, UK.

Ames BN (2006). Low micronutrient intake may accelerate the degenerative diseases of aging through allocation of scarce micronutrients by triage, *Proceedings of the National Academy of Sciences USA* **103**, 17589–17594.

Ames K (1999). Myth of the hunter-gatherer, *Archaeology* **52**, 45–49.

Ames K and Maschner HDG (2000). *Peoples of the Northwest Coast: Their Archaeology and Prehistory*, Thames and Hudson, London, UK.

Ammerman AJ, Pinhasi R and Bánffy E (2006). Comment on "Ancient DNA from the First European Farmers in 7500-Year-Old Neolithic Sites", *Science* **312**, 1875b.

An CB, Tang L, Barton L and Chen FH (2005). Climate change and cultural response around 4000 cal yr B.P. in the western part of Chinese Loess Plateau, *Quaternary Research* **63**, 347–352, available online at: www.anthro.ucdavis.edu/CARD/USPRC/pubs/An%20etal%202005.pdf.

An CB, Feng ZD and Barton L (2006). Dry or humid? Mid Holocene humidity changes in arid and semi-arid China, *Quaternary Science Reviews* **25**, 351–361, available online at: www.anthro.ucdavis.edu/card/USPRC/pubs/An%20etal%202006.pdf.

Anderson E (1967). *Plants, Man and Life*, University of California Press, Berkeley, USA.

Anderson MK (2004). Pre-agricultural plant gathering and management. In *Encyclopaedia of Plant and Crop Science*, pp. 1055–1060, Marcel Dekker Inc, New York, USA.

Anderson PC (1994). Insights into plant harvesting and other activities at Hatoula, as revealed by microscopic functional analysis of selected chipped stone tools. In: *Le gisement de Hatoula en Judée occidentale, Israël* (Lechevallier M and Ronen A, editors.). Memoires et travaux du centre de recherche Français de Jerusalem No. 8. Association Paléorient, pp. 277–293, Paris, France [in French].

Andersson JO, Sjögren AM, Davis LAM Embley MT and Roger AJ (2003). Phylogenetic analyses of diplomonad genes reveal frequent lateral gene transfers affecting eukaryotes, *Current Biology* **13**, 94–104.

Andersson L and Georges M (2004). Domestic-animal genomics: deciphering the genetics of complex traits, *Nature Reviews Genetics* **5**, 202–212.

Andrews AP, Andrews EW, and Castellanosc FR (2003). The Northern Maya Collapse and its Aftermath, *Ancient Mesoamerica* **14**, 151–156, available online at: www.faculty.ncf.edu/andrews/research/NorthMayaCollapse-2003.pdf.

Andrews JP and Bostwick TW (2001). *Desert Farmers at the River's Edge, The Hohokam and Pueblo Grande*, Pueblo Grande Museum and Archaeological Park, Phoenix, USA, available online at: http://www.ci.phoenix.az.us/PUEBLO/dfindex.html.

Angel LJ (1984). Health as a crucial factor in the changes from hunting to developed farming in the eastern Mediterranean. In: *Paleopathology at the Origins of Agriculture*, Cohen MN and Armelagos GJ editors, pp. 51–73, Academic Press, New York, USA.

Anonymous (2005). Topper site-were humans in America over 50,000 years ago? *Athena Review* **4**, 7–8, available online at: www.athenapub.com/topper.htm.

Anthony KRM, Meadley J and Röbbelen G (1993). *New Crops for Temperate Regions*, Chapman and Hall, London.

Arabidopsis sequencing initiative (2000). Analysis of the genome of the flowering plant *Arabidopsis thaliana*, *Nature* **408**, 796–815.

Araus JL, Ferrio JP, Buxo R and Voltas J (2007). The historical perspective of dryland agriculture: lessons learned from 10 000 years of wheat cultivation, *Journal of Experimental Botany* **58**, 131–145.

Arensburg B, Schepartz LA, Tillier AM, Vandermeersch B and Rak Y (1990). A reappraisal of the anatomical basis for speech in Middle Palaeolithic hominids, *American Journal of Physical Anthropology* **83**, 137–146.

Armelagos GJ and Harper KJ (2005a). Genomics at the origins of agriculture, part one, *Evolutionary Anthropology* **14**, 68–77.

Armelagos GJ and Harper KJ (2005b). Genomics at the origins of agriculture, part two, *Evolutionary Anthropology* **14**, 109–121.

Arsuaga JL (2003). *The Neanderthal's Necklace, In Search of the First Thinkers*, Wiley, Chichester, UK.

Arumuganathan K and Earl ED (1991). Nuclear DNA content of some important plant species, *Plant Molecular Biology Reporter* **9**, 211–215, table available online at: http://www.nal.usda.gov/pgdic/tables/nucdna.html.

Astill G and Langdon J, editors (1997). *Medieval Farming and Technology*, Brill, Leiden, Netherlands.

Augustin L, Barbante C, Barnes PR *et al.* (2004). Eight glacial cycles from an Antarctic ice core, *Nature* **429**, 623–628.

Austad S (1997). Postreproductive survival. In: Wachter KW, Finch CE, editors. *Between Zeus and the Salmon: the Biodemography of Longevity*, pp. 161–174, National Academy Press, Washington, DC, USA.

Ayala F and G.L. Stebbins GL (1981). Is a new evolutionary synthesis necessary? *Science* **213**, 967–971.

Bacon F (1597). *Religious Meditations. Of Heresies.* Meditationes Sacrae.

Bacon F (1626). *The New Atlantis*, available online at: http://oregonstate.edu/instruct/ph1302/texts/bacon/atlantis.html.

Bader NO (1993a). Tell Maghzaliyah: an early Neolithic site in Northern Iraq. In: Yoffee N and Clark JJ, editors, *Early Stages in the Evolution of Mesopotamian Civilization: Soviet Excavations in Northern Iraq*, pp. 7–40, The University of Arizona Press, Tucson, Arizona, USA.

Bader NO (1993b). The early agricultural settlement of Tell Sotto. In: Yoffee N and Clark JJ, editors, *Early Stages in the Evolution of Mesopotamian Civilization: Soviet Excavations in Northern Iraq*, pp. 41–54, The University of Arizona Press, Tucson, Arizona, USA.

Bader NO (1993c). Summary of The Earliest Agriculturists of Northern Mesopotamia. In: Yoffee N and Clark JJ, editors, *Early Stages in the Evolution of Mesopotamian Civilization: Soviet Excavations in Northern Iraq*, pp. 63–71, The University of Arizona Press, Tucson, Arizona, USA.

Badr A, Müller K, Schäfer-Pregl R *et al.* (2000). On the origin and domestication history of Barley (*Hordeum vulgare*)., *Molecular Biology and Evolution* **17**, 499–510, available online at: http://mbe.oxfordjournals.org/cgi/reprint/17/4/499.

Bahrani Z (2002). Peformativity and the image: Narrative, representation, and the Uruk vase. In: Ehrenberg E, editor, *Leaving no Stones Unturned: Essays on the Ancient Near East and Egypt in Honor of Donald P. Hansen*, pp. 15–22, Eisenbrauns, Winona Lake, Indiana, USA.

Balter M (2005). Are humans still evolving? *Science* **309**, 234–237.

Bakels CC (1991). The crops of the Rössen culture. In: Vytlacok S, editor, *Palaeoethnobotany and Archaeology: International Workshop for Palaeoethnobotany 8th Symposium*, Nitra Nove Vozokany 1989, pp. 23–27, Archaeological Institute of the Slovak Academy of Sciences, Nitra, Slovakia.

Bakels CC (1997). Le blé dans la culture de Cerny. In: *La culture de Cerny. Nouvelle économie, novelle societé au Néolithique, Actes du Colloque International de Nemours*, 1994, pp. 315–317, Mémoires de Musée de Préhistoire d'Ile-de-France, 6.

Baldini JUL, McDermott F and Fairchild IJ (2002). Structure of the 8200-year cold event revealed by a speleothem trace element record, *Science* **296**, 2203–2206.

Bamshad M, Wooding S, Salisbury B and Stephens JC (2004). Deconstructing the relationship between genetics and race, *Nature Reviews Genetics* **5**, 898–609.

Bancroft-Hunt N (2004). *Historical Atlas of Ancient Mesopotamia*, Checkmark Books, New York, USA.

Barber DC, Dyke A, Hillaire-Marcel C, Jennings AE, Andrews JT, Kerwin MW, Bilodeau G, McNeely R, Southon J, Morehead MD and Gagnon JM (1999). Forcing of the cold event of 8,200 years ago by catastrophic drainage of Laurentide lakes, *Nature* **400**, 344–348.

Barclay A (2004). Feral play, *Rice Today*, January 2004, 15–19, available online at: http://www.irri.org/publications/today/pdfs/3–1/feral.pdf.

Bar-Matthews M, Ayalon A and Kaufman A (1998). Late quaternary paleoclimate in the eastern Mediterranean region from stable isotope analysis of speleothems at Soreq cave, Israel, *Quaternary Research* **47**, 155–168.

Bar-Matthews M, Ayalon A, Kaufman A and Wasserburg GJ (1999). The Eastern Mediterranean paleoclimate as a reflection of regional events: Soreq cave, Israel, *Earth and Planetary Science Letters* **166**, 85–95.

Barnett WK (2000). Cardial pottery and the agricultural transition in mediterranean Europe. In: Price TD, editor, *Europe's First Farmers*, Cambridge University Press, Cambridge, UK.

Barton L (2004). Archaeobotanical analysis of early agriculture in North China, paper presented at the 69th meeting of the Society for American Archeology, Montreal, Canada, available online at: www.anthro.ucdavis.edu/CARD/barton/SAA04LWB.pdf.

Barton L, Brantingham PJ and Ji D (2006). Late Pleistocene climate change and Paleolithic cultural evolution in

Northern China: Implications from the Last Glacial Maximum. In: Madsen DB, Chen FH and Gao X, editors, *Late Quaternary Climate Change and Human Adaptation in Arid China*, Elsevier, New York, USA.

Bar-Yosef O (1996). The impact of Late Pleistocene–Early Holocene climatic changes on humans in Southwest Asia. In: Straus, LG Eriksen BV, Erlandson JM and Yesner DR, editors, *Humans at the End of the Ice Age: The Archaeology of the Pleistocene-Holocene Transition*, pp. 61–76, Plenum Press, New York, USA.

Bar-Yosef O (1998a). The Chronology of the Middle Paleolithic of the Levant. In: Akazawa T, Aoki K, and Bar-Yosef O, editors, *Neandertals and Modern Humans in Western Asia*, Plenum, New York, USA.

Bar-Yosef O (1998b). The Natufian culture in the Levant, threshold to the origins of agriculture, *Evolutionary Anthropology* **6**, 159–177, available online at: www.columbia.edu/itc/anthropology/v1007/baryo.pdf.

Bar-Yosef O (2000). The Middle and Early Upper Paleolithic in Southwest Asia and neighboring regions. In: Bar-Yosef O and Pilbeam D, editors, *The Geography of Neanderthals and Modern Humans in Europe and the Greater Mediterranean*, pp. 107–156, Peabody Museum of Archaeology and Ethnology, Harvard University Press, Cambridge, Massachusetts, USA.

Bar-Yosef O, Gopher A, Tchernov E and Kislev M (1991). Nativ Hagdud: an Early Neolithic village site in the Jordan Valley, *Journal of Field Archaeology* **18**, 405–426.

Bar-Yosef O and Meadow RH (1995). The origins of agriculture in the Near East. In: Price TD and Gebauer AB, editors, *Last Hunters-First Farmers*, pp.39–94, School of American Research Press, Santa Fe, New Mexico, USA.

Bar-Yosef O and Valla FR, editors (1991). *The Natufian Culture in the Levant*, University of Michigan Press, Ann Arbor, Michigan, USA.

Bar-Yosef O, Vandermeersch B, Arensburg B *et al.* (1992). The Excavations in Kebara Cave, Mt. Carmel, *Current Anthropology* **33**, 497–550.

Bassus C (*c.* 600 BCE). *Geoponica Sive Cassiani Bassi Scholastici*, translated by Beckh H, (1994), Sauer Verlag, Heidelberg, Germany.

Bateson W (1894). *Materials for the Study of Variation*, Macmillan, London, UK.

Baum BR (1977). *Oats: Wild and Cultivated: A monograph of the genus Avena L. (Poaceae)*, Monograph no. **14**, Biosystematics Research Institute, Canada Department of Agriculture, Ottawa, Ontario.

Baum BR and Fedak G (1985a). *Avena atlantica*, a new diploid species of the oat genus from Morocco, *Canadian Journal of Botany* **63**, 1057–1060.

Baum BR and Fedak G (1985b). A new tetraploid species of Avena discovered in Morocco, *Canadian Journal of Botany* **63**, 1379–1385.

Baumhoff MA (1963). Ecological determinants of aboriginal California, *University of California Publications in American Archaeology and Ethnology* **49**, 150–235.

Bautista NS, Solis R, Kamijima O and Ishii T (2001). RAPD, RFLP and SSLP analysis of phylogenetic relationships between cultivated and wild species of rice, *Genes and Genetic Systems* **76**, 71–79, available online at: http://www.jstage.jst.go.jp/article/ggs/76/2/71/_pdf.

Beadle GW (1978). Teosinte and the origin of maize. In: *Maize Breeding and Genetics*, Walden DB, editor, pp. 113–128, Wiley, New York, USA.

Beerling D (2007). *The Emerald Planet: How Plants Changed Earth's History*, Oxford University Press, Oxford, UK.

Beier RC (1990). Natural pesticides and bioactive components in foods, *Reviews of Environmental Contamination and Toxicology* **113**, 47–137.

Beja-Pereira A, England PR, Ferrand N *et al.* (2004). African origins of the domestic donkey, *Science* **304**, 1781.

Belderok B, Mesdag H and Donner DA (2000). *Bread-Making Quality of Wheat: A Century of Breeding in Europe*, Kluwer, Dordrecht, Netherlands.

Bell B (1971). The Dark Ages in ancient history, I. The first Dark Age in Egypt, *American Journal of Archaeology* **75**, 1–26.

Bellwood P (2005). *First Farmers: The Origins of Agricultural Societies*, Blackwell, Oxford, UK.

Bennett MD and Smith JB (1976). Nuclear DNA amounts in angiosperms, *Philosophical Transactions of the Royal Society of London*, B Biological Sciences **274**, 227–274.

Bennetzen JL, Coleman C, Liu R, Ma J and Ramakrishna W (2004). Consistent over-estimation of gene number in complex plant genomes, *Current Opinion in Plant Biology* **7**, 732–736, available online at: www.ira.cinvestav.mx:8080/papers/CurrOpinPlantBio17_732.pdf.

Benson LV, Berry MS, Jolie EA, *et al.* (2007). Possible impacts of early-11th-, middle-12th-, and late-13th-century droughts on western Native Americans and the Mississippian Cahokians, *Quaternary Science Reviews* **26**, 336–350.

Bentley RA, Lounes C and Price TD (2003). The Neolithic transition in Europe: comparing broad scale genetic and local scale isotopic evidence, *Antiquity* **77**, 112–117.

Benz BF (2001). Archaeological evidence of teosinte domestication of Guilá Naquitz, Oaxaca, *Proceedings of the National Academy of Sciences USA* **98**, 2104–2106, available online at: www.pnas.org/cgi/reprint/98/4/2104.pdf.

Bergthorsson U, Adams KL, Thomason B and Palmer JD (2003). Widespread horizontal transfer of mitochondrial genes in flowering plants, *Nature* **424**, 197–201.

Bergthorsson U, Richardson AO, Young GJ, Goertzen LR and Palmer JD (2004). Massive horizontal transfer of

Childe VG (1941). War in Prehistoric Societies, *Sociological Review* **33**, 126–138.

Childe VG (1958). *The Dawn of European Civilization*, Knopf, New York, USA.

Chilton MD (1988). Plant genetic engineering: progress and promise, *Journal of Agricultural and Food Chemistry* **36**, 3–5.

Chilton MD (2001). Agrobacterium. A memoir, *Plant Physiology* **125**, 9–14, available online at: www.plantphysiol.org/cgi/reprint/125/1/9.pdf.

Choudhury B (1995). Eggplant. In: Smartt J and Simmonds NW, editors, *Evolution of Crop Plants*, pp. 464–465, Longman Scientific and Technical, Harlow, UK.

Chrispeels MJ and Sadava DE (2003). *Plants, Genes and Plant Biotechnology*, Jones and Bartlett, New York, USA

Ciochon RL, Nisbett RA, and Corruccini RS (1997). Dietary consistency and craniofacial development related to masticatory function in minipigs, *Journal of Craniofacial Genetics Developmental Biology* **17**, 96–102.

Clark G (1999). Too much revolution: agriculture in the industrial revolution, 1700–1860. In: Mokyr J, editor, *The British Industrial Revolution: An Economic Perspective*, 2nd edition, pp. 206–224, Boulder: Westview Press, Boulder, Colorado, USA.

Clark R (2004). Agricultural Enclosures: the major phase, 1760 onwards, *The Literary Encyclopedia*, available online at: www.LitEncyc.com/php/stopics.php?rec = ture&UID = 1472.

Clark RM, Linton E, Messing J and Doebley JF (2004). Pattern of diversity in the genomic region near the maize domestication gene tb1, *Proceedings of the National Academy of Sciences USA* **101**, 700–707, available online at: http://www.pnas.org/cgi/content/abstract/101/3/700?ck = nck.

Clement CR (1999a). 1492 and the loss of Amazonian crop genetic resources I. The relation between domestication and human population decline, *Economic Botany* **53**, 188–202.

Clement CR (1999b). 1492 and the loss of Amazonian crop genetic resources II. Crop biogeography at contact, *Economic Botany*, **53**, 203–216.

Clutton-Brock J (1995). Origins of the dog: domestication and early history. In: Serpell J, editor, *The Domestic Dog: Its Evolution, Behaviour and Interactions with People*, pp. 7–20, Cambridge University Press, Cambridge, UK.

Coe MD (1999). *The Maya (Ancient Peoples and Places)* 6th edition, Thames and Hudson, London, UK.

Coe MD and Koontz R (2002). *Mexico: from the Olmecs to the Aztecs*, Thames and Hudson, London, UK.

Cohen MN (1977). *The Food Crisis in Prehistory: Overpopulation and the Origins of Agriculture*, Yale University Press, New Haven, Connecticut, USA.

Cohen MN (1987). The significance of long-term changes in human diet and food economy. In: Harris M and Ross E, editors, *Food and Evolution*, pp. 261–284, Temple University Press, Philadelphia, Pennsylvania, USA.

Cohen MN (1989). *Health and the Rise of Human Civilization*, Yale University Press, New Haven, Connecticut, USA.

Cohen NM and Armelagos GJ, editors (1984). *Paleopathology at the Origins of Agriculture*, Academic Press, New York, USA.

Cohen DJ (1998). The origins of domesticated cereals and the Pleistocene-Holocene transition in East Asia, *Review of Archaeology* **19**, 22–29.

Cohen MN (1997). Does palaeopathology measure community health? A rebuttal of 'the osteological paradox' and its implication for world history. In: Paine RR, editor, *Integrating Archaeological Demography: Multidisciplinary Approaches to Prehistoric Population*, Occasional Paper No. **24**, pp. 242–260, Center for Archaeological Investigations, Southern Illinois University, Carbondale, Illinois, USA.

Cole TW (1973). Periodicities in solar activity, *Solar Physics* **30**, 103–110.

Colledge S (1998). Identifying pre-domestication cultivation using multivariate analysis, in Damania AB, Valkoun J, Willcox G and Qualset CO, editors, *The Origins of Agriculture and Crop Domestication*, ICARDA, Aleppo, Syria, available online at: http://www.ipgri.cgiar.org/publications/HTMLPublications/47/.

Colledge SM (1994). *Plant Exploitation on Epipalaeolithic and Early Neolithic Sites in the Levant*, PhD Thesis, University of Sheffield, UK.

Colledge S (2001). *Plant Exploitation on Epipalaeolithic and Early Neolithic Sites in the Levant*, British Archaeological Reports S986, Hadrian Books, Oxford, UK.

Collinson P (1755). Letter to Linnaeus on 10 April 1755. In: Smith JE, editor, *A Selection from the Correspondence of Linnaeus and Other Naturalists*, 1821, Longman, London, UK.

Columella LJM (1st century CE). *De Re Rustica*, translated by Ash HB, Harvard University Press, Cambridge, Massachusetts, USA.

Comai L (2005). The advantages and disadvantages of being polyploid, *Nature Reviews Genetics* **6**, 836–846.

Comai L, Madlung A, Josefsson C and Tyagi A (2003). Do the different parental 'heteromes' cause genomic shock in newly formed allopolyploids? *Philosophical Transactions of the Royal Society of London B Biological Sciences* **358**, 1149–1155, available online at: http://www.comailab.genomecenter.ucdavis.edu/heterome.pdf.

Comfort NC (2001). *The Tangled Field: Barbara McClintock's Search For The Patterns Of Genetic Control*, Harvard University Press, Cambridge, Massachusetts, USA.

Conrad GW (1990). Farfan, General Pacatnamu, and the dynastic history of Chimor. In: Moseley ME and Cordy-Collins A, editors, *The Northern Dynasties: Kingship and Statecraft in Chimor*, pp. 227–242, Dumbarton Oaks Center for Pre-Columbian Studies, Washington, DC, USA.

Cook D (1984). Subsistence and Health in the Lower Illinois Valley: Osteological Evidence. In: Cohen MN and Armelagos GJ, editors, *Paleopathology at the Origins of Agriculture*, pp. 235–269, Academic Press, New York, USA.

Cook R (1937). A Chronology of Genetics, *Yearbook of Agriculture, 1937*. US Department of Agriculture, Electronic Scholarly Publishing, USA available online at: http://www.esp.org/foundations/genetics/classical/holdings/c/rc-37.pdf

Coombes P and Barber K (2005). Environmental determinism in Holocene research: causality or coincidence? *Area* **37**, 303–311, www.blackwell-synergy.com/doi/pdf/10.1111/j.1475–4762.2005.00634.x.

Cooper E (1998). The EB-MB Transitional Period at Tell Kabir, Syria. In: Fortin M and Aurenche O, editors, *Espace Naturel, Espace Habité en Syrie du Nord (10e–2e millénaires av. J.-C.)*, Maison de l'Orient, Lyons, France, pp. 271–280, (*Bulletin of the Canadian Society for Mesopotamian Studies* 33).

Cooper L (2006a). The demise and regeneration of Bronze Age urban centers in the Euphrates Valley of Syria. In: Schwartz GM and Nichols JJ, editors, *After Collapse, The Regeneration of Complex Societies*, pp. 18–38, University of Arizona Press, Tucson, Arizona, USA.

Cooper L (2006b) *Early Urbanism on the Syrian Euphrates*, Routledge, London, UK.

Cooper JS (1983). *The Curse of Agade*, Johns Hopkins University Press, Baltimore, USA.

Copley MS, Berstan R, Dudd SN, *et al.* (2003). Direct chemical evidence for widespread dairying in prehistoric Britain, *Proceedings of the National Academy of Sciences USA* **100**, 1524–1529.

Copley MS, Berstan R, Dudd SN, *et al.* (2005). Processing of milk products in pottery vessels through British prehistory *Antiquity* **79**, 895–908.

Copping LG and Hewitt HG (1998). *Chemistry and Mode of Action of Crop Protection Agents*, Royal Society of Chemistry, London, UK.

Cordain L (1999). Cereal grains: humanity's double edged sword, *World Review of Nutrition and Diet* **84**, 19–73.

Corley RHV and Tinker PB (2003). *The Oil Palm*, 4th edition, Blackwell, Oxford, UK.

Corruccini RS (1991). Anthropological aspects of orofacial and occlusal variations and anomalies. In: Kelley MA and Larsen CS, editors, *Advances in Dental Anthropology*, pp. 295–323, Wiley, New York, USA.

Costo R and Costo HJ (1995). *Natives of the Golden State: The California Indians*, Indian Historian Press, San Francisco, California, USA.

Coward H and Weaver AJ, editors (2004). *Hard Choices. Climate Change in Canada*, Wilfrid Laurier University Press, Waterloo, Canada.

Cox JD (2005). *Climate Crash*, Joseph Henry Press, Washington, DC, USA.

Cragen DC (1975). *The Boys in the Sky Blue Pants, The Men and Events at Camp Independence and Forts of Eastern California. Nevada, and Utah 1862–1877*, Pioneer Publishing Company, Fresno, California, USA.

Craig OE, Chapman J, Heron CP, Willis LH, Taylor G, Whittle A and Collins MJ (2005). Did the first farmers of central and eastern Europe produce dairy foods? *Antiquity* **79**, 882–894.

Crawford GW (1992). Prehistoric plant domestication in East Asia. In: Cowan CW and Watson PJ, editors, *Origins of Agriculture: An International Perspective*, pp. 7–38, Smithsonian Institution Press, Washington, DC, USA.

Crawford GW and Shen C (1992). The origins of rice agriculture: recent progress in East Asia, *Antiquity* **72**, 858–866.

Crawford H (2004). *Sumer and the Sumerians*, Cambridge University Press, Cambridge, UK.

de'Crescenzi P (c. 1300). *Ruralia commoda, De agricultura vulgare*, first printed in 1471, Bindoni, Venice, Italy.

Crombie AC (1953). *Robert Grosseteste and the Origins of Experimental Science 1100–1700*, Oxford University Press, Oxford, UK.

Cross SL, Baker PA, Seltzer GO, Fritz SC and Dunbar RB (2000). A new estimate of the Holocene lowstand level of Lake Titicaca, and implications for tropical paleohydrology, *Holocene* **10**, 21–32.

Crowley TJ (2003). When did global warming start? An Editorial Comment, *Climate Change* **61**, 259–260, available online at: http://stephenschneider.stanford.edu/Publications/PDF_Papers/Crowley2003.pdf.

Crown PL (1990). The Hohokam of the American Southwest, *Journal of World Prehistory* **4**, 223–255.

Crutzen PJ and Steffen W (2003). How long have we been in the Anthropocene Era? An editorial comment, *Climate Change* **61**, 250–257, available online at: http://stephenschneider.stanford.edu/Publications/PDF_Papers/CrutzenSteffen2003.pdf.

Cuffey KM and Clow GD (1997). Temperature, accumulation, and ice sheet elevation in central Greenland through the last deglacial transition, *Journal of Geophysical Research* **102**, 26383–26396.

Culbert TP, editor (1973). *The Classic Maya Collapse*, University of New Mexico Press, Albuquerque, USA.

Culbert TP (1988). *The Collapse of Classic Maya Civilization*. In: Yoffee N and Cowgill G, editors, *The Collapse of Ancient States and Civilizations*, University of Arizona Press, Tucson, USA.

Culbert TP and Rice DS, editors (1990). *Precolumbian Population History in the Maya Lowlands*, University of New Mexico Press, Albuquerque, USA.

Culbert TP (2002). *Secrets of the Maya*, Hatherleigh Press, New York, USA.

Cullen H and deMenocal PB (2000). North Atlantic influence on Tigris-Euphrates streamflow, *International Journal of Climatology* **20**, 853–863.

Cullen HM, deMenocal PB, Hemming S, Hemming G, Brown FH, Guilderson, T and Sirocko F (2000). Climate change and the collapse of the Akkadian empire: Evidence from the deep sea, *Geology* **28**, 379–382, available online at: http://research.yale.edu/leilan/cullen2000.pdf.

Currat M and Excoffier L (2004). Modern humans did not admix with Neanderthals during their range expansion into Europe, *Public Library of Science, Biology* **2** e421, available online at: http://biology.plosjournals.org/perlserv/?request = get-document&doi = 10.1371/journal.pbio.0020421.

Currat M and Excoffier L (2005). The effect of the Neolithic expansion on European molecular diversity, *Proceedings of the Royal Society B. Biological Sciences* **272**, 679–688, available online at: http://www.journals.royalsoc.ac.uk/media/mgagxktrtj297tmxcd4p/contributions/v/n/5/8/vn58861k817v11m0.pdf.

Curtis JH, Hodell DA and Brenner M (1996). Climate variability on the Yucatán peninsula (México) during the past 3500 years, and implications for Maya cultural evolution, *Quaternary Research* **46**, 37–47.

Cyranoski D (2003). Rice genome: a recipe for revolution? *Nature* **422**, 796–798.

Dalby A (2003). *Food in the Ancient World From A-Z*, Routledge, London, UK.

Damania AB (1998). Domestication of cereal crop plants and *In situ* conservation of their genetic resources in the fertile crescent. In: Damania AB, Valkoun J, Willcox G and Qualset CO, editors, *The Origins of Agriculture and Crop Domestication*, Aleppo, Syria, ICARDA, available online at: http://www.ipgri.cgiar.org/publications/HTMLPublications/47/.

Damania AB, Valkoun J, Willcox G and Qualset CO, editors (1998). *The Origins of Agriculture and Crop Domestication*, ICARDA, Aleppo, Syria, available

online at: http://www.ipgri.cgiar.org/publications/HTMLPublications/47/.

Damuth J (1993). Cope's rule, the island rule and the scaling of mammalian population density, *Nature* **365**, 748–750.

D'Andrea AC, Klee M and Casey J (2001). Archaeobotanical evidence for pearl millet (*Pennisetum glaucum*) in sub-Saharan West Africa, *Antiquity* **75**, 341–348.

Darlington CD (1963). *Chromosome botany and the origin of cultivated plants*, 2nd edition, Allen and Unwin, London.

Darwin C (1859). *The Origin of Species by Means of Natural Selection*, Penguin Books, London, UK. [This book is a reproduction of the first edition of Darwin's famous work. However, a more definitive version is contained in the 6th edition, from 1868, which is available for free online at: http://www.literature.org/authors/darwin-charles/the-origin-of-species-6th-edition/.]

Darwin C (1868). *The Variation of Animals and Plants Under Domestication*, John Murray, London, UK, available online at: http://darwin-online.org.uk.

Darwin E (1784). The Botanical Garden, *Annals of Agriculture* **1**, 380, available online at: http://www.fullbooks.com/The-Botanic-Garden4.html.

Dash M (2001). *Tulip mania: The Story of the World's Most Coveted Flower and the Extraordinary Passions It Aroused*, Random House, New York, USA.

Daunay MC, Lester RN, Gebhardt C *et al.* (2001). Genetic resources of eggplant (*Solanum melongena L.*). and allied species: a new challenge for molecular geneticists and eggplant breeders, pp. 251–274. In: van den Berg RG, Barendse GW and Mariani C, editors, *Solanaceae V*, Nijmegen University Press, Nijmegen, Netherlands.

Davies RR (1987). *Conquest, Coexistence and Change. Wales 1063–1415*, Oxford University Press, Oxford, UK.

Davis CC and Wurdack KJ (2004). Host-to-parasite gene transfer in flowering plants: phylogenetic evidence from Malpighiales, *Science* **305**, 676–678.

Davis CC, Anderson WR and Wurdack KJ (2005). Gene transfer from a parasitic flowering plant to a fern. *Proceedings of the Royal Society B. Biological Sciences* **272**, 2237–2242, available online at: http://www.journals.royalsoc.ac.uk/media/2gugtwxvvn781qrgtav0/contributions/m/2/6/4/m264w5581870v70q_html/fulltext.html.

Davis RS (1990). Central Asian hunter-gatherers at the Last Glacial Maximum. In: Gamble C and Soffer O, editors, *The World at 18,000 BP*, Volume I, pp. 267–275, Unwin Hyman, London, UK.

Davis SJM and Valla FR (1978). Evidence for the domestication of the dog 12,000 years ago in the Natufian of Israel, *Nature* **276**, 608–610.

Deacon HJ (2001). Modern human emergence: an African archaeological perspective. In: Tobias PV, Raath MA, Maggi-Cecchi J and Doyle GA, editors, *Humanity from African Naissance to Coming Millennia—Colloquia in Human Biology and Palaeoanthropology*, pp. 217–226, Florence University Press, Florence, Italy.

de Candolle A (1883). In L'Origine des plantes cultivées (Diderot, Paris) [2nd edition (1886) reprinted, 1967, Hafner, New York, USA.]

Decker DS (1988). Origin(s), evolution, and systematics of *C. pepo* (Cucurbitaceae), *Economic Botany* **42**, 4–15.

D'Egidio MG, Nardi S and Vallega V (1993). Grain, flour, and dough characteristics of selected strains of diploid wheat *Triticum monococcum* L. *Cereal Chemistry* **70**, 298–303.

Deevey ES (1960). The human population, *Scientific American* **203**, 194–204.

Delgado Salinas A, Turley T, Richman A, Lavin M (1999). Phylogenic analysis of the cultivated and wild species of *Phaseolus* (Fabaceae), *Systematic Botany* **24**, 438–460.

Delseny M (2004). Re-evaluating the relevance of ancestral shared synteny as a tool for crop improvement, *Current Opinion in Plant Biology* **7**, 126–131.

Demarest A (2004). *Ancient Maya: The Rise and Fall of a Rainforest Civilization*, Cambridge University Press, Cambridge UK.

DeMenocal P, Ortiz J, Guilderson T and Sarnthein M (2000). Abrupt onset and termination of the African Humid Period: Rapid climate responses to gradual insolation forcing, *Quarterly Science Reviews* **19**, 347–361, available online at: http://www.ldeo.columbia.edu/~peter/Resources/deMenocal.QSR.2000.pdf.

DeMenocal P (2001). Cultural responses to climate change during the late Holocene, *Science* **292**, 667–673, available online at: http://research.yale.edu/leilan/demenoca12001.pdf.

De Moulins DM (1993). Les restes de plantes carbonisées de Cafer Höyük. *Cahiers de l'Euphrate* **7**, 191–234.

De Moulins DM (2000). Abu Hureyra 2: plant remains from the Neolithic. In: Moore AMT, Hillman GC and Legge AJ, editors, *Village on the Euphrates: from Foraging to Farming at Abu Hureyra*, pp. 399–416, Oxford University Press, Oxford, UK.

Deng W, Shi B, He X, Zhang Z, Xu J, Li B, Yang J, Ling L, Dai C, Qiang B, *et al.* (2004). Evolution and migration history of the Chinese population inferred from Chinese Y-chromosome evidence, *Journal of Human Genetics* **49**, 339–348, available online at: http://library.ibp.ac.cn/html/slwj/000222645600001.pdf.

De Pamphilis , Cui L, Wall K and Lindsay B (2006). Widespread genome duplications throughout the history of flowering plants, *Genome Research* **16**, 738–749.

Descartes R (1637). *Discourse on the Method*. [The now-standard version of this and many other writings from Descartes is: Cottingham J, Stoothoff R, Murdoch D and Kenny A, translators (1991). *The Philosophical Writings of Descartes*, Cambridge University Press, Cambridge, UK.]

Desmond A (1964). How many people have ever lived on the earth? In: Mudd S, editor, *The Population Crisis and the Use of World Resources*, pp. 26–44, Dr. W. Junk Publishers, The Hague, Netherlands.

Deur D (1999). Salmon, sedentism, and cultivation: toward an environmental prehistory of the northwest coast. In: Goble DD and Hirt PW, editors, *Northwest Lands, Northwest Peoples: Readings in Environmental History*, pp. 129–155, University of Washington Press, Seattle, USA.

Deur (2002). Plant cultivation on the northwest coast: A reconsideration, *Journal of Cultural Geography* **19**, 9–35.

de Vries (1901). *The Mutation Theory*, originally published as: *Die Mutationstheorie*, Veit and Co., Leipzig, Germany.

deWet JMJ (1978). Systematics and evolution of Sorghum section Sorghum (Gramineae), *American Journal of Botany* **65**, 477–484.

deWet JMJ (1995). Minor cereals: various genera (Gramineae). In: Smartt J and Simmonds NW, editors, *Evolution of Crop Plants*, pp. 261–266, Longman Scientific and Technical, Harlow, UK.

deWet JMJ, Rao KEP, Brink DE and Mengesha MH (1984). Systematics and evolution of *Eleusine coracana* (Gramineae), *American Journal of Botany* **71**, 550–557.

Diakonoff IM (1991). *Early Antiquity* (translation of *Istoriya Drevnego Mira* Vol 1: *Rannyaya Drtevnost'*, Nauka Press, 1989), translator: Alexander Kirjanov, project editor: Philip L. Kohl, University of Chicago Press, Chicago, USA.

Diamond J (1997). *Guns, Germs and Steel*, Norton, New York.

Diamond J (2003). A new scientific synthesis of human history. In: Brockman J, editor, *Science at the Edge*, pp. 15–31, Weidenfeld and Nicholson, London, UK.

Diao X, Freeling M and Lisch D (2006). Horizontal transfer of a plant transposon, *Public Library of Science, Biology* **4**, e5, available online at: http://biology.plosjournals.org/perlserv/?request = get-document&doi = 10.1371/journal.pbio.0040005.

Dickau R, Ranere AJ and Cooke RG (2007). From the Cover: Starch grain evidence for the preceramic dispersals of maize and root crops into tropical dry and humid forests of Panama, *Proceedings of the National Academy of Sciences USA* **104**, 3651–3656.

Diehl MW (1993). Mogollon manos, metates and agricultural dependence: pithouse villages, A.D. 200–1000, *SFI*

USA **96**, 14400–14405, available online at: http://www.pnas.org/cgi/content/full/96/25/14400.

Ge S, Sang T, Lu BR and Hing DY (2001). Rapid and reliable identification of rice genomes by RFLP analysis of PCR-amplified Adh genes, *Genome* **44**, 1136–1142, available online at: www.lseb.cn/oldzjxx/ges/PDF/Ge-genome.pdf.

Gepts P (1998). Origin and evolution of common bean: past events and recent trends, *Horticultural Science* **33**, 1124–1130.

Gepts P (1999). What can molecular markers tell us about the process of domestication in common bean? In: Damania AB, Valkoun J, Willcox G and Qualset CO, editors, *The Origins of Agriculture and Crop Domestication*, pp. 198–209, ICARDA, Aleppo, Syria, available online at: http://www.ipgri.cgiar.org/publications/HTMLPublications/47/.

Gepts P (2004). Crop domestication as a long-term selection experiment, *Plant Breeding Reviews* **24**, 1–44, available online at: http://agronomy.ucdavis.edu/gepts/LTS.pdf.

Gepts P and Papa R (2002). Evolution during domestication. In: Goodman R, general editor, *Encyclopedia of Life Sciences*, Wiley, New York, available online at: http://www.agronomy.ucdavis.edu/gepts/Gepts%20and%20Papa.%202002.pdf.

Gertz A and Wricke G (1991). Inheritance of temperature-induced pseudocompatability in rye, *Plant Breeding* **107**, 89–96.

Ghatnekar L, Jaarola, M and Bengtsson BO (2006). The introgression of a functional nuclear gene from *Poa* to *Festuca ovina*, *Proceedings of the Royal Society, B: Biological Sciences* **273**, 395–399.

Ghosh A (1982). Deurbanization of Harappan civilizations. In: Possehl GL, editor, *Harappan Civilization: A Contemporary Perspective*, pp. 321–324, Oxford-IBH, New Delhi, India.

Gibbon E (1776–1788). *History Of The Decline And Fall Of The Roman Empire*, reissued by The Folio Society, 1995, London, UK.

Gibbs WW (2003). The Unseen Genome: Gems among the Junk, *Scientific American* **289**, 27–33.

Gibson McG (2000). Tell Hamoukar: early city in northeastern Syria, *Antiquity* **74**, 477–478.

Gilbert W (1600). De magnete, magneticisque corporibus, et de magno magnete tellure : physiologia noua, plurimis & argumentis, & experimentis demonstrata, Peter Short, London, UK.

Gill P, Jeffreys AJ and Werrett DJ (1985). Forensic application of DNA 'finger prints', *Nature* **318**, 577–579.

Gill RB (2000). *The Great Maya Droughts: Water, Life, and Death*, University of New Mexico Press, Albuquerque, USA.

Gimpel J (1988). *The Medieval Machine*, Wildwood House, Aldershot, UK.

Glassner JJ (2003). *The Invention of Cuneiform: Writing in Sumer*, translated by Bahrani Z and van de Mieroop M, Johns Hopkins University Press, Baltimore, USA.

Glick TF (1970). *Irrigation and Society in Medieval Valencia*, Harvard University Press, Cambridge, Massachusetts, USA.

Glick TF (1982). Agriculture and Nutrition, The Mediterranean Region. In: Strayer J, editor, *Dictionary of the Middle Ages*, I, 79–88, Scribner's, New York, USA.

Godron DA (1854). Des hybrides végétaux, considérés au point de vue de leur fécondité et de la perpétuité ou non-perpétuité de leur caractères, Annales sciences naturales, 4(e)me série, *Botanique* **19**, 135–179.

Goebel T (1999). Pleistocene human colonization of Siberia and peopling of the Americas: An ecological approach, *Evolutionary Anthropology* **8**, 208–226, available online at: www.unr.edu/cla/anthro/evolanth.pdf.

Goethe JW (1790). Versuch die Metamorphose der Pflanzen zu erklären, Carl Wilhelm Ettinger, Gotha, Germany, original available online at: http://gutenberg.spiegel.de/goethe/metamorp/metamorp.htm. [This essay is now known in English as Metamorphosis of Plants. An English translation is: Marshall A and Grotzke H (1978). *The Metamorphosis of Plants*, Biodynamic Literature, Wyoming, Rhode Island. Another English translation is available online at: http://hps.elte.hu/~zemplen/Urpflanze.doc.]

Goff SA, Ricke D, Lam TH *et al.*(2002). A draft sequence of the rice genome (*Oryza sativa* L. ssp. *japonica*)., *Science* **296**, 92–100.

Goldsmith E and Hildyard N, editors (1984). *The Social and Environmental Effects of Large Dams: Volume 1. Overview*, Wadebridge Ecological Centre, Worthyvale Manor Camelford, Cornwall, UK.

Goldschmidt WR (1951). Nomlaki Ethnography, University of California Publications. In: *American Archaeology and Ethnography* **42**, 303–443 [Facsimile version available from Coyote Press, Salinas, California, USA].

Goodman AH, Armelagos GJ and Rose JC (1984a). The chronological distribution of enamel hypoplasias from prehistoric Dickson Mounds populations, *American Journal of Physical Anthropology* **65**, 259–266.

Goodman AH, Lallo J, Armelagos GJ and Rose JC (1984b). Health Changes at Dickson Mounds, Illinois (AD 950–1300). In: Cohen MN and Armelagos GJ, editors, *Paleopathology at the Origins of Agriculture*, pp. 271–305, Academic Press, New York, USA.

Goodyear AC (2004). Evidence of pre-clovis sites in the Eastern United States. In: Bonnichsen R Lepper BT,

Stanford D and Waters MR, editors, *Paleoamerican Origins: Beyond Clovis*, Texas A and M University Press, College Station. Texas, USA.

Gomi T (1979). On dairy productivity at Ur in the late Ur III period, *Journal of the Economic and Social History of the Orient* 23, 1–42.

Gomi T (1984). On the critical economic situation at Ur early in the reign of Ibbisin, *Journal of Cuneiform Studies* 36, 211–242.

Gopher A, Abbo, S and Lev-Yadun, S (2000). The 'when', the 'where' and the 'why' of the Neolithic revolution in the Levant, *Documenta Praehistorica* 28, 49–62.

Goring-Morris AN (1991). The Harifian of the Southern Levant. In: *The Natufian Culture in the Levant*, Bar-Yosef O and Kra R, editors, pp. 173–216, International Monographs in Prehistory, Ann Arbor, Michigan, USA.

Goring-Morris AN and Belfer-Cohen A (1997). The articulation of cultural processes and Late Quaternary environmental changes in Cisjordan, *Paléorient* 23, 71–93.

Gould WA (1983). *Tomato Production, Processing and Quality Evaluation*, pp. 3–50, 2nd edition, AVI Publishing Company, Westport, Connecticut, USA.

Govaerts R (2001). How many species of seed plant are there? *Taxon* 50, 1085–1090.

Grace J and Zhang (2006). Predicting the effect of climate change on global plant productivity and the carbon cycle. In: Morison JIL and Morecroft MD, editors, *Plant Growth and Climate Change*, pp. 187–208, Blackwell, Oxford, UK.

Gradstein FM, Ogg JG and Smith AG (2004). *Geologic Time Scale 2004*, Cambridge University Press, Cambridge, UK.

GrainGenes, a compilation of molecular and phenotypic information on wheat, barley, rye, triticale, and oats, supported by the USDA-ARS Plant Genome Research Program, available online at: http://wheat.pw.usda.gov/ggpages/GrainTax/index.shtml.

Grant V (1981). *Plant Speciation*, Columbia University Press, New York, USA.

Grant-Downton RT and Dickinson HG (2005). Epigenetics and its implications for plant biology. 1. The epigenetic network in plants, *Annals of Botany* 96, 1143–1164.

Grant-Downton RT and Dickinson HG (2006). Epigenetics and its implications for plant biology. 2. The 'epigenetic epiphany': epigenetics, evolution and beyond, *Annals of Botany* 97, 11–27.

Greenberg JH (1987). *Language in the Americas*, Stanford University Press, Stanford, California, USA.

Greenberg JH, Turner CG and Zegura LZ (1986). The settlement of the Americas: a comparison of the linguistic, dental and genetic evidence, *Current Anthropology* 27, 477–497.

Greer FR, Krebs NF; American Academy of Pediatrics Committee on Nutrition (2006). Optimizing bone health and calcium intakes of infants, children, and adolescents, *Pediatrics* 117, 578–85, available online at: http://pediatrics.aappublications.org/cgi/content/full/117/2/578.

Green RE, Krause J, Ptak SE *et al.* (2006). Analysis of one million base pairs of Neanderthal DNA, *Nature* 444, 330–336.

Greilhuber J, Dolezel J, Lysak MA and Bennett MD (2005). The origin, evolution and proposed stabilization of the terms 'genome size' and 'C-value' to describe nuclear DNA contents, *Annals of Botany* 95, 255–260, available online at: http://aob.oxfordjournals.org/cgi/content/full/95/1/255.

Gregory RA (1916). *Discovery or the Spirit and Service of Science*, Macmillan, London, UK.

Gregory TR and Mable BK (2005). Polyploidy in animals. In: Gregory TR, editor, *The Evolution of the Genome*, pp. 427–517, Elsevier, New York, USA.

Gribbin J (2003). *Science: A History*, Penguin, London, UK.

Griffiths S, Sharp R, Foote T *et al.* (2006). Molecular characterization of Ph1 as a major chromosome pairing locus in polyploid wheat, *Nature* 439, 749–752.

Grimaldi D and Agosti D (2000). A formicine in New Jersey Cretaceous amber (Hymenoptera: Formicidae) and early evolution of the ants, *Proceedings of the National Academy of Sciences USA* 97, 13678–13683, available online at: http://www.pnas.org/cgi/content/full/240452097v1.

Grootes PM and Stuiver M (1997). Oxygen 18/16 variability in Greenland snow and ice with 10^{-3}- to 10^{-5}-year time resolution, *Journal of Geophysical Research* 102, 455–470.

Gupta AK (2004). Origin of agriculture and domestication of plants and animals linked to early Holocene climate amelioration, *Current Science* 87, 54–59, available online at: www.ias.ac.in/currsci/ju1102004/54.pdf.

Gupta SP (1982). The Late Harappans: A study in cultural dynamics. In: Possehl GL, editor, *Harappan Civilization: A Contemporary Perspective*, pp. 51–59, Oxford-IBH, New Delhi, India.

Haak W, Forster P, Bramanti B *et al.* (2005). Ancient DNA from the first European farmers in 7500-year-old Neolithic sites, *Science* 310, 1016–1018.

Habetha M and Bosch TC (2005). Symbiotic Hydra express a plant-like peroxidase gene during oogenesis, *Journal of Experimental Biology* 208, 2157–6, available online at: http://jeb.biologists.org/cgi/content/full/208/11/2157.

Habu J (2004). *Ancient Jomon of Japan*, Cambridge University Press, Cambridge, UK.

Hackett JD, Yoon HS, Soares MB *et al.* (2004). Migration of the plastid genome to the nucleus in a peridinin dinoflagellate, *Current Biology* **14**, 213–218, available online at: www.biology.uiowa.edu/debweb/downloads/Files/Hackett%20et%20al%20Curr%20Bio%202004.pdf.

Halstead P (1989). Like rising damp? An ecological approach to the spread of farming in southeast and central Europe. In: Milles A, Williams D and Gardner N, editors, *The Beginnings of Agriculture*, pp. 23–53, British Archaeological Reports, International Series, **496**, Oxford, UK.

Hammer K (1984). Das Domestikationssyndrom, *Kulturpflanze* **32**, 11–34.

Hammond N, Tourtellot G, Donaghey S and Clarke A (1998). No slow dusk: Maya urban development and decline at La Milpa, Belize, *Antiquity* **72**, 831–837.

Hancock JF (2004). *Plant Evolution and the Origin of Crop Species*, 2nd edition, CAB International, Wallingford, UK.

Hancock JF (2005). Contributions of domesticated plant studies to our understanding of plant evolution, *Annals of Botany* **96**, 953–963.

Hansen DP (1965). The relative chronology of Mesopotamia. Part II. The pottery sequence at Nippur from the Middle Uruk to the end of the Old Babylonian Period (3400–1600 BC). In: Ehrich RW, editor, *Chronologies in Old World Archaeology*, pp 201–214, University of Chicago Press, Chicago, USA.

Hansen J (1991). *Excavations at Franchthi cave, Greece, Fascicule 7, The Palaeoethnobotany*, Indiana University Press, Bloomington, Indiana, USA.

Hanson RE, Zhao X-P, Islam-Faridi MN, Paterson AH, Zwick MS, Crane CF, McKnight TD, Stelly DM and Price HJ (1998). Evolution of interspersed repetitive elements in *Gossypium* (Malvaceae), *American Journal of Botany* **85**, 1364–1368, available online at: www.amjbot.org/cgi/content/full/85/10/1364.

Harder B (2001). The Seeds of Malaria, Recent evolution cultivated a deadly scourge, *Science News* **160**, 296.

Hare B, Plyusnina I, Ignacio N *et al.* (2005). Social cognitive evolution in captive foxes is a correlated by-product of experimental domestication, *Current Biology* **15**, 226–230, available online at: http://www.current-biology.com/content/article/abstract?uid = PIIS0960982205000928.

Harlan HV and Martini ML (1936). Problems and results in barley breeding, *Yearbook of Agriculture 1936*, US Government Printing Office, Washington, DC, USA, available online at: http://barley.ipk-gatersleben.de/archives/Harlan&Martini1936.htm.

Harlan JR (1967). A wild wheat harvest in Turkey, *Archaeology* **19**, 197–201.

Harlan JR (1975). *Crops and Man*, American Society of Agronomy and Crop Science Society of America, Madison, USA.

Harlan JR (1981). The early history of wheat: Earliest traces to the sack of Rome. In: Evans LT and Peacock WJ, editors, *Wheat Science Today and Tomorrow*, Cambridge University Press, Cambridge, UK.

Harlan JR (1992a). Wild grass or domestication, in *Préhistorie de l'agriculture: nouvelle approaches expérimentales et ethnographiques*, Anderson PC, editor, pp. 21–28, Éditions du CNRS, Paris, France.

Harlan JR (1992b). *Crops and Man*, 2nd edition, American Society of Agronomy, Madison, Wisconsin, USA.

Harlan JR (1995). *The Living Fields: Our Agricultural Heritage*, Cambridge University Press, Cambridge, UK.

Harris DR (1967). New light on plant domestication and the origins of agriculture: a review, *Geographical Reviews* **57**, 90–107.

Harris DR (1989). An evolutionary continuum of people-plant interactions. In: Harris DR and Hillman GC, editors, *Foraging and Farming: the Evolution of Plant Exploitation*, pp. 11–26, Unwin and Hyman, London, UK.

Harris DR (1996). The origins and spread of agriculture and pastoralism in Southwest Asia. Paper given at the *International Symposium on The Origins of Agriculture* held at the International Research Center for Japanese Studies, 13–16 December 1996, Nara, Japan.

Harris DR (2003). Agriculture: why and how did it begin? *Encyclopaedia of Plant and Crop Science*, pp. 5–8, Marcel Dekker Inc, New York, USA.

Harrison P and Pearce F (2001). *AAAS Atlas of Population and Environment*, American Association for the Advancement of Science, Washington, DC, available online at: http://atlas.aaas.org/.

Harsch E (2004). Farmers embrace African 'miracle' rice: high-yielding 'Nerica' varieties to combat hunger and rural poverty, *Africa Recovery—United Nations Department of Public Information* **17**, 10–15.

Harvati K, Frost SR and McNulty KP (2004). Neanderthal taxonomy reconsidered: implications of 3D primate models of intra- and interspecific differences, *Proceedings of the National Academy of Sciences USA* **101**, 47–52, available online at: http://www.pnas.org/cgi/reprint/0308085100v1.

Harvey E and Fuller DQ (2005). Investigating crop processing through phytolith analysis: the case of rice and millets, *Journal of Archaeological Science* **32**, 739–752, available online at: www.ucl.ac.uk/archaeology/staff/profiles/fuller/pdfs/HarveyFuller05.pdf.

Harvey MH, McMillan M, Morgan MR and Chan HW (1985). Solanidine is present in sera of healthy individuals and in amounts dependent on their dietary potato consumption, *Human Toxicology* **4**, 187–194.

Haug G, Günther D, Peterson LC, Sigman DM, Hughen KA and Aeschlimann B (2003). Climate and the collapse of Maya civilization, *Science* **299**, 1731–1735.

Hawkes K, O'Connell JF, Blurton Jones NG, Alvarez H, Charnov EL (1998). Grandmothering, menopause, and the evolution of human life histories, *Proceedings of the National Academy of Sciences USA* **95**, 1336–1339, available online at: www.pnas.org/cgi/reprint/95/3/ 1336.pdf.

Hawkes JG (1998). Back to Vavilov: why were plants domesticated in some areas and not in others? in *The Origins of Agriculture and Crop Domestication*, Damania AB, Valkoun J, Willcox G and Qualset CO, editors (1998)., Aleppo, Syria, ICARDA, available online at: http://www.ipgri.cgiar.org/publications/HTMLPublications/47/.

Hayden B (1995). A new overview of domestication. In: Price TD and Gebauer AB, editors, *Last Hunters-First Farmers*, pp.273–299, School of American Research Press, Santa Fe, New Mexico, USA.

Haywood J (2005). *Historical Atlas of Ancient Civilizations*, Penguin, London, UK.

Heap D and Wirz J, editors (2002). *Genetic Engineering and the Intrinsic Value and Integrity of Animals and Plants*, Proceedings of a Workshop at the Royal Botanic Garden, Edinburgh, UK 18–21 September 2002, *Ifgene*—International Forum for Genetic Engineering, Dornach, Switzerland, available online at: http://www.ifgene.org/2002intro&summary.htm.

Hebert PDN, Penton EH, Burns JM, Janzen DH and Hallwachs (2004). Ten species in one: DNA barcoding reveals cryptic species in the neotropical skipper butterfly, *Astraptes fulgerator*, *Proceedings of the National Academy of Sciences USA* **101**, 14812–14817, available online at: www.pnas.org/cgi/reprint/101/41/ 14812.pdf.

Heckenberger MJ (1998). Manioc agriculture and sedentism in Amazonia: the Upper Xingu example, *Antiquity* **72**, 633–648.

Heckenberger MJ, Kuikuro A, Kuikuro UT *et al.* (2003). Amazonia 1492: pristine forest or cultural parkland? *Science* **301**, 1710–1714.

Hegarty MJ, Barker GL, Wilson ID, Abbott RJ, Edwards KJ and Hiscock SJ (2006). Transcriptome shock after interspecific hybridization in *Senecio* is ameliorated by genome duplication, *Current Biology* **16**, 1652–1659.

Hegarty MJ and Hiscock SJ (2005). Hybrid speciation in plants: new insights from molecular studies, *New Phytologist* **165**, 411–423.

Hegde SG, Nason JD, Clegg J, and Ellstrand NC (2006). The evolution of California's wild radish has resulted in the extinction of its progenitors, *Evolution* **60**, 1187–1197.

Heiser CB (1985). *Of Plants and People*, University of Oklahoma Press, Norman, Oklahoma, USA.

Helbaek H (1969). Plant collecting, dry-farming and irrigation agriculture in prehistoric Deh Luran. In: Hole F, Flannery KV and Neely JA, editors, *Prehistory and Human Ecology of the Deh Luran Plain: An Early Village Sequence from Khuzistan, Memoirs Museum Anthropology* No. **1**, pp. 383–426, University of Michigan, Ann Arbor, USA.

Helbaek H (1970). The plant husbandry of Hacilar. In: Mellaart J, editor, *Excavations at Hacilar*, pp. 189–244, Edinburgh: Edinburgh University Press, Edinburgh, UK.

Helbaek H (1996). Pre-Pottery Neolithic farming at Beidha, *Palestine Exploration Quarterly* **98**, 61–66.

Helm J (1963). Morphologisch-taxonomische Gliederung der Kultursippen von *Brassica oleracea*, *Kulturpflanze* **11**, 92–210.

Helms M (2003). Tangible materiality and cosmological others in the development of sedentism. In: Demarrais E, C. Gosden C and Renfrew C, editors, *Rethinking Materiality, The Engagement of Mind with the Material World*, MacDonald Institute for Archaeological Research, University of Cambridge, Cambridge, UK.

Hemming SR (2004). Heinrich events: Massive later Pleistocene detritus layers of the North Atlantic and their global climate imprint, *Reviews of Geophysics* **42**, RG 1005.

Henry DO (1989). *From Foraging to Agriculture: The Levant at the End of the Ice Age*, University of Philadelphia Press, Philadelphia, USA.

Henry DO, editor (2003). *Neanderthals in the Levant: Behavioral Organization and the Beginnings of Human Modernity*, Continuum, London, UK.

Henshilwood CS (2006). Modern humans and symbolic behaviour: Evidence from Blombos Cave, South Africa. In: Blundell G, editor, *Origins*, pp. 78–83, Double Storey, Cape Town, South Africa.

Henshilwood CS (2007). Fully symbolic sapiens behaviour: Innovation in the Middle Stone Age at Blombos Cave, South Africa. In: Stringer C and Mellars P, editors, *Rethinking the Human Revolution: New Behavioural and Biological Perspectives on the Origins and Dispersal of Modern Humans*, MacDonald Institute Research Monograph series, University of Cambridge Press, Cambridge, UK.

Henshilwood CS, d'Errico F, Marean CW, Milo RG and Yates R (2001). An early bone tool industry from the Middle Stone Age at Blombos Cave, South Africa: implications for the origins of modern human behaviour, symbolism and language, *Journal of Human Evolution* **41**, 631–678.

Henshilwood CS, d'Errico F, Yates R, Jacobs Z, Tribolo C, Duller GA, Mercier N, Sealy JC, Valladas H, Watts I and

Wintle AG (2002). Emergence of modern human behavior: Middle Stone Age engravings from South Africa, *Science* **295**, 1278–1280.

Henshilwood CS and Marean CW (2003). The origin of modern human behavior. Critique of models and their test implications, *Current Anthropology* **44**, 627–651, available online at: www.svf.uib.no/sfu/blombos/pdf/12.%20CSH%20CWM%20Origin%20MHB%20CA%202003.pdf.

Heun M, Schafer-Pregl R, Klawan D *et al.*(1997). Site of einkorn wheat domestication identified by DNA fingerprinting, *Science* **278**, 1312–1314.

Hey J (2005). On the number of New World founders: a population genetic portrait of the peopling of the Americas, *Public Library of Science, Biology* **3**, (6). e19, available online at: http://biology.plosjournals.org/perlserv/?request = get-document&doi = 10.1371/journal.pbio.0030193.

Heun M, Schäfer-Pregl R, Klawan D, Castagna R, Accerbi M, Borghi B and Salamini F (1997). Site of einkorn wheat domestication identified by DNA fingerprinting, *Science* **278**, 1312–1314.

Higham C (1984). Prehistoric rice cultivation in Southeast Asia, *Scientific American* **250**, 138–146.

Higham C (1995). The transition to rice cultivation in Southeast Asia. In: Price TD and Gebauer AB, editors, *Last Hunters-First Farmers*, pp.127–155, School of American Research Press, Santa Fe, New Mexico, USA.

Higham C and Lu TL (1998). The origins and dispersal of rice cultivation, *Antiquity* **72**, 867–877.

Hill AW (1915). The History and Function of Botanical Gardens, *Annals of the Missouri Botanical Garden* **2**, 185–240.

Hillman GC (1975). The plant remains from Tell Abu Hureyra: A preliminary report, *Proceedings of the Prehistoric Society* **41**, 70–73.

Hillman GC (1978). On the origins of domestic rye—*Secale cereale, Anatolian Studies* **28**, 157–174.

Hillman GC (1984). Traditional husbandry and processing of archaic cereals in modern times: Part I, the glume wheats, *Bulletin on Sumerian Agriculture* **2**, 114–152.

Hillman GC (1996). Late Pleistocene changes in wild plant-foods available to hunter gatherers of the northern Fertile Crescent: possible preludes to cereal cultivation. In: Harris DR, editor, *The Origins and Spread of Agriculture and Pastoralism in Eurasia*, pp. 159–203, Smithsonian Institution Press, Washington, DC, USA.

Hillman GC, Colledge SM and Harris DR (1989). Plant-food economy during the Epipalaeolithic period at Tell Abu Hureyra, Syria: dietary diversity, seasonality, and modes of exploitation. In: Harris DR and Hillman GC editors, *Foraging and Farming: the Evolution of Plant Exploitation*, pp. 240–268, Unwin Hyman, London, UK.

Hillman GC and Davies MS (1990a). Domestication rates in wild-type wheats and barley under primitive cultivation, *Biological Journal of the Linnean Society* **39**, 39–78.

Hillman GC and Davies MS (1990b). Measured domestication rates in wild wheats and barley under primitive cultivation, and their archaeological implications, *Journal of World Prehistory* **4**, 157–222.

Hillman G and Davies S (1999). Domestication rate in wild wheats and barley under primitive cultivation. In: *Prehistory of Agriculture: New Experimental and Ethnographic Approaches*, Monograph **40**, Anderson P, editor, pp. 70–102, Institute of Archaeology, University of California, Los Angeles, USA.

Hillman G, Hedges R, Moore A, Colledge, S and Pettitt P (2001). New evidence of Lateglacial cereal cultivation at Abu Hureyra on the Euphrates, *Holocene* **11**, 383–395.

Ho PT (1969). The origin of Chinese agriculture, *American Historical Review* **75**, 1–36.

Hodder I (2006). *The Leopard's Tale*, Thames and Hudson, London, UK.

Hodell DA, Curtis JH and Brenner M (1995). Possible role of climate in the collapse of Classic Maya civilisation, *Nature* **375**, 391–394.

Hodell DA, Brenner M, Curtis JH and Guilderson T (2001). Solar forcing of drought frequency in the Maya lowlands, *Science* **292**, 1367–1370.

Hofstra D and Champion P (2003). Manchurian wild rice—the alien invader can be stopped, *Protect, Winter 2003*, 15–17, National Centre for Aquatic Biodiversity and Biosecurity, New Zealand, available online at: www.biosecurity.org.nz/files/Man_Wild_Rice_screen.pdf.

Hole F (1994). Environmental instabilities and urban origins. In: Stein G and Rothman MS, editors, *Chiefdoms and Early States in the Near East: The Organizational Dynamics of Complexity*, pp. 121–143, Prehistory Press, Madison, Wisconsin, USA.

Holland T (2006). *Persian Fire*, Abacus, London, UK.

Holmes FL (1973). Liebig, Justus von. *Dictionary of Scientific Biography* **8**, 329–335, Charles Scribner's Sons, New York, USA.

Homer (*c.* 800 BCE). *The Odyssey*, translated by R. Fitzgerald R (1961), Doubleday, New York, USA.

Hopf M (1983). Jericho plant remains. In: Kenyon KM and Holland TA, editors, *Excavations at Jericho*, pp. 576–621, British School of Archaeology in Jerusalem, London, UK.

Hopf M and Bar-Yosef O (1987). Plant remains from Hayonim Cave, Western Galilee, *Paléorient* **13**, 115–120.

Hoppenbrouwers P (1997). Agricultural production and technology in the Netherlands. In: Astill G and

Langdon J, editors, *Medieval Farming and Technology*, pp. 69–88, Brill, Leiden, Netherlands.

Hossfeld U and Olsson L (2002). From the modern synthesis to Lysenkoism, and back? *Science* **297**, 55–56.

Houghton J (1681–1683). *A Collection of Letters for the Improvement of Husbandry and Trade (1681–1683)*, Houghton, London.

Houghton J (2004). *Global Warming, A Complete Briefing*, 3rd edition, Cambridge University Press, Cambridge, UK.

Howard A (1945). *Farming and Gardening for Health or Disease (The Soil and Health)*, Faber and Faber, London, UK.

Huang S, Sirikhachornkit A, Sun X *et al.*(2002). Genes encoding plastid acetyl-CoA carboxylase and 3-phosphoglycerate kinase of the *Triticum/Aegilops* polyploid wheat, *Proceedings of the National Academy of Sciences USA* **99**, 8133–8138, available online at: http://www.pnas.org/cgi/content/full/99/12/8133.

Hubbard RNLB (1990). Carbonised seeds from Tepe Abdul Hosein: results of preliminary analyses. In: Pullar J, editor, *Tepe Abdul Hosein: A Neolithic site in Western Iran, Excavations 1978*, pp. 217–221, British Archaeological Report International Series 563.

Huckell BB (1996). The archaic prehistory of the North American Southwest, *Journal of World Prehistory* **10**, 305–373.

Huke RE and Huke EH (1990). *Rice: Then and Now*, International Rice Research Institute, Los Baños, Philippines.

Hunt NB (2004). *The Historical Atlas of Ancient Mesopotamia*, Checkmark Books, New York, USA.

Huntley B (1993). Rapid early-Holocene migration and high abundance of hazel *Corylus avellana* (L.): alternative hypotheses, in Chambers FM, editor, *Climate Change and Human Impact on the Landscape*, pp. 205–215, Chapman and Hall, London, UK.

Hutchinson J, editor (1965). *Essays on Crop Plant Evolution*, Cambridge University Press, Cambridge, UK.

Hymowitz T (1970). On the domestication of the soybean, *Economic Botany* **24**, 408–421.

Hymowitz T (1995). Soybean: *Glycine max* (Leguminosae—Papilionoidae). In: Smartt J and Simmonds NW, editors, *Evolution of Crop Plants*, pp. 261–266, Longman Scientific and Technical, Harlow, UK.

Ibn-Al-Awwam (*c.* 1300 CE). *Le Livre de l'Agriculture, Kitab al-fila-hah*, translated by Clement-Mullet JJ (1862), Actes Sud, Paris, France.

Iltis H (1983). From teosinte to maize: the catastrophic sexual transmutation, *Science* **222**, 886–894.

Ingman M, Kaessmann H, Paabo S and Gyllensten U (2000). Mitochondrial genome variation and the origin of modern humans, *Nature* **408**, 708–713.

Ingram CJ, Elamin MF, Mulcare CA *et al.* (2007). A novel polymorphism associated with lactose tolerance in Africa: multiple causes for lactase persistence? *Human Genetics* **120**, 779–788 .

International Human Genome Sequencing Consortium (2004). Finishing the euchromatic sequence of the human genome, *Nature* **431**, 931–945.

International Water Management Institute (2006). *Insights from the Comprehensive Assessment of Water Management in Agriculture*, International Water Management Institute, Colombo, Sri Lanka, available online at: http://news.bbc.co.uk/1/shared/bsp/hi/pdfs/21_08_06_world_water_week.pdf.

Ishii T, Xu Y and McCouch SR (2001). Nuclear- and chloroplast-microsatellite variation in A-genome species of rice, *Genome* **44**, 658–666.

Issar AS and Zohar M (2004). *Climate Change—Environment and Civilization in the Middle East*, Springer, Berlin, Germany.

Jaaska V (1998). On the origin and *in statu nascendi* domestication of rye and barley: a review. In: Damania AB, Valkoun J, Willcox G and Qualset CO, editors, *The Origins of Agriculture and Crop Domestication*, Aleppo, Syria, ICARDA, available online at: http://www.ipgri.cgiar.org/publications/HTMLPublications/47/.

Jablonski NG and Chaplin G (2000). The evolution of human skin coloration, *Journal of Human Evolution* **39**, 57–106, available online at: http://www.bgsu.edu/departments/chem/faculty/leontis/chem447/PDF_files/Jablonski_skin_color_2000.pdf.

Jablonski NG and Chaplin G (2002). Skin deep, *Scientific American* **287**, 74–82.

Jackson RH and Castillo E (1995). *Indians, Franciscans, and Spanish Colonization—The impact of the Mission System on California Indians*, University of New Mexico Press, Albuquerque New Mexico, USA.

Jacob J (1995). Ancient Maya wetland agricultural fields in Cobweb swamp, Belize: Construction, chronology, and function, *Journal of Field Archaeology* **22**, 175–190.

Jacobsen T and Adams RM (1958). Salt and silt in ancient Mesopotamian agriculture, *Science*, **128**, 1252–1258.

Jacobsen T (1982). *Salinity and Irrigation Agriculture in Antiquity, Report on Essential Results, Diyala Basin Agricultural Project, 1957–58*, Undena Publications, Malibu, USA.

Jacquin N (1797). *Plantarum rariorum horti Caesarei Schoenbrunnensis*, Vienna, Austria.

Jaenicke-Després V, Buckler ES, Smith BD, Gilbert MTP, Cooper A, Doebley J and Pääbo S (2003). Early allelic selection in maize as revealed by ancient DNA, *Science* **302**, 1206–1208.

Jain HK and Kharkwal MC, editors (2004). *Plant Breeding: Mendelian to Molecular Approaches*, Narosa Publishing, New Delhi, India.

James MG, Robertson DS and Myers AM (1995). Characterization of the maize gene sugary1, a determinant of starch composition in kernels, *Plant Cell* **7**, 417–429, available online at: http://www.plantcell.org/cgi/reprint/7/4/417.

Janick J, editor (1999). *Perspectives on new crops and new uses, proceedings of the fourth national symposium new crops and new uses: biodiversity and agricultural sustainability*, ASHS Press, Alexandria, Virginia, USA.

Jansen K (1938). Some west Baltic pollen diagrams, *Quartar* **1**, 124–139.

Jarrige CJ, Jarrige JF, Meadow RH and Quivron G (1995). *Mehrgarh: Field Reports 1974–1985: From Neolithic Times to the Indus Civilization*, Karachi: Department of Culture and Tourism, Government of Sindh, Pakistan.

Jasny N (1944). *The Wheats of Classical Antiquity, The Johns Hopkins University Studies in Historical and Political Science*, Series LXII, No. 3, Johns Hopkins Press, Baltimore, Maryland, USA.

Jennings J (1799). *Letters and Papers on Agriculture, Planting, etc.* (Bath) **9**, 97.

Ji HS, Chu SH, Jiang W *et al.*(2006). Characterization and mapping of a shattering mutant in rice that corresponds to a block of domestication genes, *Genetics* **173**, 995–1005, available online at: http://www.genetics.org/cgi/content/full/173/2/995.

Jiménez-Espejo FJ, Martínez-Ruiz F, Finlayson C, *et al.* (2007). Climate forcing and Neanderthal extinction in Southern Iberia: insights from a multiproxy marine record, *Quaternary Science Reviews* **26**, in press.

Johnsen SJ, Clausen HB, Dansgaard W *et al.* (1997). The δ^{O-18} record along the Greenland Ice Core Project deep ice core and the problem of possible Eemian climatic instability, *Journal of Geophysical Research* **102**, 397–410.

Johnson GA (1973). *Local Exchange and Early State Development in Southwestern Iran*, Archaeological Papers 51, Museum of Anthropology, University of Michigan, Ann Arbor, Michigan, USA.

Johnson GA (1988). Late Uruk in Greater Mesopotamia: expansion or collapse? *Origini* **14**, 595–612.

Johnson GA (1982). Organizational structure and scalar stress. In: Renfrew C, Rowlands MJ and Segraves BA, editors, *Theory and Explanation in Archaeology: the Southampton Conference*, pp. 389–422, Academic Press, New York, USA.

Johnson ME, Cheng Z, Morrison VA *et al.* (2006). Eukaryotic transposable elements and genome evolution special feature: recurrent duplication-driven transposition of DNA during hominoid evolution, *Proceedings of the National Academy of Sciences USA* **103**, 17626–17631.

Johnston JS, Pepper AE, Hall AE, Chen ZJ, Hodnett G, Drabek J, Lopez R and Price HJ (2005). Evolution of genome size in Brassicaceae, *Annals of Botany* **95**, 229–235.

Johnston KJ (2003). The intensification of pre-industrial cereal agriculture in the tropics: Boserup, cultivation lengthening, and the Classic Maya, *Journal of Anthropological Archaeology* **22**, 126–161.

Jones G (2005). Garden cultivation of staple crops and its implications for settlement location and continuity, *World Archaeology* **37**, 164–176.

Jones MK, Allaby RG and Brown TA (1998). Wheat domestication, *Science* **279**, 302–303.

Jones R and Meehan B (1989). Plant foods of the Gidjingali: Ethnographic and archaeological perspectives from northern Australia on tuber and seed exploitation. In: Harris DR and Hillman GC, editors, *Foraging and Farming: The Evolution of Plant Exploitation*, pp.120–135, Unwin Hyman, London, UK.

Joshi SP, Gupta VS, Aggarwal RK, Ranjekar PK and Brar DS (2000). Genetic diversity and phylogenetic relationship as revealed by inter simple sequence repeat (ISSR). polymorphism in the genus *Oryza, Theoretical and Applied Genetics* **100**, 1311–1320, available online at: www.ccmb.res.in/publications/newpub/paps/pap412.pdf.

Jursa M (1995). Die Landwirtschaft in Sippar in Neubabylonischer Zeit, *Archiv für Orientforschung Beiheft* **25**, University of Vienna, Austria.

Jusuf M and Pernes J (1985). Genetic variability of foxtail millet (*Setaria italica* Beauv.): electrophoretic study of five isoenzyme systems, *Theoretical and Applied Genetics* **71**, 385–391.

Kain RJP, Chapman J and Oliver RR (2004). *The Enclosure Maps of England and Wales 1595–1918, A Cartographic Analysis and Electronic Catalogue*, Cambridge University Press, Cambridge, UK.

Kalyanaraman S (2003). *Sarasvati: Civilization*, Baba Saheb, Bangalore, India, available online at: http://www.hindunet.org/saraswati/heritage1.pdf#search = %22Kalyanaraman%20S%20(2003).%20Sarasvati%3A%20Civilization%2C%20Baba%20Saheb%2C%20Bangalore%2C%20India%22.

Kamen H (1997). *The Spanish Inquisition—an Historical Revision*, Yale University Press, New Haven, Connecticut, USA.

Kami J, Becerra Velásquez B, Debouck DG and Gepts P (1995). Identification of presumed ancestral DNA sequences of phaseolin in *Phaseolus vulgaris, Proceedings* of the National Academy of Sciences USA **92**, 1101–1104, available online at: www.pnas.org/cgi/reprint/92/4/1101.

Kamoun S and Smart CD (2005). Late blight of potato and tomato in the genomics era, *Plant Disease* **89**, 692–699 available online at: www.oardc.ohio-state.edu/phytophthora/pdfs/PDisease_05.pdf.

Kang MS (2002). *Quantitative Genetics, Genomics, and Plant Breeding*, CABI Publishing, Wallingford, Oxford, UK.

Kano-Murakami Y, Yanai T, Tagiri A and Matsuoka M (1993). A rice homeotic gene, OSH1, causes unusual phenotypes in transgenic tobacco, *FEBS Letters* **334**, 365–368.

Kantner J (2003). *Ancient Puebloan Southwest*, Cambridge University Press, Cambridge, UK.

Kaplan L (1986). Preceramic *Phaseolus* from Guilá Naquitz. In: Flannery KV, editor, *Guilà Naquitz: Archaic Foraging and Early Agriculture in Oaxaca, Mexico*, pp. 281–284, Academic Press, New York, USA.

Kaplan L and Lynch T (1999). *Phaseolus* (Fabaceae) in archaeology: AMS radioactive carbon dates and their significance in pre-Columbian agriculture, *Economic Botany* **53**, 262–268.

Ke Y, Su B, Song X *et al.* (2001). African origins of modern humans in East Asia: A tale of 12,000 Y chromosomes, *Science* **292**, 1151–1153.

Keeley LH (1995). Protoagricultural practices among hunter-gatherers. A cross-cultural survey. In: Price TD and Gebauer AB, editors, *Last Hunters-First Farmers*, pp.243–272, School of American Research Press, Santa Fe, New Mexico, USA.

Kellis M, Birren BW and Lander ES (2004). Proof and evolutionary analysis of ancient genome duplication in the yeast *Saccharomyces cerevisiae*, *Nature* **428**, 617–624.

Kelly RL (1995). *The Foraging Spectrum: Diversity in Hunter-Gatherer Lifeways*, Smithsonian Institution Press, Washington, DC, USA.

Kenoyer JM (1998). *Ancient Cities of the Indus Valley Civilization*, Oxford University Press, Oxford, UK.

Kennett DJ and Winterhalder B, editors (2006). *Behavioral Ecology and the Transition to Agriculture*, University of California Press, Berkeley, California, USA.

Kerr RA (2006). Climate change: yes, it's been getting warmer in here since the CO_2 began to rise, *Science* **312**, 1854.

Kershaw I (1973). The great famine and the agrarian crisis in England, *Past and Present* **59**, 3–50.

Kikkert JR, Vidal JR and Reisch BI (2005). Stable transformation of plant cells by particle bombardment/biolistics, *Methods in Molecular Biology* **286**, 61–78.

Kimber CT (2000). Origins of domesticated sorghum and its early diffusion to India and China. In: Smith CW and Frederiksen RA, editors, *Sorghum. Origin, History, Technology, and Production*, Wiley, New York, USA.

King LW (1896). *Cuneiform Texts*, Vol. **1**, British Museum, London.

King LW (1902). *The Seven Tablets of Creation: the Babylonian and Assyrian Legends Concerning the Creation of the World and of Mankind* [reprinted in 2004] Kessinger Publishing, Whitefish, Montana, USA.

Kirkbride D (1982). Umm Dabaghiyah. In: Curtis J, editor, *Fifty Years of Mesopotamian Discovery: The Work of the British School of Archaeology in Iraq 1932–1982*, pp. 11–21, British School of Archaeology in Iraq, London, UK.

Kislev ME (1988). Nahal Hemar Cave, desiccated plant remains: an interim report, *Atiqot* **18**, 76–81.

Kislev ME (1992). Agriculture in the Near East in the VIIth millennium B.C. In: Anderson PC, editor, *Préhistoire de l'Agriculture: Nouvelles Approches Expérimentales et Ethnographiques, Monographie du CRA No. **6***, Centre de Recherches Archéologiques, pp. 87–94, CNRS, Paris, France.

Kislev ME (1997). Early agriculture and paleoecology of Netiv Hagdud. In: Bar-Yosef O and Gopher A, editors, *An Early Neolithic Village in the Jordan Valley Part I: The Archaeology of Netiv Hagdud*, pp. 210–236, Peabody Museum of Archaeology and Ethnology, Cambridge, Massachusetts, USA.

Kislev ME, Hartmann A and Bar-Yosef O (2006). Early domesticated fig in the Jordan Valley, *Science* **312**, 1372–1374.

Kislev ME, Nadel D and Carmi I (1992). Epipaleolithic (19,000 BP). cereal and fruit diet at Ohalo II, Sea of Galilee, Israel, *Review of Paleobotany and Palinology* **73**, 161–166.

Kislev ME, Weiss E and Hartmann A (2004). Impetus for sowing and the beginning of agriculture: Ground collecting of wild cereals, *Proceedings of the National Academy of Sciences USA* **101**, 2692–2695, available online at: www.pnas.org/cgi/reprint/101/9/2692.pdf.

Kitching IJ, Forey PL, Humphries CJ and Williams D (1998). *Cladistics: The Theory and Practice of Parsimony Analysis*, 2nd edition, Oxford University Press, Oxford, UK.

Klein HS (1999). *The Atlantic Slave Trade*, Cambridge University Press, Cambridge, UK.

Klein RG and Edgar B (2002). *The Dawn of Human Culture*, Wiley, New York, USA.

Knapp SJ (1990). New temperate oilseed crops. In: Janick J and Simon JE, editors, *Advances in New Crops*, pp. 203–210, Timber Press, Portland, OR, available online at: http://www.hort.purdue.edu/newcrop/proceedings1990/v1–203.html.

Lu T (1998). Some botanical characteristics of green foxtail (*Setaria viridis*). and harvesting experiments on the grass, *Antiquity* **72**, 902–907.

Lu TLD (1999). The transition from foraging to farming and the origin of agriculture in China, *BAR International Series* **774**, British Archaeological Reports, Oxford, UK.

Lubbock, J (1861). The kjökkenmöddings: recent geologico-archæological researches in Denmark, *Natural History Review* **1**, 489–504.

Lucero LJ (2002). The collapse of the Classic Maya: a case for the role of water control, *American Anthropologist* **104**, 814–826.

Ludden D (2000). *An Agrarian History of South Asia*, Cambridge University Press, Cambridge, UK.

Luikart G, Gielly L, Excoffier L, Vigne JD, Bouvet J and Taberlet P (2001). Multiple maternal origins and weak phylogeographic structure in domestic goats, *Proceedings of the National Academy of Sciences USA* **98**, 5927–5932, available online at: www.pnas.org/cgi/reprint/98/10/5927.pdf.

Luikart G, Fernandez H, Mashkour M, England PR and Taberlet P (2006). Origins and diffusion of domestic goats inferred from DNA markers: Example analyses of mtDNA, Y chromosome, and microsatellites. In: Zeder MA, Bradley DG, Emshwiller E and Smith BD, editors, *Documenting Domestication: New Genetic and Archaeological Paradigms*, pp. 294–305, University of California Press, Berkeley, USA.

Lüning J (2000). *Steinzeitliche Bauern in Deutschland—die Landwirtschaft im Neolithikum*, Dr Rudolf Habelt GmbH, Boon, Germany.

Lutz W, Sanderson W and Scherbov S (2001). The end of world population growth, *Nature* **421**, 543–545.

Ma J, Devos KM and Bennetzen JL (2004). Analyses of LTR-retrotransposon structures reveal recent and rapid genomic DNA loss in rice, *Genome Research* **14**, 860–869, available online at: http://www.genome.org/cgi/reprint/1466204v1.

Mable BK (2004). 'Why polyploidy is rarer in animals than in plants': myths and mechanisms, *Biological Journal of the Linnean Society* **82**, 453–466.

Macaulay V, Hill C, Achilli A *et al.* (2005). Single, rapid coastal settlement of Asia revealed by analysis of complete mitochondrial genomes, *Science* **308**, 1034–1036.

Macdougall D (2004). *Frozen Earth. The once and future story of ice ages*, University of California Press, Berkeley, California, USA.

MacHugh DE and Bradley DG (2001). Livestock genetic origins: Goats buck the trend, *Proceedings of the National Academy of Sciences USA* **98**, 5382–5384, available online at: www.pnas.org/cgi/reprint/98/10/5382.pdf.

Mac Key J (2005). Wheat: its concept, evolution, and taxonomy. In: Royo C, Nachit MM, Di Fonzo N, Araus JL, Pfeiffer WH, Slafer GA, editors, *Durum Wheat Breeding Current Approaches and Future Strategies*, pp. 3–61, Haworth Press, Binghamton, New York, USA.

MacLuhan TC (1972). *Touch the Earth: A Self-Portrait of Indian Existence*, Abacus, London, UK.

MacNeish RS (1992). *The Origins of Agriculture and Settled Life*, University of Oklahoma Press, Norman, Oklahoma, USA.

MacNeish RS and Libby JG, editors (1995). *Origins of Rice Agriculture: the Preliminary Report of the Sino-American Jiangxi (PRC) Project SAJOR, Publications in Anthropology No. 13*, El Paso Centennial Museum, University of Texas at El Paso, El Paso, Texas, USA.

Madsen DB, Elston RG, Bettinger RL, Xu C and Zhong K (1996). Settlement patterns reflected in assemblages from the Pleistocene/Holocene transition of north central China, *Journal of Archaeological Science* **23**, 217–231, available online at: www.anthro.ucdavis.edu/card/USPRC/pubs/Madsen%20etal%201996.pdf.

Madsen DB and Elston RG (2007). Variation in late Quaternary central Asian climates and the nature of human response. In: Madsen DB, Chen F, and Gao X, editors, *Late Quaternary Climate Change and Human Adaptation in Arid China (Developments in Quaternary Sciences, Volume 9)*. Elsevier, Amsterdam, Netherlands.

Madsen DB, Chen F, Oviatt CG *et al.* (2003). Late Pleistocene/ Holocene Wetland events recorded in southeast Tengger Desert lake sediments, NW China, *Chinese Science Bulletin* **48**, 1423–1429, available online at: http://paleo.sscnet.ucla.edu/MadsenCSB2003–1.pdf.

Madsen DB, Chen F and Xing G (2007). Changing views of late Quaternary human adaptation in arid China. In: Madsen DB, Chen F, and Gao X, editors, *Late Quaternary Climate Change and Human Adaptation in Arid China (Developments in Quaternary Sciences, Volume 9)*, Elsevier, Amsterdam, Netherlands.

Maffie J (2002). Why care about Nezahualcoyotl? Veritism and Nahua philosophy? *Philosophy of the Social Sciences* **32**, 71–91.

Magny M and Haas JN (2004). A major widespread climatic change around 5300 cal. yr BP at the time of the Alpine Iceman, *Journal of Quaternary Science* **19**, 423–430, available online at: www.unige.ch/forel/PapersQG06/Magny&Haas2004_JQS.pdf.

Mahoney P (2007). Human dental microwear from Ohalo II (22500–23500 cal BP), southern Levant, *American Journal of Physical Anthropology* **132**, 489–500.

Maier U (1996). Morphological studies of free-threshing wheat ears from a Neolithic site in southwest Germany, and the history of naked wheats, *Vegetation History and Archaeobotany* **5**, 39–55.

Maloney BK, Higham CFW and Bannanurag R (1989). Early rice cultivation in Southeast Asia: archaeological and palynological evidence from the Bang Pakong Valley, Thailand, *Antiquity* **63**, 363–370.

Malville JM, Wendorf F, Mazar AA and Schild R (1998). Megaliths and Neolithic astronomy in southern Egypt, *Nature* **392**, 488–491.

Mann CC (2002). Has GM corn 'invaded' Mexico? Commentary, *Science* **295**, 1617–1618.

Mann CC (2004). *Diversity on the Farm*, Ford Foundation, New York, USA, available online at: http://www.umass.edu/peri/programs/development/Mann.pdf.

Mann CC (2006). *1491, The Americas before Columbus*, Granta, London, UK.

Mann ME, Bradley RS and Hughes MK (1998). Global-scale temperature patterns and climate forcing over the past six centuries, *Nature* **392**, 779–787.

Mann JA, Kimber CT and Miller FR (1983). *The Origin and Early Cultivation of Sorghums in Africa*, Texas Agricultural Experimental Station Bulletin 1454, Texas A and M University, College Station, Texas, USA.

Marcus GJ (1980). *The Conquest of the North Atlantic*, Boydell Press, Woodbridge, Suffolk, UK.

Marcus J and Flannery KV (1996). *The Zapotec Civilization*, Thames and Hudson, London, UK.

Markham G (1631). *Cheape and good hvsbandr[y] for the well-ordering of all beasts, and fowles, and for the generall cure of their diseases*, John Harrison, London, UK.

Marlowe FW (2005). Hunter-gatherers and human evolution, *Evolutionary Anthropology* **14**, 54–67, available online at: www.fas.harvard.edu/~hbe-lab/acrobat-files/hg%20and%20human%20ev.pdf.

Marshall F (2000). The origins of domesticated animals in eastern Africa. In: Blench RM and MacDonald KC, editors, *The Origins and Development of African Livestock*, pp. 17–44, University College of London Press, London, UK.

Martin DL and Goodman AH (2002). Health conditions before Columbus: paleopathology of native North Americans, *Western Journal of Medicine* **176**, 65–68.

Martin LJ and Cruzan MB (1999). Patterns of hybridization in the *Piriqueta caroliniana* complex in central Florida: evidence for an expanding hybrid zone, *Evolution* **53**, 1037–1049.

Martin PS (1984). Prehistoric overkill: the global model. In: Martin PS and Klein RG editors, *Quaternary Extinctions*, University of Arizona Press, Tucson, USA.

Martin W (1999). Mosaic bacterial chromosomes: a challenge *en route* to a tree of genomes, *BioEssays* **21**, 99–104, available online at: http://imbs.massey.ac.nz/bio_evol/Topic4/Martin99.pdf.

Martin W (2005). Molecular evolution: Lateral gene transfer and other possibilities, *Heredity* **94**, 565–566.

Martinez-Perez E, Shaw P and Moore G (2001). The Ph1 locus is needed to ensure specific somatic and meiotic centromere association, *Nature* **411**, 204–207.

Martinez-Perez E, Shaw P, Aragon-Alcaide L and Moore G (2003). Chromosomes form into seven groups in hexaploid and tetraploid wheat as a prelude to meiosis, *Plant Journal* **36**, 21–29.

Mason SLR (1995). Acornutopia? Determining the role of acorns in past human subsistence. In: Wilkins J, Harvey D and Dobson M, editors, *Food in Antiquity*, pp.12–24, Exeter University Press, Exeter UK.

Masson VM (1968). The urban revolution in South Turkmenia, *Antiquity* **42**, 178–187.

Masterson J (1994). Stomatal size in fossil plants: evidence for polyploidy in majority of angiosperms, *Science* **264**, 421–424.

Matsuo M, Yuki Ito Y, Yamauchi R and Obokata J (2005). The rice nuclear genome continuously integrates, shuffles, and eliminates the chloroplast genome to cause chloroplast–nuclear DNA flux, *Plant Cell* **17**, 665–675, available online at: http://www.plantcell.org/cgi/content/full/17/3/665.

Matsuoka Y (2005). Origin matters: lessons from the search for the wild ancestor of maize, *Breeding Science* **55**, 383–390.

Matsuoka Y, Vigouroux Y, Goodman MM, Sanchez J, Buckler E and Doebley J (2002). A single domestication for maize shown by multilocus microsatellite genotyping, *Proceedings of the National Academy of Sciences USA* **99**, 6080–6084, available online at: www.pnas.org/cgi/reprint/99/9/6080.pdf.

Matthiae P (1981). *Ebla: An Empire Rediscovered*, Doubleday, New York, USA.

Matthiae (1997). Ebla and Syria in the Middles Bronze Age. In: Oren E, editor, *The Hyksos: New Historical and Archaeological Perspectives*, University Museum Monographs 96, University Museum Symposium Series 8, pp.127–135, Philadelphia, USA.

Matthiae P, Pinnock F and Matthiae G (1997). Tell Mardikh 1977–1996: vingt ans de fouilles et de découvertes: la renaissance d'Ebla amorrhéenne, *Akkadica* **101**, 1–29.

Mayewski PA, Meeker LD, Morrison MC *et al.*(1993). Greenland ice core "signal" characteristics: An expanded view of climate change, *Journal of Geophysical Research* **98**, 839–847.

Mazoyer M and Roudart L (2006). *History of World Agriculture: From the Neolithic Age to the Current Crisis*, Monthly Review Press, New York, USA.

McBrearty S and Brooks AS (2000). The revolution that wasn't: a new interpretation of the origin of modern human behavior, *Journal of Human Evolution* **39**, 453–663, available online at: www.hss.caltech.edu/~steve/files/mcbrearty.pdf.

McClung de Tapia E (1992). The origins of Agriculture in Mesoamerica and Central America. In: Cowan CW and Watson PJ, editors, *The Origins of Agriculture*, pp. 143–171, Smithsonian Institution Press, Washington, DC, USA.

McCorriston J and Hole F (1991). The ecology of seasonal stress and the origins of agriculture in the Near East, *American Anthropologist* **93**, 46–94.

McCallum CM, Comai L, Greene EA, Henikoff S (2000a). Targeted screening for induced mutations, *Nature Biotechnology* **18**, 455–457.

McCallum CM, Comai L, Greene EA and Henikoff S (2000b). Targeting Induced Local Lesions IN Genomes (TILLING) for plant functional genomics, *Plant Physiology* **123**, 439–442, available online at: www.plantphysiol.org/cgi/reprint/123/2/439.pdf.

McClintock B (1987). *The Discovery and Characterization of Transposable Elements: the Collected Papers of Barbara McClintock*, Moore, JA, editor, Garland Publishing, New York, USA.

McCouch S (2004). Diversifying selection in plant breeding, *Public Library of Science, Biology* 2 e347, available online at: http://biology.plosjournals.org/perlserv?request = get-document&doi = 10.1371/journal.pbio.0020347.

McDermott F, Mattey DP and Hawkesworth C (2001). Centennial-scale Holocene climate variability revealed by a high-resolution speleothem $\delta^{18}O$ record from SW Ireland, *Science* **294**, 1328–1331.

McEwan GF (2006). Inca state origins. In: Schwartz GM and Nichols JJ, editors, *After Collapse, The Regeneration of Complex Societies*, pp. 85–98, University of Arizona Press, Tucson, Arizona, USA.

McFadden ES and Sears ER (1946). The origin of *Triticum spelta* and its free-threshing hexaploid relatives, *Journal of Heredity* **37**, 107–116.

McGee H (1984). The problems of legumes and flatulence. In: *On Food and Cooking: The Science and Lore of the Kitchen*, pp. 257–258, Unwin Hyman, London, UK.

McGrade AS (1974). *The Political Thought of William of Ockham*, Cambridge University Press, Cambridge, UK.

McIntosh SK and McIntosh RJ (1983). Current directions in West African prehistory, *Annual Review of Anthropology* **12**, 215–258.

McIntyre S and McKitrick R (2005). Hockey sticks, principal components, and spurious significance, *Geophysical Research Letters* **32**, L03710, available online at: http://www.uoguelph.ca/~rmckitri/research/trc.html.

McKenzie and Eberli (1987). Indications for abrupt Holocene climatic change: Late Holocene oxygen isotope stratigraphy of the Great Salt Lake. In: Berger H and Labeyrie D, editors, *Abrupt Climatic Change*, pp.127–136, D Reidel Publishing Company, Boston, USA.

MacNeish RS, Cunnar G, Zhao Z and Libby JG (1998). *Revised Second Annual Report of the Sino-American Jiangxi (PRC). Origin of Rice Project SAJOR*. Andover Foundation for Archaeological Research, Andover, USA.

MacNeish RS and Libby JG editors (1995). *Origins of Rice Agriculture: The Preliminary Report of the Sino-American Jiangxi (PRC). Project SAJOR*, Publications in Anthropology No. 13, El Paso Centennial Museum, University of Texas at El Paso, Texas, USA.

Meadow (1993). Animal domestication in the Near East: A revised view from the Eastern Margin. In: Possehl G, editor, *Harappan civilization*, 2nd edition, pp. 295–320, Oxford and IBH, New Delhi, India.

Meager L (1697). *The Mystery Of Husbandry or Arable, Pasture and Woodland Improved*, London, UK.

Meggers BJ (2003). Revisiting Amazonia circa 1492, *Science* **302**, 2067–2070.

Mehta KL (2002). Agricultural foundation of Indus-Saraswati civilization. In: Nene YL and Chowdhary SL, editors, *Agricultural Heritage of India, Proceedings of the National Conference*, pp. 1–21, Rajasthan College of Agriculture, University of Agriculture and Technology, Udaipur, Rajasthan, India.

Meißner B (1920). *Babylonien und Assyrien*, Carl Winters Universitätsverlag, Heidelberg, Germany.

Mekel-Bobrov N, Gilbert SL, Evans PD, Vallender EJ, Anderson JR, Hudson RR, Tishkoff SA and Lahn BT (2005). Ongoing adaptive evolution of ASPM, a brain size determinant in *Homo sapiens*, *Science* **309**, 1720–1722.

Mellars P, Gravina B, and Ramsey CB (2007). Confirmation of Neanderthal/modern human interstratification at the Chatelperronian type-site, *Proceedings of the National Academy of Sciences USA* **104**, 3657–3662.

Meltzer DJ and Mead JI (1983). The timing of the Late Pleistocene mammalian extinctions in North America, *Quaternary Research* **19**, 130–135.

Mendel K (1953). *The Development of our Knowledge on Transplantations in Plants, Actes du Septième Congres International d'Histoire des Sciences*, Jerusalem, Israel.

Merpert NY (1993). The Archaic phase of Hassuna Culture. In: Yoffee N and Clark JJ, editors, *Early Stages in the Evolution of Mesopotamian Civilization: Soviet Excavations in Northern Iraq*, pp. 115–127, The University of Arizona Press, Tucson, Arizona, USA.

Merriwether DA, Rothhammer F and Ferrell RE (1994). Genetic variation in the New World: ancient teeth, bone and tissue as sources of DNA, *Experimentia* **50**, 592–601.

Messing J, Bharti AK, Karlowski WM *et al.* (2004). Sequence composition and genome organization of maize, *Proceedings of the National Academy of Sciences USA* **101**, 14349–14354, available online at: http://www.pnas.org/cgi/reprint/101/40/14349.

Meyer P (2005). *Plant Epigenetics*, Blackwell, Oxford, UK.

Michalová A (1999). Minor cereals and pseudocereals in Europe. In: Maggioni L, complier, *Report of a Network Coordinating Group on Minor Crops*, IPGRI meeting, Turku, Finland, pp. 56–66, IPGRI, Rome, Italy, available online at: www.ipgri.cgiar.org/Publications/pdf/607.pdf.

Michalowski P (1989). *Lamentation over the Destruction of Sumer and Ur (Mesopotamian Civilizations Vol 1)*, Eisenbrauns, Winona Lake, Indiana, USA.

Miller GH, Magee JW, Johnson BJ, *et al.* (1999). Pleistocene Extinction of Genyornis newtoni: human impact on Australian megafauna, *Science* **283**, 205–208.

Miller GH, Fogel ML, Magee JW, Gagan MK, Clarke SJ and Johnson BJ (2005). Ecosystem collapse in Pleistocene Australia and a human role in megafaunal extinction, *Science* **309**, 287–290.

Miller JB, Foster-Powell K and Colaguiri S (1999). *The G.I. Factor*, Hodder and Stoughton, New York, USA.

Millon R (1973). *Urbanization at Teotihuacán*, Mexico, Vol. **1**, University of Texas Press, Austin, USA.

Milton K (1993). Diet and primate evolution, *Scientific American* **269**, 86–93.

Minc LD and Vandermeer J (1990). The origin and spread of agriculture. In: Carroll CR, Vandermeer JH and Rosset PM, editors, *Agroecology*, pp. 65–111, McGraw-Hill, New York, USA.

Mithen RF (2001). Glucosinolates and their degradation products, *Advances in Botanical Research* **35**, 213–262.

Mithen RF, Faulkner K, Magrath R, Rose P, Williamson G and Marquez G (2003). Development of isothiocyanate-enriched broccoli and its enhanced ability to induce phase 2 detoxification enzymes in mammalian cells, *Theoretical and Applied Genetics* **106**, 727–734.

Mithen S (2003). *After the Ice, A Global Human History 20,000–5,000 BC*, Phoenix, London, UK.

Moehs CP (2005). *TILLING: Harvesting Functional Genomics for Crop Improvement*, Information Systems for Biotechnology, March 2005, available online at: http://www.isb.vt.edu/news/2005/news05.Mar.html.

Mokrousov I, Ly HM, Otten T *et al.* (2000). Origin and primary dispersal of the *Mycobacterium tuberculosis* Beijing genotype: clues from human phylogeography, *Genome Research* **15**, 1357–1364, available online at: www.genome.org/cgi/reprint/gr.3840605v1.pdf.

Mokrousov I, Narvskaya O, Limeschenko E, Vyazovaya A, Otten T, and Vyshnevskiy B (2004). Analysis of the allelic diversity of the mycobacterial interspersed repetitive units in *Mycobacterium tuberculosis* strains of the Beijing family: practical implications and evolutionary considerations, *Journal of Clinical Microbiology* **42**, 2438–2444.

Molleson T (1994). The eloquent bones of Abu Hureyra, *Scientific American* **271**, 70–75.

Molleson T and Jones K (1991). Dental evidence for dietary change at Abu Hureyra, *Journal of Archaeological Science* **18**, 525–539.

Molleson T, Jones K and Jones S (1993). Dietary change and the effects of food preparation on microwear patterns in the Late Neolithic of Abu Hureyra, northern Syria, *Journal of Human Evolution* **24**, 455–468.

Money NP (2007). *The Triumph of the Fungi. A Rotten History*, Oxford University Press, Oxford, UK.

Monnin E, Indermühle A, Dällenbach A *et al.* (2001). Atmospheric CO2 concentrations over the last glacial termination, *Science* **291**, 112–114.

Montagu A (1964). Natural selection and man's relative hairlessness, *Journal of the American Medical Association* **187**, 120–121.

Mooney PR (1983). The law of the seed: another development and plant genetic resources, *Development Dialogue* **1–2**, 1–172.

Moore AMT (1978). *The Neolithic of the Levant*, D. Phil. thesis, University of Oxford, UK, available online at: http://ancientneareast.tripod.com/NeolithicLevant.html.

Moore AMT, Hillman GC and Legge AJ (2000). *Village on the Euphrates: from Foraging to Farming at Abu Hureyra*, Oxford University Press, Oxford, UK.

Moore RC and Purugganan MD (2005). The evolutionary dynamics of plant duplicate genes, *Current Opinion in Plant Biology* **8**, 122–128.

Moore T (1516). *Utopia*, Originally published in Latin as: *De Optimo Reipublicae Statu Deque Nova* insula Utopia, Louvain, English translation available online at: http://ota.ahds.ac.uk/texts/2080.html.

Moore TG (1998). *Climate of Fear: Why We Shouldn't Worry About Global Warming*, Cato Institute, Washington, DC, USA, available online at: www.cato.org/pubs/books/climate/climatepdf.html.

Moreau L, Lemarié S, Charcosset A and Gallais A (2000). Economic efficiency of one cycle of marker-assisted selection, *Crop Science* **40**, 329–337.

Morey DF (1994). The early evolution of the domestic dog, *American Scientist* **82**, 336–347.

available online at: http://www.pnas.org/cgi/content/full/97/4/1359.

Peteet DM, Daniels RA, Heusser LE, Vogel JS, Southon JR and Nelson DE (1993). Late-glacial pollen, macrofossils and fish remains in northeastern USA—the Younger Dryas oscillation, *Quaternary Science Reviews* **12**, 597–612, available online at: http://pubs.giss.nasa.gov/docs/1993/1993_Peteet_etal.pdf.

Peterson-Burch BD, Wright DA, Laten HM and Voyas DF (2000). Retroviruses in plants? *Trends in Genetics* **16**, 151–152.

Petit JR, Jouzel J, Raynaud D *et al.* (1999). Climate and atmospheric history of the past 420,000 years from the Vostok ice core, Antarctica, *Nature* **399**, 429–436.

Petrov DA, Sangster KA, Spencer Johnston J, Hartl DL and Shaw TL (2000). Evidence of DNA loss as a determinant of genome size, *Science* **287**, 1060–1062.

Pias J (1970). Les formations tertiaires et quaternaries de la cuvette tchadienne (Republique du Tchad). *Presetation de L'Esquisse Geologique au 1/1000000, Mém. Office de la Recherche Scientifique et Technique d'Outre-Mer* **43**, 425–429.

Pickersgill B (1971). Relationships between weedy and cultivated forms in some species of chili peppers (genus *Capsicum*), *Evolution* **25**, 683–691.

Piffanelli P, Ramsay L, Waugh R *et al.* (2004). A barley cultivation-associated polymorphism conveys resistance to powdery mildew, *Nature* **430**, 887–891.

Pigliucci M (2003). Species as family resemblance concepts: The (dis-).solution of the species problem? *BioEssays* **25**, 596–602, available online at: http://life.bio.sunysb.edu/ee/pigliuccilab/files/paper-BioEssays-species.pdf#search = %22Species%20as%20family%20resemblance%20concepts%3A%20The%20(dis-).solution%20of%20the%20species%20problem%3F%20%22.

Pikaard CS (2001). Genomic change and gene silencing in polyploids, *Trends in Genetics* **17**, 675–677.

Piperno DR (2001). On maize and the sunflower, *Science* **292**, 2260–2261.

Piperno DR and Flannery KV (2001). The earliest archaeological maize (Zea mays L.). from highland Mexico: New accelerator mass spectrometry dates and their implications, *Proceedings of the National Academy of Sciences USA* **98**, 2101–2103, available online at: http://www.pnas.org/cgi/reprint/98/4/2101.pdf#search = %22earliest%20archaeological%20maize%20(Zea%20mays%20L.).%20from%20highland%20Mexico%3A%20New%20accelerator%20mass%20spectrometry%20dates%20and%20their%20implications%2C%20%22.

Piperno DR and Pearsall DM (1998). *The Origins of Agriculture in the Lowland Neotropics*, Academic Press, San Diego, USA.

Piperno DR and Stothert KE (2003). Phytolith evidence for early Holocene *Cucurbita* domestication in southwest Ecuador, *Science* **299**, 1054–1057.

Piperno DR, Ranere AJ, Holst I and Hansell P (2000). Starch grains reveal early root crop horticulture in the Panamanian tropical forest, *Nature* **407**, 894–897.

Piperno DR, Weiss E, Holst I and Nabel D (2004). Processing of wild cereal grains in the Upper Palaeolithic revealed by starch grain analysis, *Nature* **430**, 670–673.

Plat H (1600). *The New and Admirable Art of Setting Corne*, London, UK.

Pliny the Elder or Caius Plinius Secundus (23–79 CE). *Historia Naturalis*, translated by Bostock J and Riley HT, 1855, available online at: http://www.perseus.tufts.edu/cgi-bin/ptext?lookup = Plin.+Nat.+toc.

Plunckett DL (1991). *Saving Lives Through Agricultural Research, Issues in Agriculture, No. 1*, Consultative Group on International Agricultural Research, Washington, DC, USA.

Póirtéir C editor (1995). *The Great Irish Famine*, Mercier Press, Dublin, Ireland.

Pollan M (2001). *The Botany of Desire: A Plant's Eye View of the World*, Random House, New York, USA.

Pollock S (1999). *Ancient Mesopotamia: The Eden That Never Was*, Cambridge University Press, Cambridge, UK.

Pollock S (2001). The Uruk Period in Southern Mesopotamia. In: Rothman MS, editor, *Uruk Mesopotamia and Its Neighbors Cross-Cultural Interactions in the Era of State Formation*, pp. 181–231, School of American Research Press, Santa Fe, New Mexico, USA.

Poncet V, Lamy F, Devos KM, Gale MD, Sarr A and Robert T (2000). Genetic control of domestication traits in pearl millet (*Pennisetum glaucum* L., Poaceae), *Theoretical and Applied Genetics* **100**, 147–159.

Poncet V, Thierry Robert T, Sarr A and Gepts P (2004). Quantitative trait locus analyses of the domestication syndrome and domestication process. In: Thomas B, Murphy DJ and Murray BG, editors, *Encyclopaedia of Plant and Crop Science*, pp. 1069–1073, Marcel Dekker Inc, New York, USA.

Pope KO, Pohl MED, Jones JG *et al.* (2001). Origin and environmental setting of ancient agriculture in the lowlands of Mesoamerica, *Science* **292**, 1370–1373.

Popkin B (2006). The nutrition transition in high and low-income countries: what are the policy lessons? *26th Conference of the International Association of Agricultural Economists*, 12–18 August, 2006, Gold Coast, Australia, see news report online at: http://news.bbc.co.uk/1/hi/health/4793455.stm.

Popovsky M (1984). *The Vavilov Affair*, Archon Books, Hamden, Connecticut, USA.

Porteres R (1956). Taxonomie agrobotanique des riz cultives *O. sativa* L. et *O. glaberrima*. S, *Journal d'Agriculture Tropicale et de Botanique Appliqué* **3**, 341–384; 541–580; 627–700; 821–856.

Possehl GL (2003). *The Indus Civilization: A Contemporary Perspective*, AltaMira Press, California, USA.

Postgate JN (1992). *Early Mesopotamia*, Routledge, New York, USA.

Postan MM (1973). *Essays on Medieval Agriculture and General Problems of the Medieval Economy*, Cambridge University Press, Cambridge, UK.

Potts D (1997). *Mesopotamian Civilization: The Material Foundation*, Cornell University Press, Ithaca, New York, USA.

Powell MA (1985). Salt, seed, and yields in sumerian agriculture. a critique of the theory of progressive salinization, *Zeitschrift für Assyriologie* **75**, 7–35.

Price KR, Lewis J, Wyatt GM and Fenwick GR (1988). Flatulence—causes, relation to diet and remedies, *Die Nahrung* **6**, 609–626.

Price TD, editor (2000). *Europe's First Farmers*, Cambridge University Press, Cambridge, UK.

Price TD, Bentley RA, Lüning J, Gronenborn D and Wahl J (2001). Prehistoric human migration in the Linearbandkeramik of Central Europe, *Antiquity* **75**, 593–603.

Price TD, Gerbauer AB and Keeley LH (1995). The spread of farming into Europe North of the Alps. In: Price TD and Gebauer AB, editors, *Last Hunters-First Farmers*, pp. 95–126, School of American Research Press, Santa Fe, New Mexico, USA.

Pringle H (1998a). The slow birth of agriculture, *Science* **282**, 1446.

Pringle H (1998b). The original blended economies, *Science* **282**, 1447.

Pryor F (2004). *Britain AD: A quest for Arthur, England and the Anglo-Saxons*, Harper Collins, London, UK.

Purseglove JW (1979). *Tropical Crops: Monocotyledons*, Longman, London, UK.

Purugganan MD, Boyles AL, Suddith JI (2000). Variation and selection at the CAULIFLOWER floral homeotic gene accompanying the evolution of domesticated *Brassica oleracea*, *Genetics* **155**, 855–862, available online at: http://www.genetics.org/cgi/content/full/155/2/855.

Purves WK, Sadava D, Orians GH and Heller C (2004). *Life. The Science of Biology*, 7th Edition, W.H. Freeman, New York, NY, USA.

Qualset CO and Shands HL (2005). *Safeguarding the Future of U.S. Agriculture: The Need to Conserve Threatened Collections of Crop Diversity Worldwide*, Report from: University of California, Division of Agriculture and Natural Resources, Genetic Resources Conservation Program, Davis, CA, USA, available online at: http://www.grcp.ucdavis.edu/publications/index.htm.

Quiros CF and Paterson AH (2004). Genome mapping and analysis. In: Pua EC and Douglas CJ, editors, *Biotechnology in Agriculture and Forestry, Volume 54, Brassica*, Springer Verlag, Berlin, Germany.

Quist D and Chapela IH (2001). Transgenic DNA introgressed into traditional maize landraces in Oaxaca, Mexico, *Nature* **414**, 541–542.

Raboin LM, Carreel F, Noyer JL, *et al.* (2005). Diploid ancestors of triploid export banana cultivars: molecular indentification of 2n restitution gamete donors and n gamete donors, *Molecular Breeding* **16**, 333–341.

Rahmstorf S (1995). Bifurcations of the Atlantic thermohaline circulation in response to changes in the hydrological cycle, *Nature* **320**, 735–738.

Rahmstorf S (2003). Timing of abrupt climate change: A precise clock, *Geophysical Research Letters* **30**, 1510–1514.

Rak Y, Ginzburg A and Geffen E (2002). Does *Homo neanderthalensis* play a role in modern human ancestry? The mandibular evidence, *American Journal of Physical Anthropology* **119**, 199–204.

Rana D, van den Boogaart T, O'Neill CM *et al.* (2004). Conservation of the microstructure of genome segments of Brassica napus and its diploid relatives, *Plant Journal* **40**, 725–733, available online at: http://www.blackwell-synergy.com/doi/full/10.1111/j.1365–313X.2004.02244.x.

Randall TA, Dwyer RA, Huitema E *et al.* (2005). Large-scale gene discovery in the oomycete Phytophthora infestans reveals likely components of phytopathogenicity shared with true fungi, *Molecular Plant Microbe Interactions* **18**, 229–243.

Rapp RA and Wendel JF (2005). Epigenetics and plant evolution, *New Phytologist* **168**, 81–91.

Rashed R, editor (1996). *Encyclopedia of the History of Arabic Science*, Routledge, London, UK.

Raven J, editor (2004). *Lost Libraries: The Destruction of Book Collections since Antiquity*, Palgrave Macmillan, Basingstoke, UK.

Raven JE (2000). *Plants and Plant Lore in Ancient Greece*, Leopard's Head Press, Oxford, UK.

Rayburn AL (1990). Genome size variation in Southwestern United States Indian maize adapted to various altitudes, *Evolutionary Trends in Plants* **4**, 53–57.

Rayburn AL and Auger JA (1990). Genome size variation in *Zea mays ssp. mays* adapted to different altitudes, *Theoretical and Applied Genetics* **79**, 470–447.

Rayburn AL, Price HJ, Smith JD and Gold JR (1985). C-band heterochromatin and DNA content in *Zea mays*, *American Journal of Botany* **72**, 1610–1617.

Redman CL (1978). *The Rise of Civilization: From Early Farmers to Urban Society in the Ancient Near East*, WH Freeman and Company, San Francisco, USA.

Reich DE and Goldstein DB (1998). Genetic evidence for a Paleolithic human population expansion in Africa, *Proceedings of the National Academy of Sciences USA* **95**, 8119–8123, available online at: www.pnas.org/cgi/reprint/95/14/8119.pdf.

Renfrew C (1979). Systems collapse as social transformation-catastrophe and anastrophe in early state societies. In: Renfrew C and Cooke KL, editors, *Transformations, Mathematical Approaches to Culture Change*, pp. 481–506, Academic Press, New York, USA.

Renssen H, Brovkin V, Fichefet T and Goosse H (2006). Simulation of the Holocene climate evolution in Northern Africa: The termination of the African Humid Period, *Quaternary International* **150**, 95–102, available online at: www.geo.vu.nl/~renh/pdf/Renssen-etal-QuatInt06.pdf.

Rich SM and FJ Ayala (2000). Population structure and recent evolution of *Plasmodium falciparum*, *Proceedings of the National Academy of Sciences, USA* **97**, 6994–7001, available online at: http://www.pnas.org/cgi/content/full/97/13/6994.

Richards TA, Dacks JB, Jenkinson JM, Thornton CR and Talbot NJ (2006). Evolution of filamentous plant pathogens: gene exchange across eukaryotic kingdoms, *Current Biology* **16**, 1857–1864.

Richardson AO and Palmer JD (2007). Horizontal gene transfer in plants, *Journal of Experimental Botany* **58**, 1–9.

Richerson PJ, Boyd R and Bettinger RL (2001). Was agriculture impossible during the Pleistocene but mandatory during the Holocene? A climate change hypothesis, *American Antiquity* **66**, 387–411.

Rick CM (1995). Tomato. In: Smart J and Simmonds NW editors, *Evolution of Crop Plants*, pp. 452–457, Longman, Essex, UK.

Rieseberg LH and Wendel JF (2004). Plant speciation—rise of the poor cousins, *New Phytologist* **165**, 411–423.

Rieseberg LH, Wood TE and Baack EJ (2006). The nature of plant species, *Nature* **440**, 524–527.

Riley R and Chapman V (1958). Genetic control of the cytologically diploid behaviour of hexaploid wheat, *Nature* **182**, 713–715.

Riley R, Chapman V and Kimber G (1959). Genetic control of chromosome pairing in intergeneric hybrids with wheat, *Nature* **183**, 1244–1246.

Rindos D (1980). Symbiosis, instability and the origins and spread of agriculture: a new model, *Current Anthropology* **21**, 751–772.

Rindos D (1984). *The Origins of Agriculture: An Evolutionary Perspective*, Academic Press, New York, USA.

Ripley G (1471). *The Compound of Alchymy*. In: *Ashmoleum Theatrum Chemicum Britannicum*, 1652.

Ristvet L and Weiss H (2005). The Habur region in the late third and early second millennium BC. In: Orthmann W, editor, *The History and Archaeology of Syria*, Vol. 1, Saarbrücken Verlag, Saarbrücken, Germany, available online at: http://research.yale.edu/leilan/RistvetWeissHAS_10.pdf.

Roaf M (1990). *The Cultural Atlas of Mesopotamia and the Ancient Near East, Facts on File*, New York, USA.

Roberts N (1998). *The Holocene: an Environmental History*, Blackwell, Oxford, UK.

Rodermel S (1999). Subunit control of rubisco biosynthesis: a relic of an endosymbiotic past, *Photosynthesis Research* **59**, 105–123.

Rodhouse JC, Haugh CA, Roberts D and Gilbert RJ. (1990). Red kidney bean poisoning in the UK: an analysis of 50 suspected incidents between 1976 and 1989, *Epidemiology and Infection* **105**, 485–491.

Rogers AR, Iltis D and Wooding S (2004). Genetic variation at the MC1R locus and the time since loss of human body hair, *Current Anthropology* **45**,105–108.

Rohling EJ and Palike H (2005). Centennial-scale climate cooling with a sudden cold event around 8200 year ago, *Nature* **434**, 975–979.

Rojas RT (1983). *La Agricultura Chinampera, Compilación Histórica*, Universidad de Chapingo, México.

Rojas RT, editor (1995). *Presente, Pasado y Futuro de las Chinampas*, Centro de investigaciones y Estudios Superiores en Antropología Social, México.

Roll-Hansen N (2004). *The Lysenko Effect: The Politics Of Science*, Humanity Books, London, UK.

Rollefson GO and Kohler-Rollefson I (1993). PPNC Adaptations in the first half of the 6th millennium BC, *Paléorient* **19**, 33–42.

Rollefson G, Simmons A, Donaldson M *et al.* (1985). Excavations at the pre-pottery Neolithic B village of 'Ain Ghazal (Jordan), 1983, *Mitteilugen der Deutschen Orient-Gesellschaft zu Berlin* **117**, 69–116.

Rosenberg M, Nesbitt M, Redding RW and Strasser TF (1995). Hallen Cemi Tepesi: some preliminary observations concerning early Neolithic subsistence in eastern Anatolia, *Antolica* **21**, 1–12.

Rossignol-Strick M (1999). The Holocene climatic optimum and pollen records of sapropel 1 in the eastern Mediterranean, 9000–6000 BP, *Quaternary Science Reviews* **18**, 515–530.

Rostovtzeff MI (1941). *Social and Economic History of the Hellenistic World*, Clarendon, Oxford, UK.

Rothman MS, editor (2001). *Uruk Mesopotamia and Its Neighbors Cross-Cultural Interactions in the Era of State*

Formation, School of American Research Press, Santa Fe, New Mexico, USA.

Roux G (1993). *Ancient Iraq*, 3rd edition, Penguin, London, UK.

Rova E and Weiss H, editors (2003). *The Origins of North Mesopotamian Civilization: Ninevite 5 Chronology, Economy, Society*, Brepols, Brussels, Belgium.

Ruddiman W (2003). The anthropogenic greenhouse era began thousands of years ago, *Climate Change* **61**, 261–293, available online at: http://stephenschneider. stanford.edu/Publications/PDF_Papers/Ruddiman20 03.pdf#search = %22Anthropogenic%20Greenhouse% 20Era%20Began%20Thousands%20of%20Years% 20Ago%22.

Ruddiman WF (2005a). How did humans first alter global climate? *Scientific American*, March 2005, pp. 46–53.

Ruddiman WF (2005b). *Plows, Plagues, and Petroleum: How Humans Took Control of Climate*, Princeton University Press, Princeton, New Jersey, USA.

Rudney JD (1982). Dental indicators of growth disturbance in a series of ancient Lower Nubian populations: changes over time, *American Journal of Physical Anthropology* **60**, 463–470.

Russo L (2004). *The Forgotten Revolution. How Science was Born in 300 BC and Why it Had to be Reborn*, Springer, Berlin, Germany.

Rust RL (1999). A brief history of chocolate, food of the gods, *Athena Review* **2**, available online at: http://www. athenapub.com/chocolat.htm.

Rutledge E (1993). Kett's rebellion. In: Wade-Morris P, editor, *An Historical Atlas of Norfolk*, Norfolk Museums Service, Norwich, UK.

Ryan W and Pitman W (1999). *Noah's Flood: The New Scientific Discoveries About the Event That Changed History*, Simon and Schuster, New York, USA.

Sabeti PC, Reich DE, Higgins JM *et al.* (2002). Detecting recent positive selection in the human genome from haplotype structure, *Nature* **419**, 832–837.

Sabeti PC, Schaffner SF, Fry B *et al.* (2006). Positive natural selection in the human lineage, *Science* **312**, 1614–1620.

Sage RF (1995). Was low atmospheric CO_2 during the Pleistocene a limiting factor for the origin of agriculture? *Global Change Biology* **1**, 93–106.

Saggs HWF (1965). *Everyday Life in Babylonia and Assyria*, Batsford, Dorset, UK, available online at: www.aina. org/books/eliba/eliba.pdf.

Salamini F, Özkan H, Brandolini A, Schafer-Pregl R and Martin W (2002). Genetics and geography of wild cereal domestication in the near east, *Nature Reviews Genetics* **3**, 429–441.

Samuel D (1996). Archaeology of ancient Egyptian beer, *Journal of the American Society of Brewing Chemists* **54**,

3–12, available online at: http://www.ancientgrains. org/samuel1996beer.pdf.

Sanders WT (1957). *A Study of the Ecological Factors in the Development of Mesoamerican Civilizations*, Harvard University Press, Cambridge, USA.

Sanjur OI, Piperno DR, Andres TC and Wessel-Beaver L (2002). Phylogenetic relationships among domesticated and wild species of *Cucurbita* (Cucurbitaceae) inferred from a mitochondrial gene: Implications for crop plant evolution and areas of origin, *Proceedings of the National Academy of Sciences USA* **99**, 5535–5540, available online at: http://www.pnas.org/cgi/reprint/99/1/535.

SanMiguel P, Gaut BS, Tikhonov A, Nakajima Y and Bennetzen JL (1998). The paleontology of retrotransposons in maize: dating the strata, *Nature Genetics* **20**, 43–45.

Sapp J (2003). *Genesis, The Evolution of Biology*, Oxford University Press, Oxford, UK.

Särkilahti E and Valanne T (1990). Induced polyploidy in *Betula, Silva Fennica* **24**, 227–234.

Sarla N and Mallikarjuna Swamy BP (2005). Oryza glaberrima: A source for the improvement of Oryza sativa, *Current Science* **89**, 955–963.

Sarnthein M., Kiefer T, Grootes PM, Elderfield H and Erlenkeuser H (2006). Warmings in the far northwestern Pacific promoted pre-Clovis immigration to America during Heinrich event **1**, *Geology* **34**, 141–144.

Sauer CO (1952). Agricultural Origins and Dispersals, *Bowman Memorial Lectures*, series **2**, American Geographical Society, New York, USA.

Sauer J (1957). Recent migration and the evolution of the dioecious amaranths, *Evolution* **11**, 11–31.

Savaolainen P, Zhang YP, Luo J, Lundeberg J and Leitner T (2002). Genetic evidence for an East Asian origin of domestic dogs, *Science* **298**, 1610–1613.

Scarano MT, Abbate L, Ferrante S, Lucretti S, Nardi SL and Tusa N (2002). Molecular characterization of Citrus symmetric and asymmetric somatic hybrids by means of ISSR-PCR and PCR-RFLP. In: Vasil IK, editor, *Plant Biotechnology 2002 and Beyond*, pp. 549–550, Kluwer, Dordrecht, Netherlands.

Scarre C (1989). *Past Worlds: "The Times" Atlas of Archaeology*, Harper Collins, London, UK.

Schimel D (2006). Climate change and crop yields: beyond Cassandra, *Science* **312**, 1889–1890.

Schlebecker JT (1975). *Whereby We Thrive: A History of American Farming*, Iowa State University Press, Ames, Iowa, USA.

Schlegel RHJ (2003). *Encyclopedic Dictionary of Plant Breeding and Related Subjects*, Haworth Press, New York, USA.

Schmandt-Besserat D (1992). *Before Writing*, Vol. 1, *From Counting to Cuneiform*, University of Texas Press, Austin, Texas, USA.

Schmandt-Besserat D (1997). *How Writing Came About* (abridged)., University of Texas Press, Austin, Texas, USA.

Schmidt K (2001). Göbekli Tepe, Southeastern Turkey. A preliminary Report on the 1995–1999 Excavations, *Palèorient* **26**, 45–54.

Schneider JL (1967). Evolution du dernier lacustre et peuplements préhistoriques aux bas-pays du Tchad (in French), Bulletin, *Associations Sénégalaise pour l'Etude du Quaternaire de l'Ouest Africain*, 18–23.

Schranz ME, Lysak MA and Mitchell-Olds T (2006). The ABC's of comparative genomics in the Brassicaceae: building blocks of crucifer genomes, *Trends in Plant Science* **11**, 535–542.

Schranz ME and Osborn TE (2000). Novel flowering time variation in the resynthesized polyploid *Brassica napus*, *Journal of Heredity* **91**, 242–246.

Schulman AH, Kalendar S and Vicient CM (2002). Dynamic DNA and genome evolution, *Plant Biology 2002* Abstract # 50004, American Society of Plant Biologists, Rockville, Maryland, USA, available online at: http://abstracts.aspb.org/pb2002/public/S05/1046.html.

Schwartz SB, editor (2004). *Tropical Babylons: Sugar and the Making of the Atlantic world, 1450–1680*, University of North Carolina Press, Chapel Hill, North Carolina, USA.

Schwartz GM and Nichols JJ, editors (2006). *After Collapse, The Regeneration of Complex Societies*, University of Arizona Press, Tucson, Arizona, USA.

Schoberth HH and Song C (2002). Chemicals and materials from coal in the 21st century, *Fuel* **81**, 15–32.

Schweinfurth G (1884). Further discoveries in the flora of ancient Egypt, *Nature* **29**, 312–315.

Second G (1982). Origin of the genic diversity of cultivated rice (*Oryza* spp.): Study of the polymorphism scored at 40 isozyme loci, *Japanese Journal of Genetics* **57**, 25–57.

Segraves KA, Thompson JN, Soltis PS and Soltis DE (1999). Multiple origins of polyploidy and the geographic structure of *Heuchera grossulariifolia*, *Molecular Ecology* **8**, 253–262.

Semino O, Passarino G, Oefner PJ *et al.* (2000). The genetic legacy of Paleolithic *Homo sapiens sapiens* in extant Europeans: a Y chromosome perspective, *Science* **290**, 1155–1159.

Sencer HA and Hawkes JG (1980). On the origin of cultivated rye, *Biological Journal of the Linnean Society of London* **13**, 299–313.

Senior L and Weiss H (1992). Tell Leilan "sila bowls" and the Akkadian reorganization of Subarian agricultural production, *Orient-Express* **2**, 16–24.

Serre D, Langaney A, Chech M *et al.* (2004). No evidence of neandertal mtDNA contribution to early modern humans, *Public Library of Science, Biology* **2**, 313–317, available online at: http://www.plosbiology.org/plosonline/?request = get-document&doi = 10.1371%2Fjournal.pbio.0020057.

Service ER (1962). *Primitive Social Organization: An Evolutionary Perspective*, Random House, New York, USA.

Service ER (1975). *Origins of the State and Civilization: The Process of Cultural Evolution*, WW Norton and Company, New York, USA.

van Seters J (1964). A date for the "Admonitions" in the second intermediate Period, *Journal of Egyptian Archaeology* **50**, 13–23.

Seyis F, Snowdon RJ, Lühs W and Friedt W (2003). Molecular characterization of novel resynthesized rapeseed (*Brassica napus*). lines and analysis of their genetic diversity in comparison with spring rapeseed cultivars, *Plant Breeding* **122**, 473–478.

Shaffer J (1993). Reurbanization: The eastern Punjab and beyond. In: Spodek H and Srinivasan DM, editors, *Urban Form and Meaning in South Asia: The Shaping of Cities from Prehistoric to Precolonial Times*, pp. 53–67, National Gallery of Art, Washington, DC, USA.

Shaffer JG and Lichtenstein DA (1995). The Cultural Tradition and Paleoethnicity in South Asian Archaeology. In: Erdosy G, editor, *The Indo-Aryans of Ancient South Asia: Language, Material Culture and Ethnicity*, pp. 126–15Walter de Gruyter, Berlin, Germany.

Shaffer JG and Lichtenstein DA (1999). Migration, Philology and South Asian Archaeology. In: Bronkhorst J and Deshpande MM, editors, Harvard Oriental Series, *Aryan and Non-Aryan in South Asia. Evidence, Interpretation, and Ideology*, pp. 239–260, Harvard University Press, Cambridge, Massachusetts, USA.

Shaked H, Kashkush K, Özkan H, Feldman M and Levy AA (2001). Sequence elimination and cytosine methylation are rapid and reproducible responses of the genome to wide hybridization and allopolyploidy in wheat, *Plant Cell* **13**, 1749–1759, www.plantcell.org/cgi/reprint/13/8/1749.pdf.

Shea JJ (1998). Neandertal and early modern human behavioral variability: A regional-scale approach to the lithic evidence for hunting in the Levantine Mousterian, *Current Anthropology* **39** (Suppl.), S45–S61.

Shea JJ (1999). Ar Rasfa, a Levantine Mousterian site from Northwest Jordan: a preliminary report, *Paléorient* **24**, 71–78.

Shea JJ (2001a). Modern human origins and neanderthal extinctions in the Levant, *Athena Review* **2**, 21–32, available online at: http://www.athenapub.com/8shea1.htm.

Shea JJ (2001b). The Middle Paleolithic: neandertals and early modern humans in the Levant, *Near Eastern Archaeology* **63**, 38–64.

Shea JJ (2003). Neandertals, competition, and the origin of modern human behavior in the Levant, *Evolutionary Anthropology* **12**, 173–187.

Sheehan H (1993). *Marxism and the Philosophy of Science: A Critical History*, 2nd edition, Humanities Press International, New Jersey. A section of this book referring to the Lysenko affair is available online at: http://www.comms.dcu.ie/sheehanh/lysenko.htm.

Shelach G (2000). The earliest Neolithic cultures of Northeast China: Recent discoveries and new perspectives on the beginning of agriculture, *Journal of World Prehistory* **14**, 363–413.

Shennan SJ (1993). Settlement and social change in Central Europe, 3500–1500 BC, *Journal of World Prehistory* **7**, 121–162.

Shingleton AW and Stern DL (2003). Molecular phylogenetic evidence for multiple gains or losses of ant mutualism within the aphid genus *Chaitophorus*, *Molecular Phylogenetics and Evolution* **26**, 26–35.

Shipek FC (1981). A native American adaptation to drought: the Kumeyaay as seen in the San Diego Mission records, 1770–1798, *Ethnohistory* **28**, 295–312.

Shipek FC (1982). Kumeyaay socio-political structure. *Journal of California and Great Basin Anthropology* **4**, 296–303.

Shipek FC (1991). *The Autobiography of Delfina Cuero*, Ballena Press, Melno Park, California, USA.

Shorto R (2005). *The Island at the Centre of the World*, Black Swan, London, UK.

Shostak M (1981). *Nisa: The Life and Words of a !Kung Woman*, Random House, New York, USA.

Slade AJ, Fuerstenberg SI, Loeffler D, Steine MN and Facciotti D (2005). A reverse genetic, nontransgenic approach to wheat crop improvement by TILLING, *Nature Biotechnology* **23**, 75–78.

Slade AJ and Knauf VC (2005). TILLING moves beyond functional genomics into crop improvement, *Transgenic Research* **14**, 109–114.

van Slageren MW (1994). Wild wheats: a monograph of *Aegilops L.* and *Amblyopyrum* (*Jaub. and Spach*). *Eig* (*Poaceae*), Papers 1994, Wageningen Agricultural University, Wageningen, Netherlands.

Sloane H (1696). *Catalogus Plantarum Quae in Insula Jamaica Sponte Proveniunt*, Brown, London, UK.

Smartt J and Simmonds NW, editors (1995). *Evolution of Crop Plants*, pp. 261–266, Longman Scientific and Technical, Harlow, UK.

Slater G (1907). *The English Peasantry and the Enclosure of Common Fields*, Archibald Constable and Co, London, UK.

Smil V (2001). *Enriching the Earth: Fritz Haber, Carl Bosch, and the Transformation of World Food Production*, MIT Press, Cambridge, Massachusetts, USA.

Smith A (1776). *An Inquiry into the Nature and Causes of the Wealth of Nations*, 2003 edition, Bantam Books, New York, USA, available online at: http://www.adamsmith.org/smith/won-intro.htm.

Smith AE and Secoy DM (1976). A Compendium of Inorganic Substances Used in European Pest Control before 1850, *Journal of Agricultural and Food Chemistry* **24**, 1180–1186, available online at: http://www.hort.purdue.edu/newcrop/history/lecture31/r_31–1.html.

Smith BD (1989). Origins of agriculture in eastern North America, *Science* **246**, 1566–1571.

Smith BD (1994). *The Emergence of Agriculture*, Freeman, New York, USA.

Smith BD (1997). The initial domestication of *Cucurbita pepo* in the Americas 10,000 years ago, *Science* **276**, 932–934.

Smith B D (1998). *The Emergence of Agriculture*, Scientific American Library, New York, USA.

Smith BD (2001a). Documenting plant domestication: The consilience of biological and archaeological approaches, *Proceedings of the National Academy of Sciences USA* **98**, 1324–1326, available online at: http://www.pnas.org/cgi/content/full/98/4/1324.

Smith BD (2001b). Low level food production, *Journal of Archaeological Research* **9**, 1–43.

Smith BD (2005). Reassessing Coxcatlan Cave and the early history of domesticated plants in Mesoamerica, *Proceedings of the National Academy of Sciences USA* **102**, 9438–9445, available online at: http://www.pnas.org/cgi/content/full/102/27/9438.

Smith BD (2006a). Eastern North America as an independent center of plant domestication, *Proceedings of the National Academy of Sciences USA* **103**, 12223–12228.

Smith BD (2006b). Seed size as a marker of domestication in squash (*Cucurbita pepo*).. In: Zeder MA, Bradley DG, Emshwiller E and Smith BD, editors, *Documenting Domestication: New Genetic and Archaeological Paradigms*, pp. 25–31, University of California Press, Berkeley, USA.

Smith CW and Frederiksen RA (2000). *Sorghum. Origin, History, Technology, and Production*, Wiley, New York, USA.

Smith P, Bar-Yosef O and Sillen A (1984). Archaeological and skeletal evidence for dietary change during the Late Pleistocene/Early Holocene in the Levant, In *Paleopathology at the Origins of Agriculture*, edited by Cohen MN and Armelagos GJ, pp. 101–136, Academic Press, New York, USA.

Smyth DR (1995). Origin of the cauliflower, *Current Biology* **5**, 361–363.

Snowdon RJ and Friedt W (2004). Molecular markers in Brassica oilseed breeding: current status and future possibilities, *Plant Breeding* **123**, 1–8, www.blackwell-synergy.com/doi/pdf/10.1111/j.1439–0523.2003.00968.x.

Sobel D (1995). *Longitude: The True Story of a Lone Genius Who Solved the Greatest Scientific Problem of his Time*, Fourth Estate, London, USA.

Sober E (1988). *Reconstructing the Past: Parsimony, Evolution, and Inference*, MIT Press, Cambridge, Massachusetts, USA.

Sollberger E and Kupper JR (1971). *Inscriptions Royals Sumériennes et Akkadiennes*, Les Éditions du Cerf, Paris, France.

Sommerfeld W, Archi A and Weiss H (2004). Why "Dada measured 40,000 liters of barley from Nagar to Sippar", *Proceedings of the 4th International Congress on the Archaeology of the Ancient Near East*, March 29-April 3, Berlin, Germany, available online at: http://research.yale.edu/leilan/publications.html.

Song K, Lu P, Tang K and Osborn TC (1995). Rapid genome change in synthetic polyploids of *Brassica* and its implications for polyploid evolution, *Proceedings of the National Academy of Sciences USA* **92**, 7719–7723, available online at: http://www.pnas.org/cgi/reprint/92/17/7719.

Sonnate G, Stockton T, Nodari, RO, Becerra Velasquez, VL and Gepts P (1994). Evolution of genetic diversity during the domestication of common-bean (*Phaseolus vulgaris* L.), *Theoretical and Applied Genetics* **89**, 629–635.

Southern R (1987). *Robert Grosseteste: The Growth of an English Mind in Medieval Europe*, Clarendon Press, Oxford, UK.

Speed A (1659). *Adam out of Eden*, London, UK.

Spencer HA and Hawkes JG (1980). On the origin of cultivated rye, *Biological Journal of the Linnean Society* **13**, 299–313.

Spielvogel J (1996). *Western Civilization: Comprehensive Volume*, 3rd Edition, West Publishing Company, Minneapolis, USA.

Spooner DM, McLean K, Ramsay G, Waugh R and Bryan GJ (2005). A single domestication for potato based on multilocus amplified fragment length polymorphism genotyping, *Proceedings of the National Academy of Sciences USA* **102**, 14694–14699, available online at: http://www.pnas.org/cgi/reprint/0507400102v1.pdf.

Spooner DM and Hetterscheid LA (2006). Origins, evolution, and group classification of cultivated potatoes. In: Motley TJ, Cross HB and Zerega NJC, editors, *Darwin's Harvest: New Approaches to Origins, Evolution, and Conservation of Crop Plants*, pp. 205–307, Columbia University Press, New York, USA.

Spring J (1997). Vertebrate evolution by interspecific hybridization: are we polyploid? *FEBS Letters* **400**, 2–8.

Stadler LJ (1928). Genetic effects of X-rays in maize, *Proceedings of the National Academy of Sciences USA* **14**, 69–75, available online at: http://www.pnas.org/cgi/reprint/14/1/69.

Stahl AB (1984). Hominid dietary selection before fire, *Current Anthropology* **25**, 151–168.

Stahl AB (1989). Plant-food processing: implications for dietary quality. In: Harris DR and Hillman GC, editors, *Foraging and Farming: The Evolution of Plant Exploitation*, pp. 171–194, Unwin Hyman, London, UK.

Stallknecht GF, Gilbertson KM and Ranney JE (1996). Alternative wheat cereals as food grains: Einkorn, emmer, spelt, kamut, and triticale, pp. 156–170. In: Janick J, editor, *Progress in New Crops*, ASHS Press, Alexandria, Virginia, USA, available online at: http://www.hort.purdue.edu/newcrop/proceedings1996/v3–156.html.

Stapf O and Hubbard CE (1934). Pennisetum. In: Prain D, editor, *Flora of Tropical Africa*, Vol. **9**, 954–1070, Balkema, Rotterdam, Netherlands.

Staubwasser M, Sirocko F, Grootes PM, and Segl M (2003). Climate change at the 4.2 ka BP termination of the Indus valley civilization and Holocene south Asian monsoon variability, *Geophysical Research Letters* **30**, 1425.

Stebbins GL (1950). *Variation and Evolution in Plants*, Columbia University Press, New York, USA.

Stebbins GL (1971). *Chromosomal Evolution in Higher Plants*, Edward Arnold, London, UK.

Steckel RH and Rose JC, editors (2002). *The Backbone of History: Health and Nutrition in the Western Hemisphere*, Cambridge University Press, Cambridge, UK.

Steele TE (2003). Using mortality profiles to infer behavior in the fossil record, *Journal of Mammalogy* **84**, 418–430, www.eva.mpg.de/evolution/pdf/Steele_JMam03.pdf.

Steenstrup JJS (1872). Sur les Kjøkkenmøddings de l'age de la Pierre et sur les faunes et la flore préhistoriques de Danmark, *Bulletins du Congrès International d'Archéologie préhistorique à Copenhague en 1869*.

Stefansson H, Helgason A, Thorleifsson G, Steinthorsdottir V, Masson G, *et al.* (2005). A common inversion under selection in Europeans, *Nature Genetics* **37**, 129–137.

Steinman H (2002). Milk allergy and lactose intolerance, *Science in Africa*, May 2002, available online at: http://www.scienceinafrica.co.za/2002/may/milk.htm.

Stiner MC (2001). Thirty years on the 'Broad Spectrum Revolution' and paleolithic demography, *Proceedings of the National Academy of Sciences USA* **98**, 6993–6996, available online at: http://www.pnas.org/cgi/content/full/98/13/6993.

Stiner MC and Kuhn SL (1992). Subsistence, technologies, and adaptive variation in Middle Paleolithic Italy, *American Journal of Anthropology* **94**, 306–339.

Stiner MC, Munro ND, Surovell TA, Tchernov E and Bar-Yosef O (1999). Paleolithic population growth pulses evidenced by small animal exploitation, *Science* **283**, 190–194.

Stinson S (2002). Early childhood health in foragers. In: *Human Diet (Its Origin and Evolution)*, pp. 37–48, Bergin and Garvey, New York, USA.

Stocker TF, Wright DG and Broecker WS (1992). The influence of high-latitude surface forcing on the global thermohaline circulation, *Paleoceanography* **7**, 529–541.

Stokstad E (2003). 'Pristine' Forest Teemed With People, *Science* **301**, 1645–1646.

Stone D (2005). *Decision-Making in Medieval Agriculture*, Oxford University Press, Oxford, UK.

Stone L (1964). The Educational Revolution in England, 1560–1640, *Past and Present* **28**, 41–80.

Stone R (2003). Peopling of the Americas: late date for Siberian site challenges Bering pathway, *Science* **301**, 450–451.

Stoskopf NC (1985). *Cereal grain crops*, Chapter 20, Reston Publishing Co., Inc., Reston, Virginia, USA.

Stringer C and Andrews P (2005). *The Complete World of Human Evolution*, Thames and Hudson, London, UK.

Stuiver M and Becker B (1993). High-precision calibration of the radiocarbon time scale AD 1950–6000 BC, *Radiocarbon* **35**, 35–65.

Stuiver M, Reimer PJ, Bard E, Beck JW, Burr GS, Hughen KA, Kromer B, McCormac G, van der Plicht J and Spurk M (1998). INTERCAL98 radiocarbon age calibration, 24,000–0 cal BP, *Radiocarbon* **40**, 1041–1083.

Surovell TA (2000). Early Paleoindian women, children, mobility, and fertility, *American Antiquity* **65**, 493–508.

Sutter NB and Ostrander EA (2004). Dog star rising: the canine genetic system, *Nature Reviews Genetics* **5**, 900–910.

Stutz HC (1972). On the origin of cultivated rye, *American Journal of Botany* **59**, 59–70.

Su B, Xiao J, Underhill P *et al.* (1999). Y-chromosome data evidence for a northward migration of modern humans into eastern Asia during the last Ice Age, *American Journal of Human Genetics* **65**, 1718–1724, available online at: www.journals.uchicago.edu/AJHG/journal/issues/v65n6/991131/991131.web.pdf.

Su B, Xiao C, Deka R *et al.* (2000). Y chromosome haplotypes reveal prehistorical migrations to the Himalayas, *Human Genetics* **107**, 582–590, available online at: http://hpgl.stanford.edu/publications/HG_2000_v107_p582.pdf.

Sweeney D, editor (1995). *Agriculture in the Middle Ages: Technology, Practice and Representation*, University of Pennsylvania Press, Philadelphia, Pennsylvania, USA.

Swisher CC, Curtis GH, Jacob T, Getty AG, Suprijo A and Widiasmoro (1994). Age of the earliest known hominids in Java, Indonesia, *Science*, **263**, 1118–1121.

Swisher CC, Rink WJ, Anton SC *et al.* (1996). Latest *Homo erectus* of Java: potential contemporaneity with *Homo sapiens* in southeast Asia, *Science* **274**, 1870–1874.

Syvanen M and Kado CI, editors, (2002). *Horizontal Gene Transfer*, Academic Press, London, UK.

Szabo VM and Burr B (1996). Simple inheritance of key traits distinguishing maize and teosinte, *Molecular and General Genetics* **252**, 33–41.

Szreter S (2004). Health, economy, state and society in modern Britain: The long-run perspective *Hygiea Internationalis* **4**, 205–227, available online at: www.ep.liu.se/ej/hygiea/ra/025/paper.pdf.

Taine HA (1878). *The French Revolution*, translated by Durand J, Liberty Fund, Indianapolis, USA, available online at: http://oll.libertyfund.org/ToC/0178–01.php.

Tanno K and Wilcox G (2006). How fast was wild wheat domesticated? *Science* **311**, 1886.

Tänzler B, Esposti RF, Vaccino P *et al.* (2002). A molecular linkage map of einkorn wheat: mapping of storage-protein and soft-glume genes and bread-making quality QTLs, *Genetical Research, Cambridge* **80**, 131–143.

Taube K (1985). The Classic Maya maize god: a reappraisal. In: Fields VM, editor, *Fifth Palenque Round Table, 1983, Volume VII* Electronic version, Pre-Columbian Art Research Institute, Monterey, USA.

Taube K (1993). *Aztec and Maya Myths* (Legendary Past Series), University of Texas Press, Austin, Texas, USA.

Tawney RH (1912). *The Agrarian Problem in the Sixteenth Century*, Longmans, Green and Co, London, UK.

Tayes N, Domett K and Nelsen K (2000). Agriculture and dental caries? The case of rice in prehistoric Southeast Asia, *World Archaeology* **32**, 68–83.

Teller JT and Leverington DW (2004). Glacial Lake Agassiz: A 5000 yr history of change and its relationship to the ^{18}O record of Greenland, *Geological Society of America Bulletin*, **116**, 729–742.

Tenaillon MI, Sawkins MC, Long AD, Gaut RL, Doebley JF and Gaut BS (2001). Patterns of DNA sequence polymorphism along chromosome 1 of maize (*Zea mays* ssp. mays L.), *Proceedings of the National Academy of Sciences USA* **98**, 9161–9166, available online at: http://www.pnas.org/cgi/reprint/98/16/9161.

Thangaraj K Chaubey G, Kivisild T *et al.* (2005). Reconstructing the origin of Andaman islanders, *Science* **308**, 996.

Tharpar R (2002). *Early India: from the Origins to AD 1300*, Allen Lane, London, UK.

Theophrastus (*c.* 310 BCE a). *De Causis Plantarum*, Books 1–6, translated by Einarson B and Link GKK, 1976, Loeb

Classical Library, Harvard University Press, Cambridge, Massachusetts, USA.

Theophrastus (*c.* 310 BCE b). *Historiae Plantarum* (Enquiry into Plants). Books 1–5, translated by Hort AF, 1916, Loeb Classical Library, Harvard University Press, Cambridge, Massachusetts, USA.

Thoen E (1997). The birth of 'the Flemish husbandry': agricultural technology in medieval Flanders. In: Astill G and Langdon J, editors, *Medieval Farming and Technology*, pp. 69–88, Brill, Leiden, Netherlands.

Thomas B, Murphy DJ and Murray BG, editors, (2003). *Encyclopedia of Applied Plant Sciences*, Elsevier/Academic Press, Oxford, UK.

Thomas ER, Wolff ER, Mulvaney R, *et al.* (2007). The 8.2 ka event from Greenland ice cores, *Quaternary Science Reviews* **26**, 70–81.

Thomas H (1995). Oats. In: Smartt J and Simmonds NW editors, *Evolution of Crop Plants*, pp. 261–266, Longman Scientific and Technical, Harlow, UK.

Thomas J (1987). Relations of production and social change in the neolithic of Northwestern Europe, *Man* **22**, 405–430.

Thomas MG, Stumpf MPH and Härke H (2006). Evidence for an apartheid-like social structure in early Anglo-Saxon England, *Proceedings of the Royal Society B: Biological Sciences*, **273**, 2651–2657, available online at: http://www.pubs.royalsoc.ac.uk/media/proceedings_b/papers/RSPB20063627.pdf#search = %22Evidence%20for%20an%20apartheid-like%20social%20structure%20in%20early%20Anglo-Saxon%20England%22.

Thompson EE, Kuttab-Boulos H, Witonsky D *et al.* (2004). CYP3A variation and the evolution of salt-sensitivity variants, *American Journal of Human Genetics* **75**, 1059–1069.

Thompson JN, Nuismer SL and Merg K (2004). Plant polyploidy and the evolutionary ecology of plant/animal interactions, *Biological Journal of the Linnean Society* **82**, 511–519.

Thompson KF (1979). Cabbages, kales etc.: *Brassica oleracea* (Cruciferae). In: Simmonds NE, editor, *Evolution of Crop Species*, pp. 49–52, Longman, London, UK.

Thompson LG (2000). Ice core evidence for climate change in the Tropics: implications for our future, *Quaternary Science Review* **19**, 19–35.

Thompson LG, Mosley-Thompson E, Brecher H, Davis M, Leon B, Les D, Lin PN, Mashiotta T and Mountain K (2006). Abrupt tropical climate change: Past and present, *Proceedings of the National Academy of Sciences USA* **103**, 10536–10543, available online at: http://www.pnas.org/cgi/reprint/0603900103v1.

Thompson RC (1924). *Dictionary of Assyrian Botany and Assyrian Herbal*, Luzak, London, UK.

Tishkoff SA, Varkonyi R, Cahinhinan N *et al.* (2001). Haplotype diversity and linkage disequilibrium at the human *G6PD* locus: Recent origin of alleles that confer malarial resistance, *Science* **293**, 455–462.

Tishkoff SA, Gonder K, Hirbo J *et al.* (2003). The genetic diversity of linguistically diverse Tanzanian populations: A multilocus analysis, *American Journal of Physical Anthropology* Supplement **36**, 208–220.

Toth N (1985). The Oldowan reassessed: A close look at early stone tools, *Journal of the Archaeological Society* **12**, 101–120.

Toogood A, editor in chief (1999). *Plant Propagation: The Fully Illustrated Plant-by-Plant Manual of Practical Techniques*, Dorling Kindersley, New York, USA.

Tradescant J (1625). Letter to E Nicholas, 31 July 1625, quoted on pp. 45 and 115 of Allen M (1964). *The Tradescants: their Plants, Gardens and Museum, 1570–1662*, M Joseph, London, UK.

Trehane P, Brickell CD, Baum BR, Hetterscheid WLA, Leslie AC, McNeill J, Spongberg SA and Vrugman F (1995). International code of nomenclature for cultivated plants, *Regnum Vegetabile* **133**, 1–175.

Trinkaus E and Shipman P (1993). *The Neanderthals*, Jonathan Cape, London, UK.

Trinkaus E and Thompson DD (1987). Femoral diaphyseal histomorphometric age determinations for the Shanidar 3, 4, 5 and 6 Neandertals and Neandertal longevity, *American Journal of Physical Anthropology* **72**, 123–129.

Trueman C, Field J, Dortch J, Charles B and Wroe S (2005). Prolonged coexistence of humans and megafauna in Pleistocene Australia *Proceedings of the National Academy of Sciences USA* **102**, 8381–8385, available online at: http://www.pnas.org/cgi/reprint/0408975102v1.

Trut L (1999). Early canid domestication: the farm-fox experiment, *American Scientist* **87**, 160–169.

Tsuneki A., Hydar J, Miyake Y *et al.* (2000). Fourth preliminary report of the excavations at Tell el-Kerkh, Northwestern Syria, *Bulletin of the Ancient Orient Museum* **21**, 1–36.

Tuberosa R and Salvi S (2006). Genomics-based approaches to improve drought tolerance of crops, *Trends in Plant Science* **11**, 405–412.

Tudge C (1988). *Food Crops for the Future*, Blackwell, Oxford, UK.

Tudge C (1998). *Neanderthals, Bandits and Farmers: How Agriculture Really Began*, Weidenfeld and Nicolson, London, UK.

Turner G (1979). *Indians of North America*, Blandford Press, Poole, Dorset, UK.

Turville-Petre F (1932). Excavations in the Mugharet Kebara, *Journal of the Royal Anthropological Institute* **32**, 271–276.

Tusser T (1560). *Five Hundred Points of Good Husbandry—Comparing Good Husbandry*, reprinted in 1984, Oxford University Press, Oxford, UK.

Twiss KC (2007). The Neolithic of the Southern Levant, *Evolutionary Anthropology* **16**, 24–36.

U N (1935). Genome analysis in *Brassica* with special reference to the experimental formation of *B. napus* and peculiar mode of fertilization, *Japanese Journal of Botany* **7**, 389–452.

Uauy C, Distelfeld A, Fahima T, Blechl A and Dubcovsky J (2006). A NAC gene regulating senescence improves grain protein, zinc, and iron content in wheat, *Science* **314**, 1298–1301.

Ugent D, Pozorski S and Pozorski T (1982). Archaeological potato tuber remains from the Casma Valley, Peru, *Economic Botany* **36**, 182–192.

Uglow J (2003). *The Lunar Men. The Friends who made the Future*, Faber and Faber, London, UK.

Ulijaszek SJ (1992). Human dietary change. In: Whiten A and Widdowson EM editors, *Foraging Strategies and Natural Diet of Monkeys, Apes, and Humans*, Proceedings of a Royal Society Discussion Meeting May 1991, pp. 111–119, Clarendon Press, Oxford, UK.

Underhill AP (1997). Current issues in Chinese Neolithic archaeology, *Journal of World Prehistory* **11**, 103–160.

Ungar PS (2006). *Evolution of the Human Diet: The Known, the Unknown, and the Unknowable*, Oxford University Press, Oxford, UK.

Ungar PS and Teaford MF (2002). *Human Diet (Its Origin and Evolution)*, Bergin and Garvey, New York, USA.

Ungar PS and Teaford MF (2002). Perspectives on the evolution of human diet. In: *Human Diet (Its Origin and Evolution)*, pp. 1–6, Bergin and Garvey, New York, USA.

Valkoun J, Waines JG and Konopka J (1998). Current geographical distribution and habitat of wild wheats and barley. In: Damania AB, Valkoun J, Willcox G and Qualset CO, editors, *The Origins of Agriculture and Crop Domestication*, Aleppo, Syria, ICARDA, available online at: http://www.ipgri.cgiar.org/publications/HTMLPublications/47/.

Vanhaeren M, d'Errico F, Stringer C, James SL, Todd JA, Mienis HK (2006). Middle Paleolithic shell beads in Israel and Algeria, *Science* **312**, 1785–1788.

Vaughan DA (1994). *The Wild Relatives of Rice*, International Rice Research Institute, Manila, Philippines.

Various (2002). *Abrupt Climate Change: Inevitable Surprises*, National Academies Press, Washington DC, available online at: http://www.nap.edu/books/0309074347/html/.

Varro MT (37 BCE). *Rerum Rusticarum de Agri Cultura*, available online at: http://www.thelatinlibrary.com/varro.html.

Varshney RK, Graner A and Sorrells ME (2005). Genomics-assisted breeding for crop improvement, *Trends in Plant Science* **10**, 621–630.

Vartanyan SL, Garutt VE and Sher AV (1993). Holocene dwarf mammoths from Wrangel Island in the Siberian Arctic, *Nature* **362**, 337–340.

Vasal IK (2002). The role of high lysine cereals in - animal and human nutrition in Asia, *Food and Agriculture Organisation Workshop*, available online at: http://www.fao.org/ag/aga/workshop/feed/papers/13vasal.doc.

Vavilov NI (1926). Centres of origin of cultivated plants, *Trudi po Prikl. Bot. Genet. Selek.* [Bulletin of Applied Botany and Genetics] **16**, 139–248 [in Russian].

Vavilov NI (1935). Theoretical basis for plant breeding, Vol. 1. Moscow. Origin and Geography of Cultivated Plants. In: Love D, translator, *The Phytogeographical Basis for Plant Breeding*, pp. 316–366, Cambridge University Press, Cambridge, UK

Veitia RA (2005). Paralogs in polyploids: one for all and all for one? *Plant Cell* **17**, 4–11, available online at: www.plantcell.org/cgi/reprint/17/1/4.pdf.

Venter JC, Adams MD, Myers EW, Li PW, Mural RJ, Sutton GG, Smith HO, Yandell Y, Evans CA, Holt RA, *et al.* (2001). The sequence of the human genome, *Science* **291**, 1304–51.

Vences FJ, Vaquero F, Garcia P and Vega MPDL (1987). Further studies on the phylogenetic relationships in *Secale*: on the origin of is species, *Plant Breeding* **98**, 281–291.

Vicente-Carbajosa J, Moose SP, Parsons RL and Schmidt RJ (1997). A maize zinc-finger protein binds the prolamin box in zein gene promoters and interacts with the basic leucine zipper transcriptional activator Opaque2, *Proceedings of the National Academy of Sciences USA* **94**, 7685–7690, available online at: http://www.pnas.org/cgi/content/abstract/94/14/7685.

Vicient CM, Jääskeläinen MJ, Kalendar R and Schulman AH (2001). Active retrotransposons are common feature of grass genomes, *Plant Physiology* **125**, 183–192, available online at: http://www.plantphysiol.org/cgi/reprint/125/3/1283.

Vietmeyer ND, editor (1989). *Triticale—a promising addition to the world's cereal grains*. Report of an Ad Hoc Panel, Board on Science and Technology for International Development, National Research Council, National Academy Press, Washington, DC, USA.

Vigouroux Y, Matsuoka Y and Doebley J (2003). Directional evolution for microsatellite size in maize, *Molecular Biology and Evolution* **20**, 1480–1483, available online at: http://mbe.oxfordjournals.org/cgi/reprint/20/9/1480.

Villareal RL, Varughese G and Abdalla OS (1990). Advances in spring triticale breeding, *Plant Breeding Reviews* **8**, 43–90.

Vilà C, Leonard JA, Götherström A *et al.* (2001). Widespread origins of domestic horse lineages, *Science* **291**, 474–477.

Vilà C, Savolainen P, Maldonado J, Amorim I, Rice JE, Honeycutt RL, Crandall KA, Lundberg J and Wayne RK (1997). Multiple and ancient origins of the domestic dogs, *Science* **276**, 1687–1689.

Vilà C, Seddon J and Ellegren H (2005). Genes of domestic mammals augmented by backcrossing with wild ancestors, *Trends in Genetics* **21**, 214–218.

Vince G (2007). Endangered languages encode plant and animal knowledge, *New Scientist* 19 February, available online at: http://environment.newscientist.com/channel/earth/dn11215-endangered-languages-encode-plant-and-animal-knowledge-.html.

Vinogradov AE (2003). Selfish DNA is maladaptive: evidence from the plant Red List, *Trends in Genetics* **19**, 609–614.

Vishnu-Mittre and Savithri R (1982). Food economy of the Harappans. In: Possehl GL, editor, *Harappan Civilization: a Contemporary Perspective*, pp. 205–221, Oxford and IBH Publishing Company, New Delhi, India.

Visser M (1986). *Much Depends on Dinner: The Extraordinary History and Mythology, Allure and Obsessions, Perils and Taboos of an Ordinary Meal*, Grove Press, New York, USA.

Vitte C and Bennetzen JH (2006). Eukaryotic transposable elements and genome evolution special feature: Analysis of retrotransposon structural diversity uncovers properties and propensities in angiosperm genome evolution, *Proceedings of the National Academy of Sciences USA* **103**, 17626–17631.

Vitte C, Lamy F, Ishii T, Brar D and Panaud O (2004). Genomic paleontology provides evidence for two distinct origins of Asian rice (*Oryza sativa* L.), *Molecular Genetics and Genomics* **272**, 504–511.

Vöchting H (1892). *Über Transplantation am Pflanzenkörper. Untersuchungen zur Physiologie und Pathologie*, Laupp'schen Buchhandlung, Tübingen, Germany.

Voight BF, Kudaravalli S, Wen X and Pritchard JK (2006). A map of recent positive selection in the human genome, *Public Library of Science, Biology* **4**, e72, available online at: http://biology.plosjournals.org/perlserv/?request = get-document&doi = 10.1371/journal.pbio.0040072.

Voigt M (1984). Village on the Euphrates. Excavations at Neolithic Gritille in Turkey, *Expedition* **27**, 10–24.

Volkman SK, Barry AE, Lyons EJ, Nielsen KM, Thomas SM, Choi M, Thakore SS, Day KP, Wirth DF and Hartl DL (2001). Recent origin of *Plasmodium falciparum* from a single progenitor, *Science* **293**, 482–484.

Vollbrecht E, Springer PS, Goh L, Buckler Iv ES and Martienssen R (2005). Architecture of floral branch systems in maize and related grasses, *Nature* **436**, 1119–1126.

Voytas DF and Naylor GJP (1998). Rapid flux in plant genomes, *Nature Genetics* **20**, 6–7.

Wade N (1999). For leaf-cutter ants, farm life isn't so simple, *New York Times*, 8 August, 1999, available online at: http://fluid.stanford.edu/~mbrennan/interests/insects/leafcutters.html.

Waines JG (1998). *In situ* conservation of wild relatives of crop plants in relation to their history. In: Damania AB, Valkoun J, Willcox G and Qualset CO, editors, *The Origins of Agriculture and Crop Domestication*, Aleppo, Syria, ICARDA, available online at: http://www.ipgri.cgiar.org/publications/HTMLPublications/47/.

Waines JG and Barnhart D (1992). Biosystematic research in *Aegilops* and *Triticum*, *Hereditas* **116**, 207–212.

Waldey G and Martin A (1993). The origins of agriculture—a biological perspective and a new hypothesis, *Australian Biologist* **6**, 88–103, available online at: http://membres.lycos.fr/xbeluga/originsofagriculture.html.

Wales HRH The prince of and Donaldson S (2007). *The Elements of Organic Gardening: Highgrove—Clarence House—Birkhall*, Weidenfeld and Nicolson, London, UK.

Walch JEI (1775). *Naturforscher* **7**, 113–116, plate 1.

Walker CBF (1987). *Reading the Past: Cuneiform*, British Museum Press, London, UK.

Wallace HA and Brown WL (1956). *Corn and its Early Fathers*, Michigan State University Press, East Lansing, Michigan, USA.

Wallace HA and Bressman EN (1949). *Corn and Corn Growing*, John Wiley and Sons, New York, USA.

Wallace JS (2000). Increasing agricultural water use efficiency to meet future food production, *Agriculture, Ecosystems and Environment* **82**, 105–119.

Walsh B (2002). Quantitative genetics, genomics, and the future of plant breeding. In: *Quantitative Genetics and Plant Breeding in the 21st Century*, Symposium of LSU Chapter of Sigma Xi, Baton Rouge, Louisiana, USA, available online at: http://nitro.biosci.arizona.edu/papers.html.

Walter of Henley (1286). Translated in 1890 as: *Walter of Henley's Husbandry, Together with an Anonymous Husbandry, Seneschaucie and Robert Grosseteste's Rules*, Longman, Green and Co, London, UK.

Walters D, Newton A and Lyon G, editors (2007). *A Sustainable Approach to Crop Protection*, Blackwell, Oxford, UK.

Walters, SD (1970). *Water for Larsa. An Old Babylonian Archive dealing with Irrigation*, Yale University Press, New Haven, Connecticut, USA.

Wang H, Nussbaum-Wagler T, Li B *et al.* (2005). The origin of the naked grains of maize, *Nature* **436**, 714–719.

Wang X, Grus WE and Zhang J (2006). Gene losses during human origins, *Public Library of Science, Biology* **4**, e52, available online at: http://biology.plosjournals.org/perlserv/?request = get-document&doi = 10.1371/journal.pbio.0040052.

Wang YJ, Cheng R, Edwards RL *et al.* (2001). A high-resolution absolute-dated late Pleistocene monsoon record from Hulu Cave, China, *Science* **294**, 2345–2348.

Wang Q and Dooner HK (2006). Remarkable variation in maize genome structure inferred from haplotype diversity at the bz locus, *Proceedings of the National Academy of Sciences USA* **103**, 17644–17649.

WARDA, *West Africa Rice Development Association* (2004). available online at: http://www.warda.org.

Ward-Perkins B (2005). *The Fall of Rome and the End of Civilization*, Oxford University Press, Oxford, UK.

Washburn S (1981). Longevity in primates. In: McGaugh JMJ, editor, *Aging, Biology and Behavior*, pp. 11–29 Academic Press, New York, USA.

Watson AM (1994). Botanical gardens in the early Islamic world. In: Robbins E and Sandahl S, editors, *The Ronald Smith Festschrift*, pp. 105–111, Tsar, Toronto, Canada.

Watson AM (1995). Arab and European agriculture in the middle ages. A case of restricted diffusion. In: Sweeney D, editor, *Agriculture in the Middle Ages: Technology, Practice and Representation*, pp. 62–75, University of Pennsylvania Press, Philadelphia, Pennsylvania, USA.

Watson PJ (1995). Explaining the transition to agriculture. In: Price TD and Gebauer AB, editors, *Last Hunters-First Farmers*, pp. 21–38, School of American Research Press, Santa Fe, New Mexico, USA.

Wayne RK, Leonard JA and Vilá CC (2006). Genetic analysis of dog domestication. In: Zeder MA, Bradley DG, Emshwiller E and Smith BD, editors, *Documenting Domestication: New Genetic and Archaeological Paradigms*, pp. 279–293, University of California Press, Berkeley, USA.

Weaver AJ and TMC Hughes (1994). Rapid interglacial climate fluctuations driven by north Atlantic ocean circulation, *Nature*, **367**, 447–450.

Webb JLA (2005). Malaria and the peopling of early tropical Africa, *Journal of World History* **16**, 269–293, available online at: http://www.historycooperative.org/journals/jwh/16.3/webb.html.

Weber SA (1999). Seeds of urbanism: palaeoethnobotany and the Indus Civilization, *Antiquity* **73**, 813–26.

Webster D (2002). *The Fall of the Ancient Maya*, Thames and Hudson, London, UK.

Weiss H (1977). Periodization, population and early state formation in Khuzistan, *Bibliotheca Mesopotamica* **7**, 347–369.

Weiss H (1983). Excavations at Tell Leilan and the origins of North Mesopotamian cities in the third millennium BC, *Paléorient* **9**, 39–52.

Weiss H (1986). The origins of Tell Leilan and the conquest of space in third millennium Mesopotamia. In: Weiss H, editor, *The Origins of Cities in Dry Farming Syria and Mesopotamia in the Third Millennium BC*, pp. 71–108, Four Quarters, Guilford, UK.

Weiss H (1997). Tell Leilan. In: Weiss H, editor, *Archaeology In Syria, American Journal of Archaeology* **101**, 97–148.

Weiss H (2000). Beyond the Younger Dryas—collapse as adaptation to abrupt climate change. In: *Confronting Natural Disaster, Engaging the Past to Understand the Future*, Bawden G and Reycraft R, editors, pp. 75–98, University of New Mexico Press, Albuquerque, New Mexico, USA, available online at: http://research.yale.edu/leilan/weissdryas2002.pdf.

Weiss H (2003). Ninevite 5 periods and processes. In: Rova E and Weiss H, editors, *The Origins of North Mesopotamian Civilization: Ninevite 5 Chronology, Economy, Society*, Brepols, Brussels, Belgium, available online at: http://research.yale.edu/leilan/Weiss2003.pdf.

Weiss H and Bradley RS (2001). What drives societal collapse? *Science* **291**, 609–610, available online at: http://research.yale.edu/leilan/weissbradley2001.pdf.

Weiss H and Courty MA (1993). The genesis and collapse of the Akkadian Empire: the accidental refraction of historical law. In: Liverani M, editor, *Akkad: The First World Empire*, pp. 131–155, Sargon, Padua, Italy.

Weiss H, Courty MA, Wetterstrom W, Senior I, Meadow R, Guichard F and Curnow A (1993). The genesis and collapse of third millennium North Mesopotamian civilisation, *Science* **261**, 995–1004, available online at: http://research.yale.edu/leilan/weisseta11993.pdf.

Weiss H, deLillis F, deMoulins D *et al.* (2002). Revising the contours of history at Tell Leilan, *Annales Archéologiques Arabes Syriennes*, Cinquantenaire, available online at: http://research.yale.edu/leilan/weisseta12002.pdf.

Weiss H, Wetterstrom W, Nadel D and Bar-Yosef O (2004). The broad spectrum revisited: Evidence from plant remains, *Proceedings of the National Academy of Sciences USA* **101**, 9551–9555, available online at: http://www.pnas.org/cgi/content/short/101/26/9551.

Weiss E, Mordechai E, Kislev ME and Hartmann A (2006). Autonomous cultivation before domestication, *Science* **312**, 1608–1610.

Index

Learning Resources
Centre